HARNESSING AUTOMATION AND MACHINE LEARNING FOR RESOURCE RECOVERY AND VALUE CREATION

HARNESSING AUTOMATION AND MACHINE LEARNING FOR RESOURCE RECOVERY AND VALUE CREATION

From Waste to Value

Edited by

KISHOR KUMAR SADASIVUNI

Center for Advanced Materials, Qatar University, Doha, Qatar

NEBOJSA BACANIN

Department of Informatics and Computing, Singidunum University, Belgrade, Serbia

JAEHWAN KIM

Creative Research Center for Nanocellulose Future Composites, Department of Mechanical Engineering, Inha University, Incheon, Republic of Korea

NEHA B. VASHISHT

Center for Advanced Materials, Qatar University, Doha, Qatar

ELSEVIER

Elsevier
Radarweg 29, PO Box 211, 1000 AE Amsterdam, Netherlands
125 London Wall, London EC2Y 5AS, United Kingdom
50 Hampshire Street, 5th Floor, Cambridge, MA 02139, United States

For accessibility purposes, images in electronic versions of this book are accompanied by alt text descriptions provided by Elsevier. For more information, see https://www.elsevier.com/about/accessibility.

Publisher's note: Elsevier takes a neutral position with respect to territorial disputes or jurisdictional claims in its published content, including in maps and institutional affiliations.

Notices

Knowledge and best practice in this field are constantly changing. As new research and experience broaden our understanding, changes in research methods, professional practices, or medical treatment may become necessary.

Practitioners and researchers must always rely on their own experience and knowledge in evaluating and using any information, methods, compounds, or experiments described herein. In using such information or methods they should be mindful of their own safety and the safety of others, including parties for whom they have a professional responsibility.

To the fullest extent of the law, neither the Publisher nor the authors, contributors, or editors, assume any liability for any injury and/or damage to persons or property as a matter of products liability, negligence or otherwise, or from any use or operation of any methods, products, instructions, or ideas contained in the material herein.

ISBN: 978-0-443-27374-2

For Information on all Elsevier publications
visit our website at https://www.elsevier.com/books-and-journals

Publisher: Candice Janco
Acquisitions Editor: Anita Koch
Editorial Project Manager: Debarati Roy
Production Project Manager: Sujatha Thirugnana Sambandam
Cover Designer: Vicky Esser Pearson

Typeset by MPS Limited, Chennai, India

Contents

5. Plastics recycling and the automation role in the recycling process

Munu Borah and Jyoti Prasad Gogoi

6. Artificial intelligence for reutilizing the plastics

Hameed Khan and Kamal Kushwah

7. Handling metals waste to salvage with automation

Izhar Ahmed, Kondaiah Seku, Sreenivasa Babu Nidamanuri and Krishna Kumar Koyyala

8. Machine learning: a better means for metal waste to reprocess

Aya Nabil Sayed, Md. Mosarrof Hossen, Tamim Al-Hasan, Mohammad Noorizadeh and Faycal Bensaali

9. Waste classification using convolutional neural networks tuned by modified metaheuristics algorithm

Ana Tasic, Luka Jovanovic, Miroslav Popovic, Nebojsa Bacanin, Miodrag Zivkovic and Milos Dobrojevic

10. Exploring the potential of lightweight computer vision YOLOv8 models for effective waste classification and management

Ana Tasic, Luka Jovanovic, Nebojsa Bacanin, Miodrag Zivkovic, Stanislava Kozakijevic, Milos Antonijevic and Miroslav Popovic

11. Waste to value added: role of automation in organic waste

Mithra Geetha, Nihal Poovadichalil, Md Nawshad, Muni Raj Maurya, Asad Ullah, Syed Faizan Haider and Anwarul Hasan

12. Future of agriculture: smart vertical farming

Mohammed Alsafran, Malik Adil Nawaz and Deepak Kasote

13. Impact of artificial intelligence for the recycling of organic waste

Malik Adil Nawaz and Deepak Kasote

14. Agriculture: the next machine-learning frontier

Al-Maadeed Somaya and Akbari Younes

15. Recycling robots to tackle electrical waste

Samia Adil, Duc Hoa Pham, Yangziaozhe Jiang, Shuaishuai Zhu and Jaehwan Kim

16. Machine learning for sustainable development in electronics

Iraklis Varlamis, Ilias Panagiotopoulos, Christos Chronis, George Dimitrakopoulos and Faycal Bensaali

17. Automated sorting of recyclable domestic waste

Jaehwan Kim, Muhammad Latif, Jongmin Song, Giseok Park and Jieun Lee

18. Intelligent waste management: a comprehensive review of machine learning and deep learning applications in advanced recycling

Adiba Tabassum Chowdhury, Muhammad E.H. Chowdhury, Md. Sakib Bin Islam, Mehrin Newaz, Abdus Salam, Faizul Sayem, Mazhar Hasan-Mia, Mohammad Nashbat, Molla Ehsanul Majid, Amith Khandakar, Saad Bin Abul Kashem and Azad Ashraf

19. Future aspects of machine learning/automation for the waste management

Nandini Gaikwad and Girish Sambhaji Gund

List of Contributors

Samia Adil Creative Research Center for Nanocellulose Future Composites, Department of Mechanical Engineering, Inha University, Incheon, Republic of Korea

Izhar Ahmed Department of Engineering, College of Engineering and Technology, University of Technology and Applied Sciences, Salalah, Oman

Tamim Al-Hasan Department of Electrical Engineering, College of Engineering, Qatar University, Doha, Qatar

Mohammed Alsafran Agricultural Research Station, Qatar University, Doha, Qatar

Azad Ashraf Department of Chemical Engineering, University of Doha for Science and Technology, Doha, Qatar

Faycal Bensaali Department of Electrical Engineering, College of Engineering, Qatar University, Doha, Qatar

Saad Bin Abul Kashem Department of Computing Science, AFG College with the University of Aberdeen, Doha, Qatar

Munu Borah Department of Physics, The Assam Kaziranga University, Jorhat, Assam, India

John-John Cabibihan Department of Mechanical and Industrial Engineering, College of Engineering, Qatar University, Doha, Qatar

Adiba Tabassum Chowdhury Department of Electrical and Electronic Engineering, University of Dhaka, Dhaka, Bangladesh

Muhammad E.H. Chowdhury Department of Electrical Engineering, Qatar University, Doha, Qatar

Christos Chronis Department of Informatics and Telematics, Harokopio University of Athens, Athens, Greece

George Dimitrakopoulos Department of Informatics and Telematics, Harokopio University of Athens, Athens, Greece

Milos Dobrojevic Department of Informatics and Computing, Singidunum University, Belgrade, Serbia

Nandini Gaikwad Department of Information Technology, Mahatma Phule Arts, Science and Commerce College, Panvel, Maharashtra, India

Mithra Geetha Center for Advanced Materials, Qatar University, Doha, Qatar

Jyoti Prasad Gogoi Department of Physics, DKD College, Dergaon, Assam, India

Girish Sambhaji Gund Department of Physics, Mahatma Phule Arts, Science and Commerce College, Panvel, Maharashtra, India

Syed Faizan Haider Mechanical Engineering Department, Qatar University, Doha, Qatar

Anwarul Hasan Mechanical Engineering Department, Qatar University, Doha, Qatar

Mazhar Hasan-Mia Department of Chemical Engineering, University of Doha for Science and Technology, Doha, Qatar

Mazhar Hasan-Zia Department of Chemical Engineering, University of Doha for Science and Technology, Doha, Qatar

Md. Mosarrof Hossen Department of Electrical Engineering, College of Engineering, Qatar University, Doha, Qatar

Md. Sakib Bin Islam Department of Electrical Engineering, Qatar University, Doha, Qatar

Yangziaozhe Jiang Creative Research Center for Nanocellulose Future Composites, Department of Mechanical Engineering, Inha University, Incheon, Republic of Korea

Deepak Kasote Agricultural Research Station, Qatar University, Doha, Qatar

Julhash U. Kazi Division of Translational Cancer Research, Department of Laboratory Medicine, Lund University, Lund, Sweden; Lund University Cancer Centre (LUCC), Lund

University, Lund, Sweden; Department of Laboratory Medicine, Lund Stem Cell Center, Lund University, Lund, Sweden

Hameed Khan Computer Science and Engineering, Jabalpur Engineering College, Jabalpur, Madhya Pradesh, India

Amith Khandakar Department of Electrical Engineering, Qatar University, Doha, Qatar

Jaehwan Kim Creative Research Center for Nanocellulose Future Composites, Department of Mechanical Engineering, Inha University, Incheon, Republic of Korea

Krishna Kumar Koyyala Department of Engineering, College of Engineering and Technology, University of Technology and Applied Sciences, Shinas, Oman

Ali Kunju Department of Chemical Engineering, University of Doha for Science and Technology, Doha, Qatar

Kamal Kushwah Department of Applied Physics, Jabalpur Engineering College, Jabalpur, Madhya Pradesh, India

Muhammad Latif Creative Research Center for Nanocellulose Future Composites, Department of Mechanical Engineering, Inha University, Incheon, Republic of Korea

Jieun Lee Creative Research Center for Nanocellulose Future Composites, Department of Mechanical Engineering, Inha University, Incheon, Republic of Korea

Luka Jovanovic Department of Informatics and Computing, Singidunum University, Belgrade, Serbia

Molla Ehsanul Majid Department of Computer Applications, Academic Bridge Program, Qatar Foundation, Doha, Qatar; Department of Computer Applications Academic Bridge Program, Qatar Foundation, Doha, Qatar

Muni Raj Maurya Center for Advanced Materials, Qatar University, Doha, Qatar

Milos Antonijevic Department of Informatics and Computing, Singidunum University, Belgrade, Serbia

Aya Nabil Sayed Department of Electrical Engineering, College of Engineering, Qatar University, Doha, Qatar

Mohammad Nashbat Department of Chemical Engineering, University of Doha for Science and Technology, Doha, Qatar

Malik Adil Nawaz Agricultural Research Station, Qatar University, Doha, Qatar

Md Nawshad Center for Advanced Materials, Qatar University, Doha, Qatar; Mechanical Engineering Department, Qatar University, Doha, Qatar

Nebojsa Bacanin Department of Informatics and Computing, Singidunum University, Belgrade, Serbia

Mehrin Newaz Department of Electrical and Electronic Engineering, University of Dhaka, Dhaka, Bangladesh

Sreenivasa Babu Nidamanuri Department of Engineering, College of Engineering and Technology, University of Technology and Applied Sciences, Shinas, Oman

Mohammad Noorizadeh Department of Electrical Engineering, College of Engineering, Qatar University, Doha, Qatar

Aghzout Otman Dept of Digitalization and Artificial Intelligence, ENSA, Morocco

Ilias Panagiotopoulos Department of Informatics and Telematics, Harokopio University of Athens, Athens, Greece

Sreedevi Paramparambath Center for Advanced Materials, Qatar University, Doha, Qatar

Giseok Park Creative Research Center for Nanocellulose Future Composites, Department of Mechanical Engineering, Inha University, Incheon, Republic of Korea

Gilroy Pereira Qatar Scientific Club, Fabrication Laboratory, Doha, Qatar

Duc Hoa Pham Creative Research Center for Nanocellulose Future Composites, Department of Mechanical Engineering, Inha University, Incheon, Republic of Korea

Nihal Poovadichalil Mechanical Engineering Department, Qatar University, Doha, Qatar

Miroslav Popovic Department of Informatics and Computing, Singidunum University, Belgrade, Serbia

Kishor Kumar Sadasivuni Center for Advanced Materials, Qatar University, Doha, Qatar

Abdus Salam Department of Electrical and Computer Engineering, Rajshahi University of Engineering and Technology, Rajshahi, Bangladesh

Faizul Sayem Department of Electrical and Electronic Engineering, University of Dhaka, Dhaka, Bangladesh

Kondaiah Seku Department of Engineering, College of Engineering and Technology, University of Technology and Applied Sciences, Shinas, Oman

Oleg Serov National Research Nuclear University, MEPhI, Moscow, Russia

Mizaj Sha Center for Advanced Materials, Qatar University, Doha, Qatar

Al-Maadeed Somaya Qatar University, Doha, Qatar

Jongmin Song Creative Research Center for Nanocellulose Future Composites, Department of Mechanical Engineering, Inha University, Incheon, Republic of Korea

Stanislava Kozakijevic Department of Psychology, Faculty of Media and Communication, Singidunum University, Belgrade, Serbia

Ana Tasic Department of Informatics and Computing, Singidunum University, Belgrade, Serbia

Rifah Tasnia Department of Computer Science and Engineering, Ahsaniya Mission University of Science and Technology, Rajshahi, Bangladesh

Asad Ullah Center for Advanced Materials, Qatar University, Doha, Qatar; Mechanical Engineering Department, Qatar University, Doha, Qatar

Iraklis Varlamis Department of Informatics and Telematics, Harokopio University of Athens, Athens, Greece

Neha B. Vashisht Center for Advanced Materials, Qatar University, Doha, Qatar

Akbari Younes Computer Science and Engineering Department, Qatar University, Doha, Qatar

Shuaishuai Zhu Creative Research Center for Nanocellulose Future Composites, Department of Mechanical Engineering, Inha University, Incheon, Republic of Korea

Mohamed Zied Chaari Qatar Scientific Club, Fabrication Laboratory, Doha, Qatar

Miodrag Zivkovic Department of Informatics and Computing, Singidunum University, Belgrade, Serbia

About the editors

Dr. Kishor Kumar Sadasivuni is an accomplished Research Assistant Professor at the Center for Advanced Materials, Qatar University. With a strong foundation in analytical chemistry, he embarked on his academic journey, earning a master's degree from Andhra University, India. His pursuit of knowledge led him to the University of South Brittany in Lorient, France, where he earned his PhD in Materials Science and Engineering in 2012 under the guidance of Professor Yves Grohens. This marked the beginning of his impressive career in academia and research. Dr. Kishor Kumar Sadasivuni's contributions to the scientific community have been exceptional. His groundbreaking work earned him recognition as one of the world's top 2% scientists, a distinction conferred by Stanford University in 2019, 2020, 2022, and 2023. Qatar University has also acknowledged his remarkable achievements with honors. Boasting over 15 years of active research experience, Dr. Kishor has made significant strides in his field. His prolific publication record includes more than 450 research articles in esteemed international peer-reviewed journals, accumulating an impressive total citation count of 18,000 and an h-index of 75. He has also authored 20 book chapters and served as the editor of 14 books, with several of his works featured in Springer's prestigious top 25 e-book downloads of 2020. Dr. Kishor's innovative mindset is evident through his patent portfolio. He holds four US patents and two Indian patents, with five patent disclosures currently under consideration. As a visionary leader, he spearheads 15 research projects, encompassing grants from the Qatar National Research Fund, including NPRP, UREP, and IRCC grants, as well as projects supported by Qatar University, collectively amounting to an impressive 5 million USD. A true collaborator, Dr. Kishor has actively engaged with researchers from diverse disciplines worldwide, boasting over 480 coauthors in fields such as computer science, biomedical sciences, industrial engineering, and electrical engineering. His global collaborations span countries like the United States, France, South Korea, Oman, Spain, Italy, Australia, and Malaysia. In summary, Dr. Kishor Kumar Sadasivuni is a distinguished figure in the world of academia and materials science, renowned for his prolific research contributions, innovative patents, and extensive collaborative efforts. His dedication to advancing knowledge and fostering international cooperation underscores his commitment to the scientific community.

Dr. Nebojsa Bacanin received his PhD degree from the Faculty of Mathematics, University of Belgrade in 2015 (study program Computer Science, average grade 1000). He was the vice-dean of the Graduate School of Computer Science and the Faculty of Informatics and Computing in Belgrade, Serbia. He currently works as a Full Professor and as a Vice-Rector for Scientific Research at Singidunum University. He is involved in scientific research in the field of computer science and his specialty includes artificial intelligence, machine learning, deep learning, stochastic optimization algorithms, swarm intelligence, soft-computing, optimization and modeling, image processing, computer vision, and cloud and distributed computing. He actively works in the domain of novel and prospective research field, hybrid methods between machine learning and metaheuristics, where metaheuristics are applied for addressing nondeterministic polynomial hard (NP-hard) challenges from machine learning domain such as hyperparameters optimization (tuning), training, and feature selection. Besides improving machine learning/deep learning models for tackling various practical tasks for classification and regression, his research also involves optimized deep learning models for univariate and multivariate time-series forecasting. Moreover, he is an expert in the area of metaheuristics, and he has been actively doing research in enhancing swarm intelligence, as well as other types of metaheuristics, by incorporating minor changes (e.g., modification in exploitation/exploration expressions, parameters' adjustments, etc.) and/or major modifications by performing hybridization with other methods (e.g., low-level and high-level hybrid metaheuristics methods). He has been applying his methods to a wide variety of practical research areas, for example, cloud computing scheduling, wireless sensor networks (WSNs) localization, coverage and energy consumption, X-ray images classification, stock price forecasting, portfolio optimization, and many others.

Jaehwan Kim serves as an Inha Fellow Professor at the Department of Mechanical Engineering at Inha University, Korea. He is also designated as the Director of CRC for Nanocellulose Future Composites. Dr. Kim is a Fellow of the Korean Academy of Science and Technology, the National Academy of Engineering of Korea, and the Institute of Physics. He is an Associate Editor of *Smart Materials and Structure* as well as *Smart Nanosystems in Engineering and Medicine* and Editor of *International Journal of Precision Manufacturing and Engineering*, *Heliyon*, and *Actuators*. He has been the Director of the Creative Research Center for EAPap Actuator funded by the National Research Foundation of Korea (NRF). Recently, he started the Creative Research Center for Nanocellulose Future Composites, sponsored by NRF. His research interests are smart materials, structures and devices, biomaterial-based smart materials, cellulose, electroactive polymers, power harvesting, biomimetic actuators, biosensors, tactile sensors, and flexible electronics. He has published more than 320 journal papers, presented 360 international conference papers, and filed more than 40 patents.

Dr. Neha B. Vashisht is working as a Researcher at Smart Nano Solution Group, Centre for Advanced Materials (CAM), Qatar University. She received her PhD in 2017 from the Department of Basic and Applied Sciences, Guru Gobind Singh Indraprastha University of Science and Technology, New Delhi, India. During her PhD, she worked on "Synthesis and Characterization of Plasmonic Noble Metal-Tin Oxide Hybrid Nanocomposites for Sensing Applications." She has many research publications in leading scientific research journals with good impact factors. Her areas of interest are the synthesis and characterization of new smart hybrid materials, nanocomposites, their modifications, and optics. She has actively participated in and presented her research work at several international conferences and seminars.

Foreword

It is my privilege to present *Harnessing Automation and Machine Learning for Resource Recovery and Value Creation*, a book that explores the transformative power of technology in addressing the global challenge of waste management. As we face increasing pressure to adopt sustainable practices, the integration of automation and machine learning has emerged as a critical approach for converting waste into valuable resources and driving a circular economy.

This book is a comprehensive guide that covers the fundamentals of machine learning and automation while showcasing their practical applications in waste classification, recycling, and resource recovery. The chapters delve into various waste streams, including plastics, metals, and organic materials, illustrating how advanced technologies like computer vision and neural networks are being used to enhance the efficiency and accuracy of waste processing.

What sets this book apart is its focus on real-world applications and the future potential of these technologies in diverse sectors such as agriculture, electronics, and smart vertical farming. The contributors, experts in their fields, provide valuable insights into how automation and machine learning are not only theoretical tools but are actively shaping the future of waste management. This work is an essential reading for researchers, practitioners, and policymakers dedicated to sustainability. I hope it inspires continued innovation and collaboration in the quest to leverage technology for environmental stewardship and resource recovery.

Dr. Milovan Stanisic,
President of Singidunum University,
Belgrade, Serbia

Preface

The increasing global emphasis on sustainability, resource conservation, and waste management has catalyzed the need for innovative solutions that can efficiently convert waste into valuable resources. As the world faces mounting environmental challenges, the integration of advanced technologies such as automation and machine learning into waste management processes has emerged as a pivotal strategy. These technologies not only enhance the efficiency of waste processing but also contribute to creating a circular economy where waste is transformed into energy, raw materials, and other valuable products.

This book, *Harnessing Automation and Machine Learning for Resource Recovery and Value Creation*, delves into the transformative potential of these cutting-edge technologies in the realm of waste management and resource recovery. The chapters presented here offer a comprehensive exploration of how automation and machine learning are being utilized to revolutionize waste classification, recycling, and management across various sectors, including plastics, metals, and organic waste.

The journey begins with an introduction to the innovative technologies driving the waste-to-energy conversion processes, followed by an in-depth discussion on the basics of machine learning and automation. Readers are then guided through the complexities of waste classification, with a particular focus on the role of automation in enhancing the efficiency and accuracy of these processes. From plastics recycling to the reutilization of metals and organic waste, each chapter provides valuable insights into how these technologies are reshaping the landscape of waste management.

This book also highlights the role of advanced computer vision models like YOLOv8 in waste classification, as well as the prospects of machine learning and automation in fields such as agriculture and electronics recycling. By examining the current state and future potential of these technologies, this book aims to inspire further innovation and collaboration in the quest for sustainable development.

In compiling this book, we have aimed to provide a valuable resource for researchers, practitioners, and policymakers who are working toward a more sustainable future. It is our hope that the insights and knowledge shared in these pages will contribute to the ongoing efforts to harness the power of automation and machine learning for the betterment of our planet.

Acknowledgments

This book was supported by the Qatar National Research Fund under Grant No. MME03-1226-210042. The statements made herein are solely the responsibility of all the editors and the authors.

Introduction

The world today is grappling with the escalating challenge of waste management. As urbanization and industrialization continue to surge, the amount of waste produced globally has reached unprecedented levels. Traditional methods of waste disposal are proving inadequate, leading to overflowing landfills and mounting environmental concerns. However, this challenge also presents a unique opportunity: to reimagine waste not as a burden, but as a valuable resource that can be recovered and reused. *Harnessing Automation and Machine Learning for Resource Recovery and Value Creation* explores how the integration of automation and machine learning is revolutionizing waste management practices. These advanced technologies have the potential to transform the way we process, classify, and recycle waste, making these processes more efficient, accurate, and sustainable. By doing so, they play a crucial role in promoting a circular economy, where resources are continuously repurposed, reducing the need for new raw materials and minimizing environmental impact. This book begins by laying a foundation with an overview of the basics of machine learning and automation. From there, it delves into specific applications, such as the classification and recycling of plastics, metals, and organic materials. The use of cutting-edge techniques like computer vision and neural networks is highlighted, showcasing how they enhance the accuracy and efficiency of waste processing. Beyond the technical aspects, this book also examines the broader implications of these technologies for sustainability and the future of resource management. It provides insights into emerging trends like smart agriculture and the recycling of electronic waste, offering a glimpse into the future of waste management. In this book, readers will find a comprehensive guide to the transformative potential of automation and machine learning in waste management, and how these technologies can lead us toward a more sustainable and resource-efficient future.

CHAPTER

1

Introduction to innovative technologies for waste-to-energy conversion using automation and machine learning

Mizaj Sha[1], Sreedevi Paramparambath[1], John-John Cabibihan[2], Mithra Geetha[1], Neha B. Vashisht[1], and Kishor Kumar Sadasivuni[1]

[1]Center for Advanced Materials, Qatar University, Doha, Qatar [2]Department of Mechanical and Industrial Engineering, College of Engineering, Qatar University, Doha, Qatar

Introduction

The world's environmental quality is jeopardized since it continues to act as a sink for toxins from various industries, from forest waste to subtle compounds in water and air settings. Furthermore, rising population growth, urbanization, and rapid industrialization have all contributed to a large increase in demand for energy and raw materials [1]. As a result, creating innovative eco-technologies for sustainable and environmentally friendly energy production from waste (resources) is critical to promoting a circular economy in the region/country and globally [2].

Recently there has been a lot of interest in applying the notion of cleaner manufacturing to diverse sectors, transforming the linear economy into a resource-conserving circular economy, and converting waste to energy or valuable commodities using a bioreactor [3]. Cleaner manufacturing is a proactive method or process used on products and services to reduce waste and health risks. As a result, transitioning from the current linear economy to the circular economy through practices such as recycling and reuse will provide answers for the efficient use of scarce resources required by industrial sectors [4]. In this context, biorefineries can help maximize the energy potential of organic wastes while also addressing waste management and greenhouse gas emissions [5].

Harnessing Automation and Machine Learning for Resource Recovery and Value Creation
DOI: https://doi.org/10.1016/B978-0-443-27374-2.00001-7

Biorefining is the process of separating the various components of a raw material using proper pretreatment procedures and biotechnologies to turn the individual components into value-added products [6]. Biorefineries provide several advantages, including resource conservation, economic feasibility, waste management, and the option to select appropriate, locally accessible feedstocks. Biorefinery is one of the fundamental concepts now being applied in a country's circular economy. Improving research and technology in this subject and implementing these principles in practice demands a strong effort from the scientific community and the commercial sectors and the active participation of various stakeholders [7].

Fundamentals of waste-to-energy conversion

Waste-to-energy (WtE) conversion is a process that involves the generation of energy from various types of waste materials. It is a sustainable approach that helps reduce the waste volume sent to landfills while producing valuable energy [8]. Here are some fundamental aspects of WtE conversion:

Types of waste material

Waste material can be classified into various types depending on its source, composition, and potential for reuse or recycling (Fig. 1.1). Here is a breakdown of some common types of waste:

1. Municipal solid waste (MSW) consists of two subtypes: Household waste and commercial waste. Household waste consists of everyday items thrown away by households,

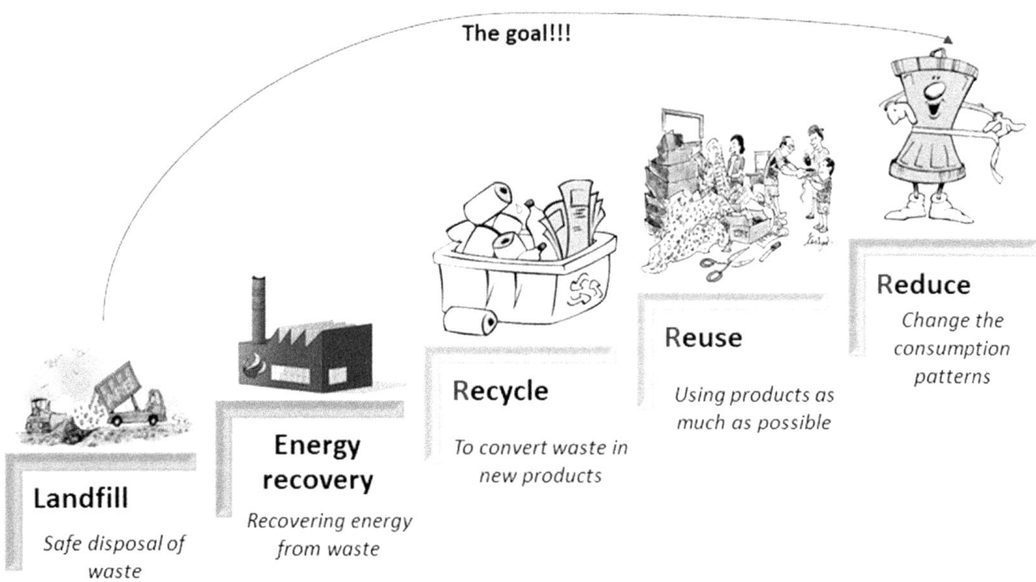

FIGURE 1.1 Waste management strategy—hierarchy [9]. Source: *From J. Lorenzo Llanes, and E. Kalogirou, Waste-to-energy conversion in Havana: technical and economic analysis, Social Sciences 8(4), 119 (2019). https://doi.org/10.3390/socsci8040119.*

businesses, and institutions, whereas establishments like offices, stores, and restaurants produce commercial waste. It often includes packaging, food waste, and paper [8].

2. Industrial waste: It consists of two subtypes. Manufacturing waste includes scrap metal, plastics, paper, and chemicals. It often originates from manufacturing processes. Construction and Demolition debris includes concrete, bricks, wood, and metal scraps from construction or demolition activities [8].
3. Hazardous waste:
   - Chemical waste Includes wastes containing harmful chemicals that can be toxic, flammable, or corrosive. Common in industrial and laboratory settings.
   - Medical waste includes discarded medical supplies like needles, gloves, and expired drugs, as well as potentially infectious materials from hospitals and clinics.
   - Electronic waste (e-waste) includes discarded electronic devices like computers, televisions, and mobile phones, which may contain hazardous substances like lead and mercury [8]. Different e-waste components are shown in Fig. 1.2.

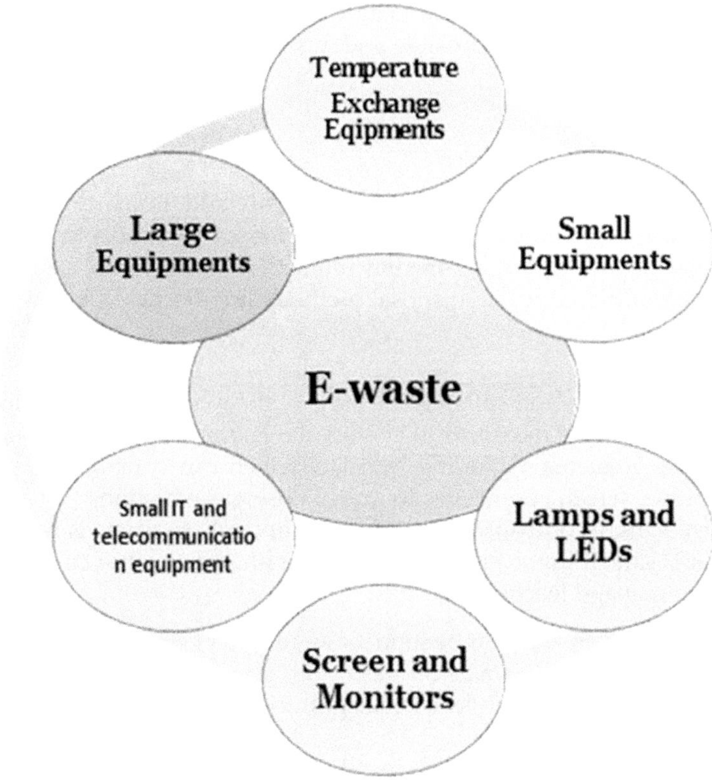

FIGURE 1.2 Different types of e-waste [10]. Source: *From H. Madkhali, S. Duraib, L. Nguyen, M. Prasad, M. Sharma, and S. Joshi, A comprehensive review on e-waste management strategies and prediction methods: a Saudi Arabia perspective, Knowledge 3(2), 163–179 (2023). https://doi.org/10.3390/knowledge3020012.*

4. Biological waste:
   - Agriculture waste includes organic waste such as crop residues, manure, and other by-products from farming and livestock production.
   - Biomedical waste includes any waste that contains infectious (or potentially infectious) materials, such as laboratory cultures and waste from surgical procedures [8].
5. Organic waste:
   - Food waste includes food scraps from households and commercial establishments.
   - Garden and Yard waste includes leaves, grass clippings, and branches [8].
6. Inert waste:
   - Dirt and rocks are typically generated from earthmoving and construction activities.
   - Concrete and rubble are often considered for recycling into construction materials [8].
7. Special waste:
   - Asbestos waste requires special handling due to its carcinogenic nature.
   - Automotive waste includes used motor oil, batteries, and tyres.
   - Batteries: Batteries must be disposed of or recycled properly due to their toxic and corrosive materials [8].

Each type of waste has specific methods and regulations for disposal, aiming to minimize its impact on the environment and human health. Proper management involves recycling and reprocessing materials wherever possible and ensuring that hazardous waste is treated and disposed of safely (Fig. 1.3).

Traditional methods of waste disposal

Traditional waste disposal methods are used historically and, in many cases, continue to be used to manage waste generated by households, industries, and other sectors. These methods vary widely regarding environmental impact, efficiency, and sustainability [12]. Some of the primary traditional waste disposal methods are discussed below:

Landfilling

Open dumping: Historically, and still in some parts of the world, waste is dumped in open areas without any treatment or measures to protect the environment. This method leads to various environmental problems, including leachate, which can contaminate groundwater, and methane generation, which contributes to greenhouse gas emissions.

Sanitary landfills: A more controlled form of landfilling where waste is isolated from the environment until it is safe. It involves layering waste with soil or other cover materials and includes measures to manage leachate and gas.

1. Incineration: This involves the combustion of organic substances contained in waste materials. Incineration converts waste into ash, flue gas, and heat. While it significantly reduces the volume of waste, it can lead to air pollution and requires high technology to manage the emissions safely.
2. Compositing: A biological process involving decomposing organic waste, such as food scraps and yard waste, into a soil-like substance called compost. This method recycles organic material and returns nutrients to the soil but is limited to biodegradable waste.

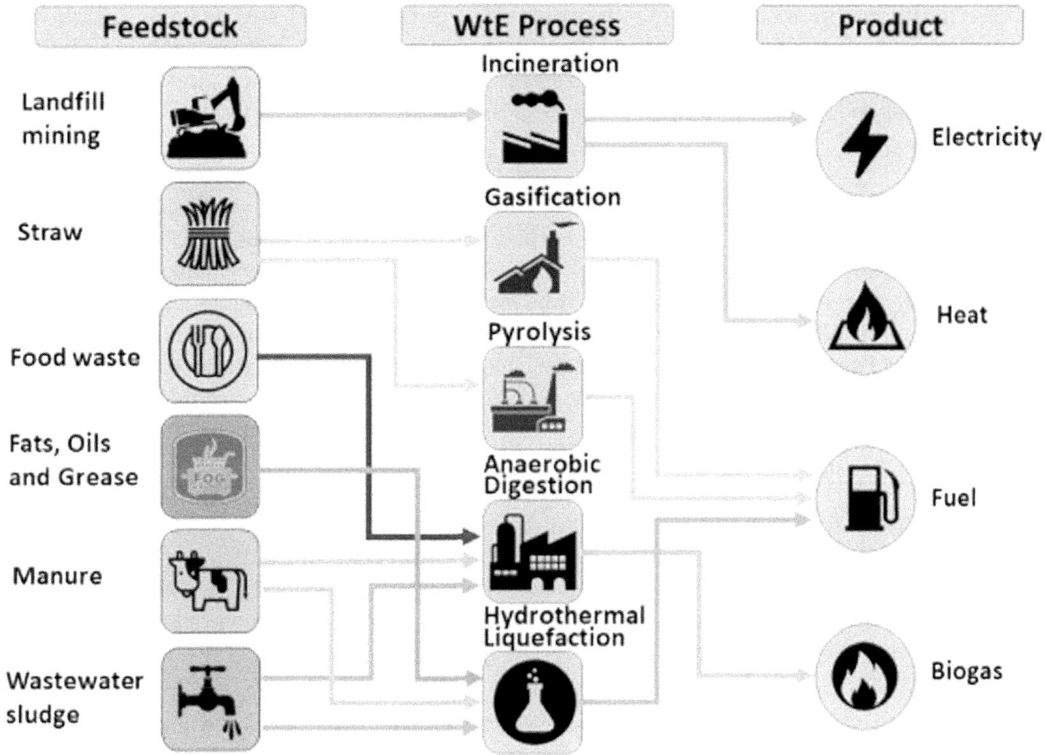

FIGURE 1.3 Schematics of waste-to-energy conversion [11]. Source: *From W. Foster, U. Azimov, P. Gauthier-Maradei, L. C. Molano, M. Combrinck, J. Munoz, J. J. Esteves, and L. Patino, Waste-to-energy conversion technologies in the UK: processes and barriers – a review, Renewable and Sustainable Energy Reviews 135 (2021). https://doi.org/10.1016/j.rser.2020.110226.*

3. Burying: In rural or less-developed areas, burying waste is common. This involves simply digging a hole and burying waste, a rudimentary form of landfilling. If not managed properly, this method can lead to soil and water contamination.
4. Burning: Often practiced in rural areas, open burning involves setting fire to waste piles. This method is quite harmful to the environment as it releases toxic fumes and particulate matter into the air, contributing to air pollution and health risks.
5. Animal feed: A traditional method where organic waste materials (e.g., food scraps) are fed to pigs and chickens. This method recycles waste but needs to be managed to prevent disease.
6. Ocean dumping: Historically, waste was sometimes disposed of by dumping it into the oceans. This method has largely been phased out in many parts of the world due to its severe environmental impacts, including harm to marine life.
7. Reuse and repurposing: Not exactly a disposal method, but traditionally many items that might be considered waste were reused or repurposed to extend their life, thereby reducing the waste stream. This includes using old containers for storage or repurposing clothing into rags.

These traditional methods vary greatly in terms of environmental sustainability and health impacts. In modern waste management, there is a significant push toward more sustainable practices such as recycling, waste minimization, and advanced technological methods like anaerobic digestion and WtE processes that aim to reduce the negative impacts of waste disposal.

Introduction to waste-to-energy technologies (principles)

WtE technologies involve converting waste materials into various forms of energy, including electricity, heat, or fuel, through various processes. This approach helps manage waste and contributes to energy production, serving dual purposes of waste reduction and renewable energy generation. WtE technologies offer a promising solution to the dual challenges of waste management and sustainable energy production. However, their success and expansion depend heavily on technological advancements, economic feasibility, and public acceptance [13].

Benefits and challenges of waste-to-energy conversion

WtE conversion technologies provide a range of benefits but also pose certain challenges. These systems can effectively reduce waste volumes while generating energy, but they must be managed carefully to mitigate potential environmental and economic impacts [14]. Here is a closer look at the benefits and challenges associated with WtE conversion:

Benefits

1. **Reduction in waste volume:** WtE processes can significantly reduce the waste volume that must be disposed of in landfills. For example, incineration can reduce waste volume by up to 90%.
2. **Energy generation**: WtE converts waste into usable forms of energy, such as electricity and heat, helping to supplement energy supplies, which can reduce reliance on fossil fuels and contribute to energy security.
3. **Low greenhouse gas emission**: By diverting waste from landfills, WtE technologies can reduce methane emissions—a potent greenhouse gas produced when organic waste decomposes anaerobically in landfills.
4. **Resource recovery:** Some WtE technologies, like incineration and gasification, can recover metals and other materials postprocess, which can be recycled and reused.
5. **Economic development:** Building and operating WtE facilities can create engineering, construction, and facility operation jobs. It also stimulates local economies through waste processing fees and the sale of generated energy.
6. **Waste management:** WtE provides an alternative to landfill disposal and can be particularly valuable in regions where land is scarce or landfill capacity is limited.

Challenges of waste-to-energy conversion

1. High capital and operational cost: WtE plants require significant initial investment for construction and advanced technology. Operating costs can also be high due to the need for continuous maintenance and management [15].

2. Emission and air quality concerns: Combustion-based WtE processes can emit pollutants like nitrogen oxides, sulfur dioxide, particulate matter, and dioxins. Modern pollution control technologies can mitigate these emissions, but they add to the cost and complexity of the plants [15].
3. Public opposition: WtE facilities often face "not in my backyard" (NIMBY) opposition due to concerns over emissions, traffic, and impacts on local property values. Public perception can be a significant hurdle even with advanced, cleaner technologies [15].
4. Waste supply dependency: WtE plants require a continuous and sufficient waste stream to operate efficiently. Waste quantity and composition changes can affect operational efficiency and economic viability [15].
5. Potential competition with recycling: There is a concern that the focus on energy recovery might detract from recycling efforts. Some materials might be burned for energy instead of being recycled, which can be less sustainable in the long run [15].
6. Regulatory challenges: WtE projects must navigate complex and stringent regulatory environments designed to protect public health and the environment. Compliance can be costly and time-consuming [15].

While WtE conversion offers valuable solutions for waste management and energy production, balancing the benefits and challenges requires careful planning, technological innovation, and public engagement. Advanced WtE technologies continue to evolve, aiming to provide more sustainable and efficient ways to manage waste and generate energy. Still, the adoption and success of these technologies depend heavily on economic, environmental, and social factors.

Automation in waste-to-energy conversion

Basics of automation: role of automation in waste sorting and preprocessing

Automation involves using technology to perform tasks with minimal human intervention. It spans various industries and applications, improving efficiency, accuracy, and productivity. Automation plays a crucial role in optimizing and enhancing the efficiency of WtE conversion processes [16]. As these facilities aim to transform waste into usable energy forms such as electricity, heat, or syngas, automation technologies ensure consistent operation, maximize output, and minimize human error and safety risks [17]. There are basic components which are used in all automation systems:

1. Sensors: Detect environmental changes and send the information to other devices. Examples include temperature sensors, motion detectors, and light sensors.
2. Controllers: Devices like PLCs (programmable logic controllers) or computer systems that receive data from sensors and make decisions to control the operation of machines.
3. Actuators: Mechanical devices that perform the action dictated by the controller, such as motors, hydraulic pistons, or solenoids.
4. Human-machine interface: Interfaces that allow humans to interact with the machine, often providing controls and data visualization.

Automated systems for waste segregation and classification

To better comprehend the various research provided and identify which type of algorithm should be utilized, the following are the different studies used, the different algorithms, and the studies where they apply. It also has strengths and disadvantages, which can be utilized to determine which algorithm is best. There are numerous ways to separate garbage into distinct groups [18].

- Intelligent waste separator (IWS)—This can replace traditional waste management methods; the prototype automatically places garbage in altered basins and accepts inbound wastes using a multimedia-embedded processor, image processing, specifically the image recognition algorithm, and machine learning (ML) to select and separate waste. It produced a prototype of a shared garbage can with additional basins using multimedia technology.
- Spot garbage is a smartphone-based app—It detects a mound of garbage and identifies its position using cell phones' location access. The program identifies waste in photographs using a convolutional neural network architecture.
- IoT-based waste collection system with infrared sensors—This automatic garbage segregator categorizes rubbish using a contemporary classification method known as convolutional neural networks. This method provides the path for improved recycling and reuse operations, resulting in more effective waste management. The project aims to build and construct a system that can successfully segregate waste using artificial neural networks and image analysis, specifically the image recognition method.
- Adaptive and interactive modeling system (AIMS) that employs an induction technique to read sensor data streams and generate an efficient description of object features, hence defining material separation strategies.
- Waste segregation system—The project aims to design and create a system for properly segregating waste using artificial neural networks. The proposed system can be constructed to correctly categorize various sorts of waste by incorporating recognition and classification ideas into artificial neural networks.
- The automatic waste segregator and monitoring system classifies waste into three categories: metal, plastic, and moist (organic). Other wastes are classified as wet waste, representing organic waste such as leftovers and vegetable peels. Table 1.1 represents the different systems that were studied, their techniques, and their focus.

Robotics and automated machinery for handling and processing

Material recovery facilities (MRFs) play an important role in waste processing since they sort and separate many materials in waste streams. In recent years using artificial intelligence (AI) robots in MRFs has emerged as a viable alternative for reducing waste sorting costs while increasing productivity [24].

AI waste-sorting robots employ AI to identify and classify waste materials. These robots are intended to help automate and optimize garbage sorting, which may be time-consuming and costly for waste management facilities [25]. Moreover, they have the ability to improve sorting efficiency over time by utilizing available data by mimicking human brain learning and decision-making (Fig. 1.4).

TABLE 1.1 Previous studies with focused systems.

System	Techniques/focuses	References
Automated waste segregation system	Image processing/garbage classification	[18]
Spot garbage	Detection of individual object	[19]
IoT-based waste collection system with infrared	Infrared obstacle line sensor/garbage collection and sort	[20]
Adaptive and interactive modeling system (AIMS)	AI sorting techniques/sorting wastes	[21]
Waste segregation system using artificial neural networks	Acquisition of images of waste material/acquiring image	[22]
Automatic waste segregator and monitoring system	Arduino controlled board/sensor sorting waste	[23]

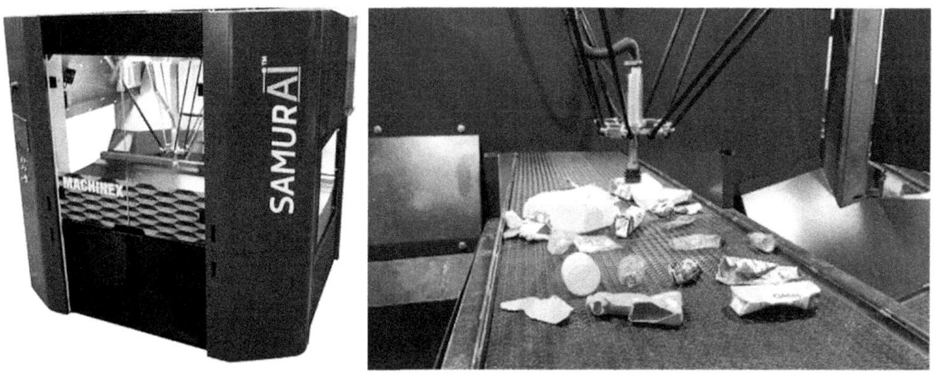

FIGURE 1.4 AI-based sorting robots.

The robots employ a computer vision system and deep ML to recognize and categorize various sorts of waste, such as plastics, metals, paper, and organic waste. This information is subsequently relayed to a robotic arm, which selects and sorts the classified items into the appropriate bins. To begin, a type of AI called computer vision can be used to identify recyclables in trash streams [26]. Then, AI robotics sort and separate the recyclables before moving on to the next stage of the recycling process. In practice, this implies that automated pickers physically transport recyclables from one site to another using information supplied by computer vision (Fig. 1.5). When paper and plastic garbage are deposited on a recycling conveyor belt, they can be scanned using computer vision, with each piece of waste detected and categorized based on its content. Then, an automated sorting solution, such as a robot, uses this data to identify which things to pick up and where to place them, and once the waste stream has been sorted by material and composition, a recycling baler can be used to prepare the recyclables for shipment. Integrating innovative technology into material

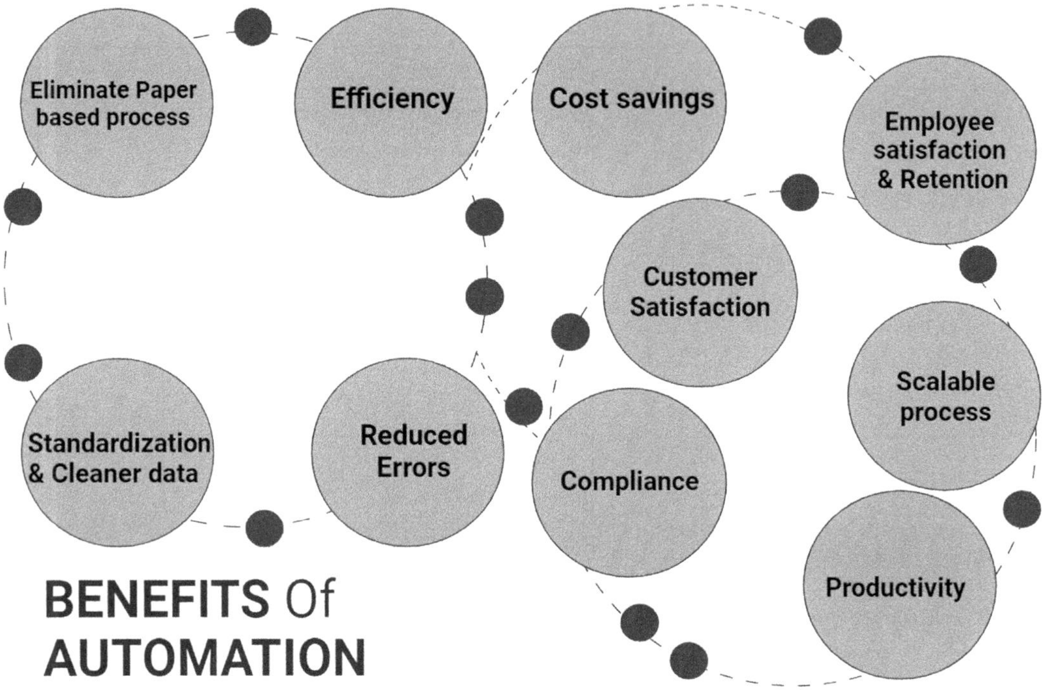

FIGURE 1.5 Various advantages of automation.

recovery facilities can improve efficiency and output [27]. Robotic sorting, in particular, provides several advantages to MRFs, including [27]:

- **Accuracy**: Accurate sorting is essential for an MRF, independent of the sorting methods or solid waste treatment equipment. Traditionally, picking and sorting at MRFs is done manually, but the accuracy provided by AI sorting robots means that many MRFs are transitioning to more technological alternatives. Using computer vision to identify recyclables allows for more precise picking. Recycleye Vision, for example, can identify many recycling classes. With this data available in near real-time, waste streams may be sorted more accurately using integrated robotics technologies.
- **Higher purity:** As one may imagine, better precision on sorting lines leads to higher purity outputs. This directly affects MRF revenue and allows sorting facilities to become more profitable. According to our recent case studies, Recycleye Vision and Recycleye Robotics boosted purity by 8% on an aluminum line, bringing output purity to 93% and improving output value by 20%. Similarly, a Recycleye waste-sorting robot improved purity on a fiber line by 12%, bringing the total purity to 97%.
- **Enhanced productivity:** Recycleye Robotics, capable of delivering 33,000 picks during 10 hours, demonstrates the significance of AI robotics in MRFs. Furthermore, the improved accuracy provided by computer vision and AI sorting technology ensures that increased efficiency translates into higher revenue.

- Bryson Recycling increased its paper line output by 10% after integrating Recycleye's waste-sorting robotic system at its MRF. With an average of 55 selections per minute and 99% purity on a split-stream plastics line, it is clear how AI robotic sorting equipment can boost efficiency and performance.
- While manual pickers normally work a 10-hour shift at MRFs, a robotic recycling sorting system can work longer. As a result, recycling facilities can work continuously to maximize productivity and outputs.
- **Reliability and consistency:** If MRFs are to create long-term advantages, they must provide optimal accuracy and higher production. Fortunately, robotic sorting provides the dependability and stability required. Once an AI robot has been integrated into an MRF's waste sorting line, it will follow defined algorithms and require little maintenance. With no ongoing training needs or significant turnover, AI sorting equipment can swiftly integrate into operations and give improved performance cheaper than alternative techniques.
- **Improved Safety:** Workplace safety is a top responsibility for any employer, but it is especially important in places such as MRFs, where industrial equipment is frequently used. Suppose MRFs use more automated equipment, such as robotic sortation systems. In that case, they can reduce their reliance on manual personnel, reducing the number of individuals who come into close contact with industrial equipment. Furthermore, robotic sorting equipment's consistency and routine operation allow for greater predictability and lower hazards in potentially hazardous situations. On an aluminum line, for example, the capacity of AI sorting robots to recognize high fire-risk canisters allowed them to be removed before processing, improving on-site fire safety.

Advantages and limitations of automation in waste-to-energy

Automation significantly improves energy recovery in WtE operations. Automated sorting and separation systems use advanced sensors and imaging technology to identify and separate waste items efficiently. This enables greater segregation of items suitable for energy generation, such as organic waste and biomass [28].

Automated systems can also improve combustion efficiency by carefully managing variables like fuel intake, temperature, and oxygen levels. This enables more efficient and thorough combustion, which leads to improved energy generation and lower emissions (Fig. 1.6). Furthermore, automated systems can monitor and alter processes in real-time, maximizing energy recovery based on waste composition and demand [29].

Automation in WtE facilities promotes sustainability in several ways. For starters, it allows for the recovery of energy from garbage that would otherwise wind up in landfills, lowering greenhouse gas emissions and lessening the environmental impact of waste disposal. Second, automation increases the quality and purity of recovered materials. Advanced sorting systems can separate recyclable elements from waste streams, including metals and plastics. These materials can then be recycled, helping conserve resources and reducing the demand for raw material extraction [31].

Furthermore, automation enables improved process monitoring and control, which leads to increased safety and lower operating and maintenance costs. Automated systems can detect and respond to problems like equipment breakdowns or abnormal process conditions, reducing downtime and increasing overall facility efficiency [32].

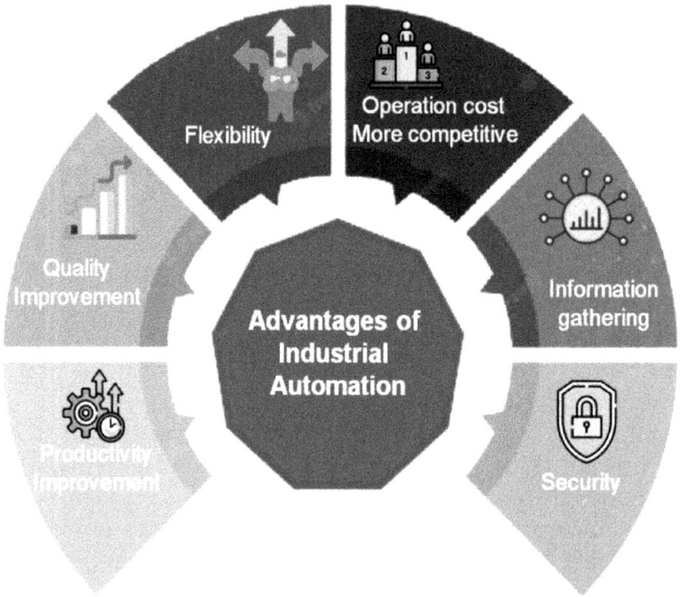

FIGURE 1.6 Advantages of automation in waste to energy [30]. Source: *From A. M. Al Shahrani, M. A. Alomar, K. N. Alqahtani, M. S. Basingab, B. Sharma, and A. Rizwan, Machine learning-enabled smart industrial automation systems using internet of things, Sensors 23(1) (2023). https://doi.org/10.3390/s23010324.*

Challenges

While automation has tremendous advantages, there are also obstacles and issues. The initial investment for deploying automation technologies can be large. However, long-term savings in operating efficiency, energy output, and reduced maintenance requirements frequently offset these expenses [33].

Additionally, incorporating automation technologies into existing WtE systems may necessitate retrofitting or modifications. Proper preparation and coordination are required to achieve successful implementation while minimizing disruptions [32].

Furthermore, workforce training and engagement are critical to reaping the full benefits of automation. Employees must be trained to operate and maintain automated systems properly. Open communication between management and employees is critical for addressing any concerns or resistance to change [32].

Machine learning applications in waste-to-energy

Most prefeasibility and case studies in underdeveloped nations have consistently demonstrated that WTE systems are technically and economically viable. However, lacking a regulatory framework, local design requirements, and limited regional research impedes the widespread adoption of such systems. Adopting cutting-edge technology like ML is crucial to meeting our sustainability objectives. It has huge promise for enhancing waste management and energy generation. The development of energy systems based on ML algorithms is progressing swiftly. This study makes substantial contributions to the field of WtE systems by conducting a detailed comparison of linear and nonlinear regression models [34].

Introduction/basics to machine learning

Classical ML is frequently defined by how an algorithm learns to make increasingly accurate predictions. There are four main types of ML: supervised learning, unsupervised learning, semisupervised learning, and reinforcement learning [34].

Data scientists choose the type of algorithm based on the nature of the data. Many methods and techniques apply to more than one of the basic ML types described below. They are frequently altered to numerous types based on the problem to be solved and the dataset. Deep learning techniques, such as convolutional neural networks and recurrent neural networks, are employed in supervised, unsupervised, and reinforcement learning tasks, depending on the problem and available data [34].

- **Supervised learning:** In supervised learning, data scientists provide algorithms with labeled training data and specify which variables the computer should look for associations. Both the algorithm's input and output are specified in supervised learning. Initially, most ML algorithms operated with supervised learning; now, unsupervised alternatives are gaining prevalent.
- **Unsupervised learning**: Unsupervised ML algorithms don't require labeling data. They sift through unlabeled data to look for patterns that can be used to group data points into subsets. Most types of deep learning, including neural networks, are unsupervised algorithms.
- **Reinforcement learning**: It creates an algorithm with a specific objective and rules for achieving that goal. A data scientist will also program the algorithm to seek positive incentives for actions that help it achieve its ultimate goal while avoiding punishments for actions that push it further away from it.

Machine learning for predictive maintenance of waste-to-energy

ML has great potential for converting WtE systems because of its ability to improve the overall process, increase efficiency, and reduce waste. With the ongoing rise in demand for renewable energy sources, WtE systems are positioned to play an increasingly important role (Fig. 1.7). ML plays an important role in optimizing the performance of these systems and facilitating their contributions to a sustainable future [35]. In light of those mentioned above, a comprehensive WtE framework is simulated, and ML methods are used to predict its efficacy in electricity generation, air heating capacity, hydrogen creation, oxygen retention, and purified water production. ML algorithms are built using both linear and nonlinear regression models, and their results are thoroughly examined for comparison [36].

Optimization of energy production using machine learning

The contribution in this field provides slight variations. Most studies focus on developing a predictive model for utilizing energy utilization regarding process inputs. Liu et al. focus on forecasting consumption by analyzing recent trends in energy usage [38]. Moreover, Kovacic and Sarler et al. included production planning concepts, thereby paving the way for optimizing energy through scheduling strategies [39]. To fulfill the objectives, models based

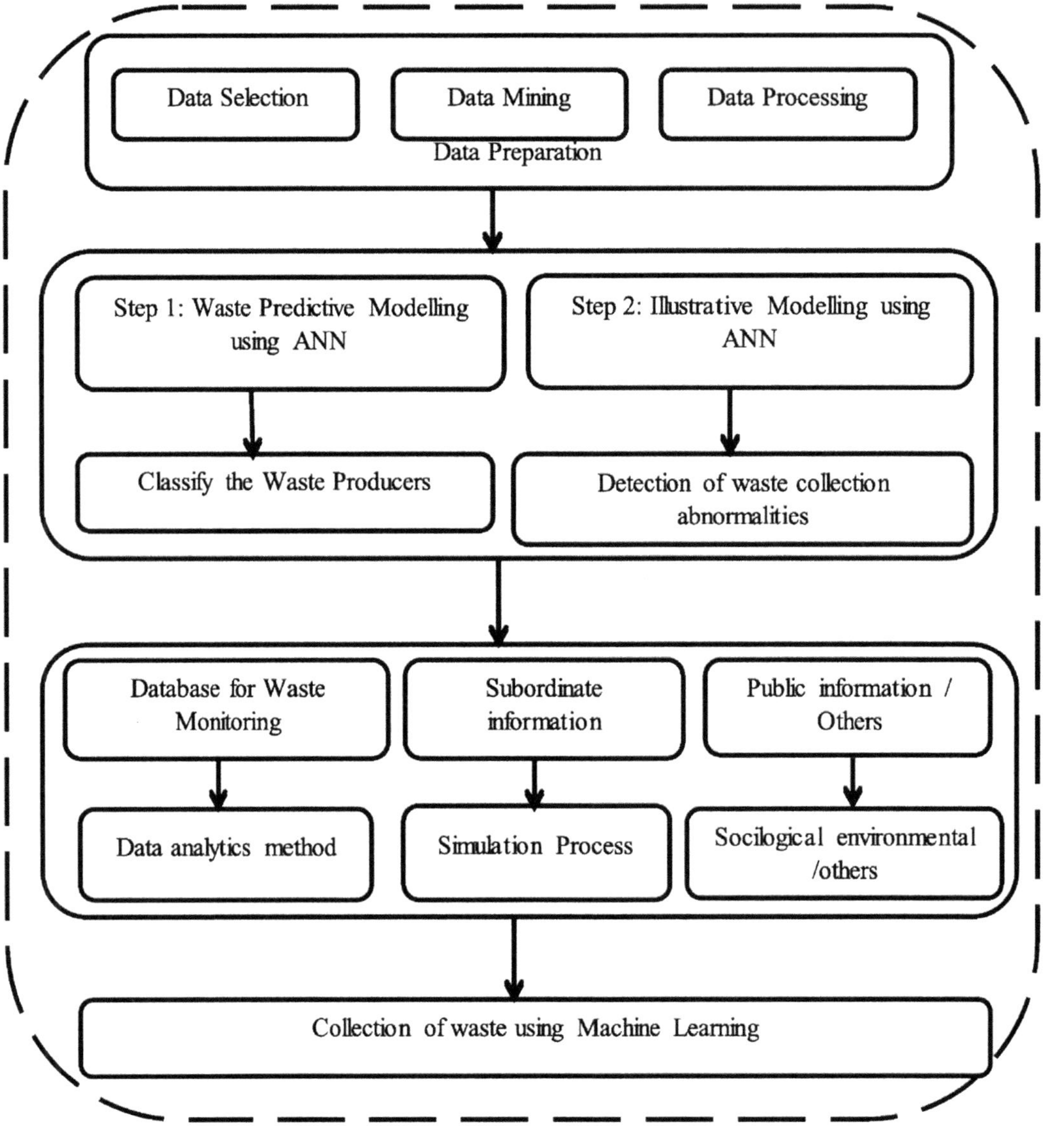

FIGURE 1.7 The machine learning model for waste management [37]. Source: *From J. Huang, and D. D. Koroteev, Artificial intelligence for planning of energy and waste management, Sustainable Energy Technologies and Assessments 47, 101426 (2021). https://doi.org/10.1016/j.seta.2021.101426.*

on ANN, genetic algorithms, signal processing, and least squares support vector machine are used. One primary objective behind this form of estimation lies in the need to align energy supply with demand closely.

The major focus in this area of study primarily focuses on defining and addressing problems in optimization, resulting in a set of inputs to improve energy efficiency. A different approach to the trend is observed by Monedero et al. where optimization seeks to identify a point adjacent to the current operating point by searching through historical data for performance enhancement [40]. All the methods share the use of a modeling phase to understand system dynamics utilizing a variety of ML instruments. The optimal resulting solutions also differ in focusing on various approaches to improve efficiency. Wang and Liang et al. focus on developing optimal scheduling configurations to enhance resource usage [41,42] and Golkarnarenji et al. aim to study the changes to input variables in continuous processes [43].

The ever-growing area of energy management (EM) has, for the last 20 years, seen that literature constantly suggesting we must find better ways to generate and use power. Every printed paper says that [44] people are interested in research that tries to solve difficult optimization problems with new methods. These methods include using ML algorithms for optimization, most of all other methods developed so far, optimization algorithms that work in real-time, heuristics based on rules of thumb rather than knowledge about the problem at hand itself; hyperheuristic approaches that combine several simpler heuristics into one complex adaptive strategy or metaheuristic procedures which are used for finding approximate solutions to difficult combinatorial optimization problems where an exhaustive search is impracticable (Fig. 1.8).

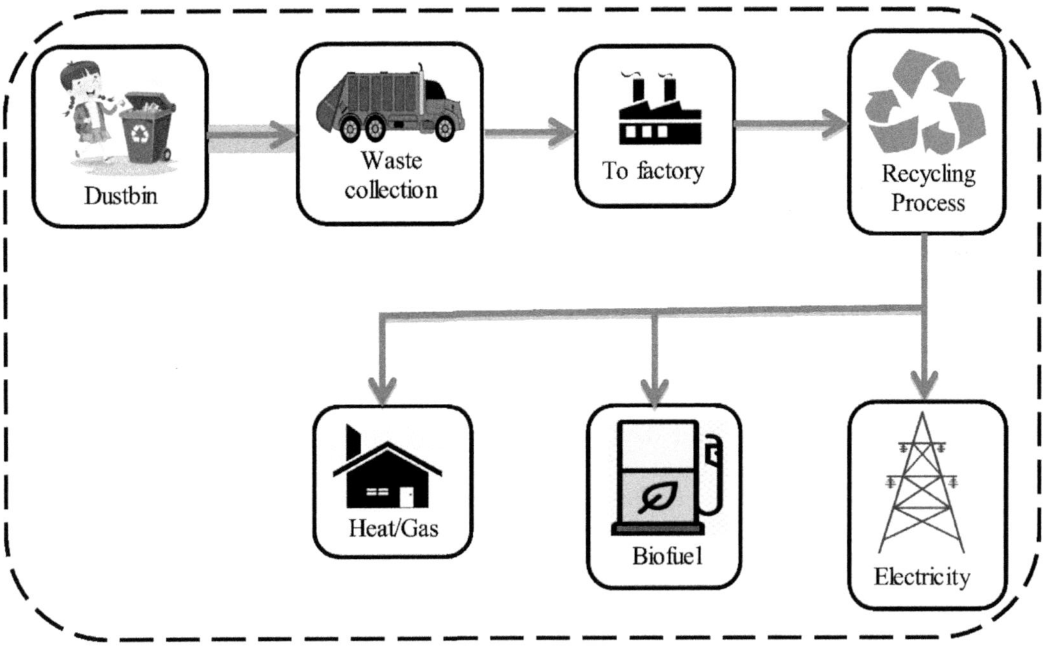

FIGURE 1.8 Waste management process and reproduction of products [25]. Source: *From B. Fang, J. Yu, Z. Chen, A. I. Osman, M. Farghali, I. Ihara, E. H. Hamza, D. W. Rooney, and P. S. Yap, Artificial intelligence for waste management in smart cities: a review, Environmental Chemistry Letters 21(4), 1959–1989 (2023). https://doi.org/10.1007/s10311-023-01604-3.*

In energy systems, the interconnection between EM and optimization is unmistakable. The exchange and formulation of innovative energy optimization and management strategies are pivotal across all sectors. Previous research in this field has covered different topics, including energy consumption management, optimization strategies to prolong equipment lifespan, enhancing the performance of system components, and optimizing renewable energy production systems like wind, solar, and hydropower [45–47]. Some studies, such as the work by Teng et al., explore tangential aspects related to energy systems, such as EM applications in electric vehicles and fuel cells, to bolster energy efficiency [48].

Various research studies have shown that waste to energy can be produced through different technologies. The studies have included or focused on optimizing the relevant processes and important technological parameters. Some research focuses on multicriteria decision analysis for WtE production.

Integration of automation and machine learning

Automation is a technology that requires minimum human intervention to perform tasks. This process reduces manual errors and enhances efficiency. Automated systems can range from simple rule-based algorithms to complex workflows orchestrated by software. ML enables the system to learn and identify data and improve over time without being programmed externally. ML algorithms identify patterns, make predictions, and learn from feedback, allowing them to perform tasks more accurately as they accumulate more data (Fig. 1.9). ML is swiftly reshaping operations and enhancing the quality of the results delivery. ML has been classified into four different categories upon their wide applications (Fig. 1.10).

Synergies between automation and machine learning in waste-to-energy

ML models are created through the iterative process of algorithms, where various parameters are optimized internally. These algorithms typically contain multiple hyperparameters, which users must set externally to control. These hyperparameters usually remain constant throughout the process. For instance, a neural network model could have millions of parameters like weights and bias values. The learning algorithm responsible for determining these parameters is governed by several hyperparameters, such as the learning rate, optimization schedule, and data augmentation techniques. Alongside selecting the appropriate algorithm, the choice of hyperparameters significantly influences the final model performance and the effectiveness of the learning process. Various sensors are designed to improve cities' infrastructure through effective waste management and classification. The physical components of the system include wastebins, grippers, and dump stations. In the beginning, the waste is collected in the smart bins and the data are being stored in the cloud. They send a notification on the web or mobile app when their limit is exceeded. This has been illustrated in Fig. 1.11. The streamlined approach where the waste bins are initially equipped with sensors to observe waste levels. The authority for waste management is responsible for collecting waste from both residential and commercial

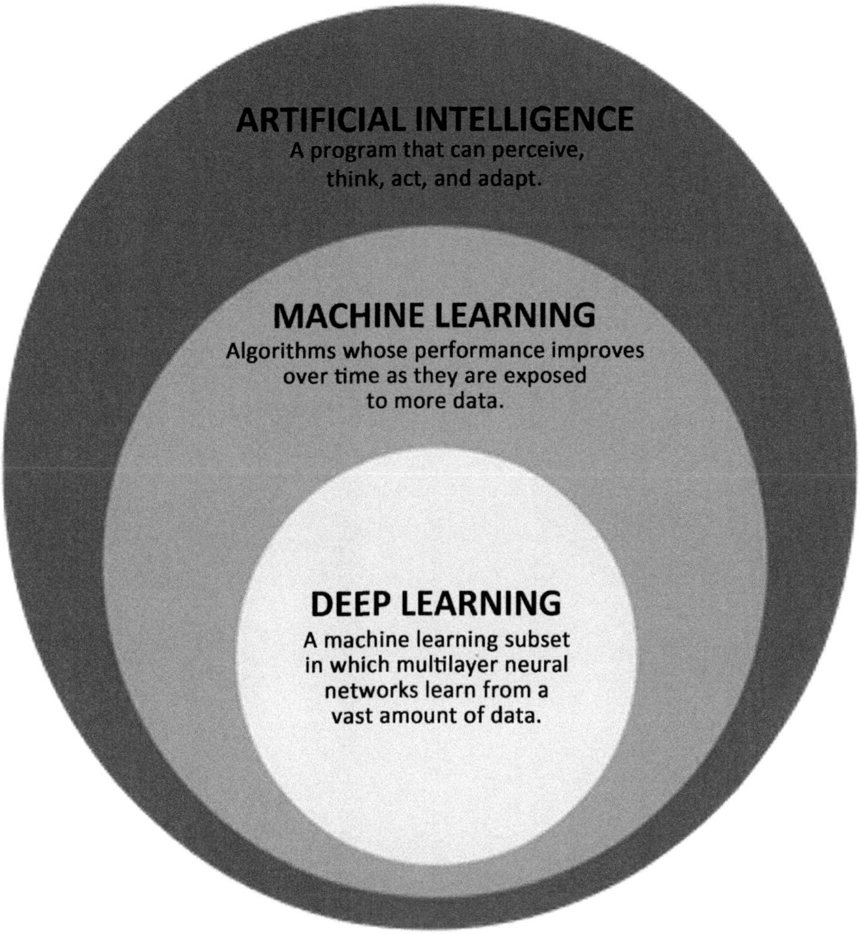

FIGURE 1.9 An overview of artificial intelligence, machine learning, and deep learning [49]. Source: *From J. Jenis, J. Ondriga, S. Hrcek, F. Brumercik, M. Cuchor, and E. Sadovsky, Engineering applications of artificial intelligence in mechanical design and optimization, Machines 11(6), 577 (2023). https://doi.org/10.3390/machines11060577.*

locations. With the help of a camera, the images of waste dump are captured and it is then segregated into grids using grid segmentation technique. The image acquired is processed with the help of a deep learning algorithm called VGG-16 for identification. Each waste is identified using a classifier and sorted into appropriate containers using a robotic arm. The overall latency is reduced by employing both edge and cloud computing. The waste classification model is organized into three main modules: edge node processing, cloud processing, and the control unit.

The strategy of grid segmentation divides the test image into grid-like cells to improve the efficiency of waste classification. This helps in reducing computational costs and enhances performance. In the first phase, the image is mapped onto a grid, converted to

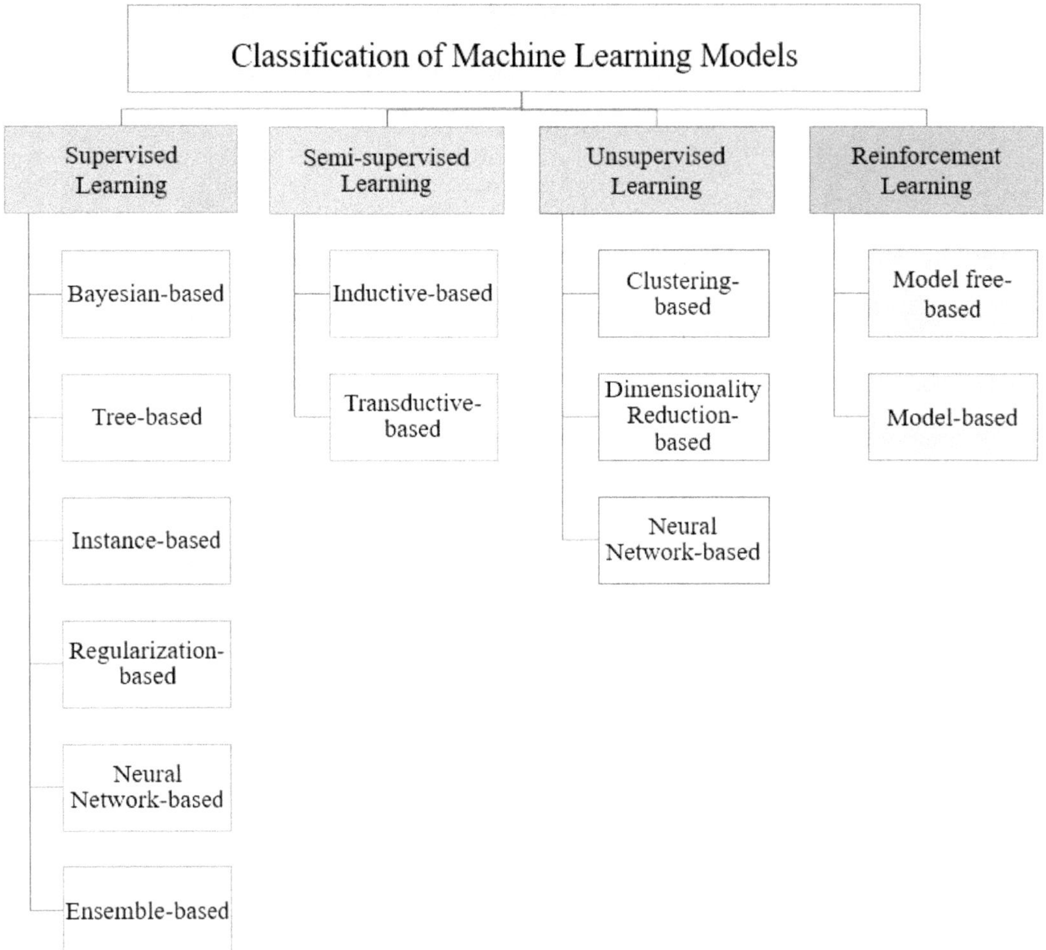

FIGURE 1.10 Different machine-learning models [50]. Source: *From T. Talaei Khoei, and N. Kaabouch, Machine learning: models, challenges, and research directions, Future Internet 15(10), 332 (2023). https://doi.org/10.3390/fi15100332.*

grayscale, and segmented based on size. In the second phase, each cell segment is classified using the VGG-16 algorithm based on texture, color, and position features. The process continues column-wise, with each identified waste object being picked by a robotic arm and placed in the appropriate bin, ensuring only correctly classified segments are included in the final result (Fig. 1.12).

There are various studies where ML plays a larger part in the automatic collection and disposal of solid waste. As the part of process, 34 categories are mapped into 10 base classes corresponding to different types of recycling and composting plants. Mapping ensures that each item is correctly assigned to appropriate recycling and composting facility (Fig. 1.13).

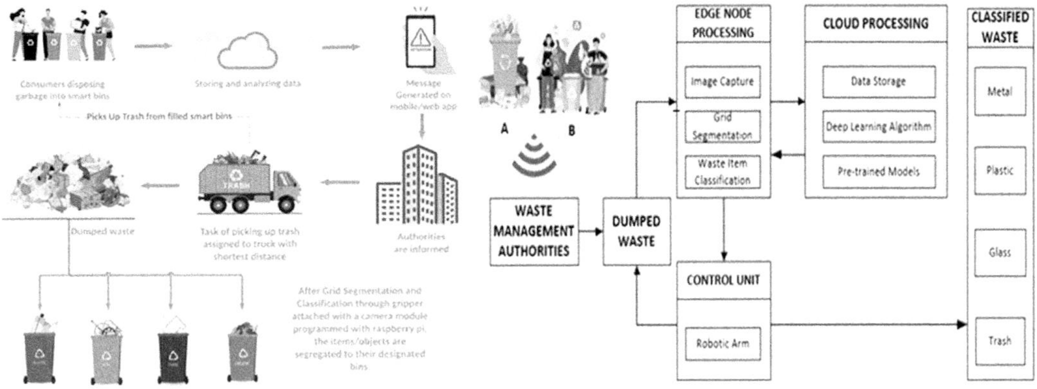

FIGURE 1.11 Mechanism of different smart waste collection and simplified system approach [18]. Source: *From S. M. Cheema, A. Hannan, and I. M. Pires, Smart waste management and classification systems using cutting edge approach, Sustainability 14(16), 10226 (2022). https://doi.org/10.3390/su141610226.*

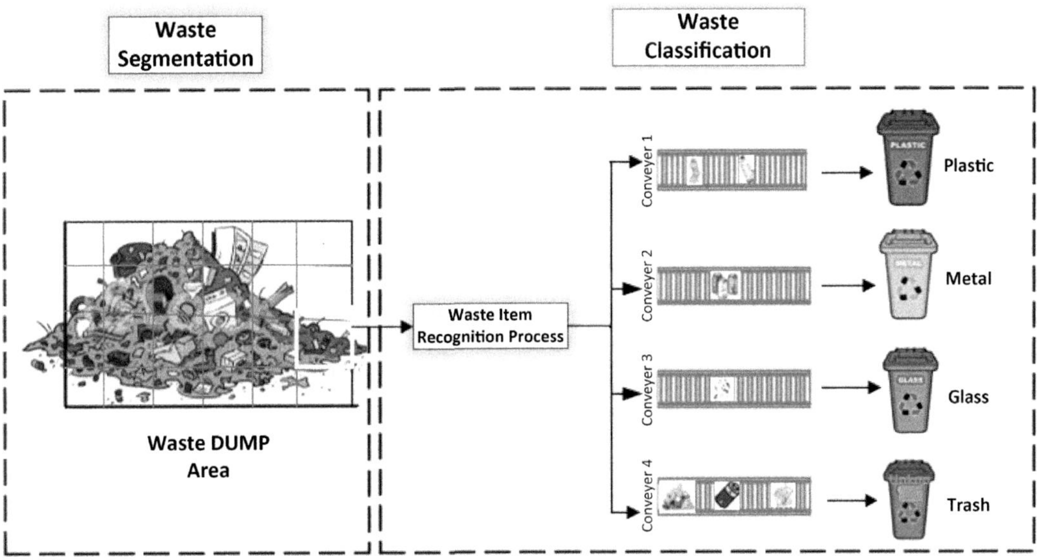

FIGURE 1.12 Grid segmentation and classification [18]. Source: *From S. M. Cheema, A. Hannan, and I. M. Pires, Smart waste management and classification systems using cutting edge approach, Sustainability 14(16), 10226 (2022). https://doi.org/10.3390/su141610226.*

AutoML aims to generate the optimal model for a given dataset by exploring various ML algorithms and their associated hyperparameter configurations. This process has two primary objectives: firstly, achieving the highest performance for a specific ML task (e.g., classification or regression accuracy) on the dataset, and secondly, ensuring robust generalization of the model on unseen data. Strategies such as hyperparameter

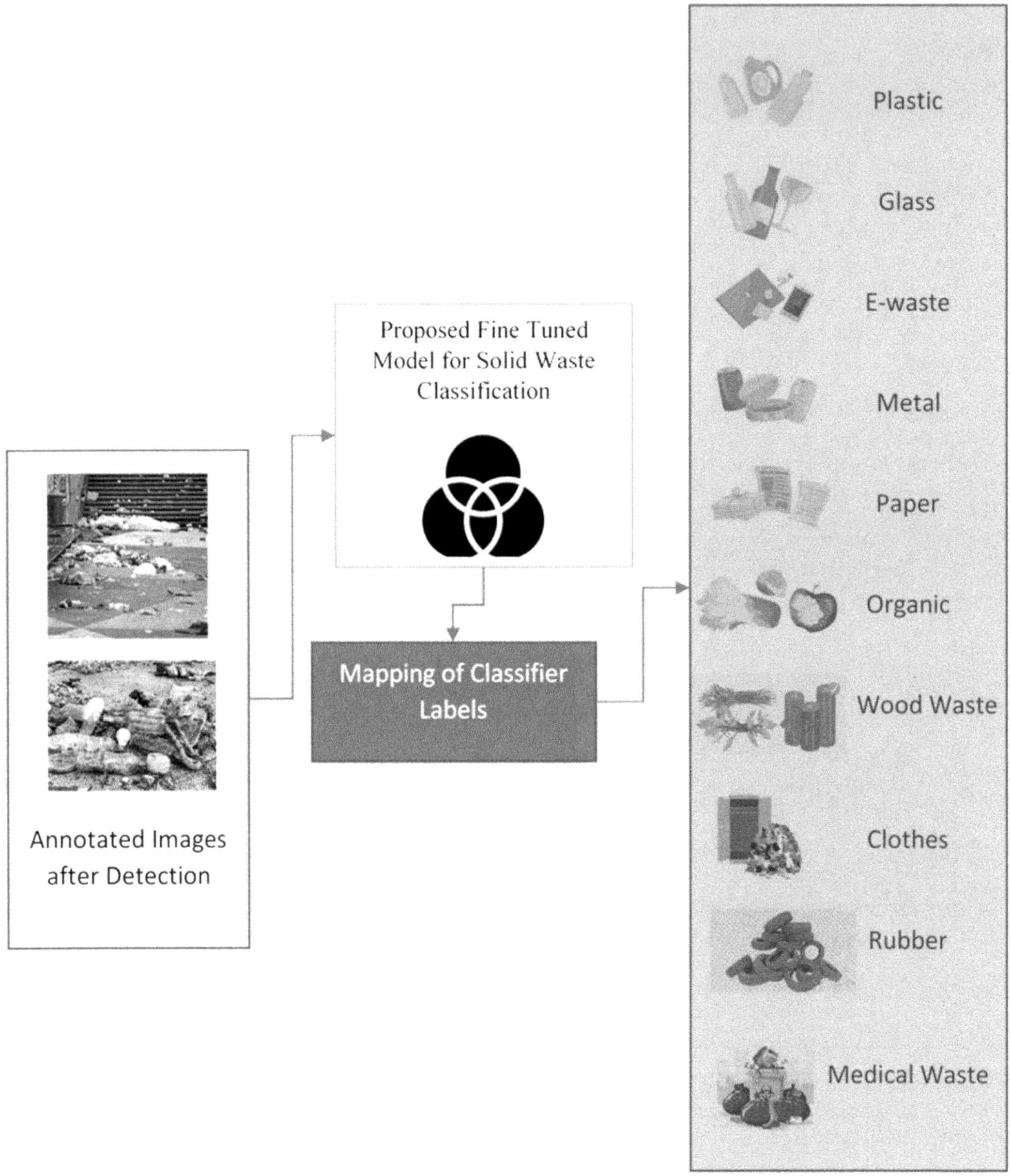

FIGURE 1.13 Solid waste classification using transfer learning model [51]. Source: *From M. Malik, S. Sharma, M. Uddin, C.-L. Chen, C.-M. Wu, P. Soni, and S. Chaudhary, Waste classification for sustainable development using image recognition with deep learning neural network models, Sustainability 14(12), 7222 (2022). https://doi.org/10.3390/su14127222.*

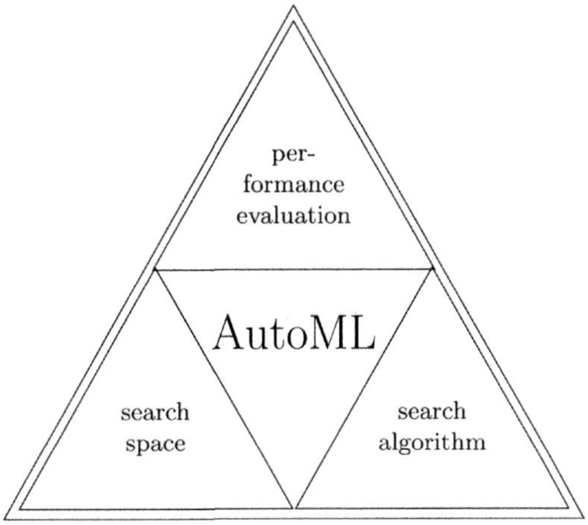

FIGURE 1.14 Components of AutoML [53]. Source: *From M. Baratchi, C. Wang, S. Limmer, J. N., van Rijn, H. Hoos, T. Bäck, and M. Olhofer, Automated machine learning: past, present and future, Artificial Intelligence Review 57(5) (2024). https://doi.org/10.1007/s10462-024-10726-1.*

optimization contribute to attaining the former goal, while techniques like cross-validation aid in assessing the model's generalization capabilities across different hyperparameter configurations (Fig. 1.14). AutoML encompasses three closely linked problems: algorithm (or model) selection, which involves evaluating multiple ML algorithms with fixed hyper-parameter settings; hyperparameter optimization, which focuses on fine-tuning a single algorithm; and combined algorithm selection and hyperparameter optimization, as outlined by Thornton et al. in 2013 [52].

Components of AutoML

It has three components: search space, search algorithm, and performance evaluation [54,55].

1. Search space: It provides all the design decisions that require consideration in the ML solution space. This also provides a range of options for selecting learning algorithms (and consequently, model classes) and their associated hyperparameters.
2. Search algorithm: The search algorithm determines how to explore the search space to enhance the performance of the resultant model. Initially, methods focused on search spaces small enough to allow exhaustive enumeration.
3. Performance evaluation: The search algorithm iteratively presents a sequence of potential solutions (such as learning algorithms or hyperparameter configurations) for tackling the specified machine-learning task. Evaluating their performance is crucial for determining the optimal choice. Moreover, assessing the final output (i.e., the model) generated by an AutoML system is necessary.

Benefits and challenges in integrating automation and machine learning

Integrating ML with optical microscopy has revolutionized the way scientists and researchers analyze and interpret various data. One such example is given in Fig. 1.15 where ML and optical microscopy techniques namely NIR or MID-IR are integrated to increase the efficiency in plastic sorting. This helps in the enhancement of real-time analysis.

Fusing ML with automation technologies fosters innovation and simplifies intricate workflows. ML combined with automation brings forth numerous advantages, such as

1. *Predictability*—ML algorithms can accurately predict potential errors or system failures and take corrective action before they occur.
2. *Cost savings*—Automation with ML can reduce operational costs by minimizing the need for human intervention in routine tasks.
3. *Efficiency*—Automating repetitive tasks reduces errors and increases efficiency, enabling businesses to focus on more strategic tasks.
4. *Scalability*—ML integrated with automation solutions can scale to meet businesses' growing demands, increasing productivity and output.
5. *Personalization*—ML solutions can personalize customer experiences by analyzing data and offering customized recommendations.

The integration of ML with automation showcases many challenges (Fig. 1.16):

1. *Job displacement*—Adopting AI-powered automation may lead to job displacement and require a transition to new types of work.

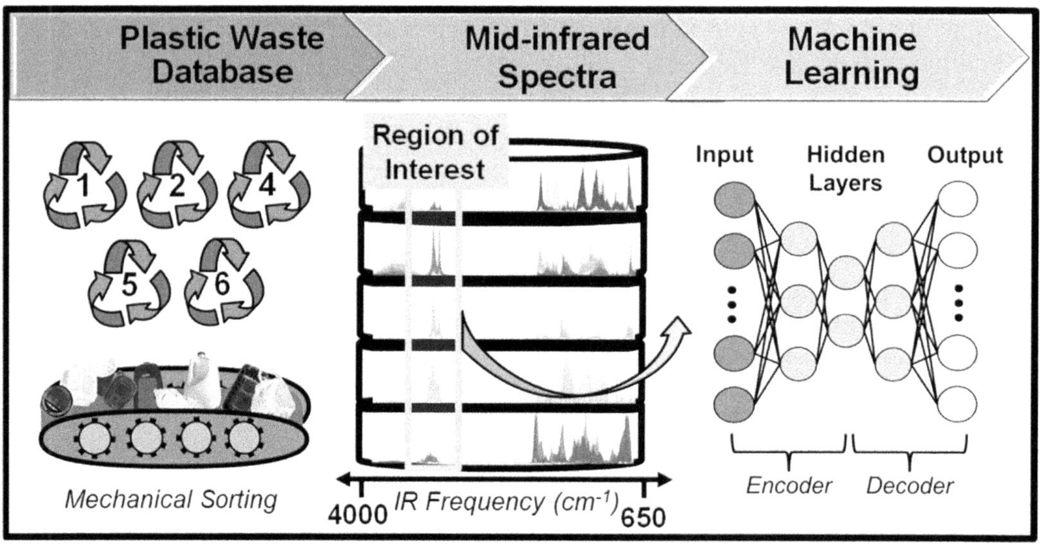

FIGURE 1.15 Sorting of plastic waste by utilizing both machine learning and optical spectroscopy [56]. Source: *From C. Lubongo, M. A. A. Bin Daej, and P. Alexandridis, Recent developments in technology for sorting plastic for recycling: the emergence of artificial intelligence and the rise of the robots, Recycling 9(4) (2024). https://doi.org/10.3390/recycling9040059.*

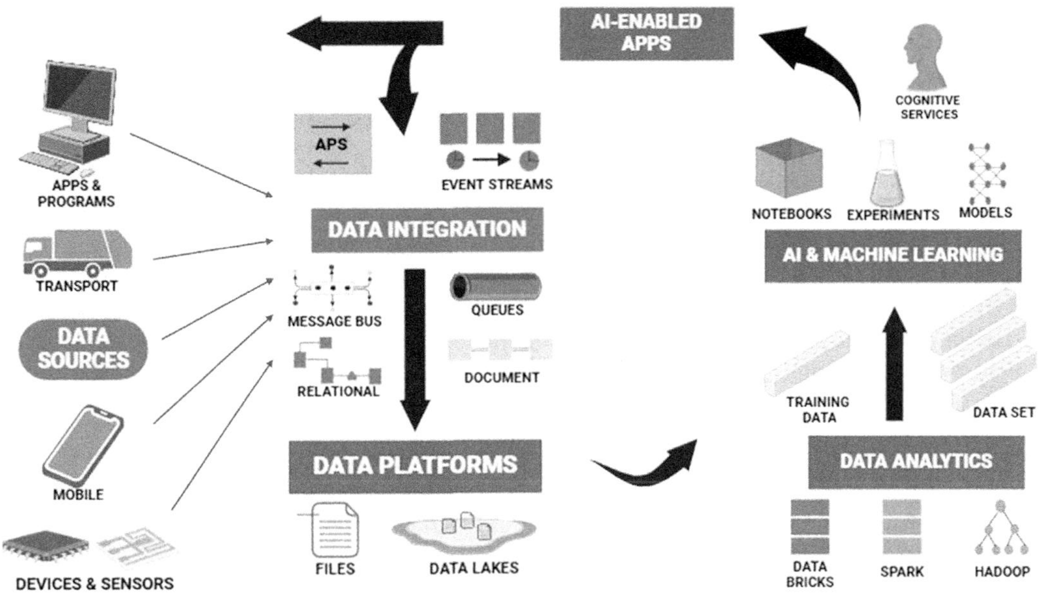

FIGURE 1.16 Integration of machine learning and automation.

2. *Ethical considerations*—The use of AI in decision-making processes raises ethical concerns, such as bias and privacy.
3. *Implementation costs*—Implementing AI-powered automation requires significant training, infrastructure, and technology investments, which may challenge smaller businesses.
4. *Security risks*—Integrating AI with automation solutions may increase security risks and vulnerabilities.

Prospects and emerging trends

The convergence of robust, fast-connecting networks, high-speed capabilities, and technologies like ML will drive the evolution of a sustainable smart energy distribution industry. As the environmental impacts of power generation and usage escalate, utility providers actively seek innovative solutions to mitigate their environmental footprint. ML is presently being incorporated by pioneers in both conventional and renewable energy sectors in various capacities to enhance the adoption of alternative energy sources, refine renewable energy predictions, promote energy conservation, and enhance usability.

1. Smart grid and distribution generation: As a crucial aspect of WtE conversion, the challenges of global warming and climate change necessitate a substantial transition toward the more sustainable utilization of resources, particularly renewable ones, encompassing their development, supply, and consumption. In this era of energy transformation, smart grids are imperative for enhancing the efficiency and effectiveness of power networks and systems [57].

2. ML transforming the conversion of waste to energy: By the year 2050, renewable energy is projected to meet approximately 70% of the global energy demand [58]. Modern ML technologies enable this restructuring of energy distribution [59].
3. Data-driven strategies: In the WtE conversion, using data-driven approaches coupled with ML facilitates discussions on implementing data-driven management and data science. This, in turn, enables the formulation of long-term energy strategies, ultimately providing a competitive edge for a sustainable future.
4. Decision-making: ML can potentially improve the efficiency of AI decision-making processes for WtE conversion. Reinforcement learning, applicable to scenarios with dynamically changing environments, can be employed in various contexts such as machine control, robotics, and optimization tasks like heating, ventilation, and air conditioning, as well as supply chain optimization [35].

Conclusion

Automation is transforming WtE operations, increasing energy recovery and overall sustainability. It makes garbage sorting and separation more efficient, optimizes combustion processes, and allows for real-time monitoring of facility operations. While there are hurdles and considerations, the benefits of automating WtE systems far surpass the initial costs. With continued technological breakthroughs, the future appears hopeful for automation to drive sustainable waste management practices while lowering our dependency on fossil fuels.

To summarize, incorporating innovative technology into WtE systems, notably through automation and ML, represents an industry-changing innovation. Automation increases operating efficiency, safety, and precision while expediting trash sorting, processing, and energy recovery. Meanwhile, machine learning algorithms provide powerful tools for predictive maintenance, process optimization, and data-driven decision-making, which result in considerable increases in system performance and energy yield. These technologies work together to overcome some of the major difficulties that traditional WtE approaches confront, such as operational inefficiencies and inefficient resource utilization. As research and development progresses, the synergy between automation and machine learning promises to drive additional breakthroughs, making WtE processes more sustainable, cost-effective, and capable of meeting rising renewable energy demand. Embracing these technologies not only improves the performance of WtE systems but also helps to achieve larger environmental and energy goals, paving the way for a more sustainable and resource-efficient future.

References

[1] S. Sasmoko, M.Z. Akhtar, H.ur R. Khan, S. Sriyanto, M.K. Jabor, A. Rashid, et al., How do industrial ecology, energy efficiency, and waste recycling technology (circular economy) fit into China's plan to protect the environment? Up to speed, Recycling 7 (6) (2022) 83. Available from https://doi.org/10.3390/recycling7060083.

[2] A. Siddiqua, J.N. Hahladakis, W.A.K.A. Al-Attiya, An overview of the environmental pollution and health effects associated with waste landfilling and open dumping, Environmental Science and Pollution Research International 29 (39) (2022) 58514–58536. Available from https://doi.org/10.1007/s11356-022-21578-z.

[3] E. Kazakova, J. Lee, Sustainable manufacturing for a circular economy, Sustainability 14 (24) (2022) 17010. Available from https://doi.org/10.3390/su142417010.

[4] U. Awan, R. Sroufe, Sustainability in the circular economy: insights and dynamics of designing circular business models, Applied Sciences 12 (3) (2022) 1521. Available from https://doi.org/10.3390/app12031521.

[5] H. Fahmi Jamal, A. Abd El-Fattah, An overview of solid waste management and privatization in Kingdom of Bahrain, Frontiers in Environmental Science 11 (2023). Available from https://doi.org/10.3389/fenvs.2023.1302711.

[6] P. Roy, A.K. Mohanty, P. Dick, M. Misra, A review on the challenges and choices for food waste valorization: environmental and economic impacts, ACS Environmental Au 3 (2) (2023) 58–75. Available from https://doi.org/10.1021/acsenvironau.2c00050.

[7] P.J. Landrigan, H. Raps, M. Cropper, C. Bald, M. Brunner, E.M. Canonizado, et al., The Minderoo-Monaco Commission on Plastics and Human Health, Annals of Global Health 89 (1), 23. https://doi.org/10.5334/aogh.4056.

[8] M.A. Alao, O.M. Popoola, T.R. Ayodele, Waste-to-energy nexus: an overview of technologies and implementation for sustainable development, Cleaner Energy Systems 3 (2022) 100034. Available from https://doi.org/10.1016/j.cles.2022.100034.

[9] J. Lorenzo Llanes, E. Kalogirou, Waste-to-energy conversion in Havana: technical and economic analysis, Social Sciences 8 (4) (2019) 119. Available from https://doi.org/10.3390/socsci8040119.

[10] H. Madkhali, S. Duraib, L. Nguyen, M. Prasad, M. Sharma, S. Joshi, A comprehensive review on e-waste management strategies and prediction methods: a Saudi Arabia perspective, Knowledge 3 (2) (2023) 163–179. Available from https://doi.org/10.3390/knowledge3020012.

[11] W. Foster, U. Azimov, P. Gauthier-Maradei, L.C. Molano, M. Combrinck, J. Munoz, et al., Waste-to-energy conversion technologies in the UK: processes and barriers – a review, Renewable and Sustainable Energy Reviews 135 (2021) 110226. Available from https://doi.org/10.1016/j.rser.2020.110226.

[12] S. Rezania, B. Oryani, V.R. Nasrollahi, N. Darajeh, M. Lotfi Ghahroud, K. Mehranzamir, Review on waste-to-energy approaches toward a circular economy in developed and developing countries, Processes 11 (9) (2023) 2566. Available from https://doi.org/10.3390/pr11092566.

[13] K.E. Okedu, H.F. Barghash, H.A. Al Nadabi, Sustainable waste management strategies for effective energy utilization in oman: a review, Frontiers in Bioengineering and Biotechnology 10 (2022) 825728. Available from https://doi.org/10.3389/fbioe.2022.825728.

[14] M.A. Nanda, W. Sugandi, A.K. Wijayanto, H. Imantho, A. Sutawijaya, L.O. Nelwan, et al., The waste-to-energy (WtE) technology to support alternative fuels for agriculture in the context of effective solid waste management in the Jabodetabek Area, Indonesia, Energies 16 (24) (2023) 7980. Available from https://doi.org/10.3390/en16247980.

[15] B.N. Kulkarni, V. Anantharama, Solid waste to energy: a prognostic for sound waste management, in: C. Baskar, S. Ramakrishna, S. Baskar, R. Sharma, A. Chinnappan, R. Sehrawat (Eds.), Handbook of Solid Waste Management: Sustainability through Circular Economy, Springer Nature, Singapore, 2022, pp. 2047–2078. Available from https://doi.org/10.1007/978-981-16-4230-2_86.

[16] J. Sousa; A. Rebelo; J. Cardoso, Automation of Waste Sorting with Deep Learning, 2019, p. 48. https://doi.org/10.1109/WVC.2019.8876924.

[17] H. Wilts, B.R. Garcia, R.G. Garlito, L.S. Gómez, E.G. Prieto, Artificial intelligence in the sorting of municipal waste as an enabler of the circular economy, Resources 10 (4) (2021) 28. Available from https://doi.org/10.3390/resources10040028.

[18] S.M. Cheema, A. Hannan, I.M. Pires, Smart waste management and classification systems using cutting edge approach, Sustainability 14 (16) (2022) 10226. Available from https://doi.org/10.3390/su141610226.

[19] Y. Himeur, M. Elnour, F. Fadli, N. Meskin, I. Petri, Y. Rezgui, et al., AI-big data analytics for building automation and management systems: a survey, actual challenges and future perspectives, Artificial Intelligence Review 56 (6) (2023) 4929–5021. Available from https://doi.org/10.1007/s10462-022-10286-2.

[20] P. Kumar, 3 - IoT-based solid waste management and collection system using infrared sensors, in: S. Roy, T.A. Tran, K. Natarajan (Eds.), Recent Advancement of IoT Devices in Pollution Control and Health Applications, Advances in Pollution Research; Woodhead Publishing, 2023, pp. 29–36. Available from https://doi.org/10.1016/B978-0-323-95876-9.00003-3.

[21] I. Gligorea, M. Cioca, R. Oancea, A.-T. Gorski, H. Gorski, P. Tudorache, Adaptive learning using artificial intelligence in e-learning: a literature review, Education Sciences 13 (12) (2023) 1216. Available from https://doi.org/10.3390/educsci13121216.

[22] M.A. Mohammed, M.J. Abdulhasan, N.M. Kumar, K.H. Abdulkareem, S.A. Mostafa, M.S. Maashi, et al., Automated waste-sorting and recycling classification using artificial neural network and features fusion: a digital-enabled circular economy vision for smart cities, Multimedia Tools and Applications (2022) 1–16. Available from https://doi.org/10.1007/s11042-021-11537-0.

[23] A. Haseeb, SmartSort—Waste Segregation System Using Machine Learning and IoT, 2024.

[24] P.G. Chavhan, V.N. Khedka, M. Gupta, K. Agarwal, Automatic waste segregator based on IoT & ML using keras model and streamlit, International Journal of Intelligent Systems and Applications in Engineering 12 (2) (2024) 787–799.

[25] B. Fang, J. Yu, Z. Chen, A.I. Osman, M. Farghali, I. Ihara, et al., Artificial intelligence for waste management in smart cities: a review, Environmental Chemistry Letters 21 (4) (2023) 1959–1989. Available from https://doi.org/10.1007/s10311-023-01604-3.

[26] W. Lu, J. Chen, Computer vision for solid waste sorting: a critical review of academic research, Waste Management 142 (2022) 29–43. Available from https://doi.org/10.1016/j.wasman.2022.02.009.

[27] M.I.B. Ahmed, R.B. Alotaibi, R.A. Al-Qahtani, R.S. Al-Qahtani, S.S. Al-Hetela, K.A. Al-Matar, et al., Deep learning approach to recyclable products classification: towards sustainable waste management, Sustainability 15 (14) (2023) 11138. Available from https://doi.org/10.3390/su151411138.

[28] A. Satav, S. Kubade, C. Amrutkar, G. Arya, A. Pawar, A state-of-the-art review on robotics in waste sorting: scope and challenges, International Journal on Interactive Design and Manufacturing (IJIDeM) 17 (2023) 1–18. Available from https://doi.org/10.1007/s12008-023-01320-w.

[29] D.B. Olawade, O. Fapohunda, O.Z. Wada, S.O. Usman, A.O. Ige, O. Ajisafe, et al., Smart waste management: a paradigm shift enabled by artificial intelligence, Waste Management Bulletin 2 (2) (2024) 244–263. Available from https://doi.org/10.1016/j.wmb.2024.05.001.

[30] A.M. Al Shahrani, M.A. Alomar, K.N. Alqahtani, M.S. Basingab, B. Sharma, A. Rizwan, Machine learning-enabled smart industrial automation systems using internet of things, Sensors 23 (1) (2023) 324. Available from https://doi.org/10.3390/s23010324.

[31] O. Mathlouthi, M. Trabelsi Ayadi, I. Ghorbel Abid, A. Nassour, Composting and mechanical biological treatment for reducing greenhouse gas emissions in Bizerte, Tunisia, Sustainability 16 (2) (2024) 694. Available from https://doi.org/10.3390/su16020694.

[32] J.T. Licardo, M. Domjan, T. Orehovački, Intelligent robotics—a systematic review of emerging technologies and trends, Electronics 13 (3) (2024) 542. Available from https://doi.org/10.3390/electronics13030542.

[33] P. Onu, A. Pradhan, C. Mbohwa, The potential of Industry 4.0 for renewable energy and materials development – the case of multinational energy companies, Heliyon 9 (10) (2023) e20547. Available from https://doi.org/10.1016/j.heliyon.2023.e20547.

[34] Y. Zhang, A.J. Aldosky, V. Goyal, M.N. Meqdad, T.U.K. Nutakki, T.R. Alsenani, et al., A machine learning study on a municipal solid waste-to-energy system for environmental sustainability in a multi-generation energy system for hydrogen production, Process Safety and Environmental Protection 182 (2024) 1171–1184. Available from https://doi.org/10.1016/j.psep.2023.12.054.

[35] Z.M. Çınar, A. Abdussalam Nuhu, Q. Zeeshan, O. Korhan, M. Asmael, B. Safaei, Machine learning in predictive maintenance towards sustainable smart manufacturing in Industry 4.0, Sustainability 12 (19) (2020) 8211. Available from https://doi.org/10.3390/su12198211.

[36] S. Onwusinkwue, F. Osasona, I. Ahmad, A. Anyanwu, S. Dawodu, O. Obi, et al., Artificial intelligence (AI) in renewable energy: a review of predictive maintenance and energy optimization, World Journal of Advanced Research and Reviews 21 (2024) 2487–2799. Available from https://doi.org/10.30574/wjarr.2024.21.1.0347.

[37] J. Huang, D.D. Koroteev, Artificial intelligence for planning of energy and waste management, Sustainable Energy Technologies and Assessments 47 (2021) 101426. Available from https://doi.org/10.1016/j.seta.2021.101426.

[38] Z. Liu, X. Wang, Q. Zhang, C. Huang, Empirical mode decomposition based hybrid ensemble model for electrical energy consumption forecasting of the cement grinding process, Measurement 138 (2019) 314–324. Available from https://doi.org/10.1016/j.measurement.2019.02.062.

[39] M. Kovačič, B. Šarler, Genetic programming prediction of the natural gas consumption in a steel plant, Energy 66 (2014) 273–284. Available from https://doi.org/10.1016/j.energy.2014.02.001.

[40] I. Monedero, F. Biscarri, C. León, J. I. Guerrero, R. González, L. Pérez-Lombard, Decision system based on neural networks to optimize the energy efficiency of a petrochemical plant, Expert Systems with Applications 39 (10) (2012) 9860–9867. Available from https://doi.org/10.1016/j.eswa.2012.02.165.

[41] Y.C. Liang, X. Lu, W.D. Li, S. Wang, Cyber physical system and big data enabled energy efficient machining optimisation, Journal of Cleaner Production 187 (2018) 46–62. Available from https://doi.org/10.1016/j.jclepro.2018.03.149.

[42] S. Wang, Y.C. Liang, W.D. Li, X.T. Cai, Big data enabled intelligent immune system for energy efficient manufacturing management, Journal of Cleaner Production 195 (2018) 507–520. Available from https://doi.org/10.1016/j.jclepro.2018.05.203.

[43] G. Golkarnarenji, M. Naebe, K. Badii, A.S. Milani, R.N. Jazar, H. Khayyam, Support vector regression modelling and optimization of energy consumption in carbon fiber production line, Computers & Chemical Engineering 109 (2018) 276–288. Available from https://doi.org/10.1016/j.compchemeng.2017.11.020.

[44] D.A.C. Narciso, F.G. Martins, Application of machine learning tools for energy efficiency in industry: a review, Energy Reports 6 (2020) 1181–1199. Available from https://doi.org/10.1016/j.egyr.2020.04.035.

[45] S. Acarer, Ç. Uyulan, Z.H. Karadeniz, Optimization of radial inflow wind turbines for urban wind energy harvesting, Energy 202 (C) (2020).

[46] A.S. Azad, A. Rahaman, M.S. Watada, J. Vasant, P. Vintaned, J.A. G, Optimization of the hydropower energy generation using meta-heuristic approaches: a review, Energy Reports 6 (2020) 2230–2248. Available from https://doi.org/10.1016/j.egyr.2020.08.009.

[47] S. Salimi, A. Hammad, Optimizing energy consumption and occupants comfort in open-plan offices using local control based on occupancy dynamic data, Building and Environment 176 (2020) 106818. Available from https://doi.org/10.1016/j.buildenv.2020.106818.

[48] T. Teng, X. Zhang, H. Dong, Q. Xue, A comprehensive review of energy management optimization strategies for fuel cell passenger vehicle, International Journal of Hydrogen Energy 45 (39) (2020) 20293–20303. Available from https://doi.org/10.1016/j.ijhydene.2019.12.202.

[49] J. Jenis, J. Ondriga, S. Hrcek, F. Brumercik, M. Cuchor, E. Sadovsky, Engineering applications of artificial intelligence in mechanical design and optimization, Machines 11 (6) (2023) 577. Available from https://doi.org/10.3390/machines11060577.

[50] T. Talaei Khoei, N. Kaabouch, Machine learning: models, challenges, and research directions, Future Internet 15 (10) (2023) 332. Available from https://doi.org/10.3390/fi15100332.

[51] M. Malik, S. Sharma, M. Uddin, C.-L. Chen, C.-M. Wu, P. Soni, et al., Waste classification for sustainable development using image recognition with deep learning neural network models, Sustainability 14 (12) (2022) 7222. Available from https://doi.org/10.3390/su14127222.

[52] C. Thornton, F. Hutter, H.H. Hoos, and K. Leyton-Brown, Auto-WEKA: combined selection and hyperparameter optimization of classification algorithms, in: Proceedings of the 19th ACM SIGKDD International Conference on Knowledge Discovery and Data Mining; KDD '13; Association for Computing Machinery, New York, NY, USA, 2013, pp. 847–855. https://doi.org/10.1145/2487575.2487629.

[53] M. Baratchi, C. Wang, S. Limmer, J.N. van Rijn, H. Hoos, T. Bäck, et al., Automated machine learning: past, present and future, Artificial Intelligence Review 57 (5) (2024) 122. Available from https://doi.org/10.1007/s10462-024-10726-1.

[54] T. Elsken, J.H. Metzen, and F. Hutter, Neural architecture search: a survey. arXiv April 26, 2019. https://doi.org/10.48550/arXiv.1808.05377.

[55] Automated machine learning: methods, systems, challenges, in: F. Hutter, L. Kotthoff, J. Vanschoren (Eds.), The Springer Series on Challenges in Machine Learning, Springer International Publishing, Cham, 2019. Available from https://doi.org/10.1007/978-3-030-05318-5.

[56] C. Lubongo, M.A.A. Bin Daej, P. Alexandridis, Recent developments in technology for sorting plastic for recycling: the emergence of artificial intelligence and the rise of the robots, Recycling 9 (4) (2024) 59. Available from https://doi.org/10.3390/recycling9040059.

[57] C. Feng, Y. Wang, Q. Chen, Y. Ding, G. Strbac, C. Kang, Smart grid encounters edge computing: opportunities and applications, Advances in Applied Energy 1 (2021) 100006. Available from https://doi.org/10.1016/j.adapen.2020.100006.

[58] M. Larsson, Global Energy Transformation, Palgrave Macmillan UK, London, 2009. Available from https://doi.org/10.1057/9780230244092.

[59] M. Ghobakhloo, Industry 4.0, digitization, and opportunities for sustainability, Journal of Cleaner Production 252 (2019) 119869. Available from https://doi.org/10.1016/j.jclepro.2019.119869.

CHAPTER

2

Basics of machine learning

Julhash U. Kazi[1,2,3]

[1]Division of Translational Cancer Research, Department of Laboratory Medicine, Lund University, Lund, Sweden [2]Lund University Cancer Centre (LUCC), Lund University, Lund, Sweden [3]Department of Laboratory Medicine, Lund Stem Cell Center, Lund University, Lund, Sweden

Introduction

Machine learning (ML) is profoundly transforming various industries, with its impact particularly pronounced in resource recovery and value creation [1]. ML facilitates optimizing material extraction and recycling processes by deploying advanced analytical tools and predictive models. This optimization is critical for enhancing sustainability and economic viability, especially in light of the escalating global demand for raw materials and the consequent imperative for efficient recycling and waste management strategies [2]. One of the pivotal ways in which ML is reshaping resource recovery is through its application in sorting and recycling operations. Sophisticated algorithms, trained on vast datasets, can accurately identify and classify diverse materials within waste streams. Such capability significantly boosts the operational efficiency of recycling facilities [3]. For example, ML algorithms are adept at distinguishing between different types of plastics, metals, and other recyclables. This precise differentiation aids in minimizing contamination and improving the separation process, thereby elevating the quality of the recycled output and increasing the economic value extracted from waste. Moreover, ML significantly optimizes energy and water consumption within resource recovery processes. By carefully analyzing consumption patterns and operational data, ML algorithms can pinpoint inefficiencies and predict the most effective settings for machinery and processes [4]. This optimization leads to a more judicious use of resources, diminished operational expenses, and a lower environmental footprint. Another critical application of ML lies in the predictive maintenance of machinery used in resource recovery. By evaluating data from sensors and operational logs, ML models can anticipate equipment failures before they are visible, thereby shortening downtime and reducing maintenance costs [5]. This predictive capability ensures more streamlined operations, enhancing both productivity and profitability. In the domain of value creation, ML offers invaluable insights into market analysis and forecasting. It empowers companies to scrutinize market trends and predict future demands for recycled materials, guiding

Harnessing Automation and Machine Learning for Resource Recovery and Value Creation
DOI: https://doi.org/10.1016/B978-0-443-27374-2.00002-9

strategic investment and production decisions [6]. Such foresight is indispensable for maintaining competitiveness in the fast-evolving resource recovery sector. Integrating ML into resource recovery and value creation represents a significant leap forward in achieving sustainable and economically viable operations. It highlights the transformative potential of technology in addressing global challenges related to resource management and environmental conservation.

Overview of machine learning

ML stands at the forefront of technological advancement, blending data analysis and artificial intelligence (AI) to foster systems capable of learning, adapting, and evolving. By extracting knowledge from data, ML enables a transformative approach to problem-solving across various domains, automating decision-making processes and uncovering insights that drive innovation. This overview describes the core aspects of ML, tracing its historical evolution, highlighting its significance in today's technology and industry landscapes, and exploring its pivotal role in resource recovery and value creation.

Definition and evolution of machine learning

ML, a critical subset of AI, fundamentally revolves around equipping machines with the ability to learn from data, recognize patterns, and execute decisions with minimal human guidance (Fig. 2.1). The origins of ML are deeply rooted in the early days of computing in the mid-20th century, evidenced by pioneering work such as Alan Turing's idea of a "learning machine" and the introduction of the perceptron in the 1950s [7,8]. Throughout its history, ML has seen a series of evolutionary leaps, transitioning from the initial symbolic AI of the 1960s and 70 s to the introduction of neural networks in the 1980s, and eventually to the contemporary surge in deep learning, driven by massive increases in data volume, computational capabilities, and algorithmic refinement.

The significance of machine learning in contemporary technology and industry

ML has emerged as an indispensable technology in the current era, catalyzing progress in numerous sectors. It has reshaped industries by refining operational processes, enhancing efficiency, and revealing new insights and possibilities. For instance, within healthcare, ML algorithms are instrumental in diagnosing diseases, predicting patient outcomes, and tailoring treatment protocols [10–12]. In finance, these models are applied for fraud detection, risk management, and algorithmic trading [13]. The advent of big data has further amplified the influence of ML, facilitating the analysis of large-scale datasets to identify meaningful patterns and predictions previously out of reach [14].

Overview of resource recovery and value creation through machine learning

ML presents transformative opportunities in the context of resource recovery and value creation. Resource recovery, the act of obtaining valuable materials or energy from waste, is significantly enhanced through the use of ML algorithms [15]. These algorithms augment the efficiency of sorting recyclable materials, determine product lifecycles, and improve supply chain management within circular economies. This aids in sustainable resource

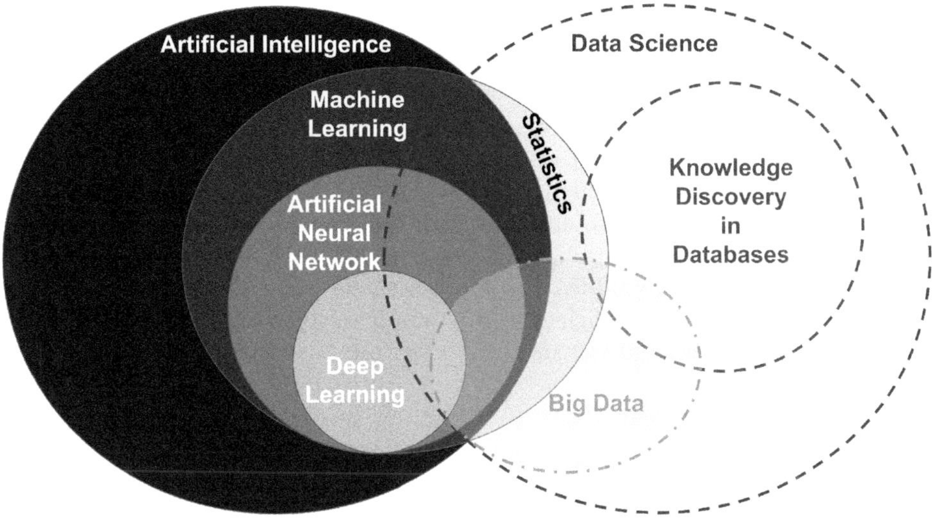

FIGURE 2.1 Interactions between artificial intelligence and data science involve the integration of advanced algorithms and data-driven methodologies to extract insights, make predictions, and automate decision-making processes. Artificial intelligence leverages data science techniques such as statistical analysis, data preprocessing, and feature engineering to enhance ML models and improve their performance. Conversely, data science benefits from artificial intelligence through the use of sophisticated algorithms for data mining, pattern recognition, and predictive analytics. Together, artificial intelligence and data science enable organizations to unlock the full potential of their data, driving innovation and improving outcomes across various domains [9]. Source: *From O. Debauche, J.B. Nkamla Penka, M. Hani, A. Guttadauria, R. Ait Abdelouahid, K. Gasmi, et al., Towards a unified architecture powering scalable learning models with IoT data streams, blockchain, and open data. Information (Switzerland), 14 (6) (2023). https://doi.org/10.3390/info14060345.*

management, an essential concern within growing environmental awareness and resource scarcity. Likewise, in the domain of value creation, ML serves as a catalyst for innovation [16]. Through analyzing market trends, consumer behaviors, and operational inefficiencies, ML algorithms assist businesses in creating enhanced value with reduced resources. They support the development of novel products and services tailored to specific consumer needs, streamline production processes, and foster sustainable practices by minimizing waste and maximizing resource utilization.

Core concepts of machine learning

Each type of ML offers unique approaches and tools for extracting insights and making predictions from data. Understanding the distinctions, strengths, and suitable application areas of these learning types is fundamental in harnessing ML's full potential in various domains. This understanding drives effective problem-solving and fosters innovation in fields as diverse as healthcare, finance, marketing, and beyond.

Types of machine learning

At the heart of ML lies a diverse set of methodologies, each with its unique approach to digesting data, learning from it, and applying the gleaned insights to make informed predictions or decisions [17]. The essence of understanding ML's core concepts revolves around grasping the different types of learning methodologies—supervised, semisupervised, unsupervised, and reinforcement learning—alongside other critical concepts such as feature selection, model evaluation, and bias-variance tradeoff (Fig. 2.2). Recognizing the nuances, strengths, and optimal application scenarios for each learning type is pivotal in effectively leveraging ML's capabilities across various fields. This deep understanding is not merely about solving problems efficiently; it is about pushing the boundaries of what is possible, driving innovation, and transforming practices in sectors as varied as healthcare, finance, marketing, and more.

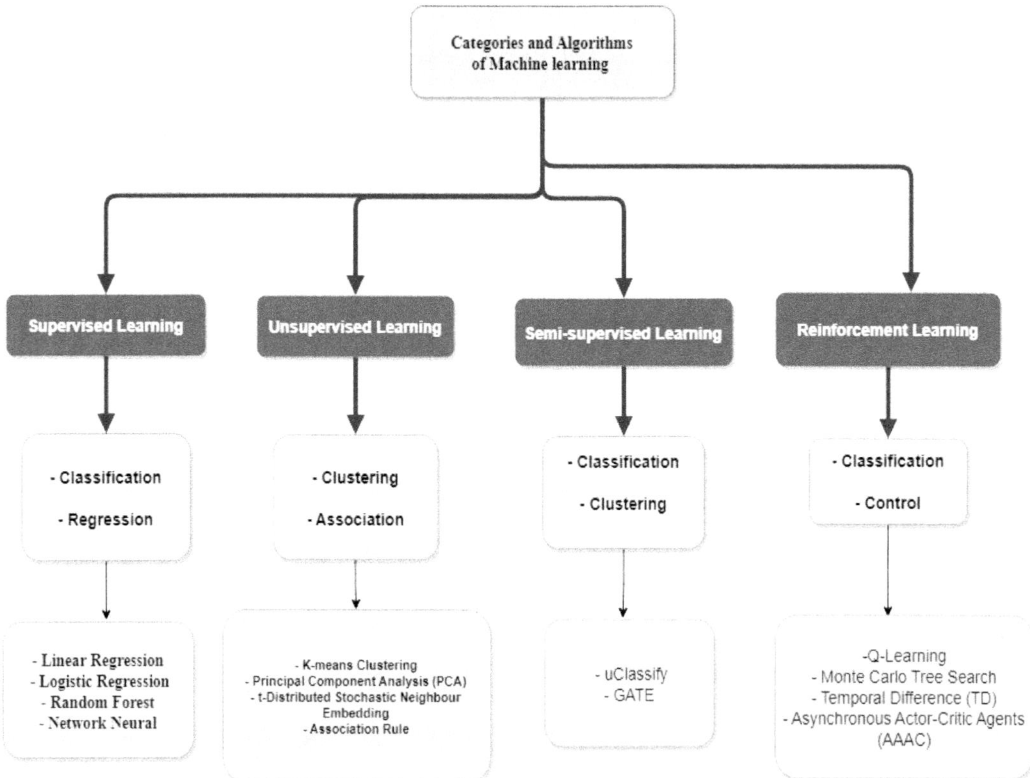

FIGURE 2.2 Overview of machine learning algorithm types and examples. A structured overview of the machine learning landscape, categorizing algorithms into four main types: supervised learning, semisupervised learning, unsupervised learning, and reinforcement learning [18]. Source: *From M.M. Taye, Understanding of machine learning with deep learning: architectures, workflow, applications and future directions. Computers 12 (5) (2023). https://doi.org/10.3390/computers12050091.*

Supervised learning

Supervised learning is defined by its use of labeled datasets to train algorithms. Such datasets comprise pairs of inputs and their corresponding outputs, allowing the model to learn the mapping from inputs to their correct outputs. This learning paradigm is indispensable for tasks like regression, which involves predicting a continuous value, and classification, which focuses on categorizing inputs into two or more distinct classes [19]. The power of supervised learning resides in its capacity to decipher complex patterns and yield precise predictions. Its versatility renders it ideal for a multitude of applications, such as disease diagnosis in healthcare based on symptoms and laboratory findings, credit scoring in finance to forecast the likelihood of loan defaults, predictive analytics in retail for projecting sales trends and customer retention, and quality control in manufacturing, where systems are trained to detect flaws in products [20].

Unsupervised learning

Unlike supervised learning, unsupervised learning utilizes unlabeled data to uncover hidden patterns or intrinsic structures. This approach is particularly advantageous for tasks such as clustering, where the aim is to group similar items, dimensionality reduction, which simplifies the number of variables considered, and association, where the objective is to discover rules that delineate relationships between various items. A key application of unsupervised learning is market segmentation, which clusters customers with similar preferences without prior labeling, thereby facilitating the creation of targeted marketing campaigns [21]. Moreover, unsupervised learning is adept at identifying the underlying structure of data without being provided with explicit instructions. Its capacity for pattern recognition extends to a wide array of uses, from market basket analysis, which interprets retail purchase patterns, to genomic sequence analysis, where it can categorize genes with comparable variation patterns. Additionally, it proves valuable in anomaly detection, pinpointing outliers that could indicate fraudulent activity, and in social network analysis, which helps reveal community structures and key influencers. These diverse applications emphasize the versatility of unsupervised learning in extracting meaningful information from complex datasets.

Semisupervised learning

Semisupervised learning operates at the intersection of supervised and unsupervised learning, harnessing both labeled and unlabeled data to construct predictive models. This hybrid approach is particularly beneficial when acquiring labeled data is costly or labor-intensive, but unlabeled data is plentiful. A small set of labeled data guides the learning process in semisupervised learning. It is then applied to a larger set of unlabeled data, allowing the model to improve its accuracy and generalize better to new, unseen data [22]. This methodology excels in scenarios where the partially labeled data can influence the structure learned from the unlabeled data, thereby enhancing the model's performance. For instance, it is effective in text classification tasks where a limited amount of text is categorized, and the learned patterns are then extrapolated to classify a larger amount of text. It is also prevalent in image recognition, where labeling is exhaustive, and the system can learn features from labeled images and apply them to categorize a vast array of unlabeled images. Semisupervised learning also finds significant applications in areas like speech analysis, where it can identify specific features from a set of labeled samples and then apply this knowledge to a broader range of sounds. In drug discovery, it aids in predicting molecular activities by learning from a small set of compounds with known properties. The versatility of semisupervised learning makes it a powerful tool in fields where the complete labeling of data is unfeasible, allowing for more efficient use of available information and enabling improved model performance with less human supervision.

Reinforcement learning

Reinforcement learning stands apart from supervised and unsupervised learning paradigms by focusing on the interaction between an agent and its environment. In this dynamic setting, the agent learns to make decisions by performing actions and receiving feedback in the form of rewards or penalties [23]. The ultimate goal of reinforcement learning is to develop a strategy known as a policy that maps situations to actions in a way that maximizes cumulative reward over time. This learning process is akin to a trial-and-error exploration, where the agent incrementally improves its policy based on the consequences of its actions. Unlike supervised learning, where a model is trained with correct input-output pairs, or unsupervised learning, which deals with finding patterns in data without any labels, reinforcement learning thrives on the concept of learning from the outcomes of actions taken, without requiring explicit examples of optimal outputs [24]. Reinforcement learning is particularly well-suited for problems where decision-making is sequential, and the consequences of an action are not immediately apparent. It has been successfully applied in various domains, such as robotics, where agents learn to perform complex tasks like walking or flying through continuous interaction with their environment. In gaming, reinforcement learning algorithms have achieved superhuman performance in various games by learning winning strategies through extensive gameplay against themselves. Moreover, reinforcement learning has practical applications in operations research, such as optimizing supply chain logistics or scheduling, where the sequence of operations affects the overall performance [25]. It can be used in finance to construct autonomous trading systems that learn to make profitable trades based on historical data [26]. Reinforcement learning's adaptability and goal-oriented nature make it a potent approach for complex problems where the learning process requires balancing the exploration of new strategies with the exploitation of known successful actions.

Comparison and contrasting of machine learning types

In the diverse ecosystem of ML, each learning paradigm—supervised, semisupervised, unsupervised, and reinforcement learning—carries its unique toolkit designed to address specific types of problems. Comprehending how these paradigms differ, their respective advantages and drawbacks, and the situations in which one might be preferred over the others, is fundamental for any ML practitioner. This discernment enables the strategic deployment of ML methods to tackle distinct challenges and achieve desired outcomes effectively. Table 2.1 below briefly outlines the advantages and disadvantages of various learning algorithms.

TABLE 2.1 Advantages and disadvantages associated with various learning algorithms.

Type	Advantages	Disadvantages
Supervised learning	High accuracy: Supervised learning models, given a sufficient amount of high-quality labeled data, can achieve high levels of accuracy in prediction tasks. Clear objectives: The learning process is directed toward a specific goal—mapping inputs to known outputs, simplifying training and evaluating labeled labeling.	The requirement of labeled data: The necessity for large amount labeled datasets can be a significant limitation, as labeling can be expensive and time-consuming. Risk of overfitting: These models can become too tailored to the training data, failing to generalize to new, unseen data.
Semisupervised learning	Utilization of unlabeled data: Semisupervised learning can leverage the abundant unlabeled data, making it more practical in many real-world scenarios. Cost-effective: Reduces the cost and effort required for data labeling while a man delivers improved performance over unsupervised methods.	Model complexity: The algorithms can be more complex than supervised models, requiring careful tuning to balance the influence of labeled and unlabeled data. Sensitivity to noise: Semisupervised methods can be sensitive to noisy or incorrect labels.
Unsupervised learning	Discovery of hidden patterns: Unsupervised learning excels at identifying underlying structures and patterns without any preexisting labels. Flexibility: It can be applied to data where the outcomes or categories are unknown, offering flexibility in exploratory analysis.	Lack of explicit outcomes: The absence of specific target variables can make the evaluation of unsupervised models more subjective and challenging. Dependence on data quality: The performance heavily relies on the inherent structure and quality of the data.
Reinforcement learning	Decision-making: Reinforcement learning is ideally suited for sequential decision-making problems and environments where feedback is available as many rewards. Adaptability: Reinforcement learning agents can adapt to environmental changes, making them suitable for dynamic contexts.	Sample inefficiency: Reinforcement learning can require a large number of trials to learn an optimal policy, which can be impractical in complex environments. Reward dependency: The design of the reward structure is crucial and can be tricky, as poor reward signals can lead the agent to undesired behaviors.

Selection criteria for machine learning approaches in different scenarios

The selection of an ML approach is governed by several criteria that align with the characteristics of the problem at hand:

Data availability: Supervised learning is often the most direct approach if labeled data are abundant. In contrast, unsupervised learning is the go-to when the data is abundant but unlabeled. Semisupervised learning finds its place when a larger amount of unlabeled data supplements a limited amount of labeled data. Reinforcement learning is chosen when the problem is modeled as a decision-making process with clear rewards.

The complexity of the task: Supervised learning is suitable for simple prediction tasks, whereas unsupervised learning is suitable for exploring the data structure. Reinforcement learning is reserved for complex tasks that require a series of decisions and can be learned from interaction over time.

Computational resources: If computational resources are limited, simpler supervised or unsupervised algorithms might be preferred. Reinforcement learning, which can be resource-intensive, is chosen when the infrastructure can support its computational demands.

Interpretability needs: If the problem requires highly interpretable results, supervised learning, especially with models like decision trees, may be preferable. Unsupervised learning can offer interpretability in the form of clusters or associations but can also lead to more ambiguous results. Reinforcement learning models are typically less interpretable due to their complexity.

Problem structure: Supervised learning is suitable for problems with a clear goal, like classification or regression. Unsupervised learning fits scenarios where the goal is to understand underlying patterns or data structure. Reinforcement learning is tailored to environments where an agent must learn to behave through trial and error.

By carefully weighing these factors the most suitable type of learning algorithm can be applied, ensuring the most effective use of resources and the highest likelihood of achieving objectives.

Data preprocessing in machine learning

The effectiveness of an ML model heavily depends on the quality of the data fed into it. Therefore careful data preprocessing is indispensable for achieving accurate and reliable model predictions. This process not only enhances the performance of the ML model but also contributes to the robustness and generalizability of the solutions it provides. Properly preprocessed data ensures that the underlying patterns within the data are accurately captured and represented, allowing the model to learn from the most relevant and clear information. This step is critical in avoiding biases and errors that could compromise the model's performance.

The importance of data quality

The foundation of any successful ML project lies in the quality of the data utilized. High-quality data ensures the reliability of the model's predictions and significantly influences the efficiency and accuracy of the learning process [27]. The careful attention to data quality—through strategic collection and management, alongside rigorous cleaning and normalization practices—is indispensable in harnessing the full potential of ML algorithms. These steps ensure that the foundation upon which models are built is robust, enabling the development of solutions that are accurate and reliable and capable of generalizing well to new, unseen data.

Data collection and management

Strategic collection: The process of data collection must be approached with strategic forethought to guarantee that the dataset is representative of the problem at hand. This necessitates a clear definition of data requirements and the careful selection of sources from which data is gathered. The aim is to capture a comprehensive and unbiased view of the problem space, ensuring that the dataset encompasses the variability and complexities of real-world scenarios that the model will encounter.

Data management: Effective data management is pivotal in maintaining data integrity and accessibility. Organizing data in a structured, easily accessible format facilitates efficient data processing and analysis. Equally important is adhering to stringent data security and privacy standards, particularly when dealing with sensitive or personal information, to comply with legal and ethical guidelines. Moreover, maintaining a detailed log of data collection, updates, and modifications enhances the transparency and reproducibility of the ML project, fostering trust and confidence in the results.

Data cleaning and normalization

Cleaning: Data cleaning is essential to rectifying errors and inconsistencies in the dataset. It involves a careful examination to identify and subsequently correct or eliminate inaccuracies, such as outliers that deviate significantly from other observations, incorrect data entries, and duplicate records [28]. Effective data cleaning enhances the quality of the dataset, leading to more accurate and reliable model performance.

Normalization: Normalization refers to the standardization of data to fit within a specific scale or format, a critical step for algorithms sensitive to data scale variations. Techniques like min-max normalization, which rescales data to a fixed range; z-score normalization, which adjusts the data based on its mean and standard deviation, and feature scaling, crucial for harmonizing the scale across different input variables, are commonly employed [29]. Normalizing data prevents the disproportionate influence of certain features over others, facilitating a balanced and fair learning process, particularly in models such as neural networks that are acutely sensitive to input scales.

Feature selection and engineering

The feature selection and engineering process plays a crucial role in developing effective machine-learning models. By carefully choosing the most relevant features and engineering new ones from existing data, practitioners can significantly enhance model performance and efficiency [28]. This section explores various techniques for feature selection, as well as the practice of feature engineering and its impact on ML models [20].

Techniques for feature selection

Learning algorithms can struggle with datasets that have too many features, many of which may be unnecessary or duplicative. Feature selection helps by reducing the number of features, and maintaining the model's decision-making quality [30].

Filter methods: These techniques employ statistical measures to evaluate the relevance of features concerning the target variable. By ranking features based on their statistical relationship with the target, filter methods help to identify the most predictive features. Common examples of filter methods include using correlation coefficients for continuous variables and the Chi-squared test for categorical variables.

Wrapper methods: Wrapper methods assess subsets of features by creating models using different combinations of features and evaluating their performance. Features are selected based on their ability to enhance model accuracy. Techniques such as forward selection, where features are incrementally added to the model, backward elimination, which involves removing features to see their effect on performance, and recursive feature elimination, which iteratively removes the least important features, are popular wrapper methods.
Embedded methods: Unlike filter and wrapper methods, embedded methods incorporate feature selection as an integral part of the learning algorithm. These methods select the most important features during the model training process itself. Decision trees, for example, inherently perform feature selection by choosing the splits that most effectively separate the data.

Feature engineering and its impact on model performance

Creation of new features: Feature engineering is the art of transforming existing data into new features that better represent the underlying problem of the predictive models [31]. This process can unveil complex patterns that were not captured by the original dataset. Techniques such as creating interaction terms between features, generating polynomial features to model nonlinear relationships, or aggregating features to capture summarized characteristics are common in feature engineering.
Impact on performance: The introduction of well-engineered features can lead to substantial improvements in model accuracy and performance. By enhancing the model's ability to understand the data, these new features can simplify the model structure, thereby reducing overfitting and computational demands [32]. Furthermore, effective feature engineering often results in more generalizable and robust models for new, unseen data, ultimately leading to more reliable predictions.

Handling imbalanced and missing data

In ML, the presence of imbalanced or missing data can significantly impact model performance. Addressing these issues requires specific techniques and strategies to ensure that the predictive models remain accurate and reliable [33].

Techniques and strategies for imbalanced data

Imbalanced datasets predominantly occur in classification problems where the instances of one class significantly outnumber those of other classes. This imbalance can make the model biased toward the majority class, resulting in poor predictive performance for the minority class [34]. Several strategies can mitigate this issue:

Resampling techniques: These techniques adjust the class distribution of the dataset before training the model. Over-sampling involves replicating instances of the minority class to increase its representation, whereas under-sampling reduces the instances of the majority class. Both methods aim to achieve a more balanced class distribution.
Algorithmic approaches: Some algorithms are inherently more robust to class imbalance. For instance, decision trees and their ensemble versions, like random forests, can often handle imbalance better due to their hierarchical structure. This allows them to focus on areas of the feature space where the minority class is more prevalent.

Cost-sensitive training: This approach involves modifying the learning algorithm to penalize misclassifications of minority class samples more than misclassifications of the majority class. By assigning a higher cost to errors involving the minority class, the model is encouraged to pay more attention to correctly classifying those instances.

Techniques and strategies for missing data

Missing data is another common challenge in ML projects. Incomplete datasets can arise for various reasons, such as errors in data collection or transmission [35]. Several strategies can be employed to handle missing values:

Deletion: This method involves removing records with missing values from the dataset. While straightforward, deletion is only advisable when the amount of missing data is minimal, as it can lead to the loss of valuable information and potentially bias the model if the missingness is not completely random.
Imputation: Imputation strategies estimate the missing values based on the available data. Simple imputation techniques include filling in missing values with the respective feature's mean, median, or mode. More sophisticated methods, like k-nearest neighbors (KNN) imputation, utilize the similarity between instances to predict missing values. In contrast, regression imputation uses the relationships between features to estimate the missing data [36].
Using algorithms that handle missing values: Certain ML algorithms can inherently deal with missing values during training. For example, random forests can maintain accuracy even when a portion of the data is missing, as they can use splits that only involve observed values [37].

Machine learning algorithms

This section provides an overview of both traditional ML algorithms and advanced neural network models, highlighting their strengths, limitations, and suitability for different types of data and problems. The choice of an appropriate algorithm is crucial. It should be based on a thorough analysis of the problem characteristics, data availability, and specific requirements of the application area, especially in fields like resource recovery, where efficiency and accuracy are paramount [38].

Traditional machine learning methods

Traditional ML methods are foundational algorithms that enable computers to learn from and make predictions based on data. These methods include supervised learning techniques like linear regression, where the model predicts an outcome based on input features, and classifiers such as support vector machines (SVMs) and decision trees, which categorize data into distinct classes. Unsupervised learning techniques like k-means clustering and principal component analysis identify patterns or groupings without prelabeled responses. Over the years, these techniques have been extensively applied to many problems ranging from spam detection to customer segmentation. They are typically characterized by their interpretability, ease of implementation, and effectiveness in handling structured data. However, they may require careful feature selection and may not scale as well to very large datasets or capture complex patterns as effectively as some more recent ML methods.

Linear and logistic regression

Linear and logistic regression are fundamental statistical methods in ML for prediction and classification [39]. Linear regression uses a linear equation to predict the value of a dependent variable based on one or more independent variables, optimizing coefficients through the least squares method. It is known for simplicity and effectiveness but may overfit complex datasets, an issue addressed by regularization techniques like ridge [40], lasso (least absolute shrinkage and selection operator) [41], and elastic net [42], which also aid in feature selection. Logistic regression, used for binary classification, predicts the probability of an input belonging to a category using a logistic function and optimizes coefficients via maximum likelihood estimation. It is valued for its simplicity and interpretability but can overfit with large or irrelevant feature sets; thus regularization is used to improve model generalization. Both models have measures to evaluate their performance and fit.

Decision trees

Decision Trees are a graphical representation of possible solutions to a decision based on certain conditions. They are structured like a tree, where each internal node denotes a test on an attribute, each branch represents the outcome of that test, and each leaf node indicates a class label or decision outcome [43,44]. This structure allows for the easy interpretation of classification rules from the root to the leaf. One of the main advantages of decision trees is their simplicity and transparency, making them easily understandable even for those with no background in data science. They require minimal data preprocessing, such as normalization or dummy variable creation, and can handle both numerical and categorical data. However, decision trees are prone to overfitting, especially when they grow complex with many branches [45]. This overfitting makes the model specific to the training data, reducing its ability to generalize to new data. Furthermore, decision trees can be sensitive to small changes in the data, leading to entirely different structures. This instability can be problematic in dynamic environments. Ensemble methods such as random forests [46] and gradient boosting [47] are utilized to combat this. These approaches involve constructing numerous decision trees and aggregating their outputs, which bolsters the predictions' stability and enhances the overall accuracy. Furthermore, decision trees have a known predisposition for favoring attributes with more distinct values. This can lead to a skewed selection process, compromising the model's balance and precision. To address this, strategies like binning, which groups a range of values into a smaller number of "bins," can be employed to alleviate the bias and contribute to a more equitable model. Despite these challenges, decision trees remain an integral part of the ML toolkit. Their capacity to handle diverse data types and provide a transparent rationale for each prediction they make renders them particularly invaluable for applications where understanding the underpinnings of the predictive outcomes is just as critical as the prediction's accuracy.

Ensemble methods

Ensemble methods enhance predictive power in ML by combining the strengths of multiple algorithms, thereby outperforming any single model. Bagging, for instance, creates multiple models where each is trained on a different subset of data, reducing overfitting by lessening the model's data peculiarities. Random forest, an advancement of bagging, forms a collection of decision trees, each on a random feature subset, which decreases the trees' errors when their predictions are aggregated [46]. Boosting, in contrast, builds models incrementally to correct its predecessors' mistakes, effectively transforming weak learners into a strong one. AdaBoost exemplifies this by adjusting the weights of misclassified instances to prioritize difficult data in subsequent models [48]. Gradient boosting machine (GBM) refines this approach by iteratively optimizing predictions to minimize errors [47], and XGBoost [49], an iteration of GBM, focuses on speed and regularized boosting to combat overfitting. LightGBM excels in handling large datasets with its memory-efficient algorithms [50], while CatBoost specializes in categorical data [51], managing it adeptly with minimal preprocessing. Stacking diverges by training base models on the full dataset and using their predictions as input for a higher-level model, thereby harnessing the unique strengths of each base model for improved accuracy. Selecting the right algorithm from this suite depends on the problem at hand, dataset specifics, and performance goals.

Support vector machines

Support vector machines (SVMs) are a powerful set of supervised learning methods useful for classification, regression, and outliers detection. The core idea behind SVM is to find the hyperplane that best separates the dataset into classes. The optimal hyperplane is the one that maximizes the margin between different classes, with support vectors being the critical elements that define this margin [52]. SVMs are particularly effective in high-dimensional spaces, where they can identify complex relationships without succumbing to the curse of dimensionality. This attribute makes SVMs robust against overfitting when dealing with high-dimensional data. Despite these strengths, SVMs face significant challenges with large datasets due to the computational complexity of training, which grows considerably as the size of the data increases. They also struggle with datasets where classes are not clearly separable or overlap significantly. The performance of SVMs is heavily dependent on the choice of the kernel function, which requires careful tuning to ensure that the model captures the underlying patterns in the data accurately.

K-nearest neighbors

K-nearest neighbor (KNN) is a straightforward and intuitive ML algorithm that classifies new cases based on the most common class among its KNN [53]. The algorithm stores all available cases and predicts the numerical target based on a similarity measure (e.g., distance functions). KNN has been utilized across various applications in both classification and regression tasks. Its simplicity lies in its nonparametric nature, meaning it makes no underlying assumptions about data distribution. Being easy to implement and understand, KNN can be a powerful tool for many predictive tasks. However, the algorithm's performance deteriorates rapidly as the volume of data increases, both in terms of the number of instances and the dimensionality of the feature space. This scalability issue is due to the need to compute the distance between the new point and each point in the dataset, which becomes computationally intensive with large datasets. Furthermore, KNN is sensitive to irrelevant or redundant features because all features contribute equally to calculating the distance between instances. Thus selecting relevant features and choosing an appropriate distance metric is crucial for the success of a KNN model. Determining the right value of k is essential to balance the bias-variance tradeoff inherent in the model's design.

Advanced machine learning methods—neural networks

Neural networks are a foundational component of modern ML. They consist of layers of interconnected nodes or neurons, each performing a specific mathematical operation. The connections between these nodes carry weights that adjust during training, enabling the network to learn complex patterns and relationships within the data.

Introduction to neural networks

Neural networks represent the forefront of developments in the field of AI, evolving from basic perceptron models to complex architectures capable of understanding complicated patterns in vast datasets. These sophisticated models include convolutional neural networks (CNNs), which excel in image and video recognition; recurrent neural networks (RNNs), particularly useful for sequential data such as language and speech; and more recent innovations like generative adversarial networks (GANs) and transformer models, which have revolutionized the generation of new content and language understanding, respectively. Each of these advanced networks leverages deep learning to automatically extract features, identify patterns, and make decisions with minimal human intervention, learning to mimic the complexity of human neural activity. As a result, they have expanded the horizons of what's achievable, pushing the boundaries in fields ranging from natural language processing (NLP) to autonomous vehicles.

Deep learning and its applications

Deep learning, an advanced branch of ML, is characterized by its use of multilayered neural networks to discover complex patterns in large datasets. These deep neural networks, consisting of multiple interconnected nodes, are adept at learning hierarchies of features, thus enabling various applications ranging from computer vision to NLP. Two primary architectures stand out within this domain: CNNs and RNNs. Some examples of deep learning have been depicted in Fig. 2.3.

CNNs are highly efficient in handling grid-like data such as images, empowering breakthroughs in facial recognition, medical diagnostics, and even creative applications like artistic style transfer [54]. Their architecture, which facilitates weight sharing and feature pooling, reduces the model's complexity, enabling it to excel in tasks that involve visual inputs while maintaining computational efficiency. Despite these advantages, CNNs require extensive training data and are prone to overfitting, necessitating careful model optimization and the application of regularization techniques.

RNNs, conversely, excel in processing sequential data, making them suitable for speech recognition, language translation, and forecasting in time series. They possess an inherent

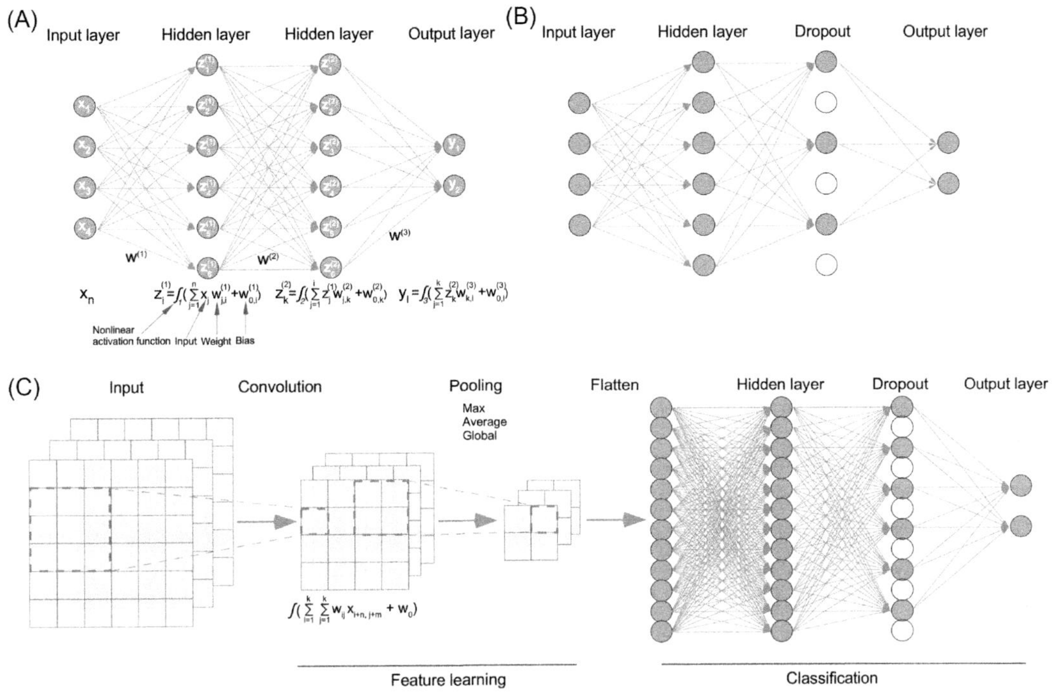

FIGURE 2.3 A brief description of deep learning architecture. (A) The deep neural network model consists of multiple interconnected layers, beginning with an input data layer where each node is fully connected to the nodes in the subsequent hidden layers. The first hidden layer processes the input data by multiplying it with corresponding weights and adding a bias term, followed by the application of a nonlinear activation function. Each subsequent hidden layer receives the output from the previous layer, repeating this process until reaching the output layer, where the final prediction or classification is made. (B) The dropout layer is a regularization technique in which a random subset of nodes is temporarily removed from the network during training, reducing the model's dependency on specific nodes and improving generalization. (C) The convolutional layer reduces the dimensionality of input data through convolution operations using a specified kernel size, such as 3x3 in this example, coupled with an activation function. Following convolution, the extracted features undergo pooling to further reduce dimensionality, after which the pooled features are flattened and passed into a fully connected DNN for final processing and output generation [11]. Source: *From R. Rafique, S.M.R. Islam, J.U. Kazi, Machine learning in the prediction of cancer therapy. Computational and Structural Biotechnology Journal 19 (2021) 4003–4017. https://doi.org/10.1016/j.csbj.2021.07.003.*

memory capability that helps them understand context and temporal sequences, which is crucial for predicting subsequent elements in a sequence [55]. However, RNNs face challenges in training due to the vanishing or exploding gradient problem and often struggle with long sequences. This has led to the development of advanced variants like Long Short-Term Memory (LSTM) networks, which are better equipped to handle long-term dependencies.

Beyond these, deep learning has significantly advanced NLP through transformer models like bidirectional encoder representations from transformers (BERT) and generative pretrained transformers (GPTs), setting new benchmarks for machine understanding of language [56]. Deep learning models interpret complex sensor data in the automotive industry to facilitate autonomous driving.

The growth of deep learning owes much to the automatic learning of representations, eliminating the need for manual feature engineering and increasing volumes of data and computational resources, especially graphics processing units (GPUs). The paper by LeCun, Bengio, and Hinton, "Deep Learning," published in Nature, extensively reviews deep learning's capabilities and the profound impact it continues to have across various fields [57].

Algorithm selection criteria

Several critical factors influence the process of selecting the appropriate ML algorithm for a specific task, reflecting the nature of the data, the complexity of the problem, available computational resources, the need for interpretability, and real-time processing requirements. Understanding these factors can guide practitioners in making informed decisions that align with their objectives and constraints.

Factors influencing the choice of algorithms

Data size and quality: The volume and quality of the available data can significantly impact the choice of algorithm. Large datasets with high-quality, well-labeled data are conducive to applying complex models such as neural networks, which require substantial amounts of data to train effectively. Conversely, smaller datasets or those with lower quality may benefit from simpler, less data-hungry models such as decision trees or SVMs, which can perform well even with limited input.

Problem complexity: The inherent complexity of the problem, particularly about high-dimensional data or complex patterns, may necessitate the use of more sophisticated algorithms. Deep learning models, known for their ability to handle complex, nonlinear relationships and high dimensionality, are often suited for such tasks.

Computational resources: Certain algorithms, especially those associated with deep learning, can have substantial computational demands. Access to powerful computing resources can therefore influence the feasibility of deploying these algorithms. More efficient algorithms that require less processing power and memory may be preferred in scenarios where computational resources are limited.

Interpretability: In applications where understanding the decision-making process is crucial, such as in healthcare or finance, the interpretability of the model becomes a significant factor. Algorithms that produce more transparent and explainable models, like decision trees, might be favored over more opaque models like deep neural networks.

Latency requirements: For real-time applications that demand quick responses, such as automated trading systems or emergency response systems, the efficiency and speed of the algorithm are paramount. In these cases, algorithms that can offer faster predictions without compromising accuracy are essential.

Examples of resource recovery

Decision trees in waste sorting: Automated waste sorting systems are integral components of modern waste management strategies, designed to enhance the efficiency and effectiveness of recycling processes [58]. These systems employ advanced decision-making algorithms to classify and separate waste materials, a task at which decision trees excel [59]. Decision trees are supervised ML algorithms that model decisions and their possible consequences, including chance event outcomes, resource costs, and utility. They are particularly useful in automated waste sorting because they can handle complex classification problems based on various attributes [60]. The Fig. 2.4 describes an example use of ML in a process line.

In the context of waste sorting, decision trees operate by breaking down a dataset into smaller subsets while, at the same time, an associated decision tree is incrementally developed. The choices made at each step of the tree depend on the best attribute to split the data to achieve the most effective classification of waste materials. This decision-making process considers various attributes of waste, such as material type (e.g., plastic, metal, organic), color, shape, size, and recyclability. The ultimate goal is to sort waste into categories that facilitate recycling, disposal, or further processing [62]. The effectiveness of decision trees in automated waste sorting systems can be attributed to several key factors, including:

Simplicity and transparency: Decision trees are relatively simple to understand and interpret, making them ideal for applications where the reasoning behind decisions needs to be transparent and explainable. This is crucial for waste sorting, where the decision-making criteria may need to be adjusted or scrutinized for efficiency and compliance with regulatory standards.
Handling of nonlinear relationships: Waste sorting involves classifying materials based on attributes that may not have a linear relationship. Decision trees can model nonlinear relationships between features without transforming them, making them highly adaptable to the complex nature of waste materials.

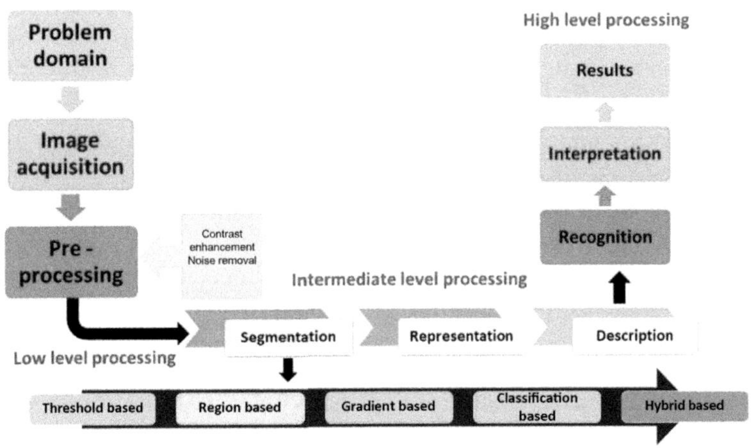

FIGURE 2.4 Machine learning application in process line. The implementation of machine learning in various stages of food processing enhances efficiency, quality control, and decision-making. Machine learning algorithms are used for tasks such as sorting and grading raw materials, predicting shelf life, monitoring and optimizing production processes, and detecting contaminants. By analyzing large datasets from sensors and production lines, machine learning models can automate quality assessments, adjust processing parameters in real time, and reduce waste. This integration helps improve product consistency, ensure food safety, and streamline operations throughout the food processing pipeline [61]. Source: *From V.K. Pandey, S. Srivastava, K.K. Dash, R. Singh, S.A. Mukarram, B. Kovács, E. Harsányi, Machine learning algorithms and fundamentals as emerging safety tools in preservation of fruits and vegetables: a review. Processes 11 (6) (2023). https://doi.org/10.3390/pr11061720.*

Versatility in handling different types of data: Decision trees can handle numerical and categorical data, an essential feature for waste sorting systems that deal with diverse material types and characteristics.

Scalability and efficiency: They can efficiently process large volumes of waste, making rapid decisions that facilitate the sorting process at a speed that matches the throughput requirements of modern recycling facilities. Research and development in this area are ongoing, with studies continuously exploring ways to optimize decision tree algorithms for better accuracy and efficiency in waste sorting. For instance, integrating decision trees with other ML techniques, like random forests or neural networks, can improve sorting accuracy by addressing the limitations of single decision tree models, such as overfitting [63].

CNNs in material recovery: CNNs represent a significant advancement in the field of ML, particularly in the processing and analysis of visual data. These deep learning models are adept at automatically and efficiently recognizing patterns within images, making them exceptionally well-suited for applications requiring high levels of visual discrimination, such as identifying and classifying materials on recycling lines [64,65]. This capability is crucial for improving the efficiency and effectiveness of material recovery processes in recycling facilities. Recycling involves several stages, from the collection of waste to its sorting, cleaning, and eventual conversion into reusable materials. The sorting stage, where different types of materials are identified and separated, is particularly challenging due to the vast diversity of waste materials. This is where CNNs come into play. They leverage their deep learning capabilities to analyze images of waste items captured on conveyor belts, recognizing and categorizing materials based on visual characteristics such as shape, size, color, and texture [66,67]. CNNs consist of multiple layers, including convolutional layers, pooling layers, and fully connected layers, each performing distinct functions that collectively contribute to the network's ability to discern and learn from complex patterns in image data:

Convolutional layers: These layers apply various filters to the input images to create feature maps that highlight specific attributes of the materials, such as edges or textural patterns. This process is critical for distinguishing between different types of recyclable materials, such as paper, plastic, glass, or metal.

Pooling layers: Pooling (or subsampling) reduces the dimensionality of each feature map while retaining the most important information, making the network more efficient and less prone to overfitting.

Fully connected layers: Toward the end, these layers interpret the feature data to make final classification decisions, effectively determining the type of material presented.

The strength of CNNs lies in their ability to learn feature representations directly from the image data, without the need for manual feature extraction. This learning ability is particularly beneficial for material recovery, where the visual characteristics of recyclable materials can be highly variable. CNNs can adapt to these variations, improving over time as they are exposed to more data, leading to higher accuracy in material identification and classification. The application of CNNs in material recovery enhances the efficiency of sorting processes. It contributes to higher recycling rates by minimizing the contamination of recyclable streams and ensuring that more materials are correctly identified and processed [68]. This technological advancement supports more sustainable waste management practices by maximizing the recovery of valuable materials and reducing the environmental impact of waste.

RNNs for predictive maintenance: RNNs are a class of artificial neural networks specialized in processing sequential data, making them highly suitable for tasks where the context or temporal sequence of data points is crucial. In the context of predictive maintenance for resource recovery facilities, RNNs offer a potent solution for anticipating equipment failures and optimizing maintenance schedules [69]. These facilities, critical in managing and repurposing waste materials, rely heavily on a wide range of mechanical and processing equipment. The ability to predict potential equipment failures before they occur can significantly enhance operational efficiency, reduce downtime, and minimize maintenance costs. Predictive maintenance involves the use of data-driven approaches to predict and prevent equipment failures. By analyzing time-series data on equipment performance, which includes variables such as temperature, vibration, sound, and pressure, RNNs can identify patterns or anomalies that precede failures [70,71]. Unlike traditional neural networks, RNNs have internal memory that allows them to remember and utilize previous information in the network's internal state, making them exceptionally well-suited for analyzing time-dependent data. The operation of RNNs in predictive maintenance can be broken down into several key steps:

Data collection: Continuous monitoring and data collection from sensors embedded in equipment to track various performance metrics.
Feature extraction: Processing the collected time-series data to extract relevant features that are indicative of the equipment's health and operational status.
Model training: Using historical data, an RNN model is trained to learn the equipment's normal operational patterns and recognize signs of potential failures.
Prediction and analysis: The trained RNN model analyzes the incoming data stream to predict potential breakdowns. The model can identify when equipment will likely fail by recognizing deviations from normal operational patterns.
Maintenance scheduling: Based on the predictions made by the RNN model, maintenance can be proactively scheduled at an optimal time to prevent the predicted failure, thereby minimizing the impact on operations.

The advantage of employing RNNs for predictive maintenance lies in their ability to process and make predictions based on sequential data, capturing temporal dependencies and patterns that other models might miss [72]. This results in more accurate predictions of equipment failure, allowing for timely maintenance actions that prevent costly downtimes and extend the lifespan of the equipment. Furthermore, using RNNs contributes to a more efficient allocation of resources. By predicting maintenance needs, facilities can prioritize maintenance activities, allocate resources more effectively, and avoid unnecessary maintenance checks, leading to significant cost savings.

The selection of an ML algorithm is a nuanced decision that must account for various factors, including data characteristics, problem complexity, computational constraints, and specific application requirements. Through carefully considering these criteria and learning from relevant case studies, practitioners can choose the most suitable algorithm to address their unique challenges, thereby maximizing the impact and effectiveness of their ML solutions.

Practical aspects of machine learning

The practical aspects of ML encompass the crucial steps of model training and validation, understanding and preventing overfitting, and evaluating the model performance [73]. These aspects are fundamental to developing robust, efficient, and accurate ML models.

Model training and validation

The model training and validation process is critical in developing effective ML models. It involves not only the application of various training techniques to build the model but also the utilization of different validation methods and metrics to evaluate its performance. This dual approach ensures that the model is accurate and generalizes well to unseen data.

Training techniques

Data splitting: A fundamental step in training an ML model involves splitting the dataset into training and test sets. A common practice is to allocate 70% of the data for training and 30% for testing. However, these proportions can be adjusted based on the project's specific requirements, such as the dataset's size and characteristics. This split allows the model to learn from a substantial portion of the data while reserving a separate subset for testing its predictions.

Cross-validation: Cross-validation techniques are employed to gain a more accurate estimate of the model's performance. K-fold cross-validation is a popular method in which the dataset is divided into "k" subsets or folds. The model is trained on "k-1" folds and tested on the remaining fold, repeating this process "k" times, with each fold serving as the test set once. This method helps mitigate the impact of the data split on model performance by systematically utilizing the entire dataset for training and testing.

Batch and epochs in training: The concepts of batches and epochs are integral, particularly in neural network training. Training data is often divided into smaller subsets or batches, allowing the model to update its weights more frequently, leading to faster convergence. An epoch refers to one complete pass of the entire training dataset through the model. Training over multiple epochs, with the data potentially reshuffled into new batches at each epoch, allows the model to learn more effectively from the data.

Validation methods and metrics

Validation techniques: Several validation techniques are utilized to assess the performance of ML models. Holdout validation is the most straightforward approach, where the dataset is simply split into a training set and a test set. Cross-validation and leave-one-out validation (a specific case of cross-validation with "k" equal to the number of data points) are more sophisticated techniques that provide a more robust evaluation by leveraging the entire dataset in the training and testing process.

Performance metrics: The choice of performance metrics depends on the nature of the ML task. For classification problems, metrics such as accuracy, precision, recall, the F1 score, and the area under the receiver operating characteristic curve (ROC-AUC) curve are commonly used to evaluate model performance [74]. These metrics provide insights into the model's ability to correctly classify instances and balance trade-offs between false positives and false negatives. In regression tasks, metrics like mean squared error (MSE), mean absolute error (MAE), and R-squared are employed to quantify the difference between the predicted and actual values, measuring the model's predictive accuracy and explanatory power.

Training and validation are indispensable stages in the lifecycle of an ML project, guiding the development of models that are not only accurate but also robust and generalizable. By carefully applying training techniques and rigorously evaluating models using appropriate validation methods and metrics, practitioners can ensure the reliability and effectiveness of their ML solutions.

Overfitting: causes and prevention

Overfitting is a common challenge in ML, marked by a model's inability to generalize to new data due to excessive training on its training dataset [75]. This section explores the phenomenon of overfitting, its causes, and various strategies to prevent it, ensuring models retain their predictive power on unseen data.

Understanding overfitting

Overfitting occurs when a model learns the details and noise in the training data to the extent that it negatively impacts the model's performance on new data. This often happens with complex models with too many parameters relative to the number of observations. Such models may excel in training data, capturing the underlying patterns and random fluctuations—essentially memorizing the data rather than learning from it [76]. When presented with new, unseen data, the overfitted model often performs poorly because it has learned to anticipate the noise and outliers of the training set rather than the general trends.

Techniques to avoid overfitting

Regularization: Implementing regularization techniques such as L1 (Lasso) and L2 (Ridge) regularization adds a penalty on the magnitude of coefficients to the loss function. This penalty discourages the model from fitting too closely to the training data by limiting the complexity of the model, thus helping to prevent overfitting.

Pruning in decision trees: This involves cutting back the branches of the tree after it has been fully grown. By removing sections of the tree that provide little power in classifying instances, pruning reduces the complexity of the final model, making it more generalizable to new data.

Dropout in neural networks: Dropout is a technique used specifically in training deep neural networks. During training, a randomly selected subset of neurons is ignored. This prevents neurons from becoming too dependent on the outputs of their neighboring nodes, thus reducing overfitting by encouraging more robust features to be learned.

Early stopping: This technique involves monitoring the model's performance on a validation set and stopping the training process once the model's performance starts to degrade (i.e., when the error on the validation set begins to increase), preventing overfitting by halting the training before the model becomes too specialized to the training data.

Balancing model complexity and generalizability

Simplifying the model: Opting for simpler models or reducing the number of features in the dataset can be an effective way to combat overfitting. Simpler models are less likely to capture noise and random fluctuations in the training data, making them more capable of generalizing to new data.
Ensemble methods: Techniques like bagging and boosting involve combining the predictions of several models to improve robustness and generalizability. By aggregating the predictions of multiple models, ensemble methods can smooth out individual models' predictions, reduce variance, and hence decrease the risk of overfitting.

Balancing a model's complexity with its ability to generalize is crucial for building effective ML models. By understanding overfitting and employing strategies to mitigate its impact, practitioners can develop models that perform well on their training data and new, unseen data, thereby achieving more reliable and robust predictions.

Model evaluation and performance metrics

Evaluating ML models involves assessing their performance using various criteria and metrics, which provide insights into their accuracy, efficiency, and general applicability to real-world scenarios. This multifaceted evaluation ensures that the developed models meet the desired standards of prediction accuracy and operational efficiency [77].

Criteria for evaluating machine learning models

Accuracy and error rates: Fundamental to evaluating ML models, these metrics measure the proportion of correct predictions made by the model (accuracy) and the rate at which it makes errors (error rate). While straightforward, these measures are critical for understanding a model's predictive capability.
Underlying assumptions: It is essential to verify that the model's theoretical assumptions are in concordance with the characteristics of the data. This alignment ensures that the model is suitable for the data it is applied to, which can significantly impact its performance.
Computational efficiency: Evaluating a model's computational efficiency involves considering the time and computational resources required to train the model and make predictions. Efficiency is particularly important in applications requiring real-time predictions or when resources are limited.
Robustness and scalability: A model's robustness refers to its ability to maintain performance despite changes or noise in the data. Scalability, however, relates to how well the model can handle increasing volumes of data. Both qualities are crucial for models intended for dynamic environments or large-scale applications.

Standard performance metrics

Classification metrics: For classification tasks, common metrics include accuracy, precision (the ratio of true positive predictions to all positive predictions), recall (the ratio of true positive predictions to all actual positives), F1 score (harmonic mean of precision and recall), and ROC-AUC (measuring the model's ability to discriminate between classes) [78].

Regression metrics: In regression tasks, performance is often measured using metrics like MSE, root mean squared error (RMSE), MAE, and R-squared (which indicates the proportion of variance in the dependent variable that is predictable from the independent variables) [79].
Clustering metrics: For tasks involving clustering, metrics such as the Silhouette Coefficient (measuring how similar an object is to its own cluster compared to other clusters), Calinski-Harabasz Index, and Davies-Bouldin Index are used to evaluate the quality of the clusters formed by the model [80–82].
Time series metrics: In time series analysis, metrics like Mean Absolute Percentage Error (MAPE) and Symmetric Mean Absolute Percentage Error (sMAPE) are utilized to assess the accuracy of forecasts, providing a percentage measure of the average deviation from the actual values [83].

Model evaluation and the selection of appropriate performance metrics are critical steps in the development of ML models. By carefully considering a range of criteria and metrics, practitioners can ensure that their models achieve high accuracy and align with the operational requirements and constraints of their specific applications, leading to more effective and reliable ML solutions.

Ethical implications and bias in machine learning

The integration of ML into various sectors has brought technological advancements and raised significant ethical considerations and concerns about bias. Addressing these issues is crucial for developing and deploying ML technologies [84].

Ethical considerations in machine learning

The integration of ML into various aspects of society has underscored the importance of ethical considerations in developing and deploying these technologies. The potential of ML systems to significantly influence human lives brings forth a set of moral obligations to ensure that these influences are positive, fair, and just [84,85].

Moral responsibility: The decisions made by ML systems can profoundly impact individuals and communities. From healthcare diagnostics to financial lending decisions, the outcomes influenced by ML models bear a moral weight, necessitating a commitment to fairness and justice in their design and implementation.
Avoiding harm: It is paramount that ML models are developed with a keen awareness of the potential for harm. This includes preventing physical harm that might arise from automated systems in critical applications, mitigating psychological impacts from personalized algorithms, and addressing social harms such as discrimination or inequality that might be perpetuated or exacerbated by biased models.
Privacy concerns: The reliance on vast datasets for training ML models raises significant privacy issues, especially when dealing with sensitive personal information. Protecting individual privacy requires robust data handling and processing protocols to ensure that personal data is not misused or exposed without consent.

Transparency and accountability: To foster trust in ML systems, it is essential to maintain transparency about how they operate and make decisions. Moreover, developers and operators of ML systems must be held accountable for their outcomes, ensuring that mechanisms are in place to address any negative impacts.

Addressing bias in machine learning models

Bias in ML models poses significant challenges to fairness, accuracy, and the ethical application of these technologies. Understanding the various types of biases that can affect ML models and adopting strategies to identify and mitigate these biases are crucial steps in developing responsible and equitable AI systems [86,87].

Types of biases in machine learning

Data bias: This form of bias emerges when the dataset used to train the ML model does not accurately represent the diversity of the real-world population or scenario it is intended to model. Data bias can lead to models that perform well on the training data but poorly in real-world applications, especially for underrepresented groups [88].
Algorithmic bias: Algorithmic bias occurs when the ML algorithm introduces systemic prejudice, often due to flawed assumptions made during the ML process [89]. This can result in models that systematically favor or discriminate against certain groups, even if the underlying data is unbiased.
Confirmation bias: This bias arises when developers' preconceptions or hypotheses influence the development and training of ML models, leading to models that are inadvertently skewed toward these preconceptions [90]. Confirmation bias can reinforce existing stereotypes and hinder the objective analysis of data.

Strategies to identify and mitigate bias

Diverse data sets: One of the most effective strategies for mitigating data bias is to ensure that the training data encompasses a broad and diverse range of inputs. This involves actively seeking out data from underrepresented groups or scenarios to create a more comprehensive and representative dataset. Ensuring diversity in training data helps prevent the model from learning and perpetuating biases present in more limited datasets.
Regular auditing: Implementing regular audits of ML models can help identify biased outcomes that may not have been apparent during initial training. Auditing involves evaluating the model's performance across different demographics and scenarios to uncover any systematic biases. If biases are detected, the model may need to be retrained with adjusted datasets or algorithms to correct these biases.
Inclusive development team: Cultivating a diverse team of developers is key to identifying and preventing biases that a more homogeneous group might not recognize. A team that reflects various perspectives and experiences is more likely to spot potential biases in datasets, model design, and overall ML processes. Inclusive development practices encourage the consideration of ethical implications and the fair treatment of all individuals, leading to more equitable and responsible ML models.

Addressing bias in ML models is an ongoing challenge that requires continuous effort and vigilance. By recognizing the types of biases that can occur and implementing strategies to mitigate these biases, developers can create ML models that are more accurate and effective, fairer and more ethical, ultimately contributing to the positive impact of AI on society.

Transparent and responsible artificial intelligence

The push toward transparent and responsible AI underscores a growing recognition of the ethical implications associated with deploying ML technologies. As AI systems increasingly influence various aspects of society, ensuring these technologies are developed and used ethically becomes paramount [91–93].

Importance of transparency in machine learning

Understanding decision-making: Transparency in ML is fundamental for comprehending the rationale behind a model's decisions. This is particularly critical in sectors with significant implications, such as healthcare, where AI-assisted decisions affect patient outcomes, or in criminal justice, where they could influence sentencing or bail decisions. Transparent models allow for a deeper understanding of their decision-making processes, facilitating identifying and correcting potential biases or errors.
Building trust: Trust is a cornerstone of technology adoption, especially for ML systems. Developers and organizations can foster trust among users and stakeholders by adopting transparent practices. Transparency reassures users that the AI systems they interact with operate fairly, responsibly, and per ethical standards, thereby encouraging broader acceptance and responsible utilization of these technologies.

Guidelines for responsible artificial intelligence practices

Ethical AI frameworks: Adherence to established ethical AI frameworks and guidelines is vital in ensuring the responsible development and deployment of AI technologies. Various organizations and regulatory bodies, including the European Union, have proposed principles and standards to guide the ethical use of AI. These frameworks often emphasize respect for human rights, nondiscrimination, fairness, accountability, and privacy.
Explainability: A commitment to explainability within ML models is crucial for responsible AI. Explainable AI (XAI) seeks to make the operations and decisions of AI systems comprehensible to human users, allowing for greater scrutiny, accountability, and trust in these technologies. By prioritizing explainability, developers can ensure that stakeholders understand how AI models derive their conclusions, which is essential for evaluating the models' fairness and accuracy.
Stakeholder engagement: Engaging diverse stakeholders in the AI development process is key to identifying and addressing the varied impacts of AI technologies. This includes those who design and deploy AI systems and the end-users and communities who may be affected by AI decisions. Involving a broad spectrum of perspectives helps anticipate potential ethical issues, societal impacts, and practical challenges, ensuring that AI technologies serve the interests and needs of all parties involved.

Tools to create transparent and responsible artificial intelligence

Developing transparent and responsible AI ensures that AI systems are ethical, fair, and accountable. Various tools and methodologies have been developed to achieve this goal. These tools aim to clarify the decision-making processes of AI systems, ensure that AI operates within ethical guidelines, and mitigate biases that could lead to unfair outcomes. Here, we explore several key tools and approaches that contribute to the creation of transparent and responsible AI:

Explainable AI (XAI) frameworks: XAI refers to techniques and models that make the outputs of AI systems understandable by humans. XAI frameworks are designed to provide insights into the reasoning behind AI decisions, making AI systems more transparent and trustworthy. Examples of XAI tools include LIME (Local Interpretable Model-agnostic Explanations) and SHAP (SHapley Additive exPlanations) [94,95], which help elucidate the contribution of each input feature to the final decision made by the model.
AI ethics guidelines and frameworks: Various organizations and governmental bodies have developed guidelines and frameworks to ensure that AI development aligns with ethical standards. These documents typically outline principles such as fairness, accountability, transparency, and respect for user privacy. The IEEE's "Ethically Aligned Design" and the EU's "Ethics Guidelines for Trustworthy AI" are prominent examples that provide comprehensive guidance for ethical AI development.
Bias detection and mitigation tools: Bias in AI can lead to unfair or discriminatory outcomes. Tools designed to detect and mitigate bias are essential for responsible AI development. AIF360 (AI Fairness 360) by IBM is an open-source toolkit that provides algorithms to detect, understand, and mitigate bias in ML models [96]. Such tools are instrumental in ensuring that AI systems treat all individuals and groups fairly.
Auditing and certification processes: Independent auditing and certification processes evaluate AI systems against established transparency, security, and ethics standards. These processes help identify potential issues and assure stakeholders of the system's compliance with ethical norms. Organizations like AI Global offer AI certification programs that assess AI products and services for responsibility, transparency, and accountability.
Privacy-preserving technologies: Ensuring user privacy is a critical aspect of responsible AI. Differential privacy and federated learning enable AI systems to learn from data without compromising individual privacy. Differential privacy introduces randomness to the data or the algorithm's outputs to prevent the identification of individuals. At the same time, federated learning allows models to be trained directly on devices, keeping sensitive data localized.
Open source and collaborative development: Open source AI projects promote transparency and collaboration in AI development. By making AI algorithms and datasets publicly available, these projects allow for greater scrutiny, peer review, and contributions from the global community. Platforms such as GitHub and initiatives like the Partnership on AI foster collaborative development, ensuring that AI technologies are developed with a wide range of perspectives and expertise.

Transparent and responsible AI practices are not just ethical imperatives but critical for sustainable advancement and integration of AI technologies into society. By prioritizing transparency, adhering to ethical frameworks, striving for explainability, and engaging with stakeholders, the AI community can navigate the complex ethical landscape of AI development and deployment, ensuring that these powerful technologies contribute positively to societal progress.

Conclusion

Summarizing key takeaways

This chapter provided a comprehensive exploration of ML, starting from its foundational concepts to the advanced techniques and ethical considerations shaping its development and application today. We introduced ML, tracing its evolution and outlining its growing significance across various sectors. Special emphasis was placed on its transformative role in resource recovery and value creation, showcasing ML's potential to drive sustainability and efficiency. The discussion then transitioned to the core concepts of ML, covering supervised, semisupervised, unsupervised, and reinforcement learning. Each learning type was dissected to reveal its unique methodologies, application areas, and the distinct challenges they address, illustrating the diversity and versatility of ML approaches. In addressing data preprocessing, the chapter underscored the paramount importance of data quality, feature selection, and strategies for managing imbalanced and missing data. These foundational steps were highlighted as critical for constructing effective ML models, ensuring they are accurate and generalizable. Exploring traditional ML methods such as decision trees, SVMs, and KNN was provided, alongside a dive into advanced neural networks, including CNNs and RNNs. This section detailed the strengths and limitations of these algorithms, offering insights into their suitability for different data types and problem scenarios. Practical aspects of ML, encompassing model training, validation, and techniques to combat overfitting, were discussed. This coverage aimed to equip readers with the knowledge needed to develop models that are not only performant but also robust and reliable under various conditions. Lastly, the chapter explored the ethical considerations and potential biases inherent in ML. The necessity for ethical ML practices, the pursuit of transparency, and the engagement in responsible AI development were stressed. This section served as a reminder of the moral responsibilities of wielding ML technologies' power. In conclusion, this chapter sought to provide a thorough overview of ML, from its theoretical underpinnings to its practical applications and ethical implications. By highlighting the key takeaways from each section, we hope to have imparted a deeper understanding of ML's potential to innovate and its challenges, paving the way for its responsible and impactful use in the future.

Future directions of machine learning in resource recovery and value creation

As ML advances, its potential applications in resource recovery and value creation are expanding, promising to significantly improve efficiency, sustainability, and innovation. Here, we explore the emerging trends and technologies in ML and their expected impact on sustainable development.

Emerging trends and technologies

AutoML (automated ML) is emerging as a powerful trend that democratizes and streamlines the ML model development process. By automating ML models' selection, composition, and parameterization, AutoML enables experts and nonexperts alike to develop effective ML solutions more quickly and with less effort. This trend will likely accelerate the adoption of ML in various sectors, including resource recovery.

Advancements in AI and IoT: Integrating AI with the Internet of Things (IoT) is set to transform resource recovery processes. By harnessing real-time data collection and analysis capabilities, AI-enhanced IoT systems can optimize resource utilization, monitor environmental conditions, and improve decision-making processes in resource management. This convergence promises to enhance operational efficiency and reduce waste.

Quantum ML: As quantum computing continues to develop, its integration with ML—known as Quantum ML—holds the promise of dramatically increasing processing capabilities. This advancement could unlock new possibilities in data analysis and modeling, enabling the handling of complex, high-dimensional datasets with unprecedented speed and accuracy. The implications for resource recovery and value creation are vast, potentially revolutionizing how we process and interpret data in these fields.

The role of machine learning in sustainable development

ML is increasingly recognized as a critical tool in promoting sustainable development, especially in the context of resource recovery and waste management:

Optimizing resource recovery: ML algorithms can significantly enhance the efficiency of resource recovery processes, from sorting recyclables with high precision to optimizing the operation of recovery facilities. By maximizing resource utilization, ML contributes directly to the principles of the circular economy, minimizing waste and extending the lifecycle of materials.

Reducing environmental impact: ML helps reduce the environmental impact of human activities through predictive analytics and optimization algorithms. By forecasting resource availability and demand, ML enables more sustainable decision-making, ensuring that resource extraction and consumption align more closely with sustainability principles.

Supporting the circular economy: ML's ability to improve waste sorting, predict maintenance needs, and optimize supply chains is vital in supporting the circular economy. By facilitating the reuse and recycling of materials, ML technologies help create more sustainable business models and reduce reliance on finite natural resources.

In summary, the future of ML in resource recovery and value creation is marked by rapid technological advancements and a growing emphasis on sustainability. As emerging trends like AutoML, AI-enhanced IoT, and Quantum ML continue to evolve, their application in sustainable development offers promising avenues for innovation. By harnessing these technologies, ML can drive significant improvements in resource efficiency, environmental protection, and sustainable growth, marking a new era in the responsible management and utilization of natural resources.

Final thoughts and reflections

The continuous evolution of machine learning

The realm of ML is inherently dynamic and characterized by constant evolution. New algorithms, techniques, and realms of application regularly emerge, broadening the scope and capabilities of ML technologies. This continuous growth demands a proactive approach to learning and adaptation from practitioners, researchers, and developers in the field. Staying abreast of the latest advancements is not merely a matter of academic interest but a requisite for effectively addressing contemporary challenges and leveraging new opportunities in technology and its applications.

Encouraging responsible and innovative use of machine learning in resource recovery

As the frontier of ML technologies expands, the imperative to cultivate a culture of responsible innovation becomes increasingly pronounced. The advancement of ML should be pursued with a holistic perspective, where technological progress is balanced with ethical considerations, sustainability goals, and the potential social impact. The application of ML in fields like resource recovery and value creation presents a significant opportunity to effect positive change, offering novel solutions to pressing environmental challenges. However, realizing this potential requires a conscientious approach, ensuring that the benefits of ML technologies are harnessed in a manner that is equitable, sustainable, and aligned with broader societal goals.

In reflecting on the journey through this chapter, we have traversed the foundational concepts of ML, explored its diverse methodologies, and considered its application in enhancing resource recovery and fostering value creation. Looking ahead, the integration of ML into efforts aimed at sustainable development appears both promising and essential. The path forward calls for a steadfast commitment to innovation, guided by principles of responsibility and inclusivity. By doing so, we can ensure that the advancements in ML not only push the boundaries of what is technologically possible but also contribute to building a more sustainable, equitable, and prosperous future for all.

Abbreviations

ML machine learning
AI artificial intelligence
KNN K-nearest neighbors
lasso least absolute shrinkage and selection operator
GBM gradient boosting machine
SVM support vector machine
CNN convolutional neural network
RNN recurrent neural network
GAN generative adversarial network
NLP natural language processing
LSTM long short-term memory

BERT	bidirectional encoder representations from transformers
GPT	generative pretrained transformer
GPU	graphics processing unit
ROC	receiver operating characteristic
MSE	mean squared error
MAE	mean absolute error
F1	score harmonic mean of precision and recall
ROC-AUC	area under the receiver operating characteristic curve
RMSE	root mean squared error
MAPE	mean absolute percentage error
sMAPE	symmetric mean absolute percentage error
XAI	explainable AI
LIME	local interpretable model-agnostic explanations
SHAP	SHapley Additive exPlanations
AIF360	AI Fairness 360
AutoML	automated ML
IoT	Internet of Things

References

[1] R. Costa-Climent, The role of machine learning in creating and capturing value, International Journal of Software Science and Computational Intelligence 14 (1) (2022) 1–19. Available from https://doi.org/10.4018/ijssci.312229.

[2] X. Chen, Machine learning approach for a circular economy with waste recycling in smart cities, Energy Reports 8 (2022) 3127–3140. Available from https://doi.org/10.1016/j.egyr.2022.01.193, http://www.journals.elsevier.com/energy-reports/.

[3] B. Fang, J. Yu, Z. Chen, A.I. Osman, M. Farghali, I. Ihara, et al., Artificial intelligence for waste management in smart cities: a review, Environmental Chemistry Letters 21 (4) (2023) 1959–1989. Available from https://doi.org/10.1007/s10311-023-01604-3, https://www.springer.com/journal/10311.

[4] N.E. Benti, M.D. Chaka, A.G. Semie, Forecasting renewable energy generation with machine learning and deep learning: current advances and future prospects, Sustainability 15 (9) (2023) 7087. Available from https://doi.org/10.3390/su15097087.

[5] A. Ucar, M. Karakose, N. Kırımça, Artificial intelligence for predictive maintenance applications: key components, trustworthiness, and future trends, Applied Sciences 14 (2) (2024) 898. Available from https://doi.org/10.3390/app14020898.

[6] Y. Kolomoyets, A. Dickinger, Understanding value perceptions and propositions: A machine learning approach, Journal of Business Research 154 (2023) 113355. Available from https://doi.org/10.1016/j.jbusres.2022.113355.

[7] D.O. Hebb, The Organization of Behavior; A Neuropsychological Theory, Wiley, 1949.

[8] A.M. Turing, Intelligent machinery, The Essential Turing (1948) 395–432.

[9] O. Debauche, J.B.N. Penka, M. Hani, A. Guttadauria, R.A. Abdelouahid, K. Gasmi, et al., Towards a unified architecture powering scalable learning models with IoT data streams, blockchain, and open data, Information (Switzerland) 14 (6) (2023). Available from https://doi.org/10.3390/info14060345.

[10] M. Nguyen, T. He, L. An, D.C. Alexander, J. Feng, B.T.T. Yeo, Alzheimer's disease neuroimaging, predicting Alzheimer's disease progression using deep recurrent neural networks, Neuroimage 222 (2020).

[11] R. Rafique, S.M.R. Islam, J.U. Kazi, Machine learning in the prediction of cancer therapy, Computational and Structural Biotechnology Journal 19 (2021) 4003–4017. Available from https://doi.org/10.1016/j.csbj.2021.07.003, http://www.csbj.org.

[12] K. Shah, A. Nasimian, M. Ahmed, L. Al Ashiri, L. Denison, W. Sime, et al., PLK1 as a cooperating partner for BCL2-mediated antiapoptotic program in leukemia, Blood Cancer Journal 13 (1) (2023). Available from https://doi.org/10.1038/s41408-023-00914-7.

[13] P.M. Mirete-Ferrer, A. Garcia-Garcia, J.S. Baixauli-Soler, M.A. Prats, A review on machine learning for asset management, Risks 10 (4) (2022). Available from https://doi.org/10.3390/risks10040084, https://www.mdpi.com/2227-9091/10/4/84/pdf.

[14] M. Saha, P. Mitra, R.S. Nanjundiah, Deep learning for predicting the monsoon over the homogeneous regions of India, Journal of Earth System Science 126 (4) (2017). Available from https://doi.org/10.1007/s12040-017-0838-7, http://www.ias.ac.in/article/fulltext/jess/126/04/0054.

[15] M. Erkinay Ozdemir, Z. Ali, B. Subeshan, E. Asmatulu, Applying machine learning approach in recycling, Journal of Material Cycles and Waste Management 23 (3) (2021) 855–871. Available from https://doi.org/10.1007/s10163-021-01182-y, http://link.springer.de/link/service/journals/10163/index.htm.

[16] D. Ni, Z. Xiao, M.K. Lim, Machine learning in recycling business: an investigation of its practicality, benefits and future trends, Soft Computing 25 (12) (2021) 7907–7927. Available from https://doi.org/10.1007/s00500-021-05579-7, http://springerlink.metapress.com/app/home/journal.asp?wasp=h83ak0wtmr5uxkah9j5m&referrer=parent&backto=browsepublicationsresults,466,533.

[17] A. Aldoseri, K.N. Al-Khalifa, A.M. Hamouda, Re-thinking data strategy and integration for artificial intelligence: concepts, opportunities, and challenges, Applied Sciences (Switzerland) 13 (12) (2023). Available from https://doi.org/10.3390/app13127082, http://www.mdpi.com/journal/applsci/.

[18] M.M. Taye, Understanding of machine learning with deep learning: architectures, workflow, applications and future directions, Computers 12 (5) (2023). Available from https://doi.org/10.3390/computers12050091.

[19] Y. Wang, Z. Cui, R. Ke, Machine Learning for Transportation Research and Applications, Elsevier, United States, 2023, pp. 1–239. Available from https://www.sciencedirect.com/book/9780323961264, https://doi.org/10.1016/C2021-0-02166-0.

[20] A. Nasimian, S. Younus, Ö. Tatli, E.U. Hammarlund, K.J. Pienta, L. Rönnstrand, et al., AlphaML: a clear, legible, explainable, transparent, and elucidative binary classification platform for tabular data, Patterns 5 (1) (2024). Available from https://doi.org/10.1016/j.patter.2023.100897.

[21] P. Wittek, Unsupervised Learning, Elsevier BV, 2014, pp. 57–62. Available from https://doi.org/10.1016/b978-0-12-800953-6.00005-0.

[22] J.E. van Engelen, H.H. Hoos, A survey on semi-supervised learning, Machine Learning 109 (2) (2020) 373–440. Available from https://doi.org/10.1007/s10994-019-05855-6, http://www.springer.com/journal/10994.

[23] A.G. Barto, Reinforcement Learning. Elsevier BV, 1997, pp. 7–30. Available from https://doi.org/10.1016/b978-012526430-3/50003-9.

[24] J. Fan, C. Xiao, Y. Huang, GDI: rethinking what makes reinforcement learning different from supervised learning. arXiv, China arXiv. 23318422 (2021). Available from https://arxiv.org, https://doi.org/10.48550/arxiv.2106.06232.

[25] Y. Yan, A.H.F. Chow, C.P. Ho, Y.H. Kuo, Q. Wu, C. Ying, Reinforcement learning for logistics and supply chain management: methodologies, state of the art, and future opportunities, Transportation Research Part E: Logistics and Transportation Review 162 (2022). Available from https://doi.org/10.1016/j.tre.2022.102712, https://www.sciencedirect.com/journal/transportation-research-part-e-logistics-and-transportation-review.

[26] B. Hambly, R. Xu, H. Yang, Recent advances in reinforcement learning in finance, Mathematical Finance 33 (3) (2023) 437–503. Available from https://doi.org/10.1111/mafi.12382, http://onlinelibrary.wiley.com/journal/10.1111/(ISSN)1467-9965.

[27] L. Budach, M. Feuerpfeil, N. Ihde, A. Nathansen, N. Noack, H. Patzla, et al., The effects of data quality on machine learning performance. arXiv, Germany arXiv. 23318422 (2022). Available from https://arxiv.org, https://doi.org/10.48550/arXiv.2207.14529.

[28] D. Chicco, L. Oneto, E. Tavazzi, F. Ouellette, Eleven quick tips for data cleaning and feature engineering, PLoS Computational Biology 18 (12) (2022) e1010718. Available from https://doi.org/10.1371/journal.pcbi.1010718.

[29] D. Singh, B. Singh, Investigating the impact of data normalization on classification performance, Applied Soft Computing 97 (2020) 105524. Available from https://doi.org/10.1016/j.asoc.2019.105524.

[30] D. Theng, K.K. Bhoyar, Feature selection techniques for machine learning: a survey of more than two decades of research, Knowledge and Information Systems 66 (3) (2024) 1575–1637. Available from https://doi.org/10.1007/s10115-023-02010-5, https://www.springer.com/journal/10115.

[31] H.N. Huang, H.M. Chen, W.W. Lin, C.J. Huang, Y.C. Chen, Y.H. Wang, et al., Employing feature engineering strategies to improve the performance of machine learning algorithms on echocardiogram dataset, Digital Health 9 (2023). Available from https://doi.org/10.1177/20552076231207589, https://journals.sagepub.com/home/DHJ.

[32] F.M. Rohrhofer, S. Saha, S. Di Cataldo, B.C. Geiger, W. von der Linden, L. Boeri, Importance of feature engineering and database selection in a machine learning model: a case study on carbon crystal structures. arXiv, Austria arXiv. 23318422. (2021). Available from https://arxiv.org, https://doi.org/10.48550/arxiv.2102.00191.

[33] Y. Liu, B. Li, S. Yang, Z. Li, Handling missing values and imbalanced classes in machine learning to predict consumer preference: demonstrations and comparisons to prominent methods, Expert Systems with Applications 237 (2024) 121694. Available from https://doi.org/10.1016/j.eswa.2023.121694.

[34] A.A. Dharmasaputro, N.M. Fauzan, M. Kallista, I.P.D. Wibawa, P.D. Kusuma Handling missing and imbalanced data to improve generalization performance of machine learning classifier. International Seminar on Machine Learning, Optimization, and Data Science (2022), ISMODE 2021. https://doi.org/10.1109/ISMODE53584.2022.9743022, http://ieeexplore.ieee.org/xpl/mostRecentIssue.jsp?punumber=9741823.

[35] X. Chen, H. Chen, S. Nan, X. Kong, H. Duan, H. Zhu, Dealing with missing, imbalanced, and sparse features during the development of a prediction model for sudden death using emergency medicine data: machine learning approach, JMIR Medical Informatics 11 (2023) e38590. Available from https://doi.org/10.2196/38590.

[36] S. Younus, L. Rönnstrand, J.U. Kazi, Xputer: bridging data gaps with NMF, XGBoost, and a streamlined GUI experience. arXiv, Sweden arXiv. 23318422 (2023), Available from https://arxiv.org, https://doi.org/10.48550/arXiv.2311.09989.

[37] A.D. Shah, J.W. Bartlett, J. Carpenter, O. Nicholas, H. Hemingway, Comparison of random forest and parametric imputation models for imputing missing data using MICE: a CALIBER study, American Journal of Epidemiology 179 (6) (2014) 764–774. Available from https://doi.org/10.1093/aje/kwt312, http://aje.oxfordjournals.org/.

[38] I.H. Sarker, Machine learning: algorithms, real-world applications and research directions, SN Computer Science 2 (3) (2021). Available from https://doi.org/10.1007/s42979-021-00592-x, https://www.springer.com/journal/42979.

[39] G. Tripepi, K.J. Jager, F.W. Dekker, C. Zoccali, Linear and logistic regression analysis, Kidney International 73 (7) (2008) 806–810. Available from https://doi.org/10.1038/sj.ki.5002787.

[40] A.E. Hoerl, R.W. Kennard, Ridge regression: biased estimation for nonorthogonal problems, Technometrics 12 (1) (1970) 55–67. Available from https://doi.org/10.1080/00401706.1970.10488634.

[41] R. Tibshirani, Regression shrinkage and selection via the Lasso, Journal of the Royal Statistical Society. Series B: Methodological 58 (1) (1996) 267–288. Available from https://doi.org/10.1111/j.2517-6161.1996.tb02080.x.

[42] H. Zou, T. Hastie, Regularization and variable selection via the elastic net, Journal of the Royal Statistical Society. Series B: Statistical Methodology 67 (2) (2005) 301–320. Available from https://doi.org/10.1111/j.1467-9868.2005.00503.x.

[43] L. Breiman, J.H. Friedman, R.A. Olshen, C.J. Stone, Classification and Regression Trees, Routledge, 1984, pp. 1–368.

[44] J.R. Quinlan, Induction of decision trees, Machine Learning 1 (1) (1986) 81–106. Available from https://doi.org/10.1007/bf00116251.

[45] J.R. Quinlan, Improved use of continuous attributes in C4.5, Journal of Artificial Intelligence Research 4 (1996) 77–90. Available from https://doi.org/10.1613/jair.279.

[46] L. Breiman, Random forests, Machine Learning 45 (1) (2001) 5–32. Available from https://doi.org/10.1023/A:1010933404324.

[47] J.H. Friedman, Greedy function approximation: a gradient boosting machine, Annals of Statistics 29 (5) (2001) 1189–1232. Available from https://doi.org/10.1214/aos/1013203451, http://projecteuclid.org/DPubS?service=UI&version=1.0&verb=Display&page=past&handle=euclid.aos.

[48] Y. Freund, R.E. Schapire, A decision-theoretic generalization of on-line learning and an application to boosting, Journal of Computer and System Sciences 55 (1) (1997) 119–139. Available from https://doi.org/10.1006/jcss.1997.1504.

[49] T. Chen, C. Guestrin, Association for computing machinery United States XGBoost: a scalable tree boosting system, in: Proceedings of the ACM SIGKDD International Conference on Knowledge Discovery and Data Mining, 2016. https://doi.org/10.1145/2939672.2939785.

[50] G. Ke, Q. Meng, T. Finley, T. Wang, W. Chen, W. Ma, et al., Neural information processing systems foundation United States LightGBM: a highly efficient gradient boosting decision tree. Advances in Neural Information Processing Systems (2017). 10495258 3147 3155.

[51] A.V. Dorogush, V. Ershov, A. Gulin, CatBoost: gradient boosting with categorical features support. arXiv, undefined arXiv. 23318422 (2018). Available from https://arxiv.org, https://doi.org/10.48550/arxiv.1810.11363.

[52] K.-R. Muller, S. Mika, G. Ratsch, K. Tsuda, B. Scholkopf, An introduction to kernel-based learning algorithms, IEEE Transactions on Neural Networks 12 (2) (2001) 181–201. Available from https://doi.org/10.1109/72.914517.

[53] T. Cover, P. Hart, Nearest neighbor pattern classification, IEEE Transactions on Information Theory 3 (1) (1967) 21–27. Available from https://doi.org/10.1109/TIT.1967.1053964.

[54] R. Yamashita, M. Nishio, R.K.G. Do, K. Togashi, Convolutional neural networks: an overview and application in radiology, Insights into Imaging 9 (4) (2018) 611–629. Available from https://doi.org/10.1007/s13244-018-0639-9, http://www.springer.com/medicine/radiology/journal/13244.

[55] R. DiPietro, G.D. Hager, Handbook of Medical Image Computing and Computer Assisted Intervention, Academic Press, 2020.

[56] D. Khurana, A. Koli, K. Khatter, S. Singh, Natural language processing: state of the art, current trends and challenges, Multimedia Tools and Applications 82 (3) (2023) 3713–3744. Available from https://doi.org/10.1007/s11042-022-13428-4, https://www.springer.com/journal/11042.

[57] Y. Lecun, Y. Bengio, G. Hinton, Deep learning, Nature 521 (7553) (2015) 436–444. Available from https://doi.org/10.1038/nature14539, http://www.nature.com/nature/index.html.

[58] S.P. Gundupalli, S. Hait, A. Thakur, A review on automated sorting of source-separated municipal solid waste for recycling, Waste Management 60 (2017) 56–74. Available from https://doi.org/10.1016/j.wasman.2016.09.015, http://www.elsevier.com/locate/wasman.

[59] I. Yani, Y. Resti, Performance of plastic-type prediction using decision tree approaches, in: AIP Conference Proceedings, American Institute of Physics Inc. Indonesia, 2023. Available from https://doi.org/10.1063/5.0172234, http://scitation.aip.org/content/aip/proceeding/aipcp 2913.

[60] S. Mishra, L. Jena, H.K. Tripathy, T. Gaber, C. So-In, Prioritized and predictive intelligence of things enabled waste management model in smart and sustainable environment, PLoS One 17 (8) (2022) e0272383. Available from https://doi.org/10.1371/journal.pone.0272383.

[61] V.K. Pandey, S. Srivastava, K.K. Dash, R. Singh, S.A. Mukarram, B. Kovács, et al., Machine learning algorithms and fundamentals as emerging safety tools in preservation of fruits and vegetables: a review, Processes 11 (6) (2023). Available from https://doi.org/10.3390/pr11061720.

[62] Y. Popova, I. Sproge, Decision-making within smart city: waste sorting, Sustainability 13 (19) (2021) 10586. Available from https://doi.org/10.3390/su131910586.

[63] H. Abu-Qdais, N. Shatnawi, E.'a AL-Alamie, Intelligent system for solid waste classification using combination of image processing and machine learning models, Journal of Experimental & Theoretical Artificial Intelligence (2024) 1–12. Available from https://doi.org/10.1080/0952813x.2024.2323043.

[64] N. Li, Y. Chen, Municipal solid waste classification and real-time detection using deep learning methods, Urban Climate 49 (2023) 101462. Available from https://doi.org/10.1016/j.uclim.2023.101462.

[65] M. Malik, S. Sharma, M. Uddin, C.L. Chen, C.M. Wu, P. Soni, et al., Waste classification for sustainable development using image recognition with deep learning neural network models, Sustainability 14 (12) (2022). Available from https://doi.org/10.3390/su14127222.

[66] K. Liu, X. Liu, Recycling material classification using convolutional neural networks, in: Proceedings—21st IEEE International Conference on Machine Learning and Applications, ICMLA 2022, Institute of Electrical and Electronics Engineers Inc., United States, 2022. Available from https://doi.org/10.1109/ICMLA55696.2022.00019, http://ieeexplore.ieee.org/xpl/mostRecentIssue.jsp?punumber=10068862.

[67] T.W. Wu, H. Zhang, W. Peng, F. Lü, P.J. He, Applications of convolutional neural networks for intelligent waste identification and recycling: a review, Resources, Conservation and Recycling 190 (2023). Available from https://doi.org/10.1016/j.resconrec.2022.106813, http://www.elsevier.com/locate/resconrec.

[68] J. Gunaseelan, S. Sundaram, B. Mariyappan, A design and implementation using an innovative deep-learning algorithm for garbage segregation, Sensors 23 (18) (2023). Available from https://doi.org/10.3390/s23187963, http://www.mdpi.com/journal/sensors.

[69] A. Rivas, J.M. Fraile, P. Chamoso, A. González-Briones, I. Sittón, J.M. Corchado, A predictive maintenance model using recurrent neural networks, 1 2020/01, Advances in Intelligent Systems and Computing 950 (2020) 261–270. Available from https://doi.org/10.1007/978-3-030-20055-8_25, http://www.springer.com/series/11156.

[70] M.A. Farahani, M.R. McCormick, R. Gianinny, F. Hudacheck, R. Harik, Z. Liu, et al., Time-series pattern recognition in smart manufacturing systems: a literature review and ontology, Journal of Manufacturing Systems 69 (2023) 208–241. Available from https://doi.org/10.1016/j.jmsy.2023.05.025, http://www.elsevier.com.

[71] R. Qi, J. Zhang, K. Spencer, A review on data-driven condition monitoring of industrial equipment, Algorithms 16 (1) (2023). Available from https://doi.org/10.3390/a16010009, http://www.mdpi.com/journal/algorithms/.

[72] S.F. Ahmed, M.S.B. Alam, M. Hassan, M.R. Rozbu, T. Ishtiak, N. Rafa, et al., Deep learning modelling techniques: current progress, applications, advantages, and challenges, Artificial Intelligence Review 56 (11) (2023) 13521–13617. Available from https://doi.org/10.1007/s10462-023-10466-8, https://www.springer.com/journal/10462.

[73] P. Gupta, N.K. Sehgal, Practical Aspects in Machine Learning, Springer Science and Business Media LLC, 2021, pp. 115–142. Available from https://doi.org/10.1007/978-3-030-71270-9_5.

[74] B.J. Erickson, F. Kitamura, Magician's corner: 9. Performance metrics for machine learning models, Radiology: Artificial Intelligence 3 (2021).

[75] X. Ying, An overview of overfitting and its solutions, Journal of Physics: Conference Series 1168 (2) (2019) 022022. Available from https://doi.org/10.1088/1742-6596/1168/2/022022.

[76] H. Li, G.K. Rajbahadur, D. Lin, C.P. Bezemer, Z.M. Jiang, Keeping deep learning models in check: a history-based approach to mitigate overfitting. arXiv, Canada arXiv. 23318422 (2024). Available from https://arxiv.org. https://doi.org/10.48550/arXiv.2401.10359.

[77] S.A. Hicks, I. Strümke, V. Thambawita, M. Hammou, M.A. Riegler, P. Halvorsen, et al., On evaluation metrics for medical applications of artificial intelligence, Scientific Reports 12 (1) (2022). Available from https://doi.org/10.1038/s41598-022-09954-8, http://www.nature.com/srep/index.html.

[78] L. Ferrer, Analysis and comparison of classification metrics. arXiv, Argentina arXiv. 23318422 (2022). Available from https://arxiv.org, https://doi.org/10.48550/arXiv.2209.05355.

[79] A. Botchkarev, Performance metrics (error measures) in machine learning regression, forecasting and prognostics: properties and typology. arXiv, Canada arXiv. 23318422 (2018). Available from https://arxiv.org, https://doi.org/10.48550/arxiv.1809.03006.

[80] L.E. Ekemeyong Awong, T. Zielinska, Comparative analysis of the clustering quality in self-organizing maps for human posture classification, Sensors 23 (18) (2023). Available from https://doi.org/10.3390/s23187925, http://www.mdpi.com/journal/sensors.

[81] Y. Lu, C.A. Phillips, M.A. Langston, A robustness metric for biological data clustering algorithms, BMC Bioinformatics 20 (2019). Available from https://doi.org/10.1186/s12859-019-3089-6, http://www.biomedcentral.com/bmcbioinformatics/.

[82] P.J. Rousseeuw, Silhouettes: a graphical aid to the interpretation and validation of cluster analysis, Journal of Computational and Applied Mathematics 20 (C) (1987) 53–65. Available from https://doi.org/10.1016/0377-0427(87)90125-7.

[83] S. Kim, H. Kim, A new metric of absolute percentage error for intermittent demand forecasts, International Journal of Forecasting 32 (3) (2016) 669–679. Available from https://doi.org/10.1016/j.ijforecast.2015.12.003, http://www.elsevier.com/locate/ijforecast.

[84] V.N. O'Reilly-Shah, K.R. Gentry, A.M. Walters, J. Zivot, C.T. Anderson, P.J. Tighe, Bias and ethical considerations in machine learning and the automation of perioperative risk assessment, British Journal of Anaesthesia 125 (6) (2020) 843–846. Available from https://doi.org/10.1016/j.bja.2020.07.040, https://www.journals.elsevier.com/british-journal-of-anaesthesia.

[85] A. Yapo, J. Weiss, Ethical implications of bias in machine learning, in: Proceedings of the Annual Hawaii International Conference on System Sciences, IEEE Computer Society, United States, 2018. Available from https://scholarspace.manoa.hawaii.edu/handle/10125/39610

[86] M.D. Abràmoff, M.E. Tarver, N. Loyo-Berrios, S. Trujillo, D. Char, Z. Obermeyer, et al., Considerations for addressing bias in artificial intelligence for health equity, NPJ Digital Medicine 6 (1) (2023).

[87] D. Dhabliya, S.S. Dari, A. Dhablia, N. Akhila, R. Kachhoria, V. Khetani, Addressing bias in machine learning algorithms: promoting fairness and ethical design, E3S Web of Conferences 491 EDP Sciences, India, 2024 2 21 2024/02/21. Available from https://doi.org/10.1051/e3sconf/202449102040, http://www.e3s-conferences.org/.

[88] M. Kumagai, Y. Ando, A. Tanaka, K. Tsuda, Y. Katsura, K. Kurosaki, Effects of data bias on machine-learning–based material discovery using experimental property data, Science and Technology of Advanced Materials: Methods 2 (1) (2022) 302–309. Available from https://doi.org/10.1080/27660400.2022.2109447.

[89] S. Akter, Y.K. Dwivedi, S. Sajib, K. Biswas, R.J. Bandara, K. Michael, Algorithmic bias in machine learning-based marketing models, Journal of Business Research 144 (2022) 201–216. Available from https://doi.org/10.1016/j.jbusres.2022.01.083, http://www.elsevier.com/locate/jbusres.

[90] E. Arazo, D. Ortego, P. Albert, N.E. O'Connor, K. McGuinness, Pseudo-labeling and confirmation bias in deep semi-supervised learning. arXiv, Ireland arXiv. 23318422 (2019). Available from https://arxiv.org. https://doi.org/10.48550/arxiv.1908.02983.

[91] S. Baker, W. Xiang, Explainable AI is responsible AI: how explainability creates trustworthy and socially responsible artificial intelligence. arXiv, Australia arXiv. 23318422 (2023). Available from https://arxiv.org, https://doi.org/10.48550/arXiv.2312.01555.

[92] N. Balasubramaniam, M. Kauppinen, A. Rannisto, K. Hiekkanen, S. Kujala, Transparency and explainability of AI systems: from ethical guidelines to requirements, Information and Software Technology 159 (2023) 107197. Available from https://doi.org/10.1016/j.infsof.2023.107197.

[93] S. Wachter, B. Mittelstadt, L. Floridi, Transparent, explainable, and accountable AI for robotics, Science Robotics 2 (6) (2017). Available from https://doi.org/10.1126/scirobotics.aan6080, http://robotics.sciencemag.org/content/robotics/2/6/eaan6080.full.pdf.

[94] S.M. Lundberg, G. Erion, H. Chen, A. DeGrave, J.M. Prutkin, B. Nair, et al., From local explanations to global understanding with explainable AI for trees, Nature Machine Intelligence 2 (1) (2020) 56–67. Available from https://doi.org/10.1038/s42256-019-0138-9, https://www.nature.com/natmachintell/.

[95] M.T. Ribeiro, S. Singh, C. Guestrin, Proceedings of the 22nd ACM SIGKDD International Conference on Knowledge Discovery and Data Mining Why Should I Trust You?: Explaining the Predictions of Any Classifier, 2016.

[96] R.K.E. Bellamy, K. Dey, M. Hind, S.C. Hoffman, S. Houde, K. Kannan, et al., Ai fairness 360: an extensible toolkit for detecting, understanding, and mitigating unwanted algorithmic bias. arXiv, United States arXiv. 23318422 (2018). Available from https://arxiv.org. https://doi.org/10.48550/arxiv.1810.01943.

CHAPTER

3

Basics of automation

Mohamed Zied Chaari[1], *Oleg Serov*[2], *Gilroy Pereira*[1], *and Aghzout Otman*[3]

[1]Qatar Scientific Club, Fabrication Laboratory, Doha, Qatar [2]National Research Nuclear University, MEPhI, Moscow, Russia [3]Dept of Digitalization and Artificial Intelligence, ENSA, Morocco

Introduction

Automation, a process that converts a work process, procedure, or machine to operate with minimal human intervention, has a rich history. The word "automation" itself was derived from the Greek word "automaton," which refers to self-moving machines. Many of the automatons developed by the Greeks were primarily toys. Still, there were a few practical automaton inventions, such as the repeating crossbow invented by Philon of Byzantium (c.280 BC–220 BC) [1]. This historical context adds a layer of intrigue to the evolution of automation, from these early inventions to the modern AI-powered systems we see today.

In the 1940s the automobile industry used automation to handle parts by machines and vehicles. It was an essential system of the car manufacturing process [2]. As textiles, metals, agriculture, and other industries were being automated in the nineteenth and twentieth centuries [3], it led to the birth of new jobs in the factory work, engineering, repair, back-office, management, and finance sectors. This growth in the usage of automation resulted in rapid worldwide development, especially in Western countries and the United States.

Automation has become a broader concept with computers and control systems. Today's machines are no longer straightforward moving machines; they are constantly processing data and adjusting their actions. Computers, whose flexibility makes them capable of driving virtually any task, are an essential part of modern automation systems [4]. Nowadays, a set of technologies resulting in industrial machines and systems operating without significant personal intervention and achieving performance superior to manual operation is considered automation.

Because of their intelligence, automated machines don't make errors; they do what they are supposed to do continuously without breaks and with high efficiency. Computers can also handle larger systems that individuals cannot operate efficiently. Therefore automation achieves better performance than manual operation.

Harnessing Automation and Machine Learning for Resource Recovery and Value Creation
DOI: https://doi.org/10.1016/B978-0-443-27374-2.00003-0

A conveyor belt was the initial industrial automation system to pull components along an assembly line [5]. However, humans were still responsible for completing most tasks on the assembly line. The use of machines has reduced the need for manual work for basic tasks and activities. Modern industrial automation comprises various actuators, devices, processors, sensors, and networks that simplify integrating industrial operations and processes. Modern industrial automation is focused on harnessing technology, including programmable logic controllers (PLCs), artificial intelligence (AI), machine learning (ML), and the Industrial Internet of Things (IIoT).

In manufacturing, automation refers to the process of producing items. It includes:

- Industrial arm robots.
- Automatic machinery tools to manufacture parts.
- Automated material handling.
- Automated assembly line machines.
- Automated storage and retrieving systems.
- Automated quality assurance systems.
- Provide feedback and control quality.
- Automated systems for creating parts from designs immediately.
- Manufacturing planning and decision-making systems.

In addition to its popular association with industrial production, automation now encompasses many unanticipated areas due to the rapid growth and reliance on technology. Automation plays a significant role in trading, environmental protection engineering, construction, pharmacology, agriculture, and food manufacturing. It is no secret that automation plays a significant role in military and strategic production lines [6]. Over the past few years, the food manufacturing industry has experienced rapid growth. Agriculture automation is helping governments and regimes worldwide address the famine crisis, especially in developing countries and the Middle East. The food industry's adoption of new automation technologies has been slow for several reasons. Considering human involvement in food production, food safety became the dominant factor. Raw material handling and delivery were at risk of contamination. Automation systems became widely adopted only after these issues were resolved [7].

The four categories of automation systems are shown in Fig. 3.1. First, the fixed automation system is distinguished by a high production volume, a dedicated piece of equipment, and fixed operation efficiency. The second category is a programmable automation system, in which

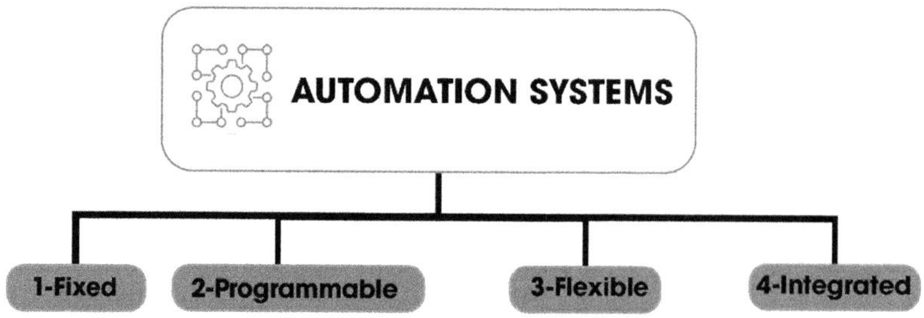

FIGURE 3.1 Four categories of automation systems.

the sequence of operations and electronic controls are programmable. Thirdly, computerized systems can be controlled by a simple program using a graphical user interface. A preprogrammed system should allow the operator to select specific settings, and the rest will be handled automatically. Finally, the integrated automation system incorporates an advanced optimization algorithm. Using complex algorithms, it performs mathematical calculations. Each of these categories of industrial automation systems has its specific benefits and drawbacks.

The term "automation" refers to a process with high quality and very little human interaction. To remain competitive in the global market, businesses must increase productivity and reduce costs per unit. Modern manufacturing relies heavily on automation, providing numerous advantages to businesses and consumers. There are several reasons why automation is essential in the manufacturing process, including:

Increasing safety factor: Automated processes can also improve the workplace's overall safety, decreasing the possibility of employee injuries.
Productivity and efficiency: Automation enables manufacturers to produce products faster, with fewer errors, and, in most cases, with zero errors.
Improved quality control: Automating processes reduces human errors, resulting in better quality control and more reliable products.
Client appreciation: By automating, manufacturers can improve customer satisfaction and brand reputation.
Labor cost reduction: Companies can drastically reduce costs by replacing employees with automated systems. However, the loss of jobs can result in societal issues, and it is a hotly debated topic in current society.

Automation is becoming increasingly necessary for companies to remain competitive in worldwide markets. Additionally, automation impacts competition between economically developed countries such as the United States and China, especially in terms of spending money on the development of automobile production and the supply chain of many products. Automation's increased efficiency, improved quality control, reduced labor costs, increased safety, and higher customer satisfaction all contribute to its importance in modern manufacturing. The chapter proposes to present the efficiency of automation in the manufacturing processes and the capability of AI in this field. Each part is discussed in this work, starting with Section 3.1, which introduces the history of automation and provides more details about it. Section 3.2 presents automation fundamentals. Section 3.3 explains the role of automation in the industry. Section 3.4 presents two case studies: the Internet of Things (IoT)-enabled drip irrigation system and drones and automation. Section 3.5 discusses automation's future and AI power. Finally, Section 3.6 concludes.

Fundamentals of automation

Automated manufacturing is focused on utilizing technology, including PLCs, artificial intelligence, sensors, actuators, and machine learning, as shown in Fig. 3.2. It uses control systems and complex equipment in a production environment or critical production industry. This includes robots, various sensors, and computers performing tasks previously done manually [8].

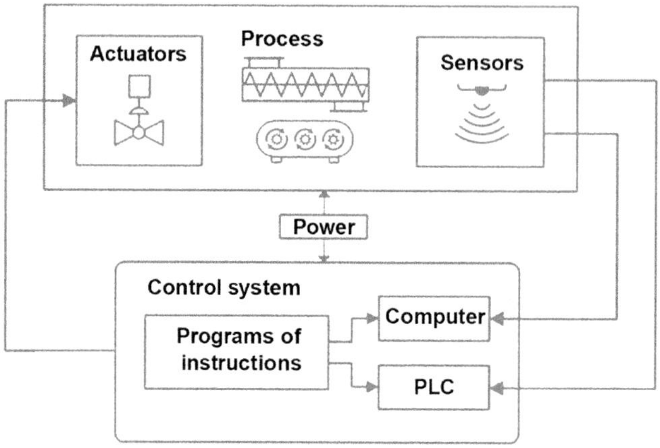

FIGURE 3.2 Industrial automation system.

TABLE 3.1 The impact of Mechanization V.s automation.

Mechanization	Automation
Saves human muscle	Saves the use of human judgment
Displaces physical employment	Displaces mental effort
The system replaces manual labor	Computers and high-tech machines replace human thinking
Provides jobs for low-skilled employees	You will create jobs for highly skilled employees by eliminating unskilled and semiskilled jobs
Impacts one or two industries simultaneously.	Multiple industries are affected at once
A short-term displacement of jobs occurs, and the process moves slowly	Displaces long-term jobs and races

These systems operate without significant individual intervention or monitoring, improving the quality and repeatability of repetitive operations. Automated platforms share many essential elements, each with its performance criteria.

Mechanization and automation are two different industrial processes. Table 3.1 shows the difference between these two different types.

Role of automation in the modern industry

.An industry's operation can be improved with the automation process. It's precious to know how these machines help the worldwide economy. Various manufacturing industries use automation at different stages; it's essential to understand what they do, and then we can identify how automation increases their income. Industries start with raw materials. Iron ore

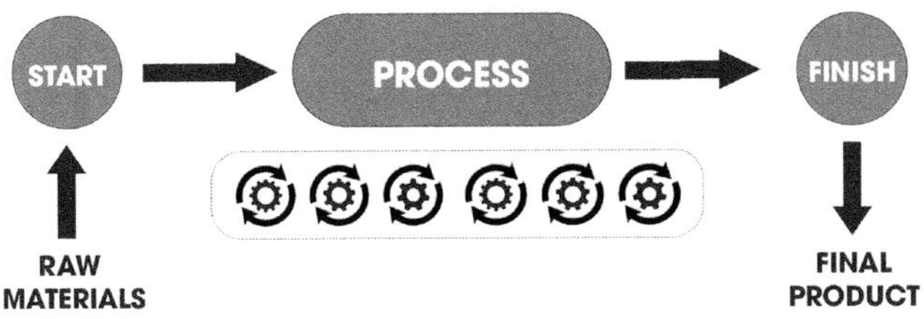

FIGURE 3.3 Industrial automation process of milk fabrication.

is a raw material in the steel industry, and the finished product is metal. In the dairy industry, a raw material is fresh raw milk, and a finished product is milk or yogurt. The raw material in the tomato pasta industry is fresh tomato, and the finished product is tomato sauce, etc.

Partially finished products are unfinished materials. For example, manufacturers assemble parts manufactured by other sectors in the automobile manufacturing industry. Factories need to use raw or unfinished materials to make finished products. It would be best to have the energy, workforce, and facilities (location, machinery, electricity, etc.) to accomplish this process. After that, raw materials are converted into finished goods. Fig. 3.3 shows the different stages in the dairy industry.

Increased productivity is one of the main benefits of automation in manufacturing. Automated manufacturing machinery can perform processes more efficiently and faster than humans, improving production times and boosting output. Manufacturing machinery can run continuously without breaks or rest. Furthermore, computerized manufacturing allows switching between products or production runs rapidly, increasing efficiency and reducing downtime. As a result, manufacturers can meet demand more effectively, reduce lead times, and boost profitability, painting a promising picture for the future. By automating repetitive or mundane tasks, workers can become more productive and efficient, allowing them to concentrate on tasks requiring more skill or complexity. Automating manufacturing generally offers manufacturers a significant global market edge by maximizing productivity.

Architecture of industrial automation

A modern industrial automation system has various types of machines, controllers, detectors, operator displays, and cabinets. As shown in Fig. 3.4, an automation system consists of a controller, a PLC, sensors, and actuators, each with its capabilities in the process field. The local environment can influence the computer of the automation system during installation, such as vibration, shock, dust, and electromagnetic interference (EMI). A temperature-controlled enclosure or essential cooling system can provide a stable temperature for the PLC and all accessories. The metal enclosure will also shield EMI. Inverters, switching power supplies, and variable-speed drives could all be included. The devices are suitable for automating operations in the same environment as computers. During the process, they will also experience things. Vibrations and mechanical stresses, such as high or low temperatures, may occur. In addition to water and mist, oil spray, UV light, sparks, and many other contaminants surround the device. Additional protection is needed if the device isn't adequately protected.

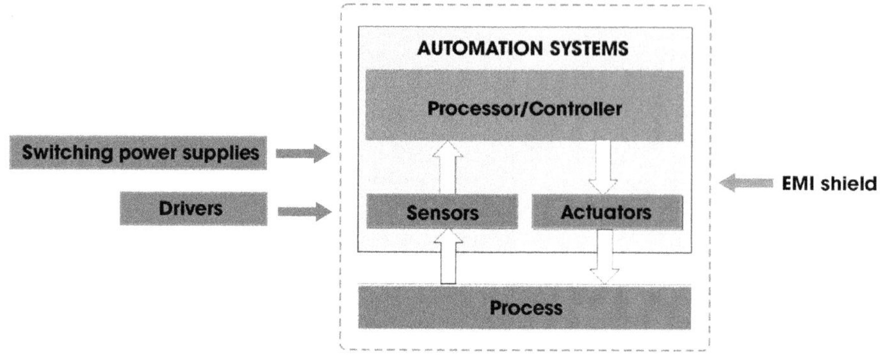

FIGURE 3.4 Essential "components" of artificial and automation systems.

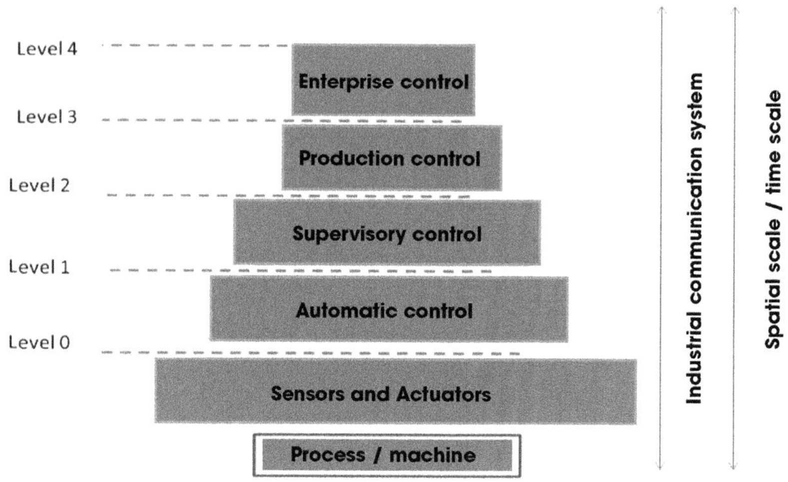

FIGURE 3.5 Automation communications levels chart.

This process is structured hierarchically in multiple levels, which helps us understand how it is structured, the elements of the process, how they relate and communicate, and their responsibilities, as shown in Fig. 3.5.

The above automation chart explains the various components of an industrial automation system and illustrates the number of devices and the speed of the components over time.

- **Sensors and actuators (level 0):** This level is closest to the processes and machines. It is used to translate signals so that signals can be derived from processes for analysis and decisions.
- **Automatic control layer (level 1):** This level consists of automatic control and monitoring systems, which drive the actuators using the process information given by sensors.
- **The supervisory control layer (level 2):** Drives the automatic control system by setting the controller's target. Supervisory control also looks after the equipment, which may consist of several control loops.

- **Production control layer (level 3):** This level solves decision problems such as production targets, resource allocation, machine task allocation, maintenance management, etc.
- **Enterprise control layer (level 4):** This level deals with less technical and more commercial activities such as supply, demand, cash flow, product marketing, new product development, etc.

Industrial control system

The PLC is a digital processor that controls automated machines by constantly monitoring input lines from detector devices. It provides action commands to react to these specific input signals. PLCs can handle multiple I/O arrangements, Fig. 3.6, and withstand various temperatures, electrical noises, vibrations, and impacts. Integrating a PLC into any production line or mechanical process is highly beneficial. Operation processes are easily changed or repeated, and critical situational information and statistics are collated and communicated. PLCs have often been a central element of equipment control, but sophistication is rising thanks to Industry 4.0 and the IIoT. PLCs generally come in two configurations: compact and modular. Compact PLCs are standalone devices that contain all the essential components. Modular PLCs start with a main module; the factory engineer can add expansion modules for additional capabilities, upgrades, or larger installations.

Here is a breakdown of the PLC's fundamental components [9]:

- The Central Processing Unit (CPU) is a microprocessor-based computer brain, usually a single board. This device reads and processes input signals before running control programs and sending appropriate responses.
- The input/output part is the real-time data interface between the machine and CPU interfaces. It's mostly a data transfer system, but it can also encode some data into signals recognized by the CPU.

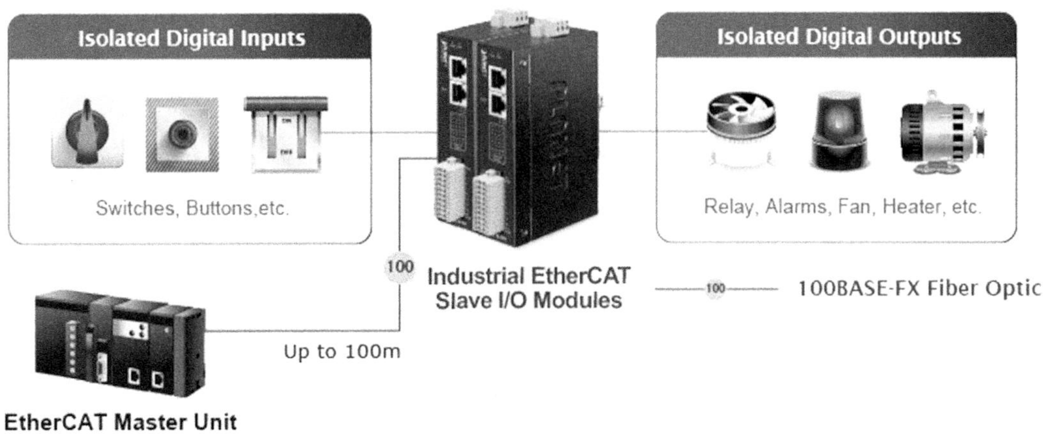

FIGURE 3.6 Modbus gateway communication.

- Programming device: This is often a separate device using a human–machine interface (HMI) software screen. The CPU is connected to this device through a communication link. Being able to re-program PLCs through an interface makes them very flexible and eliminates the need to re-wire or add additional hardware for each distinct logical configuration.
- The gateway interface enables communication with other processor-based devices or remote troubleshooting. Data transfer might be via a field bus or IP-based protocol, such as simple serial, Modbus, Profibus, Ethernet, Profinet, and CanOpen. Most of these standardized protocols were invented and implemented by specific manufacturers (Siemens, Crouzet, Alan Bradley, Omron, Mitsubishi, etc.), which will need consideration at installation.
- The power supply: For continuous operation or in areas with poor electricity availability, uninterruptible power supplies (UPS) or power generators may be required.

Input/Output Equipment

This level consists of input and output instruments closest to the processes and machinery.

Input instrument (field sensor)

This device and components provide data and control signals to a PLC (via its I/O system) or other control system. A commonly used name is a field sensor. Sensors and signals can also be directly connected to the controllers. A few examples are given below:

- HMIs, keypads, switches, and push buttons are examples of human-interaction devices that require human action.
- Various sensors that do not require human action include proximity (NPN or PNP), level, ultrasonic, inclination, and vibration sensors.
- Mechanical sensors, including limit switches, vacuum switches, vibration switches, level switches, and chock switches, are available.

Output instrument (actuator)

The controller acts on data processing results, translating the processing back into an action required by the automated machinery under control. A commonly used name is actuator. Some examples are mechanics such as pneumatic or hydraulic actuators, linear motion, valves, motor starters, control relays, pumps, and fans.

Industrial sensor system

The modern industry has sensors and instrument devices of various shapes, sizes, weights, costs, and complexity. However, they all have several similar functional elements connected in a specific manner. These elements generate a signal properly, allowing various tasks in an automation system to function.

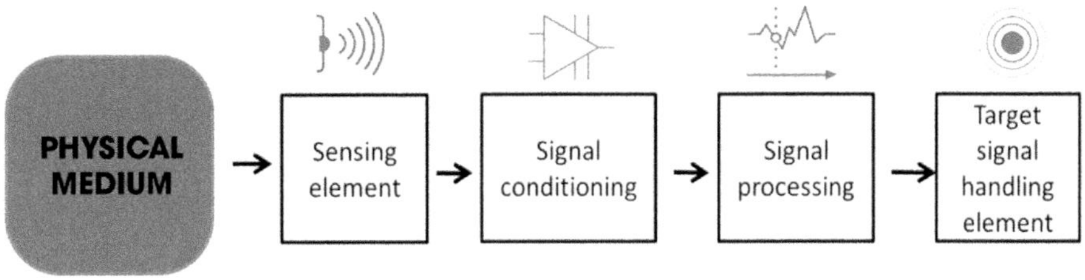

FIGURE 3.7 Functional block diagram of a sensor.

When sensors meet the medium, they usually change their physical or chemical properties in response to the measurement variable. As shown in Fig. 3.7, a sensor system is decomposed into four major components, including the medium used for measurement. The measurement of a physical variable associated with a physical medium in which a physical phenomenon occurs. Indirectly or directly, the phenomenon affects the sensing element depending on how it interacts with some component. As the first stage generates the signal, the signal processing element processes it for different purposes, like filtering to eliminate external noise, diagnostics, and linearization. Therefore signal processing systems are usually more general-purpose. The target signal-handling element may do different things depending on the target application. A process control system might contain data/ signal display modules, recording or storage modules, or feedback. Some cases may have variations, although the above description fits most cases. In a measurement system, these subsystems are often physically distinguishable from one another from a functional perspective. Most modern sensors have digital communication capabilities, either serially, parallel, or over a network. Recently, an intelligent sensor has been embedded in digital electronics, and it is called a smart sensor.

Industrial actuator system

The functional configuration, Fig. 3.8 of a typical actuator system is shown in Fig. 3.5, decomposed into its main parts. Signal-processing elements accept electrical commands from control systems and process them in various ways. In many actuators, the actuation operation is controlled by closed feedback. Because of this, the electronic signal processor also often controls the actuator. There are sometimes servo-amplifiers in the electronic power amplification element. In other cases, it may comprise power electronic drive circuits, such as motor-driven actuators. It changes the signal from electrical to nonelectrical, usually in the form of motion. Examples include electrohydraulic valves, stepper/servo motors, and current pneumatic pressure converters. Nonelectric power converters are often used when hydraulic or pneumatic mechanisms are needed to amplify power further. The actuated variable can be transformed into desired forms. For example, flow valves convert the motion to flow rate; mechanism converters convert rotary motion to linear motion; steam or other hot liquids convert the flow rate to heat, etc. In addition, there are auxiliary components for lubrication, cooling, filtering, reservoirs, prime movers, sensors for feedback, components for display, remote operation, and safety mechanisms since the power handling is high.

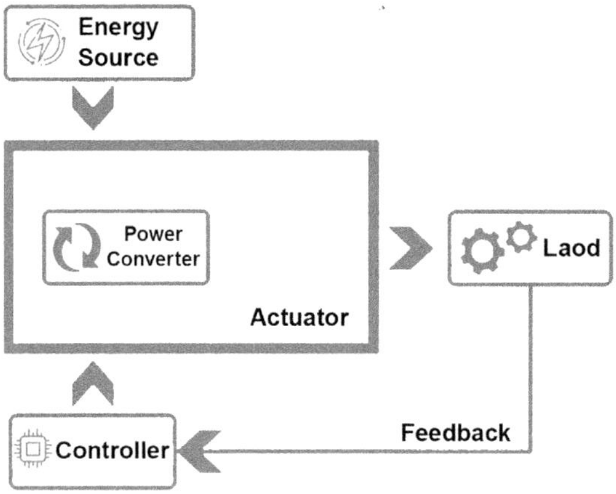

FIGURE 3.8 Functional configuration of a typical actuator system.

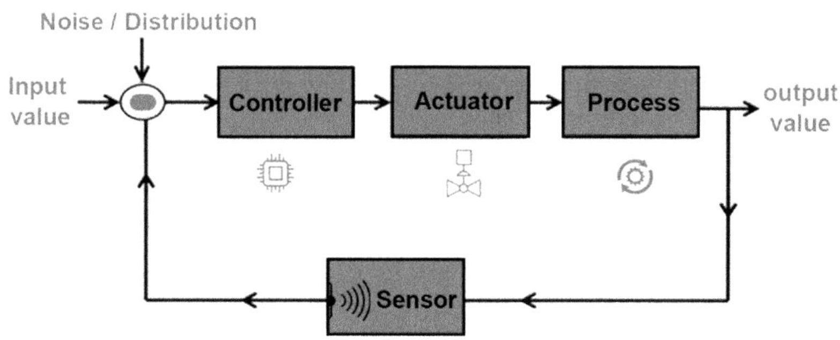

FIGURE 3.9 Industrial control system.

Automatic controller

There are four components in the standard control loop: the controller, the actuator, the sensor, and the process, as shown in Fig. 3.9. Controllers aim to keep output-measured values as closely related to input values as possible. Usually, there are two types of inputs: the command and the external inputs, which come as noise or disturbances. Achieving the performance objective is challenging because of necessary disturbances from load changes and other external factors. These disturbances include sensor noise, complexity, possible instability, uncertainty, plant dynamics variability, and actuator limitations. The most common industrial control loop signal is a continuous constant indicating desirable levels of process variables, like temperature, pressure, flow, level, etc., that ensure product quality. Because of this, the output deviates from the command signal, which results in a lesser product. In industrial machines and processes, there are four types of feedback: the feedback signals, the controller algorithm (such as P, PI, or PID), the control loop structure (standard feedback loop, cascade loop, or feedforward), or the controller gains all these types play significant roles. Most process classes have well-known control configurations. As the operating characteristics of the manufacturing process change, controller gains need to be changed occasionally.

One physical device can control one or more control loops. Adaptive control and fuzzy logic algorithms can improve controller performance and responsiveness, supplementing conventional control laws such as PID. It is essential to have a PID and self-tuning at startup. The controller should also be able to communicate with supervisory systems and on peer-to-peer networks and support manual control if automation fails. The loop controller relies on software. Software for setting up, monitoring, auto-tuning, and alarming loop controllers has become standard.

Advantages and disadvantages of industrial automation

Each technology has its advantages and disadvantages. The following paragraph presents the same benefits and drawbacks of automation.

Advantages of the automation system

- Human operators can easily replace physical tasks.
- Radiation, extreme temperatures, and pollution can create dangerous human working conditions.
- Typical human workers need help lifting heavy loads or working with small objects.
- Manual production is slower and more expensive than automated production.
- Quality control checks can ensure consistency and uniformity.
- An improved economy in the sector could raise living standards.

Disadvantages of the automation system

- Job losses.
- Manual labor is optional because machines do most of the work.
- Currently, it is possible to automate processes that can be repeated and consistent, such as those related to high-volume production and products.
- Automating an enterprise at the start of a business is expensive.
- Maintenance and service require skilled personnel.

Robotics and modern automation

In the future, robotics will play a more critical role in the global economy than it does now. The automation of industrial operations utilizes robotics to match the flexibility of humans in the same tasks and applications. The most common industrial robot has a single arm with six axes and, at the same time, nine axes. Robotic arms are commonly used in factories to automate repetitive tasks, such as welding parts, painting equipment, or picking and placing items. Many countries worldwide use robots in their economics, particularly in automation. Because arm robots are mobile, users can modify them to perform functions they were not intended to accomplish. Robot arms can replace and replicate human actions in hazardous environments. It's easy to program them to do various things and move them around a factory as capacity requires. Sometimes, a new program is all it takes.

A key feature of industrial automation leadership is the use of industrial robots, which determines a country's level of advanced technology and scale. Many people believe that China will threaten US leadership in modern industrial automation as China rises and the US reindustrializes [10]. However, that is not the case. It is important to note that although China has made significant advances in the industrial automation and robotics sector, Japan and Europe remain the global leaders in manufacturing industrial automation equipment. Chinese companies have acquired a dozen European and American industrial robotics companies since 2016, including Kuka, a German company acquired by the Midea Group. Although China has experienced notable growth in its industrial automation and robotics industries, it has yet to become a significant producer of this technology. Similarly, the United States imports many industrial robots, although not to the same extent as China. The government-sponsored acquisition campaign of China is a principal contributor to this growth, which is currently encountering resistance in Europe and the United States. The Chinese Academy of Sciences is affiliated with Siasun, China's leading industrial robot manufacturer. According to the International Federation of Robotics (IFR), China has been the world's largest industrial robot market for a decade. The distribution of industrial robot installations worldwide in 2018 was as follows: China accounted for 36% of the total, followed by Japan (13%), the United States (10%), South Korea (9%), and Germany (6%). According to this data, industrial robots are widely adopted across the globe, with China leading the way. The company's industrial robot statistics represent several applications, including handling (the movement of objects), welding, assembly, dispensing, cleanrooms, and processing machines. Asia-Pacific accounted for 67% of industrial robot installations in 2018, Europe 18%, the Americas 13%, and the rest 2%. In terms of industry, robotics were used in automotive 30%, electrical and electronics 25%, metal and machinery 10%, plastics and chemicals 5%, food and beverages 3%, and not specified 19%, which includes healthcare, construction, and textiles.

It's the latest addition to robotic automation. The padding or coverings ensure that the arm has no sharp edges. If a human or another object encounters it, it will stop or reverse its motion. According to the market research report [8], the global market for autonomous mobile robots was valued at USD 4.3 billion in 2023, and by 2032, it is expected to reach USD 18.2 billion. According to estimates, the autonomous sector is expected to grow at the most significant CAGR of 18.1% between 2023 and 2032, as shown in Fig. 3.10.

Many reports and research papers indicate that robots are used for most essential activities during the COVID-19 pandemic [11,12], specifically in China, the United States, and several other countries. China is embracing the age of intelligent manufacturing, integrating advanced technologies like AI and IoT with robotics to improve manufacturing efficiency and effectiveness. Rather than replacing human labor, Fig. 3.11 this shift adds to it, creating a synergy between the two. As human ingenuity coexists with robotic precision, manufacturing will see unprecedented levels of efficiency in China. China's first goal is to become self-sufficient in robotics and lead the industry globally. China is already the most significant

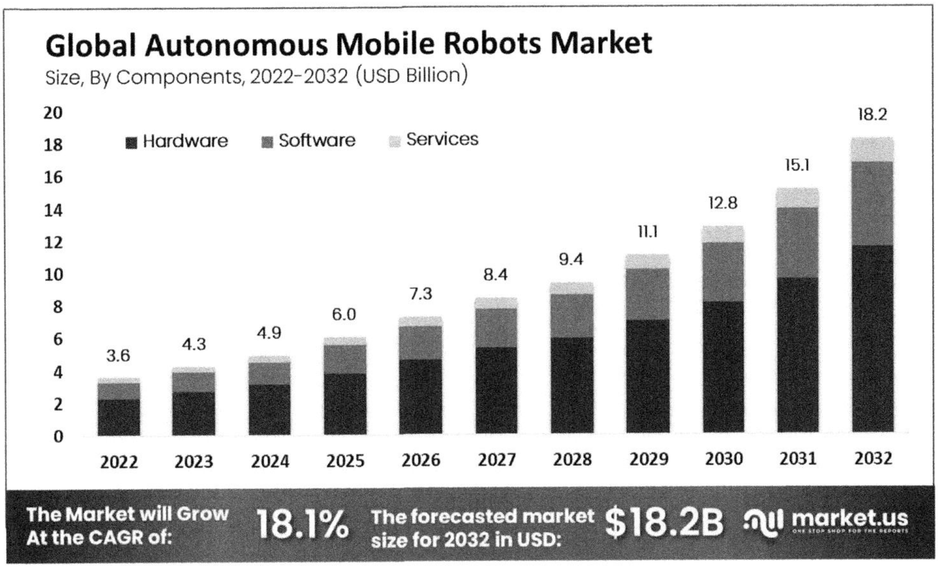

FIGURE 3.10 Global autonomous mobile robots market.

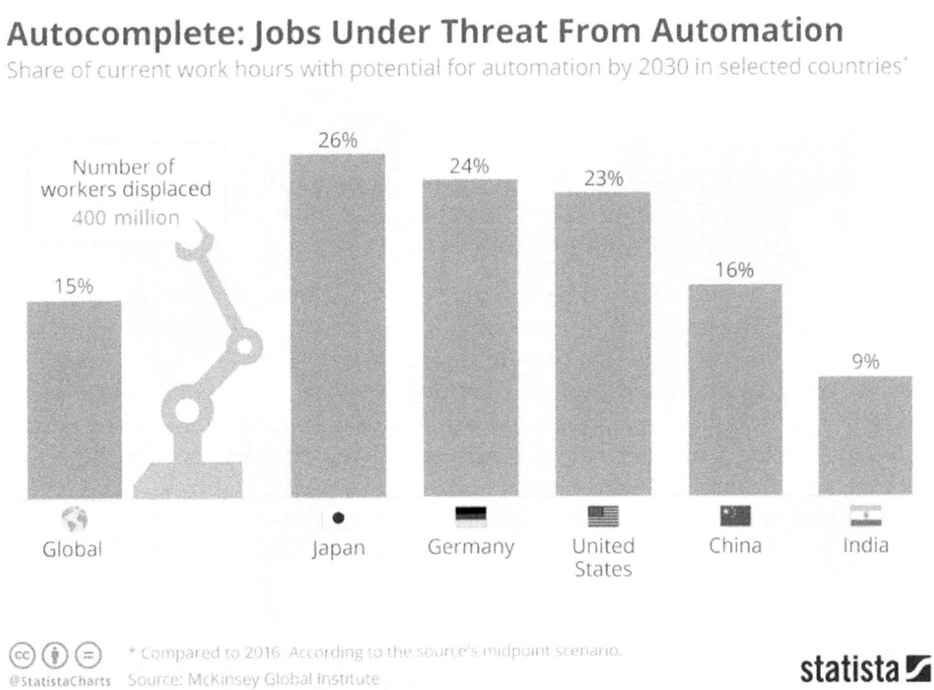

FIGURE 3.11 Autocomplete: jobs under threat from automation.

industrial robot market in the world. It accounted for 52 per cent of robot installations worldwide in 2022, giving Chinese robot producers a key advantage in their home market. The robotics market in India is estimated to generate US$694.10 million in revenue in 2024. Service robotics is expected to have a market volume of $295.60 million in 2024, making it the dominant segment in the market. According to the forecast, the market volume will reach US $639.50 m by 2028, representing a compound annual growth rate of -2.03%.

Effects of the automation and artificial intelligence in COVID-19

Automated systems have played a significant role in Coronavirus disease, which is incredible. During the COVID-19 pandemic, all the workers worldwide are inside their homes, managing the work remotely from the control room, allowing autonomous humanitarian assistance. The role of automation in fighting a pandemic and saving lives is widely acknowledged by researchers and politicians worldwide. The COVID-19 pandemic has significantly impacted global employment markets. The nature of work has changed in numerous industries [13,14]. The COVID-19 pandemic accelerates industry adoption of modern automation for technology transformation. The pandemic has affected international financial systems and business plans. The e-commerce market has increased, and AI has received more attention due to increased social distance. Automation is mainly affected by sectors impacted by mobility restrictions due to the pandemic.

As a result of the ongoing technological transformation of work and the COVID-19 pandemic shock, fewer jobs will be available to less educated and low-skilled workers. Many sources examine the long-term consequences of the pandemic on the workplace, including the impact of automation and artificial intelligence. Now, an unprecedented digitalization of the worldwide economy is underway. Several hostile forces could increase labor demand because of automation, artificial intelligence, and robotics. As a result of artificial intelligence, machines can perform some cognitive tasks better than humans.

In the following, we highlight some examples that demonstrate the capabilities of automation applied to fight COVID-19:

- A software company and a hospital in Ireland have launched an initiative called "A Robot for Every Nurse." The project aims to develop a bot that can log laboratory information in a fraction of the nurses' time, who spend approximately four hours per week entering laboratory information into hospital computer systems [15]. More tests are being conducted to prevent disease spread, and results are being announced more quickly.
- The students at the University of Massachusetts have developed a machine learning model that, when combined with a neural network, provides highly accurate and timely predictions of COVID-19 spread and the necessary lockdown measures. This advanced technology is a testament to the power of automation in healthcare and pandemic management.
- Algorithms developed by Benevolent AI company specializing in drug discovery can detect properties in medicines that may reduce the spread of viruses.
- Drone technology has been highly supported by governments worldwide, especially in China. The Chinese government utilized drone technology to respond to the outbreak

nationwide. By patrolling walkways with loudspeakers that remind people to wear safety masks or hovering over streets with bar codes [10]. They utilize signs that drivers can use to register health information so authorities can communicate data faster while maintaining a safe distance. It utilizes big-sized drones for spraying disinfectants in crowded locations, while others have been used to deliver humanitarian aid and crucial medical supplies.

Case study 1: Internet of Things-enabled drip irrigation system

The Global Report on Food Crises (GRFC) shows that nearly 281.6 million people across 59 countries are experiencing acute hunger. Due to increased stress on food systems, over 500 million people may face acute hunger next year, especially after the Russian–Ukraine matter. A global food crisis could result from the ongoing conflict between Russia and Ukraine since thirty-six of the fifty-five states are already experiencing food shortages and rely on exports from both countries. Before this conflict, Ukraine and Russia produced about 30% of the world's grain supply, a fifth of maize, and more than half of the world's sunflower oil supplies [16].

In the face of these issues, governments must act quickly, consciously, and sustainably to stabilize our global food supply based on automation processes. Several countries provide grants and rewards to encourage the adoption of agricultural robotics. To stabilize the international food supply chain, key players in agriculture must collaborate and form partnerships to run advanced automation systems based on IoT, robots, drones, and AI, especially in poor and developed countries.

AI farming is the science of cultivating soil, growing crops, and raising livestock, all controlled by autonomous machines. Since the dawn of man, agriculture has relied on technology since the first plow was fashioned from axe and hoe, as shown in Fig. 3.12. Agriculture was passed through two levels, from mechanization down to digital automation. This last level is divided into digital equipment and robotics with AI. As technology and science improved, so did the scale at which farming was possible. Since the IoT has increased in popularity and use in recent years, there are even more opportunities for technology to aid farmers and reduce hunger in every nation. The need to feed a growing global population drives technology to increase agriculture productivity. Agricultural robots collect essential crop health and yield information, allowing farmers to make data-driven decisions. As robotics, artificial intelligence, and IoT sensors advance, agricultural robots become more sophisticated.

The main benefit of automation in agriculture is its capability to improve efficiency. Technology and machines can spray, map, sense, and harvest faster, and they do not require rest or time off to perform their tasks. Farm productivity can be increased, and labor costs can be reduced. Automation in agriculture can also reduce the risk of lockdown during a pandemic, which can be controlled remotely and keep the process running without human intervention. Another benefit of automation in agriculture is its ability to boost sustainability. Precision agriculture technology, such as global positioning systems or drones, helps farmers make better use of water and fertilizer. This reduces waste and decreases the environmental impact of farming practices. Using drones and tank robots with arm robots can also reduce harmful pesticides and fertilizers since these machines can apply these

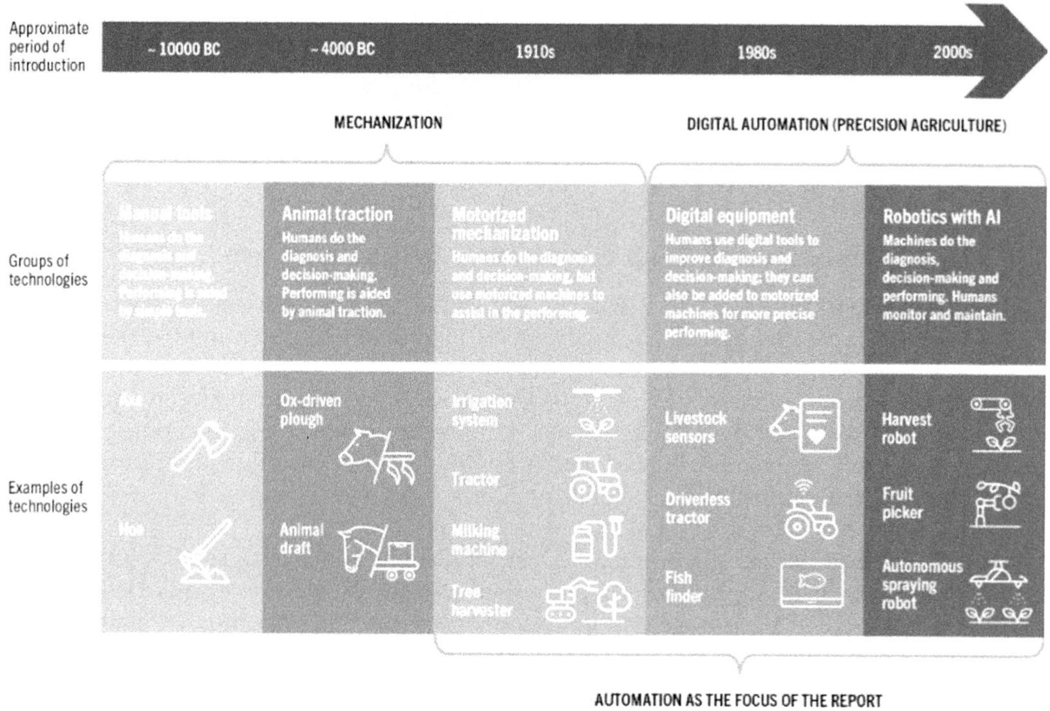

FIGURE 3.12 Evolution of agricultural automation.

products with increasing accuracy and precision based on sensor readings. Agricultural automation significantly contributes to improving productivity and reducing consumption and adverse impacts. It maximizes production by maximizing inputs and reduces consumption and impact by reducing waste. It also enhances the quality of the agricultural products produced.

The agricultural robots market is an increasingly growing industry at the intersection of agriculture and technology. Farmers can produce food quicker, cost-effectively, precisely, and sustainably with these automation machines. The agriculture automation system needs information to make accurate decisions. The automation system will make proper decisions and actions if it has correct data. Hence, the acquired data must be precise. Over the next five years (2023–2028), the global agriculture robot market is projected to grow at a CAGR of 24.3%, from \$13.5 billion in 2023 to \$40.1 billion in 2028, as shown in Fig. 3.13. Agricultural robots can automate planting, harvesting, and weeding tasks, reducing the impact of workforce shortages in Europe. Using these robots, farmers can plant, fertilize, and apply pesticides precisely, maximizing their yield and resource efficiency. Reduced fertilizer use, water management, and soil preservation contribute to sustainable agricultural practices.

In many cases, the sensors used for other applications can be used for agricultural automation. Although there are many exceptions, the requirements for agricultural automation concerning accuracy, range, and resolution usually do not exceed those of other sectors of the economy.

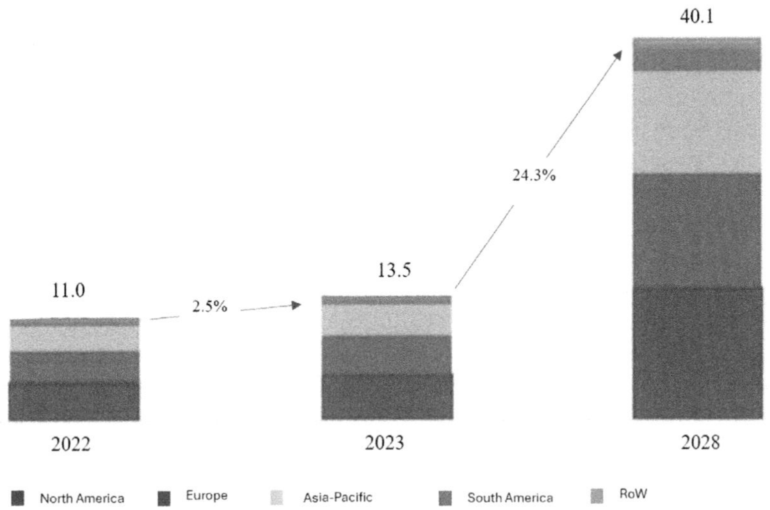

FIGURE 3.13 Agriculture robots' market global forecast to 2028 (USB BN).

The use of drones is growing rapidly in nearly every sector of the economy, but the use of drones in the agricultural industry is soaring. There is an expectation that the farm drone market will grow from a $1.2 billion industry in 2019 to a $4.8 billion industry by 2024, according to some reports. In a few short years, drones will become ubiquitous on large and small farms for various purposes, including scouting and security. The information gathered by drones on farms is often used to better inform agronomic decisions and is part of a system generally referred to as precision agriculture.

In many countries, drone use has already become essential to large-scale precision farming operations. The data collected from drone recording fields help farmers plan their planting and treatments to achieve the best possible yields. Some reports indicate that using precision farming systems can increase yields by as much as 5%, which is a sizeable increase in an industry with typically slim profit margins.

In this section, we will examine some areas where drone technologies are already being used on farms, explore some new agricultural drone technologies, and touch on some of the steps and challenges to adopting widespread drone use in agriculture.

Agricultural drones also spray fertilizer in farming areas. In contrast, others have used AI and image processing in China to see the quality of plant leaves and monitor plant growth by scanning sensors. India utilizes drone spraying to kill and stop the attack of locust swarms [17]. This feeding can destroy crops planted, causing famine and deprivation in societies that solely rely on these crops for survival. Recently, swarms of locusts have invaded several areas in India, especially Rajasthan. With nearly 90,000 Hectares of land affected across 20 districts, these growing swarms are threatening to amplify into an agricultural disaster. For this reason, India's government pushed to utilize drones to spray insecticides and map the crops planted by drones. Indian government started a program to teach new farmers to pilot drones that could be used later in their farms.

Case study 2: drone and automation

Recent technological advancements in electronics, robotics, and computing have paved the way for a new era of industrial automation, making drone technology more accessible and user-friendly. Such changes introduced significant shifts in operations paradigms across multiple sectors. These advancements have transformed traditional business models and operational processes in various industries yet brought a wide variety of new challenges associated with adopting this technology.

Recently, researchers in this area have been focusing on finding and improving practical applications of aerial imagery to support various human activities. Umar, T. (2020) showcases a wide range of industrial applications, including safety inspections in the construction industry, where they are used for photography/videography, surveying, inspections, and safety inspections. Logistics is another prospective application area that has recently received much attention; as Agatz et al. (2018) mentioned, UAVs have already been utilized in the transportation industry to support deliveries, showcasing the technology's potential to revolutionize logistics.

The industry has given much attention to aerial facility surveillance and inspection. Stroud and Weinel (2020) discuss the potential risks and benefits that may be brought to an average employee's workplace by introducing drone technologies to production facilities. Moreover, the application of drones in industrial surveillance extends to disaster relief response systems, where autonomous drones have been identified as having a significant positive impact [18].

Besides the positive impact on safety and functionality increase, cost must also be considered. The positive effect of drones on cost has been a subject of interest in various industries. For instance, drones are seen as a cost-effective solution in the logistics industry that can reduce costs and improve responsiveness [19]. Additionally, using drones in solar power plants for maintenance, surveillance, and security tasks has been identified as a competitive advantage, reducing costs, especially in large solar farms. Furthermore, the decrease in the cost of drones and the granting of licenses to commercial service providers and hobbyists have driven the fast adoption rate of drone technology, indicating its potential for cost savings [20]. The rapid adoption of the technology in multiple industries and among the general public has led to a significant decrease in cost per device. The trend is expected to continue for numerous years, Fig. 3.14, which may further boost the implementation of drones in various spheres.

Drone technology holds immense potential across various sectors, from safety inspections in construction to logistics and disaster relief efforts. These versatile aerial platforms offer unprecedented capabilities for tasks such as photography, videography, surveying, and even package deliveries. However, alongside their promise, drone operations come with many challenges.

As for most automated systems, data storage and transfer are critical points for drone-based solutions. The volume and complexity of the data collected may further result in the data processing challenge. Effective data management requires a robust infrastructure for large-scale storage, fast data retrieval, and secure transmission. Nowadays, industrial drones have sensors and cameras that collect vast data during flights, including high-resolution images, thermal imaging, and multispectral data. This data is then integrated

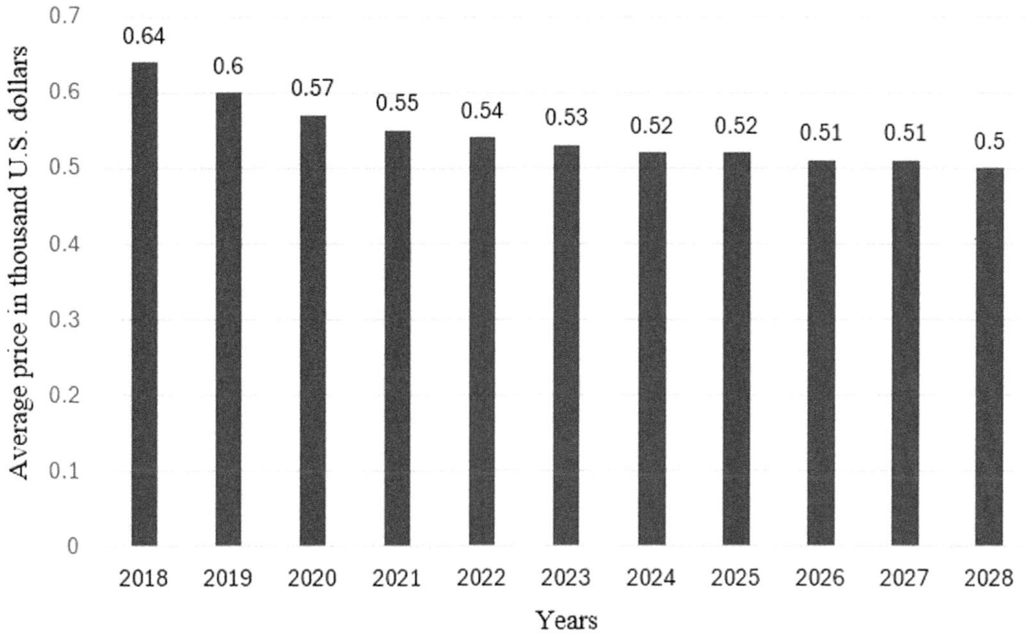

FIGURE 3.14 The dynamics of drone cost over the recent years with projections until 2028.

into advanced analytical platforms, often utilizing machine learning algorithms to process and analyze the information efficiently. For instance, combining analytics platforms with drones facilitates the development of intelligent routing systems for drone delivery, enabling efficient and optimized delivery routes [21]. Additionally, using AI and analytics platforms in conjunction with drones allows for monitoring and analyzing environmental data, such as city-scale litter monitoring and waste management, contributing to sustainability efforts [22].

Another significant obstacle to sustainable drone operations is the power source. The restricted battery life of drones creates substantial challenges in various fields. The primary issue caused by limited battery life is the reduced operational range of drones, impacting the ability to traverse long distances and carry out lengthy missions. Additionally, shortened flight times compromise the efficiency of drone operations, particularly in tasks requiring extended surveillance or delivery periods. Moreover, the carrying capacity of drones is limited by their battery life, which restricts the variety of cargo they can transport.

The challenges mentioned above must be addressed in the process of drone system design with special requirements presented by the specific application. A promising application example from the industry may be taken to demonstrate the process. In the context of high-risk and high-effort industrial applications like power line monitoring, the architecture of a drone system is meticulously designed to ensure optimal performance and reliability. The components of such a system typically include the airframe, propulsion system, navigation and control systems, communication links, and payload, which usually consists of high-resolution cameras, thermal sensors, and, sometimes, multispectral imagers Fig. 3.15). Integration and interconnection of the mentioned components are crucial for effective operation.

Each component presents its own set of requirements, such as given below:

- **Airframe and propulsion:** The airframe must be robust yet lightweight to accommodate extended flights and carry the necessary sensors and equipment. The propulsion system, often comprising electric motors powered by batteries, is designed for efficiency and reliability to handle the rigors of extended operational periods and varying atmospheric conditions.
- **Navigation and control systems**: These systems ensure that drones can navigate accurately along power lines. GPS and inertial navigation systems are combined with advanced algorithms that allow for stable flight patterns, even in challenging conditions. The control systems are often automated but can be overridden by operators from ground stations if necessary.
- **Communication systems:** Real-time data transmission is critical for immediate analysis and decision-making in industrial settings. Drones used in power line monitoring employ robust communication systems that ensure continuous data transfer between the drone and the control center, even over extended distances.
- **Sensors and payloads:** Drones can perform detailed inspections by integrating various sensors. High-resolution cameras capture visual anomalies, while thermal imaging can identify hot spots indicative of potential failures. In some cases, LIDAR (Light Detection and Ranging) technology is employed to create detailed topographical maps.

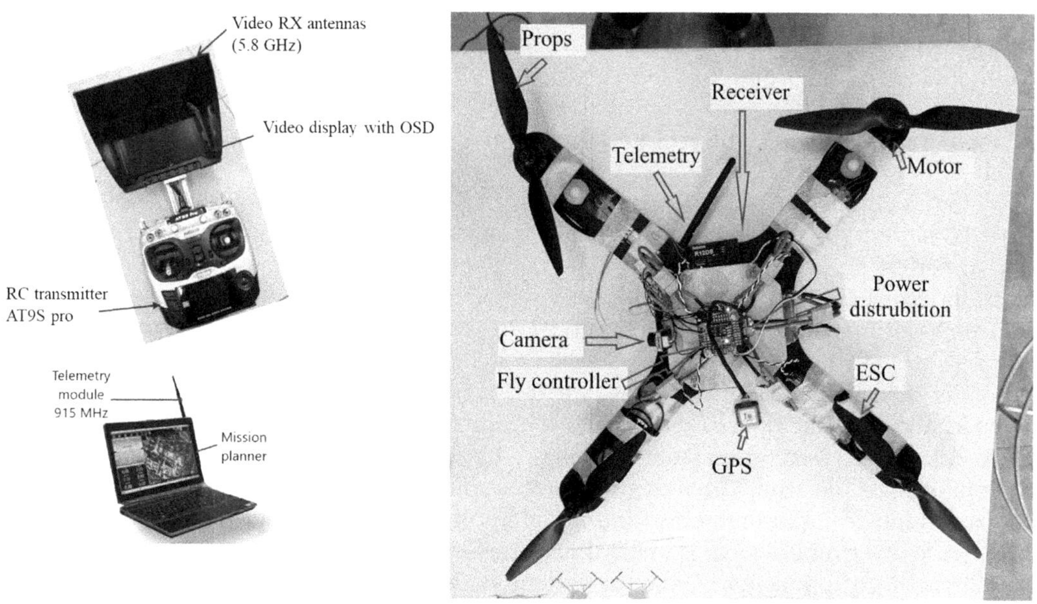

FIGURE 3.15 Typical components of a drone system.

Sophisticated software that controls the drone's flight and processes the collected data to provide actionable insights facilitates interconnection between these components. For example, AI algorithms can analyze images in real time to detect maintenance issues or potential hazards along power lines, flagging them for immediate attention. This level of integration allows for a seamless flow of information and enhances the drones' operational capabilities.

As demonstrated above, designing a drone can still be a complex and time-consuming task that requires careful consideration of numerous parameters. One beneficial aspect for today's drone design engineers is that most modern drone components can be easily purchased from any online electronics store. Additionally, many software solutions and algorithms required for drone operation are readily available in open-source libraries, simplifying the design and assembly process.

The positive impact of drones on industrial automation cannot be overstated. By automating routine inspections and data collection, drones reduce the need for manual labor, minimize human error, and significantly enhance operational safety. The ongoing improvements in drone technology, coupled with advances in AI and machine learning, promise even greater efficiencies and innovations in the future. This symbiotic relationship between drones and automation is transforming traditional business models, making operations more efficient and more adaptable to the evolving demands of the industrial landscape.

The future of automation: the power of artificial intelligence

AI is the simulation of human intelligence in machines programmed to think like humans and mimic their actions. The term was coined by John McCarthy [23] AI can be as simple as a set of rules that guide automated responses or as complex as systems that learn from data to make decisions independently. The revolutionary role of AI in enhancing automation technologies lies in its ability to process vast amounts of data quickly, learn from patterns, and make decisions with minimal human intervention.

The transformative impact of AI on various industries is profound, as it drives efficiency, accuracy, and innovation across diverse domains. AI has been shown to automate medical diagnosis and pathology in healthcare, increasing speed and accuracy in identifying diseases and conditions [24]. AI has demonstrated the potential to automate image interpretation in medical imaging, particularly in radiological applications, thereby enhancing diagnostic accuracy and efficiency. In dentistry, AI has the potential to automate routine tasks, improving the quality of work and accuracy of dental procedures.

Furthermore, AI's impact extends to the retail industry, where it has been leveraged to increase levels of automation, reduce costs, and streamline customer interactions, leading to enhanced operational efficiency and improved customer experiences [25]. In marketing, AI has revolutionized strategic planning by automating repetitive marketing functions, processing data for decision-making, and analyzing interactions and human emotions, driving innovation in marketing strategies. Moreover, AI has been identified as a critical driver of automation in the hospitality and tourism sector, leading to increased efficiency and improved customer service.

In the education sector, AI has demonstrated significant potential for automating educational processes, enhancing the efficiency of instruction, and driving innovation in personalized learning experiences [26]. Additionally, AI's impact on the manufacturing industry has been transformative, with applications in automation and optimization leading to increased efficiency, accuracy, and innovation in production processes. AI has been leveraged in supply chain management to automate and optimize various methods, improving inventory management, logistics efficiency, and accuracy.

The rapid adoption of AI across various industries has significantly influenced corporate investment strategies. As businesses recognize AI's potential to enhance efficiency, innovate products and services, and maintain competitive advantage, there has been a marked increase in investments directed toward AI research, development, and implementation (Fig. 3.16). This surge in funding is not only accelerating AI technology advancements but also fostering new opportunities for growth and transformation in the corporate sector.

Modern AI, while still far from achieving accurate artificial intelligence, can be distilled into several more straightforward concepts crucial for enhancing automation. Machine Learning involves algorithms learning from data to make predictions or decisions without

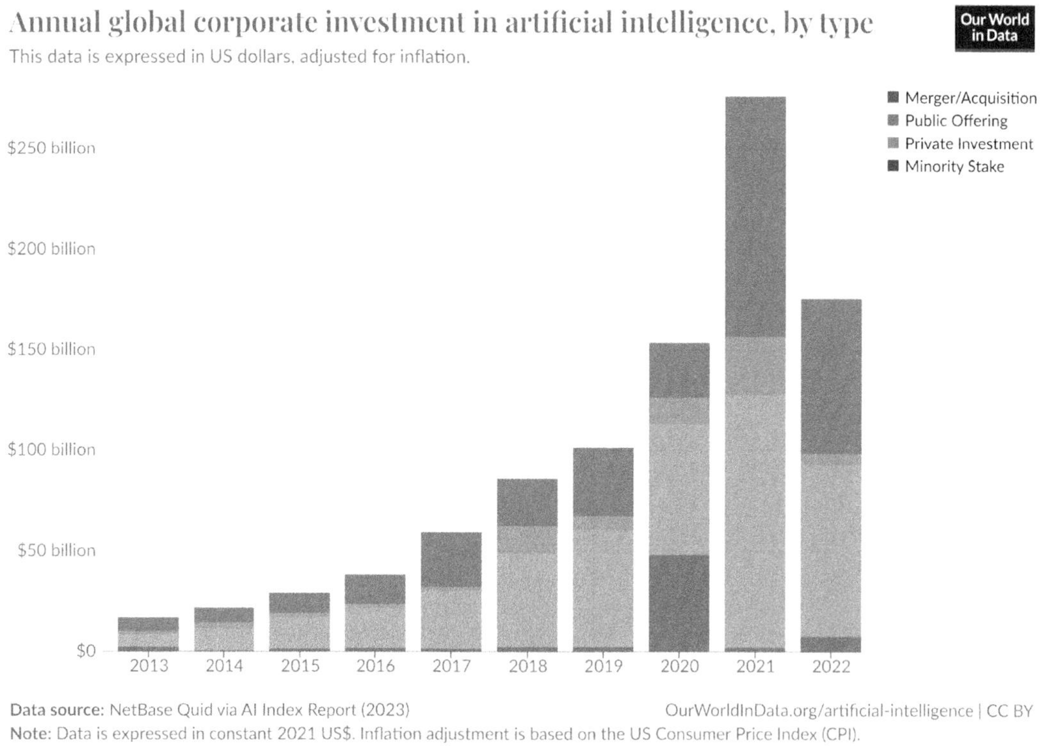

FIGURE 3.16 Dynamics of corporate investment in AI.

being explicitly programmed. Deep learning, a subset of machine learning, utilizes multi-layered neural networks to analyze complex data, making it ideal for processing images, sounds, and other unstructured data. Neural networks, inspired by the human brain, identify patterns and interpret sensory data, facilitating speech and image recognition tasks.

The historical development of AI began in the 1950s [27], rooted in early programmable computers, and has evolved significantly across decades. Key advancements include the rise of machine learning in the 1980s, the integration of the internet and computational improvements in the 1990s, and breakthroughs in deep learning during the 2000s, which enhanced image and speech recognition capabilities.

Today AI influences various sectors, including healthcare and finance, through big data analytics and innovative technologies. Future trends lean toward autonomous, self-improving systems that promise even more comprehensive problem-solving and decision-making abilities. In the sphere of smart manufacturing, AI-driven technologies are expected to fundamentally reshape production processes by integrating AI-driven robots, predictive maintenance, and advanced quality control systems. These elements will collectively enhance operational efficiency, reduce downtime, and ensure high-quality outputs.

AI-driven robots are critical in automating complex tasks that require precision and repeatability. These robots will be equipped with sensors and machine learning algorithms that allow them to adapt to new situations and improve their performance over time based on feedback from their operations. For instance, AI robots can be utilized in assembly lines to pick and place components, assemble intricate parts, and even conduct inspections. The integration of AI will enable these robots to not only execute predefined tasks but also learn from each interaction, optimizing their movements to increase speed and reduce error rates (Wan, 2021).

One of the critical components of any modern robotic automation system is the image processing module. Image processing is a powerful tool in intelligent manufacturing, offering extensive capabilities to enhance automation and improve production lines' efficiency and quality. A traditional example of an image processing application would be quality control at the production line. Cameras and sensors installed along the production line capture detailed images of products as they are manufactured. Advanced algorithms then analyze these images to identify defects such as misalignments, surface scratches, or inconsistent shapes and sizes. This automated inspection is much faster and often more accurate than manual checks, allowing for real-time quality assurance that only products meeting strict standards proceed further in manufacturing.

Regardless of the wide range of promising applications of AI in various industries, there is still a forceful drawback for the technology. Artificial intelligence in its current state is associated with challenges and considerations, including data acquisition and management issues, human resources, infrastructure, security risks, trust, and implementation challenges [28]. The need to overcome existing technical and nontechnical barriers to successfully implement and integrate AI systems in manufacturing processes has been emphasized, reflecting the complexities and considerations associated with AI adoption in manufacturing systems and processes (Kohl et al., 2021). Furthermore, the impact of AI on the cognitive load of technicians during diagnosis tasks in maintenance has been evaluated, highlighting the potential implications of AI technologies on human-machine interactions and cognitive processes in manufacturing environments [29].

Artificial Intelligence is indisputably shaping the future of smart automation, having already delivered substantial improvements in efficiency, precision, and innovation to various sectors such as healthcare, manufacturing, and education. Despite the challenges in integrating AI, such as trust concerns, high implementation costs, and technical hurdles, researchers and entrepreneurs are making active progress to overcome these obstacles. Their efforts paved the way for a new era of automation, where AI's capabilities will be fully harnessed to create more adaptive, intelligent, and efficient systems. As these challenges are addressed, AI will continue transforming industries, promising a more productive and advanced technological landscape.

Conclusion

This chapter shows that automation brings complex transformative changes to human society. Automation influence is evident in multiple sectors, altering workflow, production methods, and service delivery. Automation can lead to reduced operation costs, higher return on investment, increased production output, reduced errors, and better customer satisfaction.

Automation is being adopted in various sectors due to its efficiency in terms of output and time. While it is true that automation can lead to the displacement of specific jobs, particularly those that involve routine and repetitive tasks, they are also responsible for producing new types of jobs requiring skills such as programming, AI development, and more. Governments, educational institutions, and companies must cooperate to provide education and training to existing employees and future generations to prepare them for an automated future.

An often-overlooked positive impact of automation is that automation and AI can free the workforce from mundane, repetitive tasks. This will allow them to focus on more creative and innovative endeavors, which can lead to rapid advancements in research and development.

According to many research studies, labor automation will decrease labor demand by increasing productivity by displacing workers. Due to the productivity effect, demand for labor could decrease in the same sectors as those undergoing automation or in industries not undergoing automation. The automation process generally eliminates low-qualified workers and creates more high-quality positions, contributing to higher salary gaps by increasing the relative salaries of qualified jobs.

The adoption of automation comes with its own set of challenges and negative impacts. It is not easy to provide jobs or train the people whose jobs will be displaced due to automation. Many people will not be ready to learn new skills, especially in the later stages of their careers, and may protest automating their work environments. It is also possible that customers will prefer human interaction over the use of machines, especially in the service industry. Satisfying the workforce and customers will be a big challenge to adopting automation.

The initial costs associated with automating a facility can be incredibly high, as equipment and software must be purchased, employees must be trained, and the facility must undergo restructuring. The machines might also be too expensive and complex to repair promptly, which may result in downtime and losses. There is also a concern that over-reliance on automation systems may cause a dependency issue among the workforce, leading to a loss of human skills and expertise.

Despite the many negatives that automation brings, it is essential to remember that automation is a tool, and people determine how it is used. Even with all the challenges automation faces, it will continue to grow. The future of automation will be shaped by how it will be integrated with existing and emerging technologies and trends. AI, collaborative robotics, IoT, industry 4.0, augmented reality, cybersecurity, 3D printing, cloud computing, blockchain, and edge computing will enhance automation systems shortly.

References

[1] E. Whelan, Automation in the Ancient World: The Robots of Greece and Rome, 2020. Available from https://classicalwisdom.com/culture/history/automation-in-the-ancient-world.

[2] D.A. Hounshell, Automation, transfer machinery, and mass production in the U.S. automobile industry in the post—World War II era, Enterprise & Society 1 (1) (2000) 100–138. Available from http://www.jstor.org/stable/23699655.

[3] A Brief History of Automation in Manufacturing: Then and Now. Available from https://nanotronics.co/thinkspace/a-brief-history-of-automation-in-manufacturing/.

[4] F.G. Filip, Automation and computers and their contribution to human well-being and resilience, Studies in Informatics and Control 30 (2021) 5–18.

[5] N. Zrnić, M. Đorđević, V. Gašić, Historical background and evolution of belt conveyors, Foundations of Science 29 (2024) 225–255.

[6] R.R. Hoffman, N. Sarter, M. Johnson, J.K. Hawley, Myths of automation and their implications for military procurement, Bulletin of the Atomic Scientists 74 (2018) 255–261.

[7] Z. Wang, S. Hirai, S. Kawamura, Challenges and opportunities in robotic food handling: a review, Frontiers in Robotics and AI 8 (2022) 789107.

[8] T. Supramaniam, R. Ibrahim, T.D. Chung, S.M. Hassan, Towards the development of WirelessHART adaptor for process control applications, 2016 6th International Conference on Intelligent and Advanced Systems (ICIAS), IEEE, Kuala Lumpur, Malaysia, 2016, pp. 1–6.

[9] M.E.-S.M. Essa, A.M. El-shafeey, A.H. Omar, A.E. Fathi, A.S.A.E. Maref, J.V.W. Lotfy, et al., Reliable integration of neural network and internet of things for forecasting, controlling, and monitoring of experimental building management system, Sustainability 15 (2023) 2168.

[10] R. Kumar, A. Ray, M. Kande, WirelessHART device integration challenges and solutions in industrial automation, IEEE 18th Conference on Emerging Technologies & Factory Automation (ETFA) 2013 IEEE, Cagliari, Italy, 2013, pp. 1–4.

[11] A. Khamis, J. Meng, J. Wang, A.T. Azar, E. Prestes, H. Li, et al., AI and robotics in the fight against COVID-19 pandemic, in: A.T. Azar, A.E. Hassanien (Eds.), Modeling, Control and Drug Development for COVID-19 Outbreak Prevention, Studies in Systems, Decision and Control, Springer International Publishing, Cham, 2022, pp. 57–85. Available from https://doi.org/10.1007/978-3-030-72834-2_3.

[12] N.M. Mirza, A. Ali, Utilization of robotics during the Covid-19 pandemic, 2021 22nd International Arab Conference on Information Technology (ACIT). Presented at the 2021 22nd International Arab Conference on Information Technology (ACIT), IEEE, Muscat, Oman, 2021, pp. 1–4. Available from https://doi.org/10.1109/ACIT53391.2021.9677152.

[13] A. Chernoff, C. Warman, COVID-19 and implications for automation, Applied Economics 55 (2023) 1939–1957. Available from https://doi.org/10.1080/00036846.2022.2100870.

[14] P. Egana-delSol, G. Cruz, A. Micco, COVID-19's impact on the labor market shaped by automation: evidence from Chile, SSRN Journal (2021). Available from https://doi.org/10.2139/ssrn.3761822.

[15] I. Toni, UiPath Launches a Pro Bono Automation Project with the Mater Hospital to Free up Nurses Dealing with Infection Control Including Covid-19, 2020. Available from https://kommunikasjon.ntb.no/pressemelding/17882325/uipath-launches-a-pro-bono-automation-project-with-the-mater-hospital-to-free-up-nurses-dealing-with-infection-control-including-covid-19?publisherId=90063.

[16] W. Leal Filho, M. Fedoruk, J.H. Paulino Pires Eustachio, J. Barbir, T. Lisovska, A. Lingos, et al., How the war in Ukraine affects food security, Foods 12 (2023) 3996. Available from https://doi.org/10.3390/foods12213996.

[17] PTI, Govt Steps up Locust Control Operations; Drones to be Deployed Soon as Nagpur, UP Stay on Alert, 2020. Available from https://indianexpress.com/article/india/govt-steps-up-locust-control-operations-drones-to-be-deployed-soon-6430351/.

[18] M.F. Ardiansyah, T. William, O.I. Abdullaziz, L.-C. Wang, P.-L. Tien, M.C. Yuang, EagleEYE: aerial edge-enabled disaster relief response system, 2020 European Conference on Networks and Communications (EuCNC), IEEE, Dubrovnik, Croatia, 2020, pp. 321–325.

[19] S.S. Ali, S. Khan, N. Fatma, C. Ozel, A. Hussain, Utilisation of drones in achieving various applications in smart warehouse management, BIJ 31 (2024) 920–954.

[20] T. Czachórski, E. Gelenbe, G.S. Kuaban, D. Marek, Optimizing energy usage for an electric drone, in: E. Gelenbe, M. Jankovic, D. Kehagias, A. Marton, A. Vilmos (Eds.), Security in Computer and Information Sciences, Springer International Publishing, Cham, 2022, pp. 61–75.

[21] M. Pachayappan, V. Sudhakar, A solution to drone routing problems using docking stations for pickup and delivery services, Transportation Research Record 2675 (2021) 1056–1074.

[22] Z. Yin, M. Olapade, M. Liyanage, F. Dar, A. Zuniga, N.H. Motlagh, et al., Toward city-scale litter monitoring using autonomous ground vehicles, IEEE Pervasive Computing / IEEE Computer Society [and] IEEE Communications Society 21 (2022) 74–83.

[23] S.L. Andresen, John McCarthy: father of AI, IEEE Intelligent Systems 17 (2002) 84–85. Available from https://doi.org/10.1109/MIS.2002.1039837.

[24] F. Coppola, L. Faggioni, M. Gabelloni, F.De Vietro, V. Mendola, A. Cattabriga, et al., Human, all too human? An all-around appraisal of the "artificial intelligence revolution" in medical imaging, Frontiers in Psychology 12 (2021) 710982.

[25] L. Cao, Artificial intelligence in retail: applications and value creation logics, IJRDM 49 (2021) 958–976.

[26] J. Su (苏嘉红), W. Yang (杨伟鹏), Unlocking the power of ChatGPT: a framework for applying generative AI in education, ECNU Review of Education 6 (2023) 355–366.

[27] S. Muggleton, Alan Turing and the development of artificial intelligence, AI Communications 27 (2014) 3–10. Available from https://doi.org/10.3233/AIC-130579.

[28] S.J. Plathottam, A. Rzonca, R. Lakhnori, C.O. Iloeje, A review of artificial intelligence applications in manufacturing operations, Journal of Advanced Manufacturing and Processing 5 (2023) e10159.

[29] H. Shin, V.V. Prabhu, Evaluating impact of ai on cognitive load of technicians during diagnosis tasks in maintenance, Advances in Production Management Systems. Smart Manufacturing for Industry 4 (0) (2018) 27–34.

CHAPTER

4

Waste classification into plastics, industrial, domestic, and agriculture waste

Abdus Salam[1], Rifah Tasnia[2], Saad Bin Abul Kashem[3], Azad Ashraf[4], Ali Kunju[4], Molla Ehsanul Majid[5], Mazhar Hasan-Zia[4], Mohammad Nashbat[4], Amith Khandakar[6], and Muhammad E.H. Chowdhury[6]

[1]Department of Electrical and Computer Engineering, Rajshahi University of Engineering and Technology, Rajshahi, Bangladesh [2]Department of Computer Science and Engineering, Ahsaniya Mission University of Science and Technology, Rajshahi, Bangladesh [3]Department of Computing Science, AFG College with the University of Aberdeen, Doha, Qatar [4]Department of Chemical Engineering, University of Doha for Science and Technology, Doha, Qatar [5]Department of Computer Applications Academic Bridge Program, Qatar Foundation, Doha, Qatar [6]Department of Electrical Engineering, Qatar University, Doha, Qatar

Introduction

To achieve sustainable waste management and to protect the environment, recycling waste is an extremely important factor. Through the process of recycling, valuable materials can be recovered from waste streams such as plastic, paper, metal, and glass, and then reprocessed into new goods [1]. Because recycled materials may be used in place of raw materials such as lumber or minerals in manufacturing processes, this puts less burden on the earth's natural resources. Additionally, recycling greatly reduces the amount of waste that is delivered to landfills and incinerators for disposal. Reducing trash disposal is essential in light of the significant contributions that waste disposal has made to pollution of the air, land, and water on a worldwide scale [2]. Further, recycling may provide financial opportunities and job prospects via the collection, sorting, and reprocessing of waste items that can be recovered [3].

Harnessing Automation and Machine Learning for Resource Recovery and Value Creation
DOI: https://doi.org/10.1016/B978-0-443-27374-2.00004-2

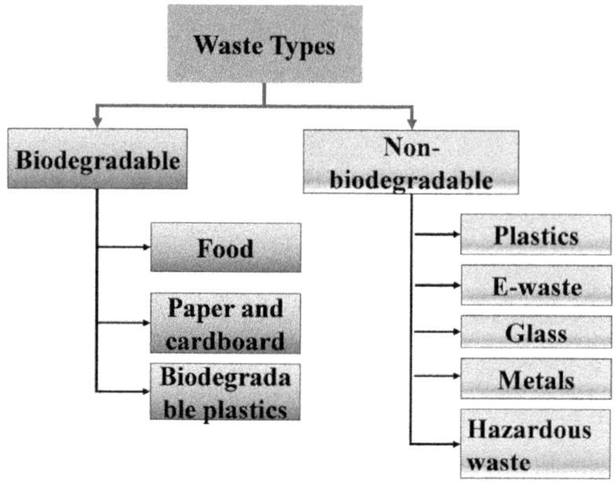

FIGURE 4.1 Typical types of wastes in our daily life.

The amount of waste produced by cities, businesses, and industries is truly astonishing. These different waste streams have unique compositions and present different environmental risks that require customized management approaches. In a comprehensive analysis, waste can be classified into different categories such as municipal solid waste, hazardous waste, electronic waste, plastic waste, and construction and demolition debris [4]. Fig. 4.1 shows some of the typical types of waste in our daily life. The municipal solid waste is made up of various components such as organic matter, paper, plastic, glass, and metal [5]. Approximately 1.47 billion metric tons (436 kg per capita per year) of municipal solid garbage are produced annually, and there is a growing trend in waste output [6]. This results in a significant amount being deposited in landfills or incinerators, causing widespread pollution of land, air, and water due to the release of leachate and ash residue. Leachates consist of dissolved organic and inorganic chemicals that have the potential to seep into the ground and contaminate groundwater sources. Incinerator ash has a tendency to accumulate heavy metals and harmful pollutants [7].

Industrial solvents, pesticides, used oils, chemicals, batteries, and medical waste are examples of hazardous waste that must be handled with utmost care due to their toxic components. Proper disposal protocols are necessary to ensure the safety of the environment and public health. When not properly treated, these can lead to significant health and environmental issues. As an example, when battery waste is not treated, it can result in contamination from lead and acid [8]. Discarded electronics and electrical equipment contain valuable metals such as copper and gold, as well as hazardous elements like mercury, cadmium, and lead. Approximately 50 million metric tons of electronic waste is produced annually on a global scale [9]. The improper disposal and basic dismantling of electronic waste for material recovery results in the release of harmful substances such as dioxins, furans, and heavy metal particulate matter [10]. Plastic waste, particularly single-use plastics, has become a significant concern because of their large quantities and limited ability to biodegrade. Every year, a staggering amount of plastic waste, exceeding 400 million tons, or 1 ton for every person on the planet, is produced worldwide [11]. Unfortunately, a considerable portion of this waste manages to evade collection systems, causing blockages in urban sewers, littering our surroundings, and polluting marine ecosystems. Microplastics, which are created when larger plastic pieces break down into tiny particles due to mechanical abrasion and photodegradation, can now be found everywhere in marine food chains [12].

Construction and demolition debris is a significant waste category that includes wood, concrete, bricks, glass, plastics, and even hazardous materials such as asbestos fibers. This rapidly expanding waste stream frequently includes components that can be recycled, such as wood and concrete aggregates. Recycling necessitates meticulous sorting and separation. Without adequate materials recovery facilities, most of the construction and demolition waste is disposed of in landfills. Poor disposal methods such as haphazard dumping, uncontrolled incineration, and improper landfilling are contributing to a significant pollution problem on a global scale. Nevertheless, it is possible to reduce their environmental impact by implementing source reduction, enforcing waste segregation, developing recycling programs, and utilizing advanced landfilling and incineration techniques. It is important to consider the waste compositions of each locality when developing region-specific integrated waste management policies. The progress in waste-to-energy, composting, and waste reclamation technologies provide prospects for sustainable management of municipal solid waste. These methods allow for the production of heat, electricity, annt years, there has been a significant advancement in the use of machine learning (ML) and deep learning (DL) for the purpose of automating classification of waste based on images.

Recent research investigate two fundamental inquiries: (1) What is the most effective method for storing wastes in dumping areas? and (2) How can waste items be categorized into distinct categories using a smart system? In order to address these obstacles, researchers developed and executed a smart waste management and classification system that capitalizes on cutting-edge technologies such as cloud computing, edge computing, fog computing, and the Internet of Things (IoT). In order to effectively manage and classify waste, the system incorporates sensors, GPS, wireless technology, cameras, and deep learning algorithms. The process commences with the collection of household refuse using smart bins that store data on the cloud and notify authorities when they are full. The waste is subsequently transported to a designated dumping area by a waste collection vehicle, where it undergoes segmentation and classification. The objective of this system is to optimize waste management and improve city infrastructure by facilitating real-time monitoring and intelligent decision-making.

The waste collection, monitoring, and classification process can be streamlined by the smart waste management system as shown in Fig. 4.2 [14], which effectively incorporates advanced technologies such as IoT, cloud computing, and AI. Consumers initially deposit waste into intelligent bins that autonomously monitor their contents and transmit this information to a cloud-based system, as illustrated in the workflow. The system notifies authorities via a mobile or web application when capacity is reached, which prompts the assignment of the nearest waste collection truck or truck. The collected waste is subsequently transported to a designated dumping site, where a gripper powered by a Raspberry Pi and equipped with a camera conducts grid segmentation and classification of waste items. This automated system effectively segregates waste into appropriate categories, including plastic, metal, glass, and organic, thereby facilitating more effective recycling and waste management practices. This system provides a scalable solution that improves both operational efficiency and environmental sustainability by addressing the increasing challenges of urban waste management through the use of cutting-edge technologies.

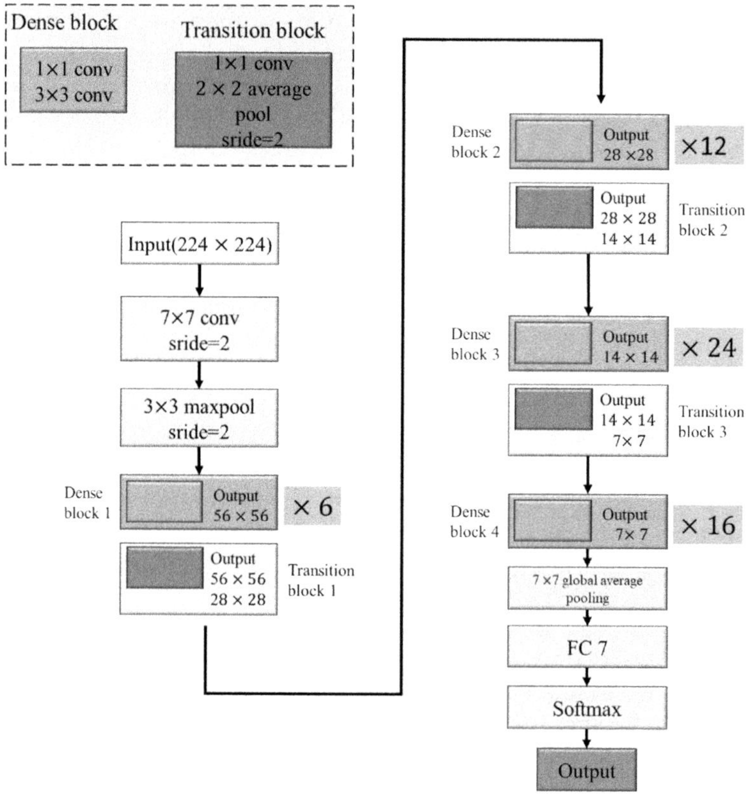

FIGURE 4.2 DenseNet121 model architecture.

Also, efficient waste management system can lead to a great economic advantage too [15]. Fig. 4.3 shows a process for turning garbage into useful goods, focusing on electronic waste. E-waste is identified and separated from normal rubbish in Step 1. E-waste is placed in cloud-monitored smart bins for data-driven decision-making (Step 2). E-waste is classified (Step 4) into plastics (Step 5) and metals (Step 6) after collection (Step 3). These materials undergo pyrolysis (Step 7) to decompose plastic waste into simpler compounds after classification. The metals are reused to make solar batteries (Step 8), promoting renewable energy. Pyrolysis produces bio-fuel (Step 9) and bio-char (Step 10), both of which are valuable resources. Bio-fuel can be utilized as an alternative energy source, while bio-char improves agricultural productivity. All steps are focused on turning waste into assets, with Step 11 emphasizing the circular economy principle of reintegrating by-products and recovered materials into the production cycle to reduce waste and maximize resource efficiency.

The classification of common waste types such as plastic, paper, metal, and glass can now be accomplished with a high degree of precision using sophisticated neural network designs. For trash categorization tasks, models that make use of convolutional neural networks (CNNs) such as AlexNet [16], ResNet-50 [17], MobileNetV3-Large [18], DenseNet-201, InceptionResnetv2 [19], and EfficientNet-B0 [20] have shown that they have the

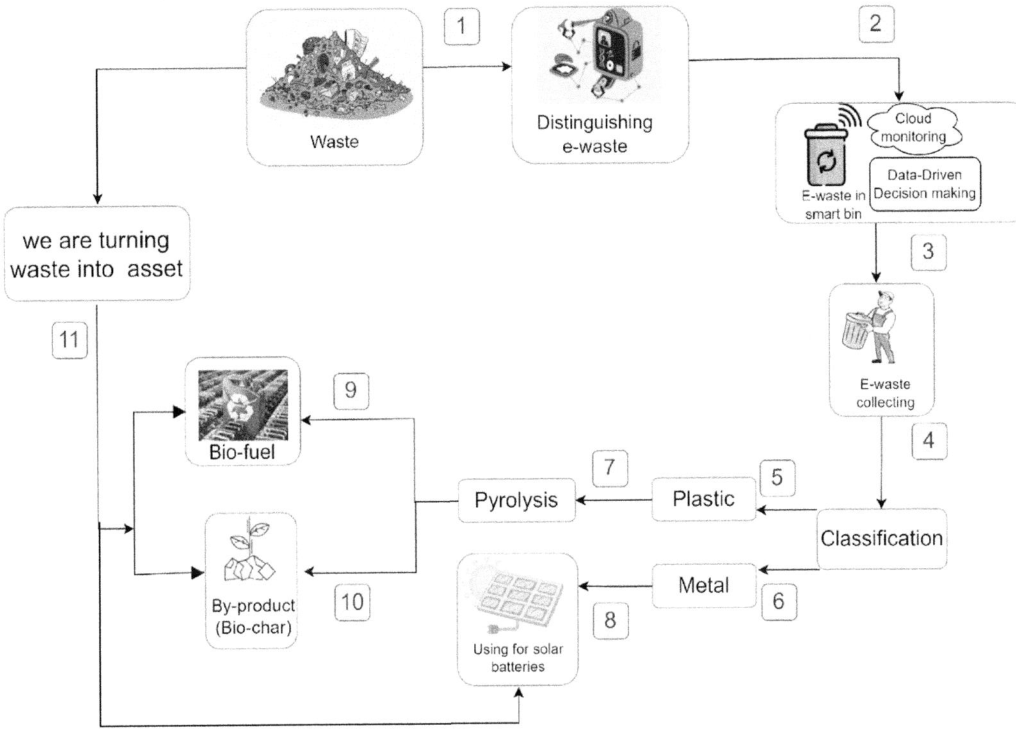

FIGURE 4.3 Effective management of waste for reuse. Source: *From M. Farjana, A.B. Fahad, S.E. Alam, M.M. Islam, An IoT- and cloud-based e-waste management system for resource reclamation with a data-driven decision-making process. IoT 4(3) (2023) 202–220. https://doi.org/10.3390/iot4030011.*

potential to be effective. Other models that have already been trained, such as VGG-16 [21] and Inception-V3 [22], have also been effectively fine-tuned. Combining convolutional neural networks (CNNs) with long short-term memory networks (LSTMs) is a hybrid strategy that can take use of both spatial and temporal correlations. In addition to the identification of waste types, machine learning, and deep learning techniques have been tailored to applications such as the prediction of trash flow via the use of sensor data, the automated quantification of garbage pile volume by stereovision, and the detection of contamination for the purpose of monitoring waste quality.

Related works

Deep learning for waste classification

Researchers aimed to use machine learning for efficient recycling processes to address waste production from population growth [23]. They used four- and five-layer deep convolutional neural networks (DCNN) to classify waste images into paper, glass, plastic, and organic materials. The primary goal of the research was to reduce environmental pollution through effective recycling and optimize recycling facility performance. The study focused on reducing human involvement in waste classification to enhance efficiency in terms of time, labor, and overall facility operations. The five-layer model achieved an impressive accuracy rate of 70%, outperforming the four-layer architecture, which attained a 61.67% accuracy rate. The study used sensitivity, specificity, precision, F-score, and accuracy metrics for comprehensive model evaluation, demonstrating the potential of machine learning-based waste classification systems in transforming traditional waste management practices. In another study, Malik et al. [24] addressed the issue of municipal solid waste (MSW) management by using advanced deep learning techniques to classify MSW. The researchers emphasized the importance of properly categorizing waste and disposing of it correctly. They employed a neural network-based approach, with a particular focus on the EfficientNet-B0 architecture. By utilizing convolutional neural networks (CNNs) and transfer learning, the team aimed to customize the pretrained EfficientNet-B0 model to specific demographic regions, ultimately improving waste classification accuracy. The study emphasized the compound-scaling approach of the model, which ensured uniform expansion in both width and depth. This approach contributed to the development of a highly efficient and fine-tuned waste classification system. Notably, the researchers mapped 34 waste categories to 10 higher-order classifications, which align with types of recycling and composting plants. The proposed system, implemented in TensorFlow over Python, showcased a remarkable accuracy of 81.2%. Interestingly, this accuracy was comparable to larger models like EfficientNet-B5, despite the significantly fewer parameters employed in their model.

Chu et al. [25] introduced a multilayer hybrid deep learning system (MHS) to improve garbage sorting accuracy. The MHS extracted visual characteristics using a high-resolution camera, sensors, and an AlexNet-based CNN. Multilayer perceptron (MLP) aggregated visual and sensor data to categorize garbage as recyclable or not. This hybrid strategy helped the MHS outperform CNN architecture with classification accuracy above 98%. The key is integrating optical and sensor data to improve garbage sorting. They presented a municipal trash management system employing sophisticated deep learning and seven CNN architectures in another work [26]. The best in categorizing garbage into nine kinds was MobileNetV3 with 94.26% accuracy. Deep learning, cloud computing, and IoT provide real-time monitoring and optimization. The whole system fosters intelligent waste management and environmental sustainability by solving categorization, monitoring, and collection issues.

In one compelling study [27], researchers developed an end-to-end waste management system using deep learning and IoT technology. At its core was a Convolutional Neural Network (CNN) that achieved 95.31% accuracy in classifying waste into digestible and indigestible categories—a promising solution to reducing recyclable material contamination. To overcome data limitations, the team utilized advanced modeling techniques for precise waste classification. The system also introduced a "smart" trash bin containing sensors and microcontrollers for real-time waste data monitoring. The integration of Bluetooth and WiFi capability enabled both short and long-range control and monitoring through a custom Android application. This comprehensive approach highlighted the power of combining deep learning and IoT to evolve waste management practices. Expanding on this, another study [28] delved deeper into algorithm performance for automated waste classification. Motivated by inefficient manual sorting processes, the researchers tested four machine/deep learning models on a dataset of 2500 waste images. The CNN decisively outperformed its machine learning (ML) counterparts, achieving 90% accuracy in categorizing glass, paper, metal, plastic, cardboard, and general waste. The study thus showcased the potential for AI-powered automation to revolutionize waste management systems. While the SVM model followed closely behind at 85% accuracy, the superiority of deep learning for image classification tasks was evident.

Innovative garbage categorization methods are being developed to automate sorting and improve disposal efficiency. One prominent work by Zhang et al. [29] presented CTR, a model built on the ResNet18 architecture. CTR has a self-monitoring module (SMM) for improved spatial feature extraction from trash photos. CTR achieved a remarkable 95.87% accuracy in classifying cardboard, glass, plastic, metal, paper, and general rubbish after extensive testing on the TrashNet dataset. The work therefore demonstrates computer vision's enormous potential for intelligent garbage sorting, particularly on mobile devices. Beyond demonstrable improvements, CTR's value includes reducing manual labor constraints and health concerns. While admitting dataset constraints, the study emphasizes future optimization and mobile adaptation to get closer to sustainable and circular waste management systems. Meanwhile, another group [30] developed a technique that combines three strong pretrained models—VGG19, DenseNet169, and NASNetLarge—to address waste diversity concerns that impede classification accuracy. Using ImageNet and transfer learning, their model achieved a fantastic performance of 96.5% on garbage datasets, outperforming previous techniques such as SVM-SIFT, CNN, Faster R-CNN, and ResNet50. The method provides a unique approach by dividing garbage into six primary and twelve subtypes, overcoming differences in material compositions and disposal regulations. The study value is in overcoming current limits by combining the benefits of diverse models via transfer learning. This improves granular categorization skills beyond previous efforts that relied primarily on deep learning approaches. Moving ahead, such multimodel solutions hold enormous potential for transforming waste management by facilitating effective classification, separation, and disposal.

Existing datasets and benchmarks

Table 4.1 summarizes some of the most popular waste classification datasets, among them, the TrashNet dataset contains 2527 images across 6 classes: plastic, glass, paper, cardboard, metal, and general waste. Prior work achieved 95.87% accuracy on this dataset using a CTR (ResNet18-based) model [27]. The Garbage Classification Dataset is larger with 5904 images but has only three high-level classes: organic, inorganic, and medical waste. The DNN-TC model based on ResNext achieved 98% accuracy on this dataset [29]. The Garbage Detection Dataset is the largest with 13,116 images across 4 common waste types: plastic, paper, metal, and glass. The YOLOv4 object detection model obtained 89.59% mAP on this dataset [30]. Finally, the Waste Classification Data contains 22,617 images spanning 6 categories similar to TrashNet: glass, paper, cardboard, plastic, metal, and trash. However, no benchmark results have been reported on this dataset. By comparing the performance of models across these varied waste classification datasets, we aim to gain insights into dataset characteristics and modeling approaches for this environmental application.

TABLE 4.1 Some available datasets on waste classification and their benchmark accuracy.

Dataset	Description	No. classes	No. images	Benchmark results	References
TrashNet	Images of various types of trash, including plastic, glass, paper, cardboard, metal, and general waste.	6	2527	95.87% accuracy with CTR (ResNet18-based) model [29]	[31]
Garbage Classification Dataset	Images of different types of trash, including organic, inorganic, and medical waste.	3	5904	98% accuracy with DNN-TC (ResNext-based) model [32]	[32]
Garbage Detection Dataset	Images of various types of trash, including plastic, paper, metal, and glass.	4	13,116	89.59% mAP with YOLOv4 model [33]	[33]
Waste Classification Data	Images of various types of waste, including glass, paper, cardboard, plastic, metal, and trash.	6	22,617	No benchmark results reported	–

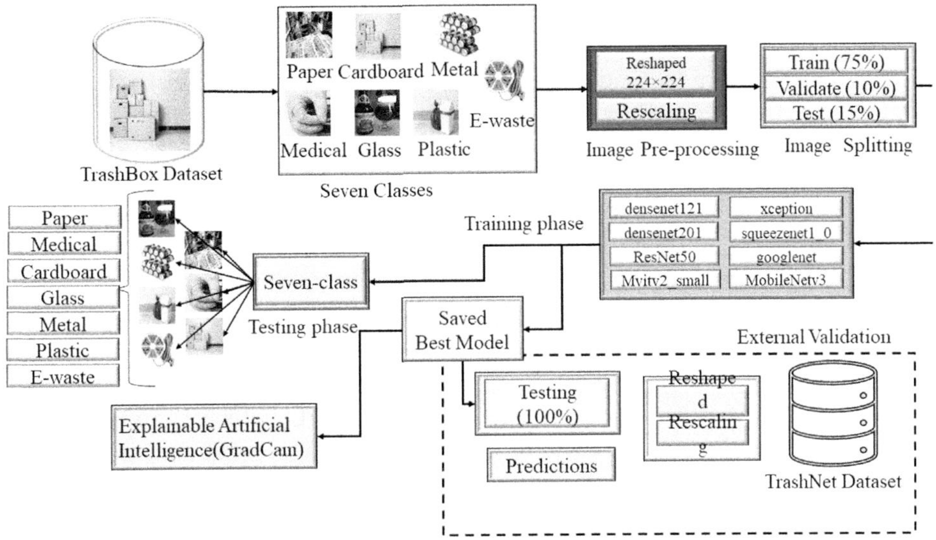

FIGURE 4.4 Proposed methodology.

Methodology

The study utilizes data from the TrashBox [34] and TrashNet [31] datasets. The TrashBox dataset comprises images of trash taken against several backdrops. The TrashBox dataset images were resized to 224×224 pixels and then standardized by mean subtraction and standard deviation normalization for our studies. The data was divided into training, validation, and test sets using a 75:15:10 ratio. The training set was used to train machine learning models for classifying photos of garbage, whilst the validation set was employed for hyperparameter tweaking and model selection. Fig. 4.4 is the illustrative representation to the proposed methodology for this study. The held-out test set was used for the ultimate assessment of the model.

Models

Various convolutional neural network architectures were evaluated for the waste classification, such as DenseNet121 [35], DenseNet201 [35], mvitv2_small [36], Xception [37], squeezenet1_0 [38], GoogleNet, MobileNetN3_small [39], ResNet101 [40], and Alexnet. The models were trained on the Trashbox training dataset using stochastic gradient descent with momentum. Data augmentation methods such as random horizontal flips, rotations, and color jittering were used to artificially increase the amount of training data. Model performance was assessed during training by monitoring the validation data. The model with the greatest accuracy on the TrashBox validation set was chosen as the optimal model. The chosen model was assessed using a separate test set from TrashBox and an external test set from TrashNet.

Evaluation: Classification model performance is assessed using evaluation measures such as accuracy, precision, recall, *F*1-score, and metrics derived from the confusion matrix. Evaluation of performance is conducted utilizing both the TrashBox test set and external TrashNet pictures to assess real-world effectiveness.

Experimental setup

The experiment is carried out in a cloud-based environment for the execution of machine learning models provided by Google Colab. It can be used as sophisticated computer capabilities, such as graphics processing units (GPUs), using Google Colab without the requirement for local hardware. For this experiment, T4 GPU, which is a kind of NVIDIA GPU intended specifically for use in data centers and cloud computing, is used. The T4 graphics processing unit (GPU) is a component of NVIDIA's Turing architecture and is designed to accelerate tasks related to machine learning. For this experiment, 24 gigabytes of GDDR6 memory is used. Which is suitable for training and inference on a wide variety of computer vision applications with its mid-range GPU. The GPU T4 is well known for its efficiency since it offers outstanding performance while using little power.

Dataset description

TrashBox dataset

The data set consists of 17,785 images of various types of garbage objects, which are categorized as follows: medical, metal, cardboard, paper, electronic waste, and glass. By further subdividing each category into subcategories, an exhaustive representation of the various varieties of refuse materials is achieved. There are 2010 images of syringes, surgical gloves, surgical masks, and medications (drugs and tablets) in the medical waste category. Comprised of electronic processors, laptops and smartphones, appliances, and electric wiring, connections, and cables, the e-waste category comprises 2883 images. The plastic category contains 2669 images of filters for plastic items, including cigarette remnants, canisters, receptacles, and bags. The paper category comprises 2695 images and contains tetra paks, newspapers, paper containers, and paper tissues, among other objects. The metal category comprises a variety of objects, including construction debris, beverage cans, food-grade cans, and aerosol cans, amounting to 2586 images in total. There are 2528 images in the glass category and 2414 images in the cardboard category. The dataset is divided into train tests and validation by splitting them by 80:10:10 accordingly. Table 4.2 shows the number of images per classes for train test and validation subsets.

The train subset consists of 14,279 images in total, which are allocated as follows among the seven classes: glass—2022 images, medical waste—1565 images, metal—2068 images, paper—2155 images, and plastic—2135 images. A total of 241 images of cardboard, 300

TABLE 4.2 Data distribution of TrashBox dataset over train test and validation subsets.

Class name	Train	Validation	Test	Total
Cardboard	1929	241	242	2412
E-waste	2405	300	302	3007
Glass	2022	252	254	2528
Medical	1565	195	197	1957
Metal	2068	258	260	2586
Paper	2155	269	270	2694
Plastic	2135	266	268	2669

images of electronic waste, 252 images of glass, 195 images of medical waste, 258 images of metal, 269 images of paper, and 266 images of plastic comprise the validation subset's total of 1781 images. Throughout the training process, this subset is utilized to validate the model, thereby aiding in the monitoring of its performance and mitigating the risk of overfitting. A total of 242 images of cardboard, 302 images of electronic refuse, 254 images of glass, 197 images of medical waste, 260 images of metal, 270 images of paper, and 268 images of plastic comprise the test subset of 1793 images. The purpose of this subset is to assess the trained model's performance on previously unseen data.

Data preprocessing

Resizing was performed on every image in the train, validation, and test subsets to 224×224 pixels. Maintaining a consistent resizing of the images facilitates standardized processing of them by the model throughout the training and inference processes. To normalize the pixel values of the RGB images to the interval [0, 1], the maximal RGB value of 255 was divided by each pixel value. By putting all pixel values within a consistent range, normalization facilitates quicker and more precise model training. The model architecture and parameters were tasked with acquiring knowledge of the visual characteristics and patterns present in the 224×224 resized images for classification.

TrashNet dataset for external validation

The dataset includes 2527 images that cover six different waste categories: glass, paper, cardboard, plastic, metal, and trash. There are different numbers of images for each class: 501 for glass, 594 for paper, 403 for cardboard, 482 for plastic, 410 for metal, and 137 for trash. Table 4.3 is the summary of a number of images per class in the TrashNet dataset.

TABLE 4.3 Number of samples per class in TrashNet dataset.

Class	Number of images
Glass	501
Paper	594
Cardboard	403
Plastic	482
Metal	410
Trash	137

The images were taken under various lighting conditions, including natural sunlight and room lighting. They show waste objects placed on a white posterboard. For the sake of consistency, all images have been resized to a resolution of 512×384 pixels. This dataset is a valuable resource for individuals interested in developing and evaluating image classification models for waste categorization. It can be of great use to researchers and practitioners in this field. To check the robustness of the trained model TrashNet dataset was used for external validation. The images were fully used for the testing set. The images were resized to 224×224 normalized and fed to the trained model as tensor.

Evaluation metrics

Accuracy

Regarding classification tasks in particular, the accuracy of a neural network model is a crucial performance metric. At its most fundamental level, accuracy assesses the ratio of samples that have been correctly classified to the overall number of samples. Accuracy is defined mathematically as follows:

$$\text{Accuracy} = \frac{\text{True Positive} + \text{True Negative}}{\text{True Positive} + \text{True Negative} + \text{False Positive} + \text{False Negative}}$$

$$= \frac{\text{True Predicted Values}}{\text{Total Observations}} \tag{4.1}$$

While accuracy is a reasonable initial criterion for evaluating the performance of a model, it does not consistently offer a comprehensive perspective. A model may accurately predict the majority class for all inputs in unbalanced datasets, for example, while failing to identify the minority class, which may be of greater interest [35].

Precision

Precision is measured mathematically as

$$\text{Precision} = \frac{\text{True Positive}}{\text{True Positive} + \text{False Positive}}$$

$$= \frac{\text{True Positive}}{\text{Total Predicted Positive}} \tag{4.2}$$

In simple terms, precision measures how many of the things marked as positive are truly positive. It lets us know how likely it is that the model's good class predictions will come true. False alarms are less likely to happen with a model that is very accurate.

Recall

Recall, which is also called sensitivity, true positive rate, or hit rate, is another important metric used to measure how well a classification model works, especially when the false negative rate is a major worry. Recall answers the question, "How many of the actual positive cases did the model get right?"

Mathematically, here's how to figure out recall:

$$\text{Recall} = \frac{\text{True Positive}}{\text{True Positive} + \text{False Negative}}$$

$$= \frac{\text{True Positive}}{\text{Total Actual Positive}} \tag{4.3}$$

*F*1 score

The *F*1 score is a measure of precision and recall, ranging from 0 to 1, with 1 indicating perfect precision and recall and 0 indicating neither. It provides a summary of a model's evaluation but doesn't consider true negatives. The *F*1 score is the harmonic mean of precision and recall, giving more weight to lower values. It is a robust measure for balancing precision and recall and addressing uneven class distribution.

The mathematical definition of the *F*1 score is the harmonic mean of accuracy and recall

$$F1\text{score} = 2 \times \frac{\text{Precision} \times \text{Recall}}{\text{Precision} + \text{Recall}} \tag{4.4}$$

Receiver Operating Characteristic curve

A model that can classify things perfectly will have an Receiver Operating Characteristic (ROC) curve that goes through the upper-left corner, or the point (0, 1), which means it has 100% Sensitivity (no false negatives) and 100% specificity (no false positives). The area under the curve (AUC-ROC), which is calculated from the ROC curve, is an important summary statistic that measures the model's general ability to tell the difference between the positive and negative classes [35].

True positive rate: It is another name of Recall which can be expressed as

$$TPR = \frac{\text{True Positive}}{\text{True Positive} + \text{False Negative}} \tag{4.5}$$

False positive rate: Can be expressed as follows:

$$FPR = \frac{\text{False Positive}}{\text{False Positive} + \text{True Negative}} \tag{4.6}$$

Hyperparameters and training

The experimental methodology utilized a deep learning approach for image classification. The models were initialized with pretrained weights from the ImageNet dataset to leverage knowledge transfer and facilitate faster convergence. Table 4.4 represents the hyperparameters used to train the models The experiments were carried out using a batch size of 32, carefully considering both computational efficiency and the need for stable training. The

TABLE 4.4 Hyperparameters used to train the models.

Hyperparameter	Value
ImageNet	True
q_order	3
batch_size	32
Optimization function	Adam
Learning rate	0.0001
stop_criteria	Loss
Number of epochs	100
Epochs patience	80
Learning factor	0.1
max_epochs_stop	80

input data comprised of RGB images that were resized to a standardized size of 224x224 pixels prior to being inputted into the network. The input data was normalized by subtracting the mean and dividing by the standard deviation for each input channel, ensuring that it was within a consistent range. A researcher would typically select the Adam optimizer to train the fully connected layers, starting with an initial learning rate of 0.0001.

The training process incorporated stopping criteria that relied on validation loss. Additionally, early stopping mechanisms were used to protect against overfitting. If the validation loss did not decrease for 80 consecutive epochs, the learning rate would decrease by a factor of 0.1. In addition, if the validation loss did not show any improvement within a maximum of 80 epochs, the training was stopped prematurely. The model was trained without using pretrained weights and without employing any data augmentation techniques.

Results and discussion

Quantitative results

The deep learning models were evaluated on a comprehensive test dataset using key classification performance metrics including accuracy, precision, sensitivity, F1-score, and specificity.

Accuracy measures the overall effectiveness of a model at correctly classifying both positive and negative cases. As you can see in Table 4.5, with an accuracy of 97.08%, the DenseNet121 model demonstrated superior classification performance on the test dataset. Following closely were the mvitv2_small and xception models, which achieved strong accuracies of 95.47% and 95.42%, respectively. Assessing precision involves measuring the model's capacity to provide accurate positive classification, calculated as the ratio of true positives to all predicted positive cases. As in Fig. 4.5 from overall model performance comparison in all metrices, Again, the DenseNet121 model demonstrated superior performance compared to other models, achieving a precision of 89.95%. This indicates that almost 90% of the cases classified as positive were accurately identified. The high precision of DenseNet121 showcases its ability for accurate positive classification while minimizing false positives.

TABLE 4.5 Model performance over several metrices.

Model	Accuracy (%)	Precision (%)	Sensitivity (%)	F1 score (%)	Specificity (%)
mvitv2_small	95.47	85.85	84.16	83.89	97.34
xception	95.42	84.89	84.10	84.02	97.28
DenseNet121	97.08	89.95	89.85	89.84	98.30
densenet201	95.23	84.48	83.49	83.32	97.16
squeezenet1_0	90.35	70.48	67.04	66.24	94.07
googlenet	94.09	81.22	79.53	79.20	96.47
alexnet	89.55	67.17	63.97	63.58	93.72
mobilenetv3_small	95.98	86.17	86.06	86.07	97.64

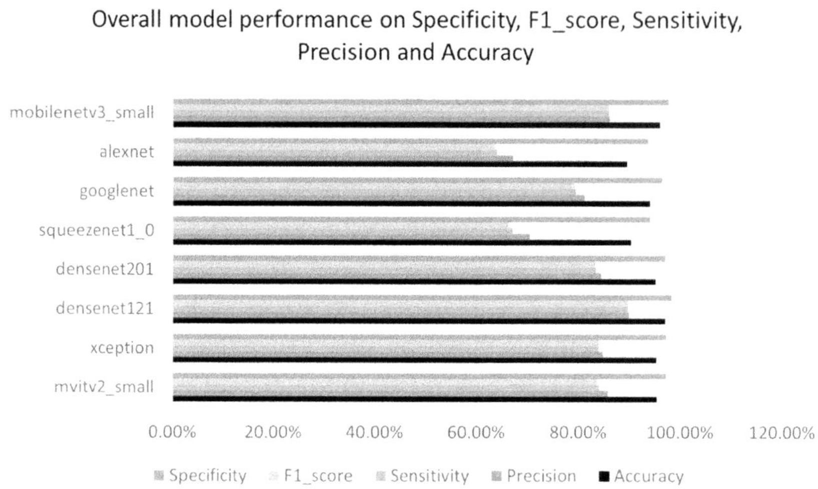

FIGURE 4.5 Overall model performance comparison in all metrices.

Sensitivity involves evaluating how accurately positive cases are identified. Comparing the sensitivities of DenseNet121, Xception, and MViTv2_small, it can be observed that DenseNet121 and Xception had very similar sensitivities of 89.85% and 84.10%, with MViTv2_small slightly lower at 84.16%. All three models demonstrated proficiency in detecting and classifying the target condition in test samples, which is crucial for diagnostic and detection tasks.

The *F*1 score strikes a balance between precision and sensitivity by calculating their harmonic mean. With an impressive *F*1 score of 89.84%, the DenseNet121 model demonstrates exceptional precision and sensitivity. Mvitv2_small and xception were the next best performers, achieving accuracy rates of 83.89% and 84.02%. And specificity assesses the model's ability to accurately identify negative cases. DenseNet121 showed an impressive specificity of 98.30%, indicating rare misclassification of healthy, normal cases as positive. By doing this, resources are not wasted on investigating false positives. Mvitv2_small and xception also demonstrated strong performance with specificities exceeding 97%. Summarizing, the DenseNet121 architecture demonstrated outstanding performance with an accuracy of 97.08% and high precision, sensitivity, F1 metrics, and specificity on the test dataset. The DenseNet121 achieves a perfect equilibrium by accurately identifying positive cases and minimizing false alarms. The DenseNet121 model is highly recommended for future deployment and inference due to its superior performance across all evaluation metrics.

Summarizing, the DenseNet121 architecture demonstrated outstanding performance with an accuracy of 97.08% and high precision, sensitivity, *F*1 metrics, and specificity on the test dataset. The DenseNet121 achieves a perfect equilibrium by accurately identifying positive cases and minimizing false alarms. The DenseNet121 model is highly recommended for future deployment and inference due to its superior performance across all evaluation metrics. Table 4.6 compares the accuracy of recent waste image classification techniques and datasets, demonstrating a wide range from 61.67% to 97.08% accuracy. The highest accuracy is achieved in this work by using the large public TrashBox dataset combined

TABLE 4.6 Accuracy comparison with recent works.

Dataset Used	Models	Accuracy	References
Custom waste image dataset with 400 RGB images	4-layer DCNN, 5-layer DCNN	61.67%, 70%	[23]
Not specified	EfficientNet-B0	81.2%	[24]
Not specified	InceptionV3, ResNet50, MobileNetV3, etc.	94.26%	[26]
Not specified	CNN	95.31%	[27]
2500 waste images	SVM, Random Forest, Decision Tree, CNN	85%, 90%	[28]
TrashNet (2527 images)	CTR (ResNet18 based)	95.87%	[29]
Two waste image datasets	TL-VGG19, TL-DenseNet169, etc. (transfer learning)	96.5%, 94%	[30]
TrashBox and TrashNet combined dataset	ResNet152, DenseNet169, and MobileNetV2	93.10%	[41]
TrashBox Dataset	DenseNet121	97.08%	This work

with a DenseNet121 deep transfer learning model, beating the next best result of 96.5%. The comparison shows deep convolutional neural networks can achieve over 97% accuracy given substantial labeled image data, significantly outperforming simpler methods like SVMs.

External validation

In general, the models demonstrated an overall reduction in performance when compared to the TrashBox training distribution, which highlights the difficulty of dealing with domain changes seen in the actual world. Both DenseNet121 and DenseNet201 demonstrated the highest level of accuracy, with an accuracy of 93.24% on TrashNet. As you can see from the illustrative representation of the overall model Accuracy on TrashNet dataset in Fig. 4.6 the models achieved a consistently high level of accuracy that was greater than 78%, which reflects consistency in positive categorization. Approximately 78%–79% of their sensitivity remained the same as that of the TrashBox dataset, which indicates that there was only a little decline in the identification of true positives. A high specificity of 96% was also reported by us, which indicates that just 4% of the negative instances were incorrectly identified.

The AlexNet and SqueezeNet networks, on the other hand, had discernible decreases in accuracy and sensitivity, which demonstrated their limited transferability. In terms of precision, GoogleNet and Xception models were not as accurate as DenseNet models, but they did give comparable accuracy estimates close to 92%. Verification of DenseNet's superiority in

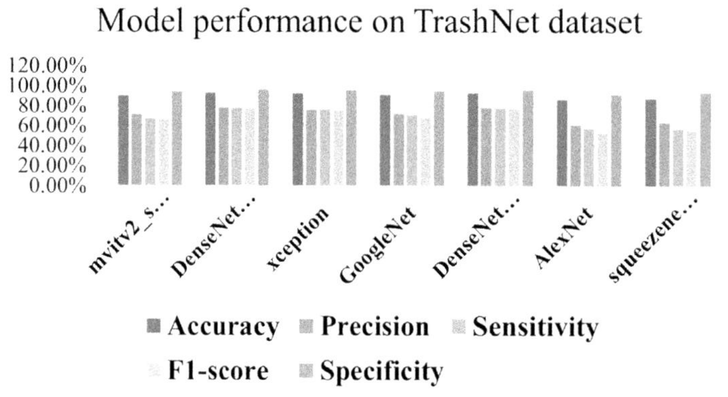

FIGURE 4.6 Overall model performance on TrashNet dataset in External validation.

TABLE 4.7 Quantitative evaluation of generalizability to TrashNet.

Model	Accuracy (%)	Precision (%)	Sensitivity (%)	F1 score (%)	Specificity (%)
mvitv2_small	90.16	71.29	67.46	66.53	94.33
DenseNet121	93.24	78.64	78.33	77.06	96.06
xception	92.78	76.36	76.95	75.62	95.70
GoogleNet	91.21	72.10	70.79	68.80	95.02
DenseNet201	93.24	78.64	78.33	77.06	96.06
AlexNet	86.66	60.81	57.67	53.33	91.90
squeezenet1_0	87.45	63.51	57.15	55.97	93.41

real-world trash categorization from photos is provided by the assessment, which highlights the significance of doing thorough benchmarking on a wide variety of target data.

When domain shift was encountered, the validation revealed that there was a large amount of performance variability across several models. In the face of distributional changes, the DenseNet models demonstrated a remarkable level of robustness, with DenseNet121 and DenseNet201 reaching the greatest accuracy of 93.24%. This indicates Table 4.7 that DenseNet's learning can be generalized even when the subtleties of new test cases differ from those of previous test cases.

Qualitative analysis

Fig. 4.7 shows the confusion matrix over 4 different models used in this study. A confusion matrix shows the categorization result on the hold-out test set of ground truth examples. Misclassifications between actual and predicted classes are off diagonals, whereas

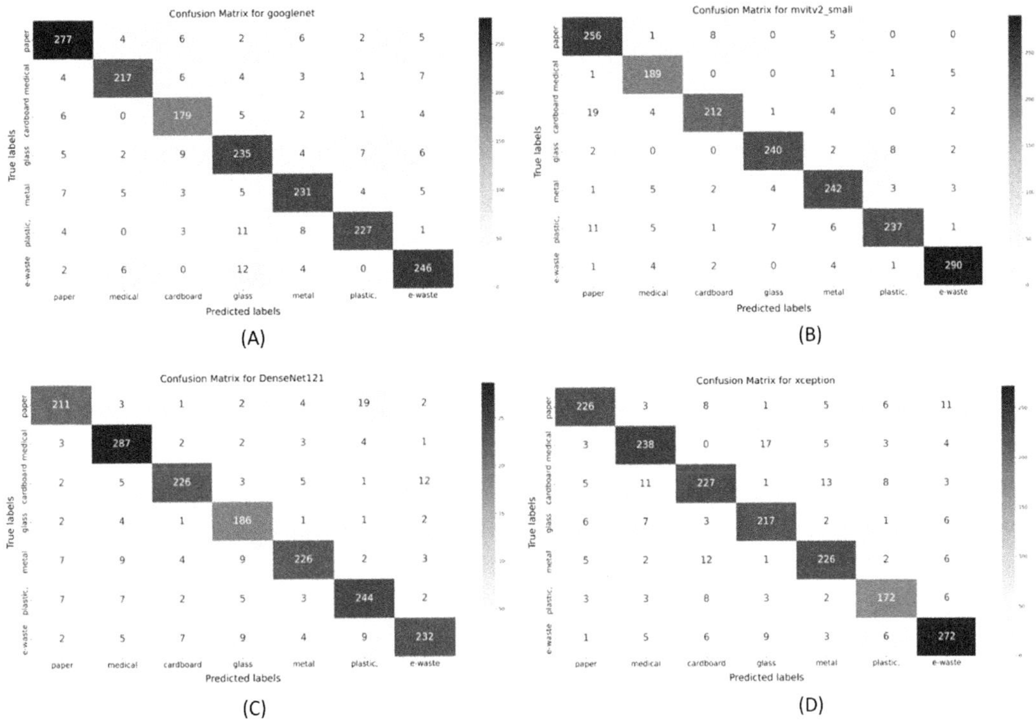

FIGURE 4.7 Confusion matrix for (A) GoogleNet, (B) mvitv2_small, (C) DenseNet121, and (D) Xception.

accurate classification counts are diagonal. DenseNet121 classified paper well with 211 true detections but confused 19 with plastics. Medical waste was classified correctly with 287 metals and 7 false. Interestingly, 12 medical cases were labeled e-waste. The model handled cardboard (226 right) well but mixed it with paper, metal, and e-waste, indicating room for improvement. Glass categorization was accurate (186), with few off-target mistakes. Metal has 226 correct categorizations, but plastic mix-ups continue. Finally, 244 examples show correct plastic identification, while misreads into other categories show optimization gaps. The grid helps identify priority areas—plastic vs. e-waste and medical vs. metal may benefit from training refinement highlighting differentiating traits.

Discussion and conclusion

A thorough evaluation of eight SOTA CNN architectures for automated waste classification was performed in this study. Using the TrashBox dataset, which contains 17,000 images spanning six waste classes, the models were trained. The real-world generalization capabilities of the diverse TrashNet dataset were rigorously assessed through external validation. Undoubtedly, the DenseNet121 architecture demonstrated superior performance, attaining a substantially higher accuracy of 97.08% in comparison to the

MobileNetV3 Small, the next best model, which achieved 95.98%. It is worth noting that DenseNet121 exhibited remarkable dependability, as evidenced by its class-leading specificity of 98.30% and precision of 89.95%; this equates to a mere 1.7% of negative cases being erroneously classified as waste. This permits deployment with high confidence and minimal false alarms. Additionally, DenseNet121 achieved an optimal trade-off between sensitivity (89.85%) and false positive reduction when detecting waste. Although the specificity of the majority of models was satisfactory, DenseNet121 performance was unparalleled. Unexpectedly, advanced architectures such as Xception were unable to surpass DenseNet121 in critical reliability metrics. In general, our exhaustive benchmarking process revealed that DenseNet121 outperformed all others across all metrics and was exceptionally effective at discerning subtle visual distinctions among intricate urban waste classifications.

These results validate DenseNet121 suitability for immediate implementation in automated waste segregation in the real world. Its equilibrium between precision and dependability can facilitate the development of scalable infrastructure for waste monitoring and management, thereby increasing recycling pipeline efficiency. Our research establishes deep learning as a revolutionary advancement in refuse classification, surpassing the capabilities of conventional computer vision methods.

Limitations and future directions

This study demonstrated a waste classification accuracy of 97.08% with DenseNet121 model, but there are significant challenges for real-world implementation. Initially, the testing was limited to carefully selected waste datasets with minimal background interference. However, real-world systems would have to function in environments filled with obstacles and varying lighting conditions. Therefore it is probable that the current performance metrics are overestimating the practicality in real-world scenarios. Thoroughly assessing the models using a variety of real-world waste images is a crucial upcoming task. Additionally, the testing has been limited in scope with only internal validation on the training set and one external dataset. Extensive testing on more locale-specific waste categories is necessary to ensure the models' generalizability. Although achieving accuracy is commendable, the misclassification rate indicates that further enhancements are needed before widespread implementation.

The techniques have primarily focused on waste classification, but practical automated sorting requires waste detection and segmentation capabilities as well. Creating multitask models and evaluating their performance on additional tasks can significantly enhance the practicality of real-world systems. Considering these constraints, upcoming research should focus on conducting more extensive testing on actual waste images, utilizing extra labeled datasets to further enhance classification accuracy, assessing the ability for detection and segmentation tasks, and exploring techniques to enhance model interpretability. Optimizing efficiency on edge devices commonly found in waste sorting infrastructure through model compression could enhance real-time operational viability.

By emphasizing augmented real-world testing, expanding to new waste categories, incorporating additional waste tasks such as detection and segmentation, and improving model explainability and efficiency, this research direction aims to create automated waste sorting systems that have significant sustainability impacts using cutting-edge deep learning techniques. The current models demonstrate impressive accuracy scores on controlled datasets as a foundation. Nevertheless, overcoming the specified constraints by conducting thorough real-world assessments and developing multitask models is crucial for moving this technology from academic potential to a practical and implementable solution. In the future, researchers should work on making the model work in a wider range of real-world settings and adding more jobs, like finding waste and separating it into different groups. For real-time uses in waste sorting infrastructure, it will also be important to make these models work best when deployed on edge devices. The results of this study show that using cutting-edge AI methods in waste management could lead to even more progress in being environmentally friendly.

Conclusion

In the end, this study shows a strong way to sort trash using cutting edge deep learning methods, mainly the DenseNet121 design. We got a great classification accuracy of 97.08% by using the TrashBox dataset and testing it against the TrashNet dataset. This was better than several advanced models. The DenseNet121 model had a great mix of precision, sensitivity, and specificity, which meant it could be used with confidence in waste management systems. Our method, which included a lot of data preprocessing, hyperparameter tuning, and the use of advanced CNN models, shows that deep learning can handle difficult classification tasks in the real world. The study's results show that deep learning models, especially DenseNet121, have the ability to automatically sort trash into categories with high accuracy and few false positives, which is very important for recycling to work well. Additionally, DenseNet121 excellent performance and ability to reliably handle a wide range of complex waste types make it a top choice for future use in automated garbage sorting systems. This work not only adds to what is known about how deep learning can be used in garbage management, but it also lays the groundwork for the creation of scalable and effective waste classification systems that can make recycling much easier and have less of an effect on the environment.

Acknowledgment

We extend our heartfelt gratitude to ChatGPT and ClaudeAI for their invaluable assistance in refining the language and structure of this research paper. Special thanks to Quillbot and Grammarly for Their contributions that have greatly enriched the quality of this work.

References

[1] S.L. Blum, Tapping resources in municipal solid waste, Science (New York, N.Y.) 191 (4228) (1976) 669–675. Available from https://doi.org/10.1126/science.191.4228.669.

[2] R.S. Stein, Polymer recycling: opportunities and limitations, Proceedings of the National Academy of Sciences of the United States of America 89 (3) (1992) 835–838. Available from https://doi.org/10.1073/pnas.89.3.835.
[3] Z. Xu, A. Elomri, S. Pokharel, Q. Zhang, X.G. Ming, W. Liu, Global reverse supply chain design for solid waste recycling under uncertainties and carbon emission constraint, Waste Management 64 (2017) 358–370. Available from https://doi.org/10.1016/j.wasman.2017.02.024.
[4] W. Schubert, J. Tomas, Liberation of valuables embedded in particle compounds and solid waste, In: Handbook of Powder Technology 12 (2007) 989–1018. Available from https://doi.org/10.1016/S0167-3785(07)12027-3.
[5] D.B. Nirmala, M.S. Sunil Kumar, R. Rajendra Prasad, Feasibility study on mechanical properties of concrete using construction waste as the partial replacement of coarse and fine aggregates, International Journal of Civil Engineering 9 (3) (2022) 24–29. Available from https://doi.org/10.14445/23488352/IJCE-V9I3P103.
[6] A.U. Zaman, A comprehensive study of the environmental and economic benefits of resource recovery from global waste management systems, Journal of Cleaner Production 124 (2016) 41–50. Available from https://doi.org/10.1016/j.jclepro.2016.02.086.
[7] P. Kjeldsen, M.A. Barlaz, A.P. Rooker, A. Baun, A. Ledin, T.H. Christensen, Present and long-term composition of MSW landfill leachate: a review, Critical Reviews in Environmental Science and Technology 32 (4) (2002) 297–336. Available from https://doi.org/10.1080/10643380290813462.
[8] M. Li, J. Liu, W. Han, Recycling and management of waste lead-acid batteries: a mini-review, Waste Management & Research 34 (4) (2016) 298–306. Available from https://doi.org/10.1177/0734242X16633773.
[9] A. Islam, T. Ahmed, M.R. Awual, A. Rahman, M. Sultana, A.A. Aziz, et al., Advances in sustainable approaches to recover metals from e-waste: a review, Journal of Cleaner Production 244 (2020) 118815. Available from https://doi.org/10.1016/j.jclepro.2019.118815.
[10] M.H. Wong, S.C. Wu, W.J. Deng, X.Z. Yu, Q. Luo, A.O.W. Leung, et al., Export of toxic chemicals: a review of the case of uncontrolled electronic-waste recycling, Environmental Pollution 149 (2) (2007) 131–140. Available from https://doi.org/10.1016/j.envpol.2007.01.044.
[11] C. Symeonides, M. Brunner, Y. Mulders, P. Toshniwal, M. Cantrell, L. Mofflin, et al., Buy-now-pay-later: hazards to human and planetary health from plastics production, use and waste, Journal of Paediatrics and Child Health 57 (12) (2021) 1795–1804. Available from https://doi.org/10.1111/jpc.15777.
[12] A.L. Andrady, Microplastics in the marine environment, Marine Pollution Bulletin 62 (8) (2011) 1596–1605. Available from https://doi.org/10.1016/j.marpolbul.2011.05.030.
[13] P. Prajapati, S. Varjani, R.R. Singhania, A.K. Patel, M.K. Awasthi, R. Sindhu, et al., Critical review on technological advancements for effective waste management of municipal solid waste—updates and way forward, Environmental Technology & Innovation 23 (2021) 101749. Available from https://doi.org/10.1016/j.eti.2021.101749.
[14] H. Wang, Garbage recognition and classification system based on convolutional neural network VGG16, Proceedings of the 2020 3rd International Conference on Advanced Electronic Materials, Computers and Software Engineering (AEMCSE), AEMCSE, 2020, pp. 252–255. Available from https://doi.org/10.1109/AEMCSE50948.2020.00061.
[15] J.W. Feng, X.Y. Tang, Office garbage intelligent classification based on Inception-v3 transfer learning model, Journal of Physics: Conference Series 1487 (2020) 012008. Available from https://doi.org/10.1088/1742-6596/1487/1/012008.
[16] A. Altikat, A. Gulbe, S. Altikat, Intelligent solid waste classification using deep convolutional neural networks, International Journal of Environmental Science and Technology 19 (2) (2022) 1285–1292. Available from https://doi.org/10.1007/S13762-021-03179-4.
[17] M. Ragazzi, I. Katsoyiannis, E. Magaril, P. Viotti, H.H. Al-Kayiem, M. Schiavon, et al., Waste classification for sustainable development using image recognition with deep learning neural network models, Sustainability 14 (14) (2022) 7222. Available from https://doi.org/10.3390/SU14127222.
[18] Y. Chu, C. Huang, X. Xie, B. Tan, S. Kamal, X. Xiong, Multilayer hybrid deep-learning method for waste classification and recycling, Computational Intelligence and Neuroscience 2018 (2018) 5060857. Available from https://doi.org/10.1155/2018/5060857.
[19] C. Wang, J. Qin, C. Qu, X. Ran, C. Liu, B. Chen, A smart municipal waste management system based on deep-learning and Internet of Things, Waste Management 135 (2021) 20–29. Available from https://doi.org/10.1016/J.WASMAN.2021.08.028.
[20] M.W. Rahman, R. Islam, A. Hasan, N.I. Bithi, M.M. Hasan, M.M. Rahman, Intelligent waste management system using deep learning with IoT, Journal of King Saud University - Computer and Information Sciences 34 (8) (2022) 2072–2087. Available from https://doi.org/10.1016/J.JKSUCI.2020.08.016.

[21] K.N. Sami, Z.M.A. Amin, R. Hassan, Waste management using machine learning and deep learning algorithms, International Journal of Perceptive Cognition and Computation* 6 (2) (2020) 97–106. Available from https://doi.org/10.31436/IJPCC.V6I2.165.

[22] G.L. Huang, J. He, Z. Xu, G. Huang, A combination model based on transfer learning for waste classification, Concurrency and Computation: Practice and Experience 32 (20) (2020) e5751. Available from https://doi.org/10.1002/CPE.5751.

[23] M. Yang, G. Thung, Classification of trash for recyclability status (n.d.).

[24] A.H. Vo, L. Hoang Son, M.T. Vo, T. Le, A novel framework for trash classification using deep transfer learning, IEEE Access 7 (2019) 178631–178639. Available from https://doi.org/10.1109/ACCESS.2019.2959033.

[25] K.A.P. Saputra, Waste object detection and classification using deep learning algorithm: YOLOv4 and YOLOv4-tiny. Turkish Journal of Computer and Mathematics Education 12(14), 1666–1677. https://doi.org/10.17762/TURCOMAT.V12I14.10504.

[26] N.V. Kumsetty, A. Bhat Nekkare, S. Kamath, M.A. Kumar, TrashBox: trash detection and classification using quantum transfer learning, in: Proceedings of the Conference on Open Innovation Association FRUCT, April 2022, pp. 125–130. https://doi.org/10.23919/FRUCT54823.2022.9770922.

[27] G. Huang, Z. Liu, L. Van Der Maaten, K.Q. Weinberger, Densely connected convolutional networks, Proceedings of the IEEE Conference on Computer Vision and Pattern Recognition (CVPR) (2017) 2261–2269. Available from https://doi.org/10.1109/CVPR.2017.243.

[28] Y. Li, C.Y. Wu, H. Fan, K. Mangalam, B. Xiong, J. Malik, et al., MViTv2: improved multiscale vision transformers for classification and detection, in: Proceedings of the IEEE Conference on Computer Vision and Pattern Recognition (CVPR), 2021, pp. 4794–4804. https://doi.org/10.1109/CVPR52688.2022.00476.

[29] F. Chollet, Xception: deep learning with depthwise separable convolutions, in: Proceedings of the 30th IEEE Conference on Computer Vision and Pattern Recognition (CVPR), 2016, pp. 1800–1807. https://doi.org/10.1109/CVPR.2017.195.

[30] F.N. Iandola, S. Han, M.W. Moskewicz, K. Ashraf, W.J. Dally, K. Keutzer, SqueezeNet: AlexNet-Level Accuracy with 50x Fewer Parameters and 0.5MB Model Size 2016 (2016).

[31] A. Howard, Searching for MobileNetV3 2019 (2019).

[32] K. He, Deep Residual Learning for Image Recognition 2015 (2015).

[33] D. Das, A. Sen, S.M.M. Hossain, K. Deb, Trash image classification using transfr learning based deep neural network, Lecture Notes in Networks and Systems 569 (2023) 561–571. Available from https://doi.org/10.1007/978-3-031-19958-5_53/COVER.

[34] G.E. Veselov, A. Tselykh, B.-G. Kim, S.M. Cheema, A. Hannan, I.M. Pires, Smart waste management and classification systems using cutting edge approach, Sustainability 14 (16) (2022) 10226. Available from https://doi.org/10.3390/su141610226.

[35] M. Farjana, A.B. Fahad, S.E. Alam, M.M. Islam, An IoT- and cloud-based e-waste management system for resource reclamation with a data-driven decision-making process, IoT 4 (3) (2023) 202–220. Available from https://doi.org/10.3390/iot4030011.

[36] H. Zhang, H. Cao, Y. Zhou, C. Gu, D. Li, Hybrid deep learning model for accurate classification of solid waste in the society, Urban Climate 49 (2023) 101485. Available from https://doi.org/10.1016/j.uclim.2023.101485.

[37] K. Lin, Y. Zhao, L. Wang, W. Shi, F. Cui, T. Zhou, MSWNet: a visual deep machine learning method adopting transfer learning based upon ResNet 50 for municipal solid waste sorting, Frontiers in Environmental Science and Engineering 17 (2023) 1–12. Available from https://doi.org/10.1007/s11783-023-1677-1.

[38] Y. Zhao, H. Huang, Z. Li, H. Yiwang, M. Lu, Intelligent garbage classification system based on improved MobileNetV3-Large, Connexions Science 34 (8) (2022) 1299–1321. Available from https://doi.org/10.1080/09540091.2022.2067127.

[39] R.A. Aral, S.R. Keskin, M. Kaya, M. Haciömeroğlu, Classification of TrashNet dataset based on deep learning models, in: Proceedings - 2018 IEEE International Conference on Big Data (Big Data 2018), 2018, pp. 2058–2062. https://doi.org/10.1109/bigdata.2018.8622212.

[40] A. Masand, S. Chauhan, M. Jangid, R. Kumar, S. Roy, ScrapNet: an efficient approach to trash classification, IEEE Access 9 (2021) 130947–130958. Available from https://doi.org/10.1109/access.2021.3111230.

[41] D. Das, A. Sen, S.M.M. Hossain, K. Deb, Trash Image Classification Using Transfer Learning Based Deep Neural Network, in: Vasant, P., Weber, GW., Marmolejo-Saucedo, J.A., Munapo, E., Thomas, J.J. (eds) Intelligent Computing & Optimization. ICO 2022. Lecture Notes in Networks and Systems, vol 569, Springer, Cham, 2023. https://doi.org/10.1007/978-3-031-19958-5_53.

CHAPTER

5

Plastics recycling and the automation role in the recycling process

Munu Borah[1], *and Jyoti Prasad Gogoi*[2]

[1]Department of Physics, The Assam Kaziranga University, Jorhat, Assam, India [2]Department of Physics, DKD College, Dergaon, Assam, India

Introduction

Background

Plastics are so useful, strong, and affordable that they have become essential to contemporary life. But because of its extensive use, serious environmental problems like pollution, resource depletion, and ecological degradation are now illustrated in Fig. 5.1. As a result, there is a growing focus on the necessity of efficient plastic management plans, of which recycling is an essential element. An extensive examination of the environmental effects of plastics over their whole lifecycle, the difficulties associated with plastic pollution, the advantages and restrictions of plastic recycling, and the search for novel ways to improve recycling effectiveness and lower plastic waste are all included in this analysis of the role that plastics and plastic recycling play in environmental protection. Acknowledging that in order to address the plastic challenge and promote sustainable plastic management practices, policy initiatives, public awareness campaigns, and technology breakthroughs are critical.

Purpose of the chapter

Artificial intelligence (AI) is a potent instrument that is transforming processes and decision-making across a wide range of sectors. AI has the potential to significantly improve sustainability, accuracy, and efficiency in the field of plastic recycling. Using machine learning (ML), computer vision, robotics, and data analytics, AI tools are applied to plastic recycling processes, such as sorting, quality control, material identification, and predictive analysis, to address plastic recycling issues and advance the circular economy. This chapter will cover a thorough examination of recent studies and case studies to demonstrate the revolutionary role that AI is playing in promoting environmentally friendly plastic recycling methods.

Harnessing Automation and Machine Learning for Resource Recovery and Value Creation
DOI: https://doi.org/10.1016/B978-0-443-27374-2.00005-4

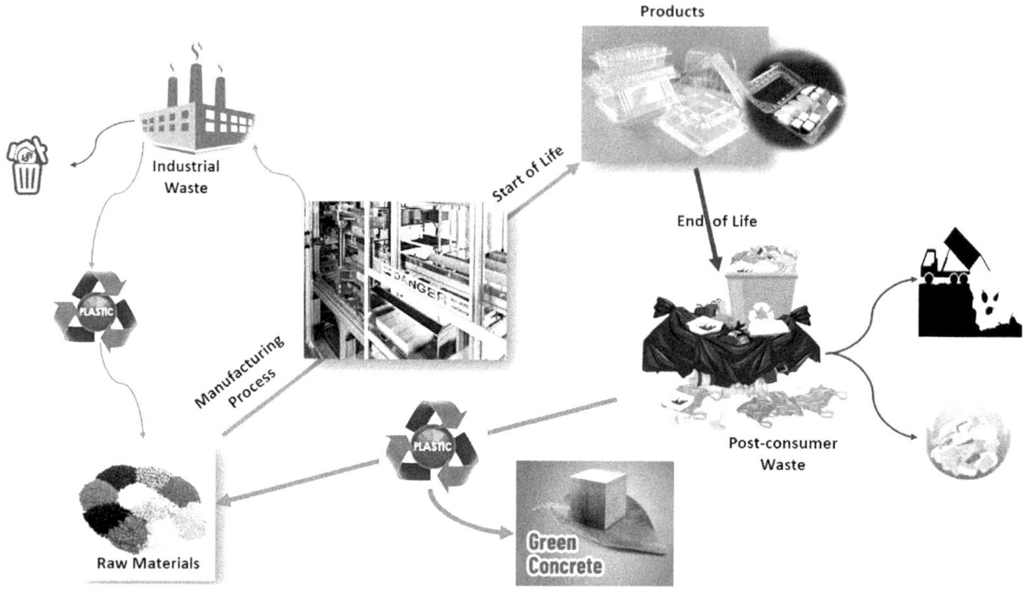

FIGURE 5.1 The lifecycle of polymer-based food trays and packaging [1]. Source: *From H. Mohammadhosseini, R. Alyousef, M. Md. Tahir, Towards sustainable concrete composites through waste valorisation of plastic food trays as low-cost fibrous materials, Sustainability 13 (4) (2021) 2073. https://doi.org/10.3390/su13042073.*

Scope and limitations

The recycling of plastic trash has benefited greatly from automation, which has raised uniformity, efficiency, and scalability. In this particular setting, it is imperative to identify the scope and limitations of automation. Automation can be used to efficiently sort plastic trash according to numerous parameters like type, color, and content. Sensor-equipped optical sorting machines can identify distinct types of plastics, allowing for successful segregation. Processing automation makes it easier to shred, grind, and pellet trash made of plastics. Modern machinery can turn plastic waste into homogenous pellets or flakes that may be recycled back into production. Quality control involves checking recycled plastics for contaminants, and data management offers insights into resource optimization, material utilization, and efficiency. Additionally, automated solutions can reduce the dangers to worker safety that come with human labor in recycling plants. Automation improves workplace safety by reducing the time workers spend on potentially dangerous tasks like sorting and processing. its implementation is limited due to the complexity of sorting, significant initial investment, and regular maintenance costs. However, at the same time, it suffers limitations in implementation due to the complexity of sorting, significant initial investment, regular maintenance cost, adaptability to new materials, and restrictions on use in contaminated waste streams. Although automation has many advantages for recycling plastic trash, such as increased efficiency and safety, it is important to understand its limitations and the need for ongoing investment and innovation to overcome obstacles and improve recycling procedures.

Plastics recycling overview

Types of plastics

Plastics come in a wide variety, each with unique constructions and varying degrees of recyclable nature [2]. Building with recyclable, sustainable materials derived from recycled resources is crucial for an eco-friendly enterprise. In 1988 the Society of the Plastics Industry (SPI) developed a classification scheme to help recyclers and consumers distinguish between various kinds of plastic. Every manufactured plastic product has an SPI code, or number usually molded into the bottom. The most common plastic types with SPI codes fall into the following categories as shown in Fig. 5.2.

Polyethylene terephthalate: (Society of the Plastics Industry Code 1)

Polyethylene terephthalate (PETE or PET) is the common plastic material used for food packaging, pharmaceutical containers, and bottles of water and soft. It is readily recyclable and is frequently made into carpets, fleece, and stings. Its qualities include a 255°C melting point, strong heat resistance, toughness, and solvent resistance. In general, it doesn't release any chemicals into the food or beverages that are kept inside. On the other hand, prolonged use may cause it to release harmful substances.

High-density polyethylene: (Society of the Plastics Industry Code 2)

High-density polyethylene (HDPE) is safe plastic since it emits no hazardous substances. Shampoo, detergent, and motor oil containers are frequently made using it. It is easily recyclable into furniture, plastic timber, and new HDPE bottles. Transparent, flexible, and durable, HDPE plastic frequently has a waxy texture. Its melting point is between 120°C and 180°C, and it has a strong heat resistance and solvent resistance. This implies that it can be repeatedly sterilized. It doesn't release any toxins that could harm people's or animals' health.

Polyvinyl chloride: (Society of the Plastics Industry Code 3)

Polyvinyl chloride (PVC) is mostly utilized in pipes and tiles. It can be recycled and utilized to make walls, flooring, and materials of an industrial grade. It cannot be recycled. Reusable, but not recommended for use with food or by little children. PVC is a white,

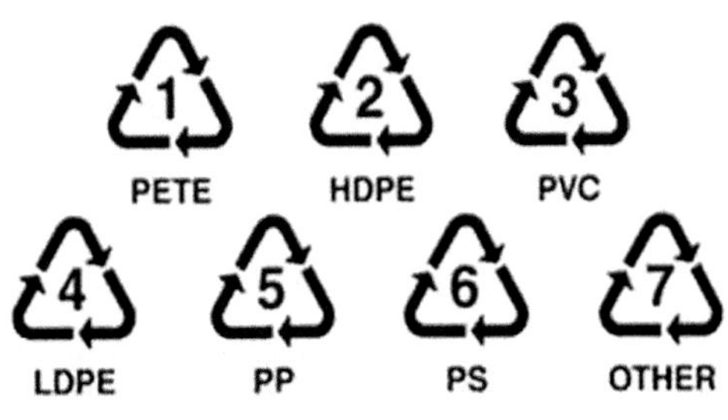

FIGURE 5.2 Plastic types with Society of the Plastics Industry codes.

brittle, long-lasting plastic that is gradually replacing conventional building materials. Melting begins at 100°C, and heat distortion can begin as early as 60°C. It has significant concentrations of hazardous substances that, if consumed for prolonged use, can kill living organisms.

Low-density polyethylene: (Society of the Plastics Industry Code 4)

Low-density polyethylene (LDPE) is a flexible, long-lasting plastic used to make Ziploc bags and is safe for storing food. Although LDPE is not frequently recyclable, it can be used to make furniture, garbage cans, and other everyday home items. More recycling programs are starting to accept LDPE plastics, which can be used to make waste bins, flooring, and bubble wrap, among other things. Processing it is simple, powerful, and resilient. It has a low point of melting. It is extremely secure and not known to generate any dangerous substances.

Polypropylene: (Society of the Plastics Industry Code 5)

Polypropylene (PP) is the plastic material that makes up Tupperware, stadium cups, and prescription bottles. Compared to materials like HDPE or PETE, it is not as commonly accepted for recycling; nonetheless, it can be utilized for sturdy products like rakes and ice scrapers. It is exceptionally resistant to heat, chemicals, oil, and grease, robust and hardy, with a high melting point. Extremely secure and not known to release any dangerous substances.

Polystyrene: (Society of the Plastics Industry Code 6)

The structure of polystyrene (PS), or Styrofoam, cannot tolerate recycling, and the substance becomes more poisonous when heated and treated with liquids. Although recycling Styrofoam takes a lot of energy, the material may be utilized to make items like license plate frames, rulers, and insulation. It is affordable and multipurpose, a good barrier against heat. Simple to form. Glassy texture, elevated melting point. It has been demonstrated to impair neuronal function and induce illness.

Other plastics: (Society of the Plastics Industry Code 7)

Any plastic that doesn't fit into one of the preceding six categories—such as polycarbonates (PCs) and polylactides, which are difficult to recycle—is classified as SPI Code 7. The most frequent products made of these materials are infant bottles, storage containers, and compact disks. Plastic timber can be made from the other recyclable materials in this group. Very seldom is it recycled. Verify whether the product has the polylactic acid code next to the number; if so, it can be composted. Its detrimental effects on the endocrine system have been demonstrated. Making a note of the SPI codes of the anticipated plastic materials might help ensure that environmentally friendly, sustainable materials are used.

Environmental impact of plastics

Plastics are ubiquitous and tenacious in the environment; reports of their presence reach the highest and most distant mountain summits and the deepest recesses of the ocean. Numerous studies have examined the ecotoxicological and broader ecological effects of plastic consumption and accumulation, in addition to the sources, breakdown, and transportation of plastics. Numerous effects are linked to the buildup of plastic trash (macro, micro, and nanoplastics) in the environment. The implications can be broadly classified into three categories: (1) effects on the environment's fauna and flora, (2) economic ramifications, and (3) effects on people's health and well-being [3]. Impacts on fauna and flora within the environment: It can impede an organism's capacity to breathe, move, and feed. When animals eat plastic, it can cause obstructions in their digestive tracts, which can cause false satiation. These are only a few of the physical effects. Plastic-related impacts could operate as a conduit for potentially hazardous contaminants. By being consumed at various trophic levels, these substances may bioaccumulate up the food chain. Reduced physical size, decreased nutrition, and changed blood chemistry have all been related to exposure. Additional effects may include faster melting rates of polar snow and ice, warmer soil and less suitable habitat, a higher risk of disease in coral reefs, easier movement of invasive species between habitats, and sediment deposition that could affect animals.

Loss of associated ecological services and biodiversity worldwide. Decrease in the value of natural capital: less provision of surface water, less control over flooding, and less climate regulation, effects on food availability, the detoxification of waste, and the advantages of culture, especially leisure time, effects on fisheries, tourism, aquaculture, and agriculture. Impacts on health and well-being: Litter can jeopardize food security, livelihoods, income, and overall health, as well as the psychological advantages that the environment offers. Human feces, human placenta, human breast milk, human gut microbiota cultured in vitro, and human Caco-2 cells have all been found to contain plastic. There is strong evidence that the chemicals used in the manufacturing of plastics, as well as long-term exposure during the process and uncontrolled waste disposal techniques such as open burning, can harm human health. It is unclear, therefore to what degree environmental plastic waste poses toxicological risks. Plastics have many beneficial effects on our civilization. Plastics can raise living standards, lessen suffering, and lessen our human imprint on the environment when they are used wisely. The problem can be solved by combining approaches, including reduction, reusing, and recycling with technological innovations.

Need for recycling

The use and disposal of plastic at current levels causes several environmental issues. About 4% of the nonrenewable resource produced worldwide—oil and gas—is used as feedstock for plastics, and an additional 3%–4% is utilized to generate energy for the process of making them. Most plastic generated annually is used to create packaging that is thrown away after a year of usage or other short-lived goods. Furthermore, large amounts of wasted end-of-life plastics are building up as garbage in landfills and natural environments around the world due to the polymers' endurance. One of the most significant steps that can be taken to lessen these effects is recycling, with the most prominent reasons including environmental preservation, resource conservation, resource conservation, reduction of pollution, economic benefits and community engagement [2]. Recycling plastics keeps plastics out of landfills, and less garbage enters the environment, protecting ecosystems and natural habitats. Since plastics are made from nonrenewable resources like crude oil and natural gas, recycling plastics could help protect these priceless resources. Generally speaking, recycling plastic uses less energy than creating new plastics from raw materials. Pollution from plastic waste in the environment can damage marine habitats and wildlife. Recycling lessens the quantity of plastic debris that ends up in the ocean and waterways, which helps to alleviate environmental pollution. Industries involved in the gathering, sorting, processing, and manufacturing of recycled materials benefit from recycling. Recycling initiatives promote community involvement and increase knowledge of the value of resource conservation and trash reduction. The typical stages involved in plastic recycling are shown schematically in Fig. 5.3

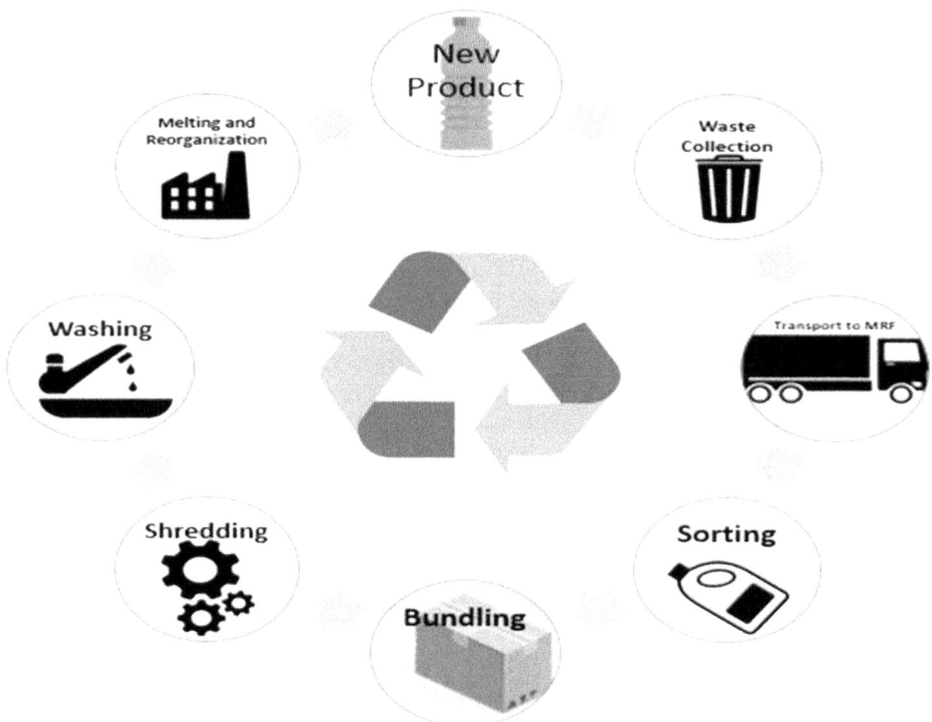

FIGURE 5.3 The typical stages involved in plastic recycling [4]. Source: *From S. Khadke, P. Gupta, S. Rachakunta, C. Mahata, S. Dawn, M. Sharma, D. Verma, A. Pradhan, A.M.S. Krishna, S. Ramakrishna, S. Chakrabortty, G. Saianand, P. Sonar, S. Biring, J.K. Dash, G.K. Dalapati, Efficient plastic recycling and remolding circular economy using the technology of trust–blockchain, Sustainability 13 (16) (2021) 9142. https://doi.org/10.3390/su13169142.*

Current state of plastics recycling

Collection and sorting

Polyolefins (POs) typically consist primarily of polymers made from resources like ethylene, propylene, 1-butene, and other olefin monomers. PP and Polyethylene (PE) are the two POs that are most frequently utilized. HDPE, LDPE, and linear low-density polyethylene are all members of the PO family of polymers. Plastic polymers are employed to make consumer goods and are utilized in many other applications to create synthetic fibers, foams, coatings, adhesives, and sealants. When a product (such as plastic packaging, used electronics, window frames, etc.) is thrown away by the end user after it has fulfilled its intended function and is no longer useful, it is called postconsumer plastic trash. Plastic garbage is created in large numbers due to the short lifespan of many plastic items and materials. 40% of plastic products are believed to endure less than a month. Some products, like filters, have an active lifespan of just 1–2 years, but other products, like automotive parts, have an active lifespan of 10 years or more. When it comes to plastic components, there are two different ways to look at them: the first is that certain items are made up of only one component. The second is that some products comprise a system of multiple components (like an entire automobile). Plastic materials are more challenging to recycle in the latter scenario [5]. In postconsumer waste (PPW), plastic types include PET (such as soda or water bottles); PP (such as meal trays for microwaves, ice cream containers and bottles for cleaning products for the kitchen and bathroom); PE (such as milk or most shampoo bottles); film (such as carrier bags and packaging foils); a combination of hard plastics (such as PS and PVC; nonbottle PET; and falsely sorted PE, PP, and PET). Apart from PVC, the majority of the remaining five polymers (DPE, LDPE, PP, and PET), are predominate in the composition of PPW. It is feasible to examine the kind of polymer and the elemental makeup of the separate parts of various plastic packaging, with the most prevalent elements being C, H, N, S, and O, but also metals and halogens. Due to the multiple immiscible polymer types that makeup PPW and the product designs (such as color and polymer separability), significant physical losses may occur throughout the sorting process. The most often used additives in polymeric packaging are flame retardants, plasticizers, acid scavengers, antioxidants, thermal, light, heat stabilizers, antistatic agents, pigments, lubricants, and slip compounds. Additives are crucial since they improve a plastic product's final functional qualities and provide them. The functional additives (stabilizers, antistatic agents, flame retardants, plasticizers, lubricants, slip agents, curing agents, foaming agents, biocides, etc.), colorants (pigments, soluble azo-colorants, etc.), fillers (mica, talc, kaolin, clay, calcium carbonate, barium sulfate, etc.), and reinforcements (carbon fibers, glass fibers, etc.) are the different categories of additives that can be identified.

Challenges in collection

Source separation and postseparation are the two methods of collecting postconsumer plastic. Plastics should be separated from other waste by households prior to collection in source separation, and waste should be separated after collection, typically at treatment and recovery centers, in postseparation. According to reports, there are several ways to collect postconsumer plastic garbage. Two common methods are mixed collection with municipal solid refuse waste (MSRW) or independent collection of waste. However, there are challenges to be overcome in collecting the waste plastics [6]: Ensuring frequent pickups, scheduling, and route planning are necessary to establish an effective collecting system. This can be especially difficult in isolated or crowded places. Many areas lack the infrastructure required for effective garbage management. This covers recycling facilities, garbage collection facilities, and transportation networks. This infrastructure can be expensive and time-consuming to build and maintain. It is essential to encourage people and communities to dispose of their plastic garbage responsibly. However, education, awareness-raising efforts, and incentives are needed to change people's attitudes and behaviors toward trash disposal. When plastic garbage is contaminated with other substances, such as food or chemicals, it becomes less recyclable and requires more money to process. To reduce contamination, proper sorting and segregation at the source are crucial. There might not be as many financial incentives in other places for gathering and recycling plastic debris. Businesses and individuals may be less inclined to participate in rubbish collection programs if the right incentives are not provided. Pollution and ecological damage can result from collection techniques that are not environmentally sustainable, such as careless dumping or cremation. To lessen these effects, ecologically friendly collection techniques must be used. Globally, there is an enormous amount of plastic waste produced, and gathering it all is extremely difficult. To properly solve this issue, cooperation between nations and regions is required. Inadequate or badly implemented waste management regulations may make collection operations more difficult. Robust regulations that penalize inappropriate disposal of waste, prohibit the use of single-use plastics, and encourage recycling are required.

Researchers have studied different schemes of collecting plastic waste. Plastic garbage is gathered using either mixed collection with MSRW or separate collection. Mechanical recycling is then often completed. When plastic packaging garbage is separated at the source, it is collected by curbside services or drop-off containers. Curbside collection carried out in tandem with a bring collection was examined by Pires et al. [7]; they dubbed this method a mixed collection system. Comparing this approach to other collection systems, the economic metrics it shows are superior. The authors did note, however, that curbside collection was not optimized within the system. Martinho et al. [8] discovered that because the previous system included a curbside component, the mixed collection system produced lower contamination rates and higher rates of material separation and recycling than the drop-off-only system. Curbside collection operates less efficiently in the mixed system than it does in the drop-off collection and exclusive drop-off system. Postseparation is more convenient for citizens (fewer containers), less infrastructure is needed for collection, and it is technically straightforward, as demonstrated by Cimpan et al. [9]. It is difficult to compare garbage collecting systems, in part due to the wide range of performance measures utilized in the literature. Gallardo et al. [10] employed the fractioning rate, separation rate, and quality in container rate as metrics to assess the level of efficiency of four Municipal Solid Waste (MSW) collection systems. Within a five-fraction collection system—which includes mixed and organic waste placed in curbside bins, as well as paper/cardboard, glass, and light packaging collected from drop-off locations—the chosen waste was maintained in optimal condition from the point of separation in homes to the recycling facilities. The

Hahladakis group [11] examined three primary collection schemes: curbside collection, bring sites/banks (BSs), and household waste recycling centers (HWRCs), commonly referred to as "civic amenity sites." Packaging plastics recovery in curbside collection was higher than in HWRCs and BSs, with weight percentages (wt.%) of 90%, 9%, and 1%, respectively. Curbside collection yielded a better return when mixed plastic recyclables were collected alternate weeks in wheeled garbage cans. Reprocessors received only 16% of the mixed plastic bottles and mixed plastics. According to Chruszcz and Reeve's [12] estimation, pots, tubs, trays (27.2%), film (15.7%), and bottles made up half of the curbside collected plastics (50.8%). Toys and pipes, for instance, are classified as nonpackaging plastics and makeup 4.4% of all plastics. Approximately 1.9% of the materials gathered could not be identified. For recycling, three different kinds of hard plastic polymers were gathered: PET, HDPE, and PP. With 40.3% of the total composition, PET was the most common among them, followed by HDPE (21.6%) and PP (10.2%). Other forms of polymers, such as PS (1.5%), Expanded Polystyrene (EPS) (0.4%), and PVC (0.1%), made up very tiny portions of plastics; black plastics accounted for 3.7%.

Sorting technologies

Recyclables can be recovered using two very different methods: separation at the source (in individual homes) and separate collection systems, or recovery in centralized facilities that process and sort mixed residual waste mechanically in order to handle a bigger waste stream. For recycling material, there are three setups for collection and sorting. The first is mixed municipal garbage sorting, which mainly entails producing a significant proportion of fuel obtained from refuse in addition to recycling metals, plastics, and glass. Mixed dry recyclables, or the sorting of paper, plastic, glass, and metal for use or additional sorting, is the name given to the second configuration. The third, known as source-separated recyclables, involves producing various amounts and quality of source material by fine-sorting individual material fractions using various sorting processes. Waste management and recycling systems depend heavily on centralized sorting and separation systems, such as material recovery facilities (MRFs), package sorting facilities, and mechanical–biological treatment plants. The centralized sorting of recyclables frequently found in household garbage was assessed by Cimpan et al. [9]. The methods created for separating mixed recyclables from different collections have also been effectively applied to the improvement of recycling facilities for residual trash. In addition to source separation, recyclables, polymers, and metals are recovered from residual waste. The centralized sorting of residual trash is pertinent in regions where source separation poses challenges. It has the potential to serve as a supplement to source separation or a replacement for the separate collection of certain items, including metals and plastics. Centralized sorting will undoubtedly play an increasingly bigger role in the future given the increased complexity of materials, the fast urbanization that is occurring, which complicates conventional collection systems, and the expanding need for secondary raw materials worldwide. Evidence suggests that recycling plastic trash becomes more cost-effective when postseparation is selected, as demonstrated by Dijkgraaf and Gradus [13]. 25% of the collected home plastics are still being used for energy recovery, according to the 75% recycling rate for these mixed plastics. Postseparation has the benefit of not requiring separation of these "contaminated" and eventually burned polymers, while building a suitable plant requires a significant financial outlay. It is recommended to combine postseparation with a decrease in the quantity of mixed plastics. The technical viability of packaging waste fractions straight from mixed municipal waste has been examined in a number of extensive trials. The recovery of PET bottles (three colors), HDPE, and occasionally PP has been the main emphasis of current food packaging sorting systems, which are mostly based on near-infrared (NIR) technology [14]. In order to collect as many mono-materials as possible from the extended street collection, which is mostly implemented in Belgium, sorters who receive trash mixed plastics must separate the waste into 11 distinct parts. Chruszcz and Reeve [12] examined the plastic waste generated by Plastic Reclamation Facilities (PRFs) and MRFs. According to estimates, the MRFs and PRFs generated 72.2% of the output in the form of bottles (MRF only), 13.2% of rigid packaging (MRF only), and 7.4% of plastic films (MRF only). 3.7% of nonplastic components and 1.0% of fines were also discovered in the output. For PET bottles with jazz, the percentage of nonplastic materials was zero; for mixed plastics, it was 9.6%. Certain material fractions, including 1.5% nonpackaging plastics (varying from 0% for the PET bottle jazz outputs to 9.8% for the pot, tub, and tray outputs), could be more challenging to process. To separate garbage into different materials, two technical methods are used: human sorting and automatic/mechanical sorting. Since some steps in the process are best completed by hand, sorting facilities typically combine automatic and manual sorting. However, targeted, safer, and less expensive automation is advantageous for other phases. For example, different techniques, such as waste screening (disk, trommel, or oscillating screen), air separation (rotary, zigzag, and crossflow classifiers, or suction hood), ballistic separation, magnetic sorting eddy current separation, film grippers, sensor technology (NIR), visual spectrophotometry (VIS), X-ray fluorescence (XRF), X-ray transmission, electromagnetic sensor, far-infrared (FIR), and manual sorting depend on the type of waste stream. Combinations of these technologies are common [15]. Numerous types of robots have been documented in the literature: stationary robots that are utilized in MFRs to detect and sort waste, as well as mobile robots that can be used to collect garbage in difficult situations (like beaches). Robotic arms, grippers, RBG cameras, actuators, proximity sensors, tracks, track belts encircling wheels, a conveyor to transfer gathered debris, and more can be added to mobile robots (Fig. 5.4). The most urgent issues in polymer sorting and recycling, aside from handling metal contaminants in sorting, are the separation of packaging and composite films. To make matters worse, PET trays and beverage cartons cannot be recycled. Therefore it is necessary to enhance both the methods for sorting and separating these fractions as well as their recycling routes [6]. AI techniques and digitization can have a significant impact on how well plastic garbage is sorted. According to Chidepatil et al. [16], research is presently being done on a blockchain platform for a circular economy integrating plastic trash as well as multisensor-based AI for plastic waste segregation. The multisensor-based separation procedure makes use of three different kinds of sensors. Their operating frequencies determine which category they fall into: VIS, NIR, and FIR. These sensors record several kinds of data (attributes) regarding waste plastic, including color, composition (both chemical and physical), shape, and so on. Closing the loop on the life cycle of plastic items is feasible when combined with blockchain technology, which acts as a trust-based platform between recyclers, purchasers of recovered raw materials (manufacturers), and separators of plastic waste. AI-powered robot sorting can take the role of manual sorting. The application of an AI-based sorting method for mixed MSW was presented by Wilts et al. [17]. They demonstrated that it is possible to achieve up to 100% purity of the sorted fractions and separated materials based on practical data, which is much beyond early expectations. Waste recovery, however, did not work out. There is compelling evidence that substantial benefits could be achieved by further optimising and adjusting a real waste treatment plant's operating conditions, integrating the growing digitization of waste management in all of its forms, and taking digitization into account from more than just material recovery rates [15]. Physical and computational infrastructures are becoming more coordinated because to emerging infrastructure and the potential provided by blockchain technology, the Internet of Things (IoT), and cyber-physical systems. Using distinctive fluorescent markers (tracers) to label resins composed of polymers is one of the potential strategies for reaching high sorting purity. The markers are added to fresh resins in the manufacturing process at ppm quantities. The levels are just high enough to be picked up by devices that use fluorescence emissions to detect them. The polymers' mechanical characteristics and optical appearance are unaffected, nevertheless.

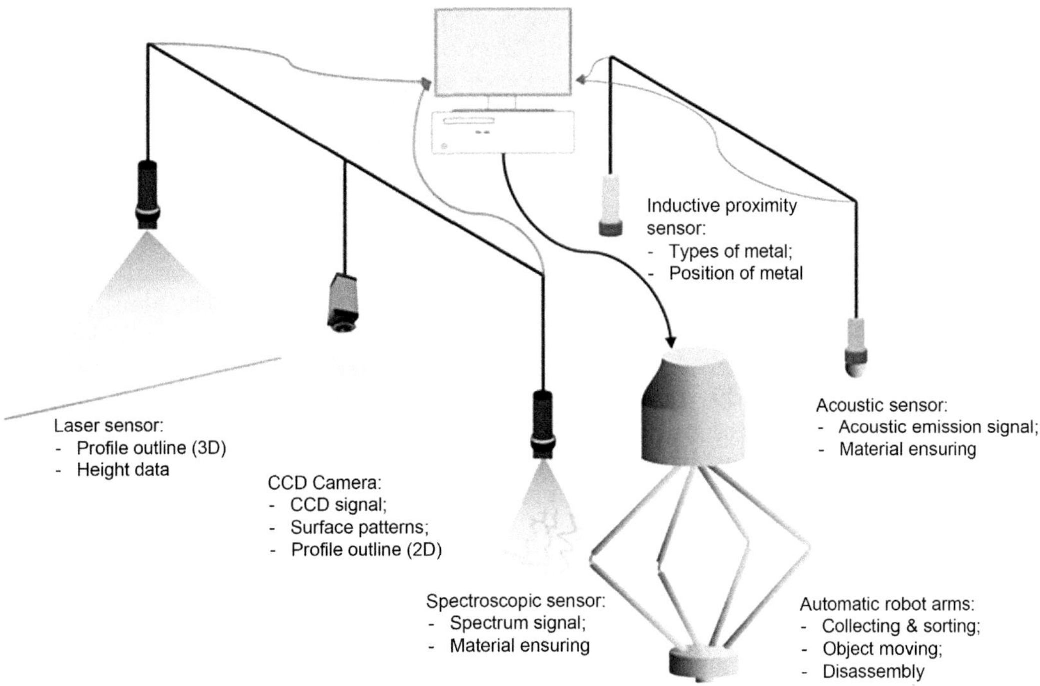

FIGURE 5.4 Schematic of different types of sensors used in robots to sort plastic waste [18]. Source*: From C. Lubongo, M.A.A. Bin Daej, P. Alexandridis, Recent developments in technology for sorting plastic for recycling: the emergence of artificial intelligence and the rise of the robots, Recycling 9 (4) (2024). https://doi.org/10.3390/recycling9040059.*

Processing and reprocessing

Various stages are involved in processing and reprocessing plastic recycling, from product development to collecting. Reprocessing denotes the use of recycled plastic pellets in the production of new goods, frequently combined with virgin materials to improve their qualities. Collecting and sorting, cleaning and shredding, melting and reprocessing, etc., are important steps and techniques that involve

Mechanical recycling

Mechanical recycling is a key method of plastic recycling that involves the physical processing of plastic waste into new products [1]. Mechanical recycling involves physical processes such as sorting, crushing, washing, melting and pressing to convert plastic waste into recyclable materials. The first step in mechanical recycling is to sort plastic waste into different categories according to polymer type, color and other properties. This can be done manually or using automated sorting techniques such as optical sensors, magnets and air sorters. The aim is to separate the different types of plastic to ensure the purity and quality of the recycling process. After sorting, plastic waste is crushed into smaller pieces using shredders or granulators. Crushing increases the surface area of the plastic material, making it easier to process and handle in later stages. Depending on the desired end product, shredded plastic is usually reduced to small flakes or pellets. The crushed plastic flakes or pellets are then washed to remove dirt, debris, tags, adhesives, or food residues. Washing is usually done with water and detergents in special equipment called washing lines. Purity is crucial to ensure the quality and purity of the recycled plastic material.

Washed plastic flakes or pellets are melted in an extruder or melting furnace. Heating the plastic above its melting point allows it to melt and flow. This molten plastic can then be molded into different shapes using molds or extrusion dies. Melting helps homogenize the plastic material and remove any impurities or moisture. In the re-extrusion step, molten plastic is pressed through a die to form new plastic products such as sheets, films, fibers or profiles. Extrusion is a common manufacturing process used to produce various plastic products with uniform dimensions and properties. Recycled plastic material can be mixed with virgin plastic or additives to meet certain performance requirements. Different chemical properties and properties of different plastics viz. PET, HDPE, LDPE, PP, PVC require special processing techniques in mechanical processing. PET is commonly used in beverage bottles, food packaging, and polyester fibers. Processing involves sorting, shredding, washing, and melting. However, due to its sensitivity to heat, PET may undergo solid-state polymerization to improve its molecular weight before reprocessing. Mechanical recycling of PET often results in recycled PET, which can be used in new bottles, fibers for textiles, and other packaging applications. HDPE is widely used in milk jugs, detergent bottles, and plastic pipes. HDPE is relatively easier to recycle mechanically than other plastics due to its high melting point and chemical resistance. Processing typically involves sorting, shredding, washing, and melting. HDPE can be melted and re-extruded into various products such as bottles, pipes, and containers. LDPE is commonly found in plastic films, grocery bags, and squeeze bottles. Mechanical recycling of LDPE is challenging due to its low melting point and tendency to degrade during processing. Techniques such as aggregation (melting and compression) or compatibilization (blending with other polymers) may be employed to improve the processability of LDPE in mechanical recycling. PP is used in packaging, automotive parts, and consumer goods. PP has a high melting point and good chemical resistance, making it suitable for mechanical recycling. Processing involves sorting, shredding, washing, and melting. Recycled PP can be used in various applications, including packaging, automotive components, and household products. PVC is used in pipes, window frames, and flooring. Mechanical recycling of PVC is challenging due to the presence of additives such as plasticizers and stabilizers, which can affect the material's properties. PVC recycling often involves specialized processes to remove contaminants and additives before shredding, washing, and melting. The specific processing techniques may vary depending on each plastic material's chemical properties, characteristics, and end-use applications. Adapting recycling processes to suit the unique requirements of different plastics is essential for maximizing material recovery and ensuring the quality of recycled products. Advantages of mechanical recycling are: Reduces the demand for new plastic production, saving natural resources and energy. Helps divert plastic waste from landfills and reduces environmental pollution. Provides financial benefits by creating opportunities for material reuse and reducing disposal costs. Challenges related to mechanical recycling of plastic are: Pollution: Impurities such as food scraps, dirt or other materials can affect the quality of recycled plastic. Degradation: Mechanical recycling can cause plastic degradation due to repeated use, limiting the recycled material's quality. Sorting complexity: Sorting plastics by type and color can be tedious and difficult, affecting the efficiency of the recycling process. Applications of mechanically recycled plastics are: Recycled plastics from mechanical recycling are commonly used in various applications such as packaging, construction materials, automotive components, textiles and consumer goods. The quality and suitability of recycled plastic for various applications depends on factors such as cleanliness, strength, appearance etc. Advances in sorting technology, material recovery techniques and process optimization should improve the effectiveness and efficiency of mechanical recycling of plastic materials. Increasing consumer awareness and government regulations on promoting the recycling of plastic and the use of recycled materials will likely stimulate the growth of mechanical recycling in the near future.

Chemical recycling

Chemical recycling, also known as advanced recycling or raw material recycling, is a process that breaks down plastic waste into molecular components or turns it into raw materials to make new plastics, fuels or other chemicals [19]. The processes involved in chemical recycling can be discussed below.

Chemical recycling typically involves depolymerization, a process in which the long polymer chains of plastics are broken into smaller molecules or monomers. This can be achieved by various methods such as pyrolysis, gasification, solvolysis, hydrolysis or enzymatic degradation. Conversion to raw materials: monomers or other chemical intermediates obtained after depolymerization can be used as raw materials for producing new plastics, chemicals or fuels. Depending on the particular method, purification steps may be required to remove impurities and isolate the desired products from the reaction mixture. Chemical recycling involves pyrolysis, gasification, solvolysis and hydrolysis. During pyrolysis, plastic waste is heated without oxygen to break down the polymers into smaller molecules, usually in the form of gases, liquids (pyrolysis oil), and carbon. These products can be further processed to obtain fuels or chemical raw materials. Gasification involves the conversion of plastic waste into synthesis gas (Syngas) consisting of hydrogen and carbon monoxide through a high-temperature reaction with controlled amounts of oxygen and/or steam. Natural gas can be used to produce chemicals, fuels or electricity. Solvolysis uses solvents or supercritical fluids to break down plastic polymers into smaller molecules. This method is particularly suitable for plastics that are difficult to recycle using traditional mechanical methods. Hydrolysis involves breaking down plastics using water and heat to break the polymer chains into monomers or oligomers. This process is often used to recycle polyester-based plastics such as PET.

Advantages of chemical recycling of plastics

Chemical recycling can handle mixed plastics and contaminated plastic waste, providing a solution for materials that are difficult to recycle mechanically. This recycling processes can achieve high conversion yields and recycle a variety of valuable products from plastic waste. Chemical recycling enables a closed approach to plastic recycling, turning plastic waste into raw material to produce verities of new plastic materials for different applications.

Challenges related to chemical recycling of plastics

Many chemical recycling processes require high temperature and (or) pressure, making them energy-intensive and potentially less environmentally friendly than mechanical recycling. Chemical recycling technologies can be capital-intensive and require significant infrastructure and treatment facilities investment. Ensuring the quality and purity of recycled products can be difficult, especially with complex chemical reactions where side reactions and impurities can occur. The main applications of chemical recycling of plastic materials are: chemical recycling can produce raw materials for producing new plastics, including virgin and recycled plastics. It can also produce fuels or chemicals that can be used in various industrial processes or as a source of energy. Technology and process optimization advancements can improve chemical recycling methods' efficiency and profitability. Increasing regulatory pressure and consumer demand for sustainable solutions will likely increase investment and development of chemical recycling technologies. In general, chemical recycling offers a complementary approach to mechanical recycling and offers a solution to plastic waste that is difficult to recycle using traditional methods. Although ongoing research and innovation face challenges such as energy intensity and cost, they promise the continued development of chemical recycling as a sustainable solution for plastic waste management.

Automation in plastics recycling

Shredding and size reduction

Automating the shredding and size-reduction processes of plastic recycling has significant potential to increase efficiency, reduce costs and improve the overall sustainability of the recycling industry. Automation can become revolutionizing in this regard. Automation streamlines the crushing and size-reduction process, enabling continuous operation without constant human interference. This results in higher throughput and shorter processing times, ultimately increasing the overall efficiency of the recycling operation. Automated crushing systems can precisely control factors such as particle size and consistency, resulting in more consistent material in output. Consistent quality is critical in downstream processes such as extrusion or injection molding, where differences in material properties can affect product quality. By automating shredding and size reduction operations, recycling companies can reduce their reliance on manual labor, which is often one of the most significant costs of the recycling process. Automated systems can be programmed to optimize resource use by adjusting disintegration parameters based on feed material type and properties. This ensures efficient use of energy and other resources, resulting in cost savings and environmental benefits. Automatic crushing systems can be integrated with advanced sorting technologies such as NIR sensors and optical scanners. These technologies make it possible to identify and separate different types of plastic based on their chemical composition or physical properties, further increasing the recycling process's efficiency and effectiveness. Automation enables real-time shredding and size reduction monitoring and provides valuable information on factors such as processing speed, energy consumption and equipment performance. This information can be analyzed and open further optimization and identification of continuous improvement in the recycling process. Automatic crushing systems are designed with safety features to minimize the risk of accidents and ensure compliance with relevant regulations and standards. By reducing the need for manual procedures for potentially hazardous tasks, automation helps create a safer work environment for recycling plant workers. Automatic crushing systems can be designed to meet a wide range of feed materials and processing requirements, making them highly scalable and adaptable to changing market needs. This flexibility is also essential for recycling companies to remain competitive and meet changing customer needs.

Sorting and separation

Automating sorting and separation processes is key to improving the efficiency and effectiveness of plastic recycling. The different concerns of automation in sorting and separation of plastic recycling are discussed as follows: Automated sorting systems utilize advanced technologies such as NIR sensors, optical scanners, and AI algorithms to accurately identify and separate different types of plastics based on their chemical composition, color, and other characteristics. This level of precision ensures that each type of plastic can be properly sorted for recycling, maximizing the value of the recovered materials. Automated sorting systems can quickly process large volumes of material, greatly increasing throughput compared to manual sorting methods. This allows recycling facilities to process larger amounts of incoming plastic waste, resulting in higher overall recycling rates and better use of resources. By automating the sorting process, recycling facilities can minimize the risk of pollution from mixed plastics or nonrecyclables. Automated systems can quickly and accurately identify contaminants and remove them from the recycling stream, ensuring that the recycled plastic meets quality standards for further processing. Automatic sorting systems enable the recovery of a wider range of plastics, including more complex materials such as black plastic or multilayer packaging. By effectively sorting these materials, recycling facilities can increase the overall yield of high-quality recycled products and maximize the value of recovered resources. Although the initial investment in automated sorting technology can be significant, it can result in long-term savings due to improved efficiency and reduced labor costs. Automated systems require less manual intervention and inspection, allowing recycling facilities to allocate resources more efficiently and achieve economies of scale. Automated sorting systems can have sensors and monitoring devices that provide real-time information on key performance metrics such as sorting efficiency, material cleanliness and equipment availability. Automatic sorting systems can be configured to handle a variety of plastic materials, shapes and sizes, making them highly adaptable to changing raw material composition and market demands. Automatic sorting systems can be seamlessly integrated with downstream processes such as crushing, washing and granulation, creating a fully automated recycling line from sorting to final product. Hence, the automation of sorting and separation processes is necessary to promote the circular economy and increase the sustainability of plastic recycling. By utilizing advanced technologies and data optimization, recycling facilities can achieve higher performance, lower operating costs and produce higher-quality recycled products, ultimately contributing to an efficient and greener recycling ecosystem in the future.

Automation in washing and cleaning

Automation in washing and cleaning is crucial to improve the overall process of plastic recycling as follows: automated washing systems can provide consistent cleaning quality by precisely controlling parameters such as temperature, pressure, and detergent concentration. It can help remove dirt, residues, and impurities from recycled plastic, resulting in higher-quality recycled materials suitable for various applications. Automation allows washing and cleaning processes to be performed continuously and quickly, significantly increasing performance compared to manual or semiautomatic methods. This allows recycling facilities to process larger amounts of plastic waste efficiently, resulting in higher overall recycling rates and resource utilization. Automatic washing systems can be designed to optimize the use of water and chemicals and minimize the consumption of resources and the generation of waste. This can significantly save the environment by precisely controlling the amount of water and detergent required to clean recycling plants. Automatic washing systems can incorporate energy-efficient technologies such as heat systems, speed fluctuations, and insulation measures to minimize energy consumption during cleaning. Automation reduces the need for manual labor, as discussed earlier. Enable real-time process monitoring and optimization. Automatic washing systems can seamlessly integrate downstream sorting and separation processes, creating a fully automated recycling line from sorting to the final product. Automated washing systems can help recycling facilities comply with environmental regulations and industry standards, ensuring that recycled plastics meet quality and purity standards. Automatic washing systems can be designed to handle a variety of plastic materials, shapes and sizes, making them highly adaptable to different compositions and recycling requirements. This flexibility allows recycling companies to efficiently process different types of plastic waste and respond quickly to changes in market demand.

Automation in drying process

Automated drying systems remove excess moisture from washed plastics to prepare them for further processing. Proper drying is essential to prevent defects in recycled plastic products and maintain material quality. Modern drying systems are facilitated with advanced sensors and control systems. These sensors monitor real-time parameters such as temperature, humidity, airflow, and humidity. Advanced control algorithms dynamically adjust drying parameters to optimize performance and energy efficiency ensuring consistent drying results. Energy-efficient drying methods have become increasingly important. The development of insulation materials, heat exchange systems, and heat recovery technologies helps to minimize energy consumption and reduce operating costs. Some systems use waste heat from other processes in a recycling plant or use renewable energy sources to improve sustainability. Advanced drying systems can use adaptive algorithms that adapt to changing raw materials, environmental conditions, and production requirements. These algorithms optimize drying parameters based on real-time data, maximizing yield and quality while minimizing energy consumption and processing time. Optimized airflow design plays a vital role in improving drying efficiency and consistency. Computational fluid dynamics simulations and experimental studies are used to design drying chambers, air distribution systems, and smoke assemblies for optimal airflow patterns. This ensures a thorough removal of moisture from all surfaces of plastic materials, leading to a higher quality end result. Hybrid systems can integrate hot air drying with infrared radiation or microwave drying to achieve faster drying rates and better process control. These hybrid methods allow for more precise temperature control and shorten drying times while maintaining product integrity. Remote monitoring and control functions allow users to monitor and control drying processes from anywhere using computer interfaces or mobile devices. It enables real-time performance monitoring, troubleshooting, and optimization, improving operational efficiency and reducing downtime. Integration with Industry 4.0 technologies such as IoT connections, cloud computing, and data analysis highlights technology in a new era. Automated drying systems equipped with IoT sensors can collect vast amounts of operational data that can be analyzed to identify patterns, optimize processes and predict maintenance needs. This data-driven approach can improve plastic recycling operations' reliability, productivity and sustainability. Advanced drying systems can provide customizable drying profiles adapted to the specific characteristics of different types of plastic materials. By adjusting parameters based on the properties of the input materials, such as temperature, airflow and residence time, these systems can achieve optimal drying results while minimizing energy consumption and preventing plastic degradation. Together, these advanced automation technologies have the potential to contribute to the improvement of drying systems for recycling plastics, making them more efficient, reliable, and environmentally friendly.

Automated pelletizing systems

Automated pelletizing can play a key role in plastic recycling by facilitating the conversion of plastic waste into reusable granules or pellets. Plastic waste collected for recycling is initially sorted to separate different types of plastics according to resin type and color. After sorting, the plastics are crushed or granulated into smaller pieces to improve processability and ensure a uniform melt during granulation. Any impurities such as labels, adhesives, or residual materials are removed by washing or cleaning. After cleaning, plastic flakes or pellets residual moisture must be removed to prevent defects in the final product. Automatic drying systems are often integrated into the methods such as heat drying or desiccant drying to completely remove moisture from the plastic feed. The dried plastic flakes or pellets are then fed into an automated granulation system, usually via a conveyor belt or feed system. In a pelletizing machine, the plastic is melted in an extruder, which is heated to the melting temperature and converted into melt. The molten plastic is then pushed through a die that molds it into uniform granules or pellets of a predetermined size and shape. Accurate temperature control during the melting and extrusion process is critical to ensure optimal plastic melting without degradation. Automated granulation systems are equipped with temperature sensors and controls to maintain the desired temperature profile during extrusion. This can ensure consistency of required pellet quality. Die design is vital in determining pellet size, shape, and uniformity. Advanced pelletizing systems have customizable nozzles that allow flexibility in producing pellets of different sizes and configurations to meet specific customer requirements. After extrusion, the newly created pellets pass through a cooling system to solidify them. This may involve submersion in a water bath or contact with chilled rollers to quickly cool the plastic and maintain the desired shape and size. Proper cooling is important to prevent warping and ensure pellet size and shape uniformity. After cooling, the pellets can be further processed to ensure a uniform size distribution and to remove over or undersized particles. Screening equipment or a classifier separates the pellets by size, ensuring consistent quality, and efficiency. Quality control measures are carried out during the granulation process to ensure that the granules meet the required characteristics such as size, shape, density and melt flow. Automated systems may include sensors, cameras, and quality control mechanisms that detect defects and deviations from quality standards. The finished pellets are then packed into bags, containers or bulk shipments for distribution to producers. Automated granulation systems are an integral part of the plastic recycling process. They enable plastic waste to be transformed into valuable raw materials for the production of new products while boosting resource efficiency and environmental sustainability, as discussed in Section Automation in drying process.

Automated robotic sorting and handling

Revolutionized automation technologies can greatly improve efficiency, accuracy, and performance, reduce labor costs, and improve safety in recycling plastics. Automated robotic sorting and handling can also impact this regard, as discussed below: Automated robotic sorting systems can use advanced sensors, cameras, and AI algorithms to identify and classify different types of plastics based on resin type, color, shape etc. These systems can accurately separate various plastic materials, including PET, HDPE, PVC, PP, and others, enabling precise sorting and separation. Once plastic materials are identified, a robotic arm equipped with grippers, suction tubes, or other specialized tools is used to select and sort the items. These robotic arms are programmed to perform specific tasks, such as picking individual plastic items from a belt, moving them to designated sorting bins or conveyor lines, and handling materials of different shapes, sizes, and weights. Automated robotic sorting systems are often integrated with conveyor systems to facilitate continuous material movement through the recycling facility. Conveyor belts transport plastic goods from the inbound stage, receiving them from collection points or sorting locations and finally to robotic sorting stations. Robotic sorting systems are equipped with sensors and cameras that provide real-time feedback on the location, direction, and condition of processed plastic materials. Some robotic sorting systems have ML and adaptive algorithms that continuously learn from experience and improve their performance over time. These systems can adapt to changes in the composition or properties of incoming plastic materials and optimize sorting strategies for greater efficiency and accuracy. Safety is paramount in automated robotic sorting systems, especially in environments where humans work alongside robots. These systems are equipped with safety sensors, barriers, and safety locks to prevent accidents and ensure safe operation. Collaborative robots that can work safely with humans are increasingly being used in recycling plants. Automated robotic sorting systems require regular maintenance and technical support to ensure optimal performance and reliability. Manufacturers offer maintenance schedules to troubleshoot potential problems. Robotic sorting systems are designed to be scalable and adaptable to changing performance requirements and handling needs. Additional robotic sorting stations can be added or reconfigured to meet increasing demand in material composition. By improving the efficiency of plastic recycling processes, automated robotic sorting systems help reduce the environmental impact of plastic waste. These systems help increase plastic recovery and recycling rates, divert them from landfills and incineration, and conserve natural resources. Automatic robotic sorting and handling systems play an important role in plastic recycling because they simplify the sorting process, improve accuracy, and increase throughput. These advanced technologies are necessary to achieve sustainable waste management goals and circular economy principles in the plastic recycling industry.

Integration of artificial intelligence and machine learning in plastic recycling

Integrating AI and ML into plastics recycling has transformed the industry by offering innovative solutions in many ways as discussed below: AI and ML algorithms are used to identify different types of plastic based on their resin type, color, shape, and other properties. Deep learning (DL) algorithms such as convolutional neural networks (CNN) can accurately classify plastics in real-time, enabling automated sorting systems to sort plastic waste efficiently. AI algorithms analyze data from various sensors and monitoring devices to optimize the operation of recycling equipment such as shredders, washers, dryers, and extruders. ML models predict equipment failures and maintenance needs based on historical data, thereby reducing downtime and improving overall equipment reliability. Optimized algorithms powered by AI adjust processing parameters such as temperature, pressure, and residence time to maximize the yield and quality of recycled plastic products. AI-based systems inspect recycled plastic materials for defects, contamination, and impurities. ML models detect deviations from quality standards and enable early intervention to prevent substandard materials from entering production. Automatic quality control systems ensure that recycled plastic pellets meet the required characteristics such as size, shape, color, and cleanliness. To optimize logistics and resource allocation, AI algorithms analyze supply chain data, including material flows, inventory levels, transportation routes, and market demand. ML models predict future demand for recycling plastics and raw materials, enabling manufacturers to decide on production planning and supply strategies. Moreover, AI-based decision support systems can identify opportunities to optimize the supply chain, reduction of costs, and minimize the environmental impact associated with it. AI and ML tools assess the environmental impact of plastic recycling processes, considering factors such as energy consumption, greenhouse gas (GHG) emissions and resource use. Life cycle assessment (LCA) models use ML techniques to analyze the environmental footprint of different recycling strategies and identify opportunities for improvement. AI algorithms analyze market data, consumer trends, and regulatory developments to provide insights into the demand for recycled plastics and new opportunities in the recycling industry. ML models predict future market trends, allowing recyclers and manufacturers to anticipate changes in demand and adjust their business strategies accordingly. AI-powered market monitoring platforms provide real-time updates on market dynamics, competitive landscape, and regulatory changes that can help stakeholders to make decisions about investments, partnerships, and product development initiatives in plastic recycling industries.

Remote monitoring and control

Remote monitoring and control systems in the plastic recycling landscape allow users to monitor and control recycling processes from almost anywhere, thereby increasing the efficiency, safety, and sustainability of recycling. Remote monitoring systems collect real-time data from various sensors, gauges, and monitoring devices installed throughout the recycling facility. These sensors measure parameters such as temperature, pressure, flow rate, humidity, and energy consumption. Data acquisition modules transmit collected data to a central control system or cloud-based platform using wired or wireless communication protocols such as Ethernet, Wi-Fi, or cellular network, etc. Collected data is processed and visualized in a centralized control interface that operators can access through web-based dashboards, desktop applications or mobile devices. The control interface provides real-time information on the performance, status, and efficiency of recycling equipment and processes, allowing users to monitor operations and identify potential problems or anomalies remotely. Remote monitoring systems use algorithms to analyze incoming data and detect unusual conditions or equipment malfunction. When anomalies are detected, the system generates alerts and notifications that are sent to operators via email, SMS, or push notifications to their mobile devices. Users can quickly respond to alarms, remotely troubleshoot, and initiate corrective actions to minimize downtime and prevent disruptions to recycling operations. In addition to monitoring, the remote-control functions allow users to adjust equipment settings remotely, parameters, and operating modes. Remote control interfaces allow users to start, stop, or stop recycling processes, adjust process parameters such as temperature or pressure, and switch between different operating systems as needed. This remote-control function increases operational flexibility and responsiveness, allowing users to optimize recycling processes in real-time and adapt to changing conditions or production requirements. Remote monitoring systems analyze historical data to assess the effectiveness and efficiency of recycling processes over time. Advanced analytical tools identify trends, patterns, and optimization opportunities that enable users to improve process efficiency, reduce energy consumption, and optimize resource usage. Performance reports and analytics dashboards provide information on key performance indicators such as yield, performance, energy efficiency, and material recovery, facilitating informed decision-making and continuous improvement. Remote monitoring systems are often integrated with maintenance management systems to monitor equipment health, schedule preventive maintenance, and manage spare inventory. Predictive maintenance algorithms analyze equipment data to identify early warning signs of potential failure or damage, allowing users to address maintenance issues and avoid unplanned downtime proactively. Remote monitoring systems include strong security features to protect sensitive information and ensure compliance with privacy regulations. Encryption, authentication, and access control mechanisms protect communication channels and prevent unauthorized access to system resources. In general, remote monitoring and control systems are important in optimizing plastic recycling operations by providing real-time visibility, control, and information about recycling processes. By utilizing advanced data analysis and remote management capabilities, operators can improve operational efficiency, maximize equipment performance and minimize environmental impact, contributing to a more sustainable and resource-efficient recycling industry.

Innovations in plastics recycling technologies

Numerous strategies have already been put out to improve the circularity of plastics and lessen the negative effects of disposed plastic products on terrestrial and aquatic environments. Effective recycling is difficult since plastic trash differs chemically and physically from other types of waste. This difficulty increases if the separation process is not done correctly in the beginning (ideally at the source). The quality of the waste material (as a raw material) is crucial for plastic recycling [20]. Plastic recycling is hampered by several reasons, including the complexity, high cost, and inefficiency in classifying and separating plastics [21]. Using coatings and additives, contamination of plastics, and thermoset plastics, in particular, lead to hardly recyclable materials [22]. Impurities account for a sizeable portion of plastic waste (28%), and the majority of the sample plastic trash is classified as "Low Quality," according to a study that characterized and evaluated the recyclability of plastic waste from recycling centers [23]. A suitable separation process must be used to identify the waste with the most potential for recycling, maximizing circularity and reducing the amount of garbage going to landfills and fuel created from refuse [24]. For compostable plastics, effective sorting and separation are also essential to reducing pollution [25]. Typically, a series of stages are used to sort the materials: manually by operators, automatically by taking advantage of variations in the material qualities, or both (i.e., manual and automated sorting). According to the study [26], many MRFs rely on manual separation to produce high-purity, differentiated products. On the other hand, it can be laborious, prone to human mistakes, and harmful to people's health. Smart waste management systems integrate various IoT technologies with sensors, computer vision, and ML to facilitate automatic trash separation. This could reduce occupational risk or health concerns and result in a more accurate separation of waste items [27]. According to most cost evaluations, automation offers cheaper long-term operational costs than manual garbage sorting, despite the initial investment being more [28]. Furthermore, medium-to high-volume plants are better off using automation [26].

Advanced sorting technologies

Two distinct approaches are being used in the field of automated garbage sorting and segregation research: sorting by sensor technology and AI via ML. Different sorting systems could be coupled to increase product output or sorting efficiency [26]. Tracer-based sorting, digital watermark-assisted sorting, and other novel sorting technologies are being developed and showcased [29]. Numerous research on trash management has been conducted, including ones describing advancements in automated garbage sorting [30] and aiming to sort and categorize plastic waste streams into multiple types [31]. The majority of scholarly investigations concerning automated trash separation were conducted through the use of either sensor or camera technologies. It should be highlighted, nonetheless, that there isn't a simultaneous application of sensor-based separation based on material composition and visible separation using cameras. Research mainly focuses on integrating spectral pictures from sensors with DL models to increase separation efficiency. A set of ML techniques and a compact, inexpensive triad spectroscopy sensor module were investigated by Martinez-Hernandez et al.[32] to identify six distinct types of plastic. It looks at using the low-cost OSRAM multispectral chipset AS7265x. Three sensor devices (AS72651, AS72652, and AS72653) that are capable of spectral identification in the visible to NIR spectrum make up this chipset. With a full width at half maximum bandwidth of 20 nm, each of these three sensors has six optical filters that can measure light intensity in the range of 410 nm (ultraviolet—UV) to 940 nm (infrared—IR). To create an AS7265x chipset with 18 multispectral wavelengths, these three sensors were integrated. To light and assess surfaces for light spectroscopy, these three multispectral sensors were combined into a single package called the triad spectroscopy sensor, which was created by the American company SparkFun Electronics. The package included visible, UV, and IR LEDs.

The suggested experimental setup for plastic recognition makes use of this one package. The complete triad spectroscopy sensor costs approximately GBP 56, which is roughly 20 times less than high-priced spectroscopy sensors. The single unit cost of the AS7265x sensor is approximately GBP 15. The configuration of the visible, infrared, and ultraviolet chips and associated LEDs is displayed in Fig. 5.5 of the triad spectrum sensor module.

Every stage of the waste management process, including monitoring, collecting, transportation, processing, disposal/recycling, and waste material analysis, has extensively used IoT and sensor technology [33]. Numerous studies on intelligent waste management systems that combine sensor and IoT technologies to recover more secondary materials of greater quality have been published. The aim is to accomplish this goal. There are several smart systems that can be used on a smaller scale for the primary separation of municipal solid waste in what are known as smart waste bins [34] or on a larger scale for the secondary separation of waste, which takes place in waste treatment plants and MRFs. Smart waste bins can distinguish between dry and wet waste [35] biodegradable and nonbiodegradable waste [36], different waste classes [37], the dimensionality of waste items, and other characteristics with the use of sensor technology. Different gas sensors are utilized to detect gases, such as CO2, CH4, H2S, and NH3, and humidity is used to detect biodegradable substances that may be present in packaging waste [38]. Laser sensors could assess the waste item profile in order to ascertain the dimensionality of packing materials (either 2D or 3D) [39]. Using various optical sensors,

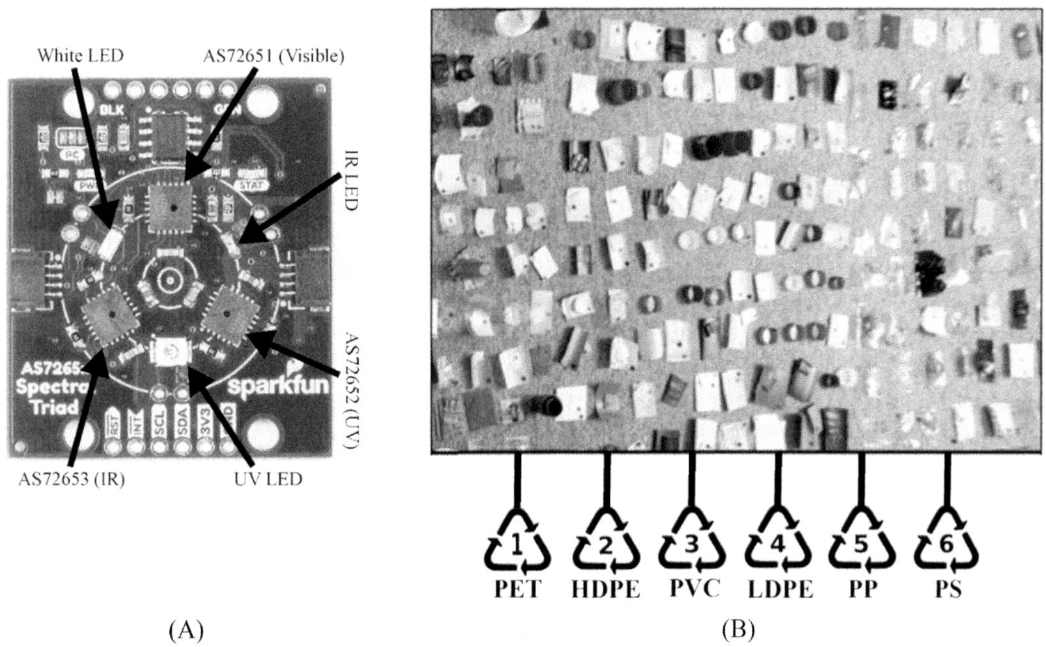

FIGURE 5.5 Low-cost spectroscopy sensor and examples of plastic waste. (A) Triad spectral sensor module from SparkFun Electronics. (B) Examples of household plastic waste used for data collection and recognition processes [32]. Source: *From U. Martinez-Hernandez, G. West, T. Assaf, Low-cost recognition of plastic waste using deep learning and a multi-spectral near-infrared sensor, Sensors 24 (9) (2024). https://doi.org/10.3390/s24092821.*

polymers that make up various kinds of plastic packaging materials can be classed and classified. NIR reflectance sensors are extensively used optical sensors for classifying plastic kinds. These sensors detect the amount of light that is reflected off an object. The NIR area, which spans from 750 to 2500 nm [40], is caused by an overtone or a mixture of various chemical bond vibrations (C–H, O–H, and others) that cause light to be absorbed and reflected [41]. Different absorption patterns result from the many elements and chemicals that make up polymers. The common plastic varieties, such as PET, HDPE, PVC, LDPE, PP, PS, and PC, could be successfully recognized using various NIR spectral properties [42]. Similar to this, plastic waste from electronics and home appliances that comprise polymers including PP, PS, ABS/PC, and acrylonitrile butadiene styrene (ABS) might be separated [43]. Bhati and Bhattacharya [44] presented an IoT system that uses affordable NIR sensors to identify various types of plastics. The device can accurately separate PET, PP, HDPE, and PS plastics by using various machine-learning methods. Moreover, plastics can be separated at the source using a portable NIR spectrometer or in smart trash cans that have built-in tiny NIR sensors. There has been a steady increase in the number of publications implementing AI in waste management systems [45]. Artificial neural networks, K-nearest neighbor, decision trees, support vector machines, adaptive neuro-fuzzy inference system, and many other models have been created and deployed in waste management systems. CNN algorithms, which belong to the DL branch of AI, shown exceptionally strong performance in picture recognition and classification, according to a study [46]. Transfer learning, a technique that leverages a pretrained model from another task, has been suggested as a way to avoid training CNN models from scratch. Transfer learning leverages the benefit of employing a pretrained model, transferring the information to a new dataset, where it is then retrained for the new target domain [47]. This is advantageous because DL models often require a large amount of data and time for training. Various models, including ResNet-50, MobileNet, VGG-16, AlexNet, and others, were employed to distinguish and arrange waste materials into distinct categories, primarily consisting of plastic, metal, paper, and biological materials. A sizable image dataset must be included in order for CNN models to be successfully trained [48]. It is feasible to distinguish between distinct plastic packaging objects based only on their look with the use of image recognition technology. It is possible to use this technology at any size. On a smaller scale, the bin itself might make it possible to separate single-use plastic materials into distinct categories, as in the case of smart garbage bins. DL models for camera detection are more frequently employed in MRFs on a greater scale. Agarwal et al. [49] provided a model of an image-based classification approach for plastic waste dubbed the one-shot learning technique, which allowed for the accurate separation of plastic waste into five categories: PET, HDPE, PP, PS, and others. While computer vision-based technologies are excellent at categorizing waste based on visual qualities, spectroscopic methods have been shown to be useful in classifying waste based on its molecular properties. To get the best separation performance, combining these two strategies shows promise For example, Maliks and Kadikis [50] suggested a combination method utilizing NIR spectroscopy and a CNN model to effectively separate plastics into five polymer classes under a range of circumstances (dry, wet, scratched, and dirt-covered). Likewise, a mixed strategy was used to improve worn polymer sample separation effectiveness [51]. A further study [52] analyzed 159 plastic samples—half of which were black in color—using the combined approach. Various spectroscopy methods, including NIR combined with DL, also showed excellent accuracy in the plastic component separation process. A CNN model that had been trained on particular spectra was fed with the spectrum data gathered from several infrared (IR) sensors. The materials were then categorized by the CNN model based on similarities between the obtained spectra and a database. Rok Pužcnik et al. [53] propose a simple intelligent waste sorting model to compare the accuracy of plastic packaging material separation using sensor technology and DL models. The study offers fresh methods to support ongoing advancements in the trash sorting of plastic packaging. Here, it is assumed that the sorting is done by placing the waste plastic on a (small) conveyor belt, then using sensors or cameras to identify and separate the plastic. Together with an RGB camera, six sensors including NIR, humidity, temperature, CO_2, CH_4, and a laser profile sensor are used to classify packaging materials according to their make up, dimensions, cleanliness, and look. Such a separation system might be used as a central postsorting system at a larger size, or at a smaller scale (households or communities) as an on-site smart waste bin system. Six distinct waste kinds, mostly single-use plastic packaging materials, were included in a case study of 60 products. Other nonplastic trash types, as well as unclean and mixed plastics, were also taken into consideration. Four CNN models—ResNet-50, MobileNet-v2, Inception-v3, and DenseNet-201—that were pretrained on a custom database and downloaded from the Deep Network Designer app on MatLab® (MathWorks, 2023) were utilized and compared for camera-based detection. The goal of the basic optimization model was to: (1) reduce the amount of items that wind up in the dirty and residual trash bins; and (2) reduce the quantity of placements that the camera incorrectly recognizes. A machine-learning-based method is established by Roman-David Kulko et al. to categorize various plastic granulate and ascertain the percentage of each component in granule combinations (Fig. 5.6). The VIS and NIR spectrometers are part of the experimental equipment. The Vis29 uses a grating monochromator with 148–1161 nm measuring ranges and a silicon line sensor with 512 pixels. The Nir22 uses a grating monochromator and an InGaAs line sensor with 128 pixels. The wavelength range under measurement is 877–1902 nm. The usage of VIS-NIR spectroscopy and the given fusion model could be used as a concept for creating analyzers for examining binary mixes of 25 distinct PGs.

AI-powered sorting algorithms

The several applications and integrated strategies where AI technologies improve the sorting procedures for plastic trash are referred to as AI manifolds in plastic sorting algorithms. The precision, efficacy, and general efficiency of recycling operations are all increased by this technology. An overview of AI in plastic sorting is as follows: ML and computer vision: Image Recognition: AI systems employ computer vision to recognize various plastic types according to their visual attributes. Photos of plastic objects are taken using high-resolution cameras and ML algorithms trained to identify particular shapes, colors, and textures associated with various types of plastic and then analyze the photos. Spectral analysis: By combining AI algorithms with hyperspectral imaging, plastics may be distinguished according to their chemical makeup. This method captures and analyzes a wide range of visible light beyond visible wavelengths to determine the different types of plastic polymers. Robotic sorting: Automated sorting robots: AI-driven robots with

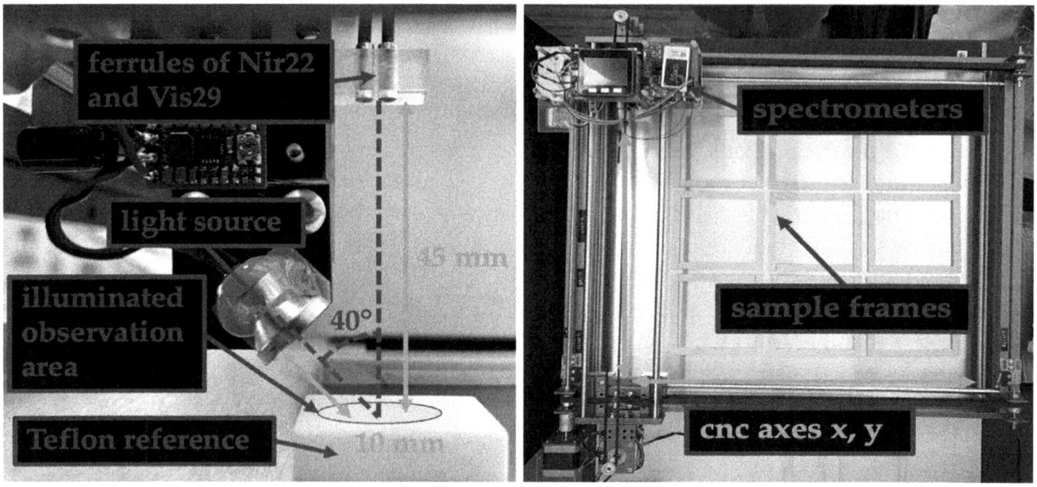

(A) Probe **(B)** Measurement setup

FIGURE 5.6 (A) The probe consisting of ferrules and lamp with teflon reference block. (B) The top view of the experimental setup [54]. Source: *From R.D. Kulko, A. Pletl, A. Hanus, B. Elser, Detection of plastic granules and their mixtures, Sensors 23 (7) (2023). https://doi.org/10.3390/s23073441.*

sophisticated grippers and sensors sort plastic items on conveyor belts. These robots recognize and collect particular types of plastics using real-time data from computer vision systems, then recycle them in bins that are allocated for that purpose. Adaptive learning: These robots can continually increase the accuracy of their sorting thanks to ML algorithms. With every sorting operation, they gain knowledge and refine their algorithms to minimize mistakes and better manage novel varieties of plastic materials. Multisensor integration, or sensor fusion, combines data from several sensors (such as RGB cameras, infrared sensors, and X-ray fluorescence) with AI algorithms to improve sorting and identification. This method makes it possible to distinguish between various plastic types and other materials more precisely. AI-driven decision making: In order to make decisions on the sorting of plastics in real-time, AI systems examine data from these sensors. For instance, they can identify the kind of polymer, degree of contamination, and suitability for recycling of every item on the conveyor belt. Analytics of data and predictive models: Pattern recognition: To find patterns and trends, AI examines enormous databases of plastic waste properties. Using this data, sorting procedures may be made more efficient, waste stream composition can be predicted, and recycling activities can run more smoothly overall. Predictive maintenance minimizes downtime and ensures continuous operation by using AI models to forecast when sorting equipment may break down or need maintenance. Cloud-based AI systems and the IoT combine to create connected sorting systems. IoT devices in sorting facilities gather and send data to cloud-based AI systems. By analyzing the data, these systems can offer insights and suggestions for enhancing the efficiency and accuracy of sorting. Remote monitoring and control: Through cloud-based platforms, operators can keep an eye on and manage sorting operations remotely. They can make real-time modifications based on AI-driven insights. Some viable application of robotics includes Autonomous Manipulation and

Perception (AMP) Robotics employing robotics and AI to precisely sort plastics in recycling facilities, lowering contamination rates and raising the caliber of recycled plastics. ZenRobotics; utilizing cutting-edge ML and sensor technology, the company creates robotic systems driven by AI that sort garbage, including plastics. Tomra Sorting Solutions; improves the recognition and segregation of various plastic polymers by integrating AI and ML into their sorting devices. AI's advantages in plastic sorting enhanced accuracy: AI systems are highly accurate in differentiating between various plastic types, which lowers contamination and raises the caliber of recycled materials. Enhanced efficiency: By managing plastic garbage more quickly and efficiently, automated sorting systems save labor expenses and processing delays. Scalability: huge volumes of plastic garbage can be handled by AI-powered sorting systems, which can be scaled up to meet the demands of huge recycling operations. Sustainability: By lowering the quantity of plastic debris that ends up in landfills or the ocean, better-sorting results in higher recycling rates, which support environmental sustainability. Still, it has challenges and considerations like initial investment: Some recycling operations may find it prohibitive to employ AI-powered sorting technologies due to the high upfront expenditures. Data quality: The caliber and variety of the training data determine how successful AI algorithms are. Sorting errors might result from low-quality data. Integration: To guarantee smooth operation, meticulous planning, and management are needed when integrating AI technology with the current infrastructure. Maintenance: To maintain AI systems operating at their best and able to adjust to new kinds of plastic materials, they require constant maintenance and updates.

Some of the latest research on AI powered plastics sorting are highlighted. Cai et al. [55] studied multiscale feature fusion method for RGB and hyperspectral images based on segmenting objects by locations, which enhances the accuracy of identifying transparent PP, blue PET, and transparent PET bottles on a black conveyor belt using multisensor fusion technology, Fig. 5.7 RGB and hyperspectral images are acquired concurrently using a line-scan camera and a NIR hyperspectral camera, which covers the spectral range from 935.9 to 1722.5 nm. Images of discarded plastic bottles were simultaneously captured in RGB and hyperspectral modes using the twin camera acquisition platform mode of image acquisition. These samples were regularly collected after being arranged randomly on the conveyor belt.

In a paper published in 2020, Chowdhury et al. [56] showed how well CNNs performed in classifying different types of plastic from photos. They achieved high accuracy rates by

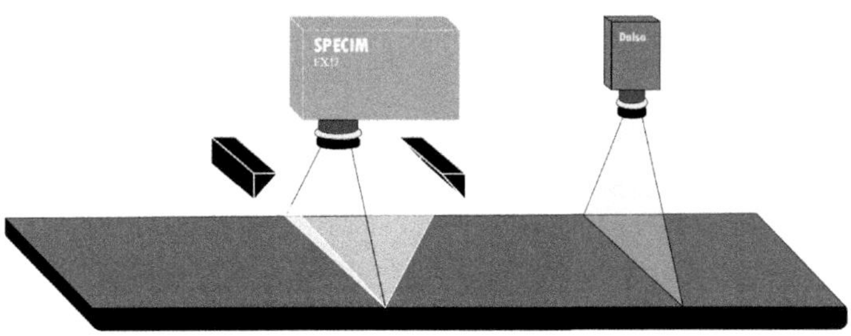

FIGURE 5.7 Dual camera platform mode of image acquisition [55]. Source: *From Z. Cai, J. Yang, H. Fang, T. Ji, Y. Hu, X. Wang, Research on waste plastics classification method based on multi-scale feature fusion, Sensors 22 (20) (2022). https://doi.org/10.3390/s22207974.*

training the models on big datasets of waste photographs with labels. Adebayo et al.'s [57] study investigated the sorting of plastics using hyperspectral imaging and showed that AI models could reliably identify plastics according to their spectral characteristics. The development of robotic systems that employ reinforcement learning to increase their sorting efficiency over time by learning from their actions and results was emphasized in the work of Zeng et al. [58]. In their discussion of applying adaptive learning algorithms in robotic sorters, Yang et al. [59] emphasized the ongoing enhancement of sorting efficiency and accuracy. The work by Liu et al. [60] showed how sensor fusion, which uses AI to integrate data from many sensor modalities, may effectively improve the accuracy of plastic sorting. García et al.'s [61] research on real-time AI processing for sorting systems highlighted how sorting times could be shortened and throughput could be increased. Strong AI models for garbage sorting require a variety of high-quality datasets, according to a review by Zhang et al. [62]. The practical difficulties of implementing AI sorting algorithms at scale were examined in the work of Smith et al. [63], which offered insights into scalability issues and integration tactics. The ethical and environmental aspects of AI in garbage management were examined in research by Jones et al. [64], which recommended sustainable methods and legislation to reduce adverse effects. Chidepatil et al. [16] covered Blockchain Tools and MultiSensor-Driven AI. As illustrated in Fig. 5.8 various AI-enabled sensor types are supported during the segregation stage. High-definition optical sensors, or cameras, are used to obtain the texture, color, and form data. We can use this, for instance, to distinguish between different colored or shaped plastic bottles. Another option is to use (NIR) laser diodes. Plastics' light absorption spectroscopy, especially in the 300–3000 nm wavelength region, revealed new opportunities for optical sensing of plastics. The study of various

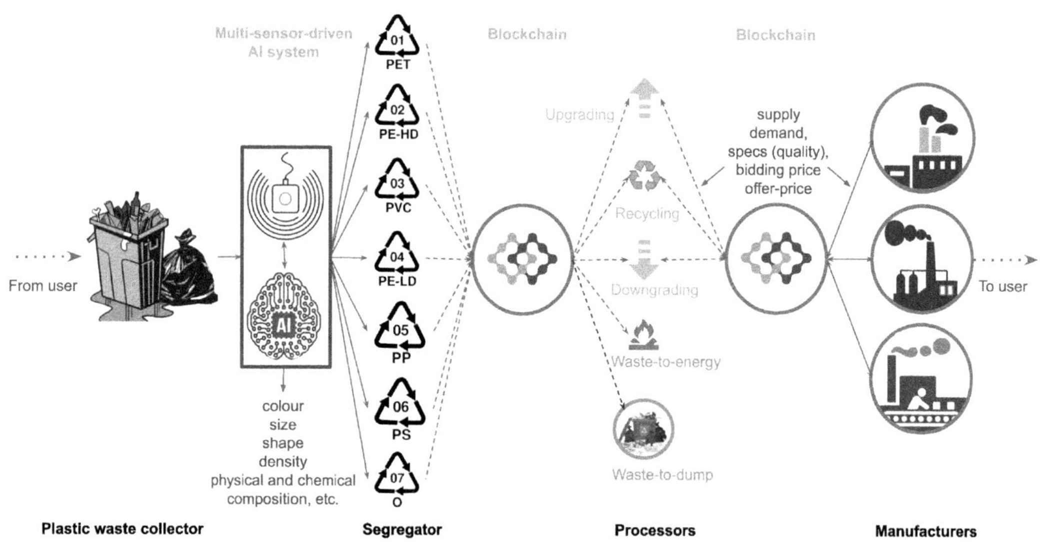

FIGURE 5.8 Blockchain and multisensor-powered artificial intelligence interfaces for plastic waste [16]. Source: *From Chidepatil, Aditya, Bindra, Prabhleen, Kulkarni, Devyani, Qazi, Mustafa, Kshirsagar, Meghana & Sankaran, Krishnaswamy. (2020). From Trash to Cash: How Blockchain and Multi-Sensor-Driven Artificial Intelligence Can Transform Circular Economy of Plastic Waste? Administrative Sciences, 10(2). https://doi.org/10.3390/admsci10020023.*

polymers' resonance frequencies is also becoming increasingly dependent on NIR laser diodes. The six types of detectable plastics that these laser sensors can distinguish between are ABS, PET, PE, PVC, PP, and PS. These sensors use each type of plastic's resonance frequency to distinguish between different grades of plastic. Resonant frequencies are those that are closest to the peak of the plastic's absorption spectrum. Resonant frequencies vary amongst polymeric molecules. We can determine the resonance frequency of the plastic wastes by utilizing these laser detectors to scan them over a range of frequencies. The kind of plastic can be determined by its resonance frequency. In this multisensor platform, AI reduces uncertainty and facilitates effective and thoughtful segmentation.

Robotics and machine learning

Recycling facilities are using robotic devices more often to automate the sorting process. These systems recognize and sort various types of plastics using robotic arms and conveyor belts fitted with sensors and cameras. Robotic arms: Research like that done by Petersen and Nygaard [65] shows how robotic arms can be used to sort plastics according to sensor and visual data. These systems greatly increase sorting efficiency by operating at a high pace and maintaining precision. Conveyor belt systems: Huang et al. [66] show how robotic arms can be integrated with sophisticated conveyor belt systems. These systems use sensors to identify different types of plastic and route them to the proper sorting stations. Compared to hand sorting, robotic sorting systems provide the following advantages: Robots are fast and efficient because they can work nonstop, which increases sorting facilities' throughput [67]. Compared to manual sorting, robots' high level of precision results in a lower error rate [68]. Safety: Automation improves workplace safety by lowering human exposure to potentially dangerous materials [69].

ML in plastic sorting: CNNs are a ML technique used to identify various types of plastics from photographs of waste items. A work by Li et al. [70] investigates the use of CNNs in plastic identification, obtaining high accuracy rates in identifying different forms of plastic based on visual characteristics. Zhao et al. [71] conducted a study that employs NIR spectroscopy and ML to identify plastics according to their chemical composition. The results show notable improvements in sorting accuracy.

Plastic sorting plants also employ ML algorithms for process optimization and predictive maintenance. Wang et al.'s paper [72] from 2021 offers a framework for predictive maintenance that uses ML to detect equipment breakdowns, reducing downtime and guaranteeing continuous operation. Research by Kumar et al. [73] emphasizes how ML may be used to evaluate operational data and optimize sorting procedures, which can lead to increased productivity and lower expenses. Real-time data analytics from ML systems allow operators to keep an eye on sorting line performance and make well-informed judgments. Real-time Monitoring: Johnson et al.'s work [74] from 2021 shows how ML may be used to monitor sorting operations in real-time, allowing for quick modifications and improving overall performance. Over time, sorting accuracy can be continuously improved using ML algorithms by learning from sorting performance data [75]. Robotics and machine intelligence come together to form a potent sorting system for plastic. Robots quickly and accurately carry out ML algorithms' fast decisions on the type of plastic by analyzing data from cameras and sensors. Companies using AI-powered robots to improve plastic sorting include AMP robots and ZenRobotics. Numerous recycling facilities have shown notable increases in sorting accuracy and efficiency thanks to these technologies [76,77].

Future studies and advancements in this area will concentrate on advanced sensors. More precise identification and sorting of a greater variety of plastics by robots and ML systems will be possible with improved sensor technology [78]. Better algorithms: As ML research continues, more advanced algorithms that can identify between various plastic kinds and increase sorting accuracy will be developed [79]. Scalability: More recycling facilities will use these sophisticated sorting systems as technology costs come down and efficiency rises, which will result in a wider deployment and a bigger impact on the management of plastic trash [80].

Emerging chemical recycling technologies

Chemical recycling technologies are starting to show promise as a way around the drawbacks of mechanical recycling of plastic waste, namely, contamination problems and polymer property degradation. This literature review examines the state-of-the-art in chemical recycling technologies, with an emphasis on the benefits, drawbacks, applications, and mechanisms.

Pyrolysis is a type of thermal decomposition in which waste plastic is heated in an oxygen-free atmosphere to produce a mixture of hydrocarbons. According to research by Lopez et al. [81], pyrolysis may be used to handle mixed and polluted plastic streams with flexibility, yielding products like syngas and synthetic crude oil that can be processed further to create fuels and chemicals. It has advantages, such as a high potential for energy recovery and versatile feedstock tolerance. Nevertheless, there are drawbacks, such as high energy consumption and the necessity for complex emission management. Gasification is a process that creates syngas by reacting plastic waste with regulated amounts of steam and oxygen. Its capacity to process various plastic kinds and produce syngas—which may be utilized as a chemical feedstock or to generate power—is highlighted in a study conducted by Arena [82]. The advantages are a wide range of feedstock compatibility and valuable syngas output and challenges faces high initial investment expenditures and intricate syngas purification procedures. Depolymerization is the process by which polymers are broken down into their monomers by chemical processes such hydrolysis, glycolysis, and methanolysis. Al-Sabagh et al. [83] state that this process is especially successful in recycling PET because it produces high-purity monomers that can be repolymerized into new polymers. Using catalysts, catalytic cracking breaks down polymeric polymers into smaller hydrocarbons. Although catalyst deactivation is still a major obstacle, Aguado et al. [84] emphasize the potential of this method to produce high-value compounds and fuels at lower temperatures compared to pyrolysis. Solvolysis refers to a group of processes in which polymers are broken down into monomers or oligomers using solvents and reactants. Kember et al. [85] concentrate on the effectiveness of solvolysis for polyurethanes and PET, generating outputs with high purity that can be used for polymerization. Enzymatic recycling breaks down plastics using enzymes; it works especially well with PET, under a seminal research, Tournier et al. [86] showed how designed enzymes could efficiently and selectively break down PET into its monomers under moderate environments. High-energy plasma arcs are used in plasma pyrolysis to break down polymers into syngas and other useful byproducts. Although it requires a lot of energy, Panda et al. [87] claim that this approach is quite effective at decomposing complicated and mixed plastic garbage. The circular economy relies on chemical recycling technology to turn plastic waste back into basic materials for new products. As recommended by Ragaert et al. [88], combining several chemical recycling techniques can increase overall effectiveness and increase the variety of plastics that can be recycled. Reducing the environmental impact of these technologies through sustainable chemical management and renewable energy sources is essential to their sustainability.

Technologies for chemical recycling are greatly aided by government laws and policies. Adopting and driving investment through supportive legislation can promote innovation and scale-up. Hopewell et al. [89] conducted studies that highlight the significance of regulatory frameworks that guarantee environmental safety and encourage chemical recycling.

Pyrolysis

One option for recycling plastic solid waste (PSW) that is becoming increasingly common is thermochemical conversion. Thermochemical recycling is more scalable and capable of handling polluted and unsorted trash than biological recycling. Pyrolysis is one of the thermochemical processes that has attracted increasing interest recently due to its ability to convert PSW into highly valuable fuels and chemicals in an environmentally friendly manner [90]. Pyrolysis is done without oxygen, as opposed to gasification and incineration, which reduces the amount of harmful pollutants and CO2 emissions. Particularly, under pyrolytic conditions—that is, below 600°C and without oxygen—polychlorinated dibenzo-para-dioxins and polychlorinated dibenzofurans can be significantly reduced [91,92]. Additionally, pyrolysis has distinct benefits in terms of product value, operating costs, environmental management, and alternative energy generation. The plastic pyrolysis mechanism involves long-chain polymers that are broken down into smaller fractions during plastic pyrolysis, and the resulting oils and gases are both extremely useful building blocks for chemicals and fuels [90]. The major result of pyrolysis of PP and PE at 500°C–600°C, without the use of any catalyst, is C20+hydrocarbons, which at room temperature solidify as wax. More gas and oil products, which are made up of C1–C4 olefin blends, C5–C20 olefins, and a small quantity of aromatics, can form at higher temperatures. For instance, when PE is heated to a crack, a variety of hydrocarbons are produced, the main ones being gases (C1–C4), mono-aromatics, C5–C10 nonaromatic hydrocarbons, and waxes (>C11). The temperature increases, yet the wax production decreases linearly while the gas yield rises dramatically. It seems that running at a higher temperature could somewhat promote the aromatics' production. Generally speaking, PE pyrolysis can be represented as follows: random scission and chain-end scission (secondary breakdown), followed by recombination of various chains (termination). This is the process of free radical initiation. Similar to this, PS pyrolysis is thought to proceed through types of free radical reactions, with the major products being aromatic monomers, dimers, and trimers.

These processes include β-scission, hydrogen abstraction, hydrogen transfer, radical recombination, and disproportionation [93–95]. Most aromatic monomers are created by an unzipping process, in which the C–C bond between two aromatic rings breaks and the terminal aromatic ring separates. PVC is completely different since it can still produce a high yield of liquid oil made of polycyclic aromatic hydrocarbons by PVC pyrolysis without the need for catalysts [96]. It is commonly acknowledged that the pyrolysis of PVC can be identified as a three-phase process, comprising of three stages: (1) dechlorination with inner cyclization; (2) scission of aromatic chains; and (3) release of two to four rings aromatics [96]. The primary thermal degradation route during PET pyrolysis is thought to involve □-scission and retro-hydroalkoxylation, resulting in the formation of benzoic acid and vinylic compounds. This is where the bridging glycol O C link can cleave, transferring a β-H atom to the carbonyl O [97]. Furthermore, in line with the previously published PET pyrolysis product distribution, the ester groups will be liberated to create a significant amount of CO2, CO, and ethylene [98,99]. By lowering reaction temperature and/or processing time, a catalyst can reduce the activation energy of pyrolysis reactions, which is crucial for minimizing pyrolysis's energy consumption. Additionally, the catalyst can lessen or prevent the creation of wax and speed up the breakdown of long-chain polymers [100]. A catalyst can efficiently reduce the energy needed to break the C–C and C–H bonds in the aforementioned PE pyrolysis process, producing more carbonium ions as a result. Pyrolysis can be carried out in a variety of reactor types, each with unique benefits: Fixed bed reactor: This straightforward device heats plastic trash by placing it on a bed; it is inexpensive and simple to use but has limited scalability and poor heat transfer. In a fluidized bed reactor, waste plastic is fluidized with a hot inert gas (for example, nitrogen) to guarantee even heating. It is more sophisticated and has greater running costs, but it has superior temperature control, high throughput, and good heat transfer. Reactor with rotary kiln, a revolving cylinder that is fed and pyrolyzes plastic trash continually. Good heat transmission and continuous operation compatibility are required, although moving parts wear and mechanical complexity are also present. Ablative reactor: This device heats and scrapes waste plastic using a heated surface. Excellent for big plastic components and has high heat transfer rates, but its mechanical complexity and scalability are limited.

Solvent-based recycling

Solvent-based recycling, also known as dissolution–precipitation or solvent extraction, has emerged as a promising alternative to plastic recycling. The elimination of contaminants (such stacking fault energy, Microwave Assisted Extraction, Advanced Semiconductor Engineering, and other plastic additives), dissolution (homogeneous or heterogeneous dissolution), and reprecipitation or devolatilization are all included in the general framework of plastic recycling by solvent extraction. In other words, each polymer is crystallized in a chosen way after dissolving the polymer(s) in the solvent(s). A solvent can ideally be used for selective dissolution if it has the ability to dissolve the target polymer as well as all other polymers save the target one. Choosing a selected solvent is obviously essential to the dissolution process. Dissolution temperature [101], concentration kinds of polymers and solvents, polymer size, and molecular weight of polymer [102], and polymer size. Two transport processes—solvent diffusion and chain disentanglement—are engaged in a thermodynamic polymer. According to Martini et al. [103] the polymer's self-diffusion is crucial to chain disentanglement. The dissolution becomes disentanglement-controlled rather than diffusion-controlled as the molecular weight of the polymer increases. In general, the plasticization of the polymer during dissolution causes an enlarged layer that resembles gel and has two distinct surfaces. In other words, distinct interfaces emerge between the solvent and the gel layer as well as between the glassy polymer and the gel layer. The solvents force the swollen polymer into the solvent, increasing the swollen surface layer until a quasistability condition is obtained, due to surface lay formation process. Solvent dissolution is kinetically dependent on both the solvent size and the gel phase's free volume [104]. Because of the enhanced mass transfer rate, the dissolving times decrease as the particle size decreases and the diffusivity of polymers increases. For instance, the addition of a small-molecule nonsolvent boosted the rates at which Polymethyl methacrylate (PMMA) dissolved [105]. Glassy polymers are composed of four layers: a liquid, a gel, an infiltration layer, and a solid swelling layer. The polymer has molecular-sized pores and channels when it is glassy. Without leaving any new holes behind, the solvent molecules fill these voids and initiate the diffusion process. While the gel layer vanishes as a result of the stress energy decreasing, all layers form in "normal dissolution," signifying the change from normal dissolution to cracking. By reducing the diffusion of both solvents and polymers, the insoluble polymers are believed to be the extra layer. Normal dissolving usually entails the solvent penetrating the polymers, the polymers expanding, and the solvent diffusing into the polymers. In actuality, a variety of factors influence the dissolution of polymers, including the solvent's composition and size as well as the polymer's molecular weight, content, and structure. For instance, due of benzene's smaller size and higher vapor pressure, cracks in PMMA and PS happen earlier with benzene than with toluene [106]. As a result, benzene dissolves PS and PMMA more quickly than toluene. A tiny amount of nonsolvent in the solvent mixture can enhance the dissolution of polymers in terms of solvent composition [106]. Moreover, swelling is also reduced in some ranges by adding low-molecular-weight nonsolvent. This suggests a mixed solvent design that is more quantitative, it seems. Heating can reduce the swelling of polymers caused by the high pressure of SCFs at supercritical circumstances. As a result, it is imperative that the temperature and pressure be balanced. Fuchs and Suhr's summary of solvents and nonsolvents for the polymers employed in the dissolution/reprecipitation process was released in 1989. The process of choosing the best solvents involves adhering to a number of important guidelines. High polar polymers dissolve in high polar solvents, whereas low polar polymers dissolve in low polar solvents, in accordance with the solubility rule "like dissolves like." (2) When the interaction (explained below) between the polymer and solvent takes precedence over the cohesion of the polymers, the polymer is readily dissolved in the solvent. The easier it is to dissolve the polymer in the solvent, the closer in polarity the solvent and polymer are. (3) The polymer dissolves more easily in the solvent if the solubility parameters of the solvent and polymer are comparable or equal. To assess the solvation ability, a number of solubility parameter systems have been created, including the ET(30) system. For a given recycle, the solubility parameter systems should be chosen carefully. One of the solubility parameters, the prediction of Hazardous and Special Products values, is used to choose appropriate solvents. Some excellent examples of solvent-based recycling for plastic recovery. PS is soluble in solvents that are hazardous [101] as well as ecologically benign [107]. The best solvents for recycling X-ray photoelectron spectroscopy were found to be limonene, terpinene, cymene, and phellandrene. In a similar vein, limonene functions well as an antioxidant and solvent for EPS even when it is heated. Systems for recycling PS foam that worked well were MEK/methanol, MEK/n-hexane, and p-xylene/n-heptane. The model PS is recycled using conventional solvents (such as xylene/methanol) at various dissolving temperatures. Within a specific temperature range, a higher temperature was associated with a higher PS recovery. To recover 100% of the EPS, a more environmentally friendly solvent called dlimonene—extracted from the rinds of citrus fruits—was utilized, taking into account the potential effects on the environment. Because of its exceptional chemical and physical qualities, PC is one of the most used engineered thermoplastics. Achilias et al. [101] concentrated on the recycling of three types of polymers from e-waste, namely, PC, ABS, and PS, utilizing the dissolved/reprecipitation process. Three pairs of solvent/nonsolvent systems—acetone/methanol, chloroform/methanol, and dichloromethane (DCM)/methanol—were used to recover these model polymers. With the DCM/methanol combination, PC recovery was 98% even at low temperatures. The outcome of recycling plastic from e-waste was the same as the model polymer for the same solvent combination. The two main thermoplastic forms, LDPE and HDPE, are widely utilized in the production of toys, bags, pipes, containers, housewares, and industrial wrappings, among other products. Toluene/acetone has been demonstrated by Poulakis and Papaspyrides [108] to be a possible solvent or nonsolvent for recycling PE. Grain size, mechanical performance, melt flow index, molecular weight, and crystallinity were all strongly retained in the recyclates. Achilias et al. [101] examined the effects of solvent types and dissolution temperature on recovery values, as well as the use of toluene/n-hexane and xylene/n-hexane (v:v2 1:3) as a solvent system for recycling POs using the dissolution/reprecipitation process. There was no discernible structural variation between the recycled polymers and the model. PET is used in packaging, textiles, films for overhead projectors, and aluminum-coated reflective materials. The recovered PET was purified using a selective dissolving procedure (Vane and Rodriguez [109]). In the dissolution/reprecipitation process for recycling PET, NMP/n-octane ÷ n-hexane was utilized as a particularly appropriate solvent/nonsolvent system. Processing and additives may be to blame for the minor drop in the number-average molecular weight during consecutive recycling cycles. Poly(styreneco-acrylonitrile) copolymer, which is an amorphous continuous phase, and scattered butadiene or butadiene copolymer, which is a rubbery phase, are components of ABS, an engineering thermoplastic. A dissolution-based technique using room-temperature acetone (0.25 g/mL) was presented by Arostegui et al. [110]. ABS is not degraded by the solvent-based approach, according to Fourier Transform Infrared Spectroscopy, Differential Scanning Calorimetry, Melt Flow Index. However, because of the removal of stabilizers, the decrease in molecular weight, and the breakdown of butadiene, the dissolution and injection molding may hasten the degradation.

Economic and environmental implications

Due to growing environmental concerns and the global movement toward sustainable practices, the economic consequences of recycling plastic have received a great deal of attention. This study of the literature looks at a number of economic factors related to plastic recycling, such as cost reductions, market expansion, employment creation, investment prospects, and regulatory consequences. When compared to virgin plastic, recycled plastic can drastically lower production costs, according to several studies. In comparison to making new plastics from raw materials, recycling plastics can result in energy savings of up to 75%, according to a 2016 analysis from the Ellen MacArthur Foundation [111]. Additionally, utilizing recycled materials can lower raw material prices, giving businesses a competitive edge, according to a study by Lazarevic et al. [112]. Recycling contributes to lower landfill and incinerator expenses. By directing plastic garbage to recycling facilities, municipalities can reduce their waste management expenses, according to the European Environment Agency (EEA) [112,113]. This can result in significant savings on trash management costs, particularly in areas with high landfill levies and stringent waste management laws. One of the main industries contributing to employment growth is recycling. The Institute for Local Self-Reliance [114] states that the collection, sorting, and processing of recyclable materials generates 10 times as many jobs as landfilling and incineration put together. In developing nations, where labor costs are lower and manual sorting methods are more common, the recycling industry has grown significantly [115]. Recycled plastics are becoming more and more popular across a range of industries. The packaging, construction, and automotive industries are expected to drive the global recycled plastics market, which Grand View Research [116] estimates would reach USD 66.74 billion by 2025. Growing environmental consciousness and governmental pressure to utilize sustainable materials are blamed for this industry rise. Investments in recycling technology and infrastructure have increased noticeably. According to the World Economic Forum [117] increasing capacity and creating effective recycling procedures depend heavily on green investments. Startups and innovations aimed at raising recycling rates and raising the caliber of recovered plastics are receiving more and more support from investors. Businesses are spending money on recycled plastics in order to comply with legislation and achieve their environmental goals. A McKinsey & Company [118] study found that businesses that include recycled materials in their goods can sometimes even save money while also improving their brand image and appealing to eco-conscious consumers. The consumer goods and packaging industries are seeing particularly strong trends in these areas. All around the world, governments offer financial incentives to boost the recycling sector. For example, significant funding has been set aside for recycling technology research and innovation under the European Union's Horizon 2020 program (European Commission [119]). The initial high expenses of setting up recycling facilities must be mitigated in part by these subsidies and grants. Extended Producer Responsibility (EPR) laws that force manufacturers to handle their products' end-of-life concerns have been successful in encouraging recycling [120]. EPR programs encourage manufacturers to create goods that are easier to recycle and to employ more recycled materials in manufacturing, according to the OECD [120]. Both wealthy and developing nations are adopting these measures more frequently. The quality of recycled plastics can be diminished and processing costs raised dramatically if there is contamination in the recycling stream. According to Hopewell et al. [89], contamination is a significant obstacle to effective recycling, highlighting the need for improved garbage sorting and consumer education. Plastic recycling has a wide range of economic effects, including reduced costs, the creation of jobs, market expansion, investment opportunities, and policy effects. Even though recycling has many advantages, issues like pollution and market volatility must be resolved to increase recycling's economic feasibility. Subsequent investigations ought to concentrate on formulating more effective recycling methodologies and regulations that bolster the recycling industry's steady economic expansion.

Cost-effectiveness of automation

A vital step in the recycling process is automated sorting. The precision and speed of sorting have been greatly increased by technologies including machine vision systems, XRF, and NIR spectroscopy [121,122]. Automated systems can have a significant upfront cost. Nevertheless, these upfront costs may be compensated for over time by reducing operating costs. Depending on the size and complexity of the system, the cost of deploying automated sorting systems can vary from several hundred thousand to several million dollars. The use of manual labor, which is a major expense in conventional recycling operations, is reduced by automation. Automation can save labor costs by as much as 70%, according to studies [123–129]. Automated systems can run constantly with less downtime, increasing efficiency and throughput. When compared to hand sorting, automated systems can process more material in an hour. For example, an automated plant can sort five tonnes of plastic in an hour as opposed to one or two tonnes by hand. Improved sorting accuracy results in recycled plastics of a higher caliber, which can command higher market prices. In addition to increasing cost-effectiveness, automation in plastic recycling has major environmental advantages, such as a decrease in landfill use and energy savings. Automation can provide significant long-term economic benefits. Because to labor cost savings, higher quality recycled materials, and faster throughput, many plants record a return on investment in less than 3–5 years. Automation enables facilities to create recycled plastics of greater quality at cheaper costs, increasing their competitiveness in the market.

Reduction in greenhouse gas emissions

Recycling plastic has become a key tactic in minimizing environmental effects, especially when cutting GHG emissions [130–133]. This review of the literature looks at several research and publications that measure and talk about the decrease in GHG emissions that comes from recycling plastic. Plastic production is a significant source of GHG emissions. The extraction of raw materials, polymerization processes, and energy-intensive manufacturing contribute to high levels of carbon dioxide (CO_2) and other GHGs. Recycling plastic has become a key tactic in minimizing the effects on the environment, especially when it comes to cutting GHG emissions. This review of the literature looks at several research and publications that measure and talk about the decrease in GHG emissions that come from recycling plastic. More than 850 million metric tonnes of GHGs will be released into the atmosphere in 2019 alone as a result of the production and burning of plastic, according to the Centre for International Environmental Law. This amount equals the emissions from 189 coal-fired power plants running at full capacity. A thorough lifecycle analysis reveals that about 6 kg of CO2 is released for every kilogram of plastic produced. In order to decrease GHG emissions through more plastic recycling, the European Union has put strict recycling objectives and laws into place. According to data from the EEA, plastic recycling in the EU decreased CO2 emissions by about 3.8 million metric tonnes in 2018. According to estimates from the U.S. Environmental Protection Agency (EPA), recycling plastics in the country lowers GHG emissions by around 1.1 million metric tonnes of CO2 equivalent each year. China's efforts to raise the standard of recyclable materials imported from abroad have, inadvertently, reduced GHG emissions by encouraging more effective recycling techniques. Japan has significantly reduced GHG emissions due to its high recycling rate, fueled by strict rules and cutting-edge recycling technologies.

Job impact and workforce transition

The field of automation in plastic recycling is expanding and has important ramifications for worker transitions and the labor market. The goal of incorporating cutting-edge technology like robotics, AI, and ML into the recycling process is to increase efficacy and efficiency. But this transition also affects workforce dynamics, skill needs, and job availability, changing the employment environment [134–136]. Worker displacement resulting from automation may occur when workers undertake manual and routine jobs. The once labor-intensive task of sorting is increasingly being handled by robots. This displacement is especially important in areas where human labor is a major component of recycling operations. However, automation also opens up new career paths in developing, managing, and upkeep cutting-edge recycling systems. Experts in robotics, AI, and machine maintenance are in greater demand. The skill sets needed for employment in the recycling industry must change due to this change.

Several tactics can be used to lessen the negative effects of job relocation and take advantage of new chances. offering training courses to current employees to learn new abilities related to automated recycling technologies. This includes receiving technical training to operate robots, comprehend AI systems, and maintain sophisticated technology. creating educational programs with the demands of the recycling sector in mind. Collaborations between academic institutions and businesses can guarantee that workers have the right skills from the start. Governments may make a significant contribution by implementing policies that facilitate worker transfers. This might include social safety nets for displaced workers, incentives for businesses to invest in staff development, and subsidies for training programs. Working together, industry stakeholders may exchange best practices and expertise about managing workforce migrations. Consortia within the industry can collaborate to create uniform training courses and certifications. AI-driven robots, used by AMP Robotics in recycling plants, have greatly increased sorting efficiency and accuracy. AMP Robotics has made training program investments for its staff in order to mitigate the effects on the workforce and prepare them for positions in technology management and maintenance. ZenRobotics, a multinational company, sorts garbage using robotics and AI. They have collaborated with academic institutions to create programmes that educate aspiring employees for jobs in automated recycling. By being proactive, they can guarantee a consistent supply of qualified workers to support their operations.

Regulatory landscape

Government policies and incentives

Due to the detrimental effects that plastic pollution has on the environment, managing plastic waste—especially recycling—has gained international attention. To encourage plastic recycling and decrease plastic waste, governments have implemented several policies and incentives, such as financial incentives for consumers and businesses, regulatory frameworks and legislation, economic instruments, public awareness and education, and more. Producers must take responsibility for the full lifespan of their products, including PPW disposal, under the terms of EPR. According to studies, implementing EPR programmes in nations such as Germany and Japan has resulted in a notable rise in recycling rates and a decrease in plastic waste. Plastic bans and restrictions: Single-use plastics are the subject of bans or restrictions in many nations. For instance, the Single-Use Plastics Directive of the European Union forbids the use of some single-use plastic products to eliminate marine litter. To promote greater recycling rates, governments set recycling targets. Taxes on the production or consumption of plastic can encourage recycling and reduction. Governments also contribute money to the recycling of technologies and infrastructure. To improve recycling habits, public awareness campaigns and educational programs are essential. Customers who return discarded plastic bottles and containers can receive financial compensation through deposit return schemes (DRS). Companies that use recycled materials or invest in recycling infrastructure may be eligible for tax breaks or rebates. Governments can encourage the use of recycled plastics by prioritizing products that contain recycled materials in their procurement strategies.

International collaboration and standards

Plastic pollution affects oceans and ecosystems globally, crossing national boundaries. Working together internationally is essential to solving this global challenge in its entirety. Coordinated actions can improve programs' and policies' efficacy, having a greater overall influence on the environment. Shared technology, expertise, and resources can be advantageous to nations. Developed countries can provide poor nations with modern recycling technologies and infrastructure. Building capacity through international cooperation might help less developed nations enhance their recycling programs and reduce plastic waste. Among the important international initiatives one important international convention that aims to reduce the flow of hazardous waste, including plastics, is the Basel Convention on the Control of Transboundary Movements of Hazardous Wastes and Their Disposal. In order to improve control and lower plastic pollution, the convention included plastic waste in its regulatory framework in 2019. According to this amendment, exporting polluted or mixed plastic trash requires prior informed approval from the exporting nation. The World Economic Forum launched the Global Plastic Action Partnership, which encourages cooperation between public and private sectors as well as civil society in the fight against plastic pollution. Its main objectives are to develop national action plans, encourage creative thinking, and mobilize funding. The possible influence of such cooperative efforts on national plastic waste management plans is exemplified by nations such as Ghana and Indonesia. A number of plastic recycling standards have been created by the International Organization for Standardization (ISO), including ISO 15270, which offers instructions for the recovery and recycling of plastics. By allowing the international trade of recycled materials and encouraging global best practices, these standards guarantee uniformity and quality in recycling processes. ASTM International creates technical standards for a variety of materials, including plastics. For recycling purposes, ASTM D7611, for instance, standardizes the tagging of plastic products. By assisting in identifying and classifying different types of plastic, these guidelines enhance the efficacy and efficiency of recycling initiatives worldwide.

Successful implementation of automation

Robots powered by AI have been created by Colorado-based AMP Robotics for use in recycling plants. Using computer vision and ML, these robots recognize, classify, and handle recyclables. High-speed cameras, sensors, and AI algorithms are all combined in AMP's technology. The robots are placed on conveyor belts, where they recognize and classify different kinds of plastics according to their kind and color. improved operating effectiveness, decreased contamination rates, and improved sorting accuracy are the results. Greater recyclability is achieved by sorting polymers with higher purity levels. Another example of automation in action is the use of sensor-based sorting technology in recycling, which was first introduced by the Norwegian business Tomra Sorting Solutions. It makes use of X-rays, visual spectrometers, and NIR sensor technologies. It significantly reduces waste and sorting errors. enhanced material quality that makes higher-value recycling streams possible. An automated system has been put in place by UK-based green recycling to better handle and sift plastic garbage. automated sorting devices with AI and sensors. A Canadian company called Plastrec Inc. recycles PET plastic bottles into premium flakes and pellets with the application of sophisticated robotics.

Challenges faced in real-world applications

Plastic recycling faces many challenges in real applications due to both technical and logistical factors. These factors are briefly described below:

Pollution: One of the biggest challenges in plastic recycling is pollution. Contaminants such as food scraps, dirt, or nonrecyclable materials can reduce the quality of recycled plastics and reduce their value and usability. Sorting and separation: There are several different types and classes of plastic, each requiring specific recycling processes. Efficient and accurate sorting and separation of different plastics is a major challenge, especially in large recycling plants. The complexity of plastic types: Plastics are of different types, available as discussed above, and each with its own chemical composition and properties. Some plastics are easier to recycle than others, and some types of plastic are not economically viable to recycle at all. Lack of infrastructure: Many areas lack plastic recycling infrastructure. This includes both collection systems and recycling facilities. Without adequate infrastructure, collected plastic can end up in landfills or incinerators instead of being recycled with environmental and future risks. Market demand and economics: Demand for recycled plastic varies and is influenced by many factors. For example, the price of virgin plastic and consumer preferences. If there is little market demand for recycled plastic, it can be difficult to make recycling economically viable. Technological limitations: Although recycling technologies have advanced, there are still limitations. For example, certain types of plastics, such as multilayer packaging or plastic blends, may be difficult or expensive to recycle and challenging one to be consider. Consumer and education: Encouraging consumers to recycle correctly and educating them about the importance of recycling is always an important task. Without proper training and awareness, contamination rates can remain high, and valuable recyclables can end up in landfills.

Globalization and trade: The global nature of the plastics industry means that recycled plastics are subject to the dynamics of international trade, including tariffs, regulations, and market fluctuations. These factors can affect the economics of recycling and the viability of recycled plastic as a commodity. Addressing the above challenges with technological innovation, infrastructure development, policy initiatives, consumer education, and cross-sector collaboration could improve the efficiency and effectiveness of plastic recycling in real-world applications in the near future.

Future prospects

The future outlook for plastic recycling presents both challenges and opportunities as our own society is struggling with the environmental impacts of plastic waste. Technological Innovations with continuous research and development of recycling technologies promise to improve the efficiency and effectiveness of plastic recycling. Innovations such as improved sorting techniques, chemical recycling processes, and the development of new, more easily recyclable materials can significantly improve the recyclability of plastics. Governments and regulators are increasingly implementing policies and regulations to encourage recycling and reduce plastic waste. Measures such as EPR, plastic bans, DRS, and landfill charges can create an enabling environment for plastic recycling. Increased investment and funding in the development of plastic recycling infrastructure and technology is necessary to increase recycling capacity and improve recycling rates. Public–private partnerships and collaboration between industry stakeholders, governments, and nongovernmental organizations can facilitate investment in recycling initiatives. Moreover, educating consumers about the importance of recycling and proper waste management can help reduce pollution levels and increase the quantity and quality of recycled plastics. Marketing campaigns, labeling schemes, and community engagement programs can raise awareness and encourage behavior change. Similarly, the collaboration between stakeholders in the plastic value chain, including manufacturers, retailers, waste management companies, recyclers, and policymakers, is critical to promoting plastic recycling. Partnerships that promote innovation, knowledge sharing, and resource mobilization can accelerate the transition to a more sustainable plastics ecosystem. Highlighting the demand for recycled plastics from producers and consumers can drive investment in recycling infrastructure and create market opportunities for recycled plastics. Initiatives such as green procurement policies and corporate sustainability commitments can increase the demand for recycled plastic.

Technological advancements

Technological advances in plastic recycling are necessary to effectively overcome the challenges of plastic recycling. Automated sorting technologies such as optical sorting systems and NIR sensors enable more accurate identification and separation of different types of plastics. These technologies improve the efficiency of recycling processes by reducing pollution and increasing the purity of recycled materials. Chemical recycling, also known as advanced or tertiary recycling, involves breaking down plastic waste into chemical components to produce new plastics or other valuable materials. Processes such as pyrolysis, depolymerization, and gasification offer new opportunities for recycling plastics that are difficult to recycle using traditional mechanical methods. The development of biodegradable and compostable plastics offers an alternative to traditional plastics that are derived from renewable resources and can be removed by composting. Although these materials may not always be suitable for traditional recycling processes, they reduce the environmental impact of plastic waste. Additive manufacturing technologies, such as 3D printing can offer opportunities for plastic recycling, enabling the production of new products from recycled plastic raw materials. These technologies enable more flexible and decentralized recycling processes allowing local communities to turn plastic waste into useful items. Mechanical recycling technologies, such as melt filtration systems and extrusion processes, improve the quality and properties of recycled plastics. These advances make it possible to produce high-performance recycled plastics that can be used in a variety of applications, including food packaging and automotive parts. As discussed earlier also the robotics and Automation play an important role in simplifying and optimizing recycling operations such as sorting, crushing, and granulation. Automated systems increase throughput, reduce labor costs, and improve the safety of recycling facilities, increasing efficiency and productivity. Similarly, IoT-enabled sensors and monitoring devices, provide real-time information on waste generation, collection, and recycling processes. These solutions enable predictive maintenance, route optimization, and data-driven decision making that can improve the efficiency and sustainability of plastic recycling initiatives. Also, using these technological advances, the plastics recycling industry can overcome existing challenges and move toward a more circular and sustainable economy where plastic waste is minimized and precious resources are conserved.

Integration of circular economy principles

Integrating circular economy principles into plastic recycling practices is essential to creating a more sustainable and resource-efficient system. These principles can be applied to plastic recycling as follows: Design for reuse: The recyclability of plastic products can be improved by adopting design principles that prioritize recyclability and facilitate disassembly and separation of parts. Designing products with fewer materials and using cash or easily separated materials simplifies the recycling process. EPR: The implementation of EPR policies places responsibility for the management of end-of-life products on producers, and encourages them to design recyclable products, use recycled materials, and invest in recycling infrastructure. EPR systems encourage producers to take responsibility for the entire life cycle of their products, including collection, recycling, and disposal. Resource efficiency and closed-loop systems: Maximizing resource efficiency by reusing and recycling materials from end-of-life products promotes a closed-loop system, where materials are continuously recycled and reintegrated into the production process. Closed-loop systems reduce the demand for new materials, minimize waste production, and reduce environmental impact. Value chain cooperation: Cooperation in the plastic value chain, including between producers, retailers, recyclers, and consumers, is crucial for the effective implementation of circular economy principles. Partnerships and supply chain integration facilitate the collection, sorting, and recycling of plastic waste and the development of the market for recycled materials. Reverse logistics and collection system: Creating effective reverse logistics and collection systems to collect and transport plastic waste from consumers and businesses to recycling sites is essential to closing the circle and maximizing the use of resources. Convenient collection points, encouraging return systems, and innovative collection methods encourage participation and increase recycling rates. Market demand for recycled materials: Increasing demand for recycled materials through purchasing policies, incentives, and consumer preferences creates market opportunities for recycled plastics and encourages investment in recycling infrastructure. By creating an attractive force for recycled materials, circular economy principles guide market-based solutions for handling plastic waste. News and technology: Investment in research, development, and innovation in recycling technologies, materials science, and product design will drive continuous improvement in plastic recycling processes. Innovations such as advanced sorting technologies, chemical recycling, and biodegradable materials are expanding recycling options and creating new recycling opportunities. Education and awareness: Educating consumers, businesses, and policymakers about the principles and benefits of the circular economy promotes a culture of sustainable development and responsible consumption. Information campaigns, training programs, and stakeholder engagement initiatives raise awareness of the environmental impact of plastic waste and empower people to take action. By integrating circular economy principles into plastic recycling practices, stakeholders can move toward a more sustainable and sustainable system that conserves resources, reduces waste, and minimizes environmental pollution.

Potential barriers to progress

Despite the importance of plastic recycling, several barriers hinder progress in this area. Here are some potential barriers: Complexity and diversity of plastic types: Plastics come in various types and compositions, each requiring different recycling processes. This complexity makes sorting and recycling challenging and may lead to contamination and reduced recycling efficiency.

Contamination: Contamination of recycled plastics with nonrecyclable materials, food residues, or other contaminants can compromise the quality of recycled materials, reducing their value and usability. Lack of infrastructure: Many regions lack adequate infrastructure for plastic recycling, including collection systems, sorting facilities, and recycling plants. The absence of infrastructure limits the capacity to collect, process, and recycle plastic waste effectively. Economic viability: The economics of plastic recycling can be challenging, particularly when the cost of recycling exceeds the value of recycled materials. Limited market demand for recycled plastics and fluctuations in commodity prices can further impact the economic viability of recycling operations.

Technological limitations: While advancements have been made in recycling technologies, certain plastics, such as multilayered or mixed plastics, remain difficult to recycle economically. Technological limitations may restrict the types of plastics that can be effectively recycled.

Consumer behavior: Inadequate consumer awareness and participation in recycling programs can contribute to low recycling rates and contamination of recyclable materials. Education and outreach efforts are needed to encourage proper recycling practices among consumers.

Regulatory challenges: Inconsistent or inadequate regulations related to plastic recycling may pose barriers to progress. Regulatory frameworks that lack enforcement mechanisms, standards for recycled content, or incentives for recycling may hinder investment in recycling infrastructure and innovation. Global trade dynamics: The global nature of the plastics industry means that recycled plastics are subject to international trade dynamics, including tariffs, regulations, and market fluctuations. Uncertainty and instability in global markets can impact the economics of recycling and trade in recycled materials. Limited end markets: The availability of end markets for recycled plastics is crucial for the success of recycling programs. However, limited demand for recycled materials in certain industries or regions may constrain the growth of plastic recycling initiatives. Environmental concerns: Despite its environmental benefits, plastic recycling is not without environmental concerns, such as energy consumption, emissions, and the release of microplastics. Addressing these concerns while promoting recycling requires careful consideration of the environmental impacts throughout the recycling process. Addressing these barriers requires a coordinated effort involving stakeholders across the plastics value chain, including governments, industry, academia, and civil society. Strategies to overcome barriers may include investing in infrastructure, promoting innovation, improving regulations, raising awareness, and fostering collaboration among stakeholders.

Conclusion

Plastic recycling has huge potential as a critical solution to reduce environmental pollution, save natural resources, and move toward a more sustainable and circular economy. Despite many challenges such as technological limitations, financial constraints, and regulatory barriers, the importance of plastic recycling cannot be overemphasized. Overcoming these challenges requires collaboration between governments, industry stakeholders, researchers, and consumers. Investing in innovative recycling technologies, improving infrastructure, implementing effective policies, and raising consumer awareness are important steps to realizing the full potential of plastic recycling. Ultimately, concerted efforts to promote plastic recycling are not only necessary but necessary to solve the global plastic pollution crisis and create a more sustainable and greener future. According to the circular principle, we can transform plastic waste into valuable resources, minimize its environmental impact and create a more sustainable world for future generations.

Recap of key findings

A key prescription of plastic recycling technology is the importance of advances in sorting, chemical recycling, and additive manufacturing processes. These innovations are necessary to improve the efficiency and effectiveness of plastic recycling, enable better sorting accuracy, reuse plastics that were previously difficult to recycle, and create new products from recycled materials. For the improvement of Plastic Recycling technology, we can highlight the importance of technological innovation, political support, and consumer engagement in improving recycling rates, reducing pollution and promoting a circular economy.

Recommendations for future research

Future research on plastic recycling should prioritize the following areas

1. Develop improved sorting techniques to improve accuracy and efficiency.
2. Improve chemical recycling processes for a wider range of plastic types and scalability.
3. Create biodegradable plastics compatible with existing recycling infrastructure.
4. Conduct a LCA to better understand environmental impacts.
5. Explore new markets and applications for recycled plastics.
6. Review and improve the policy framework, including EPR.
7. Explore strategies to increase consumer participation and awareness.
8. Integrate circular economy principles into plastic recycling practices.

Abbreviations and acronyms

ABS acrylonitrile butadiene styrene
AI artificial intelligence
CNN convolutional neural network
DL deep learning
FIR far-infrared
HDPE high-density polyethylene
IoT Internet of Things
LDPE low-density polyethylene
MRFs material recovery facilities

MSRW municipal solid refuse waste
NIR near infrared
PET polyethylene terephthalate
PRFs plastic reclamation facilities
POs polyolefins
PPW postconsumer waste
PP polypropylene
PVC polyvinyl chloride
SPI Society of the Plastics Industry
VIS visual spectrophotometry
XRF X-ray fluorescence

References

[1] H. Mohammad Hosseini, R. Alyousef, M. Md. Tahir, Towards sustainable concrete composites through waste valorisation of plastic food trays as low-cost fibrous materials, Sustainability 13 (2021) 2073. Available from https://doi.org/10.3390/su13042073.

[2] J. Hopewell, R. Dvorak, E. Kosior, Plastics recycling: challenges and opportunities, Philosophical Transactions of the Royal Society B: Biological Sciences (2009). Available from https://doi.org/10.1098/rstb.2008.0311.

[3] I.E. Napper, R.C. Thompson, Plastics and the environment, Annu. Rev. Environ. Resour. 48 (1) (2023) 55–79. https://doi.org/10.1146/annurev-environ-112522-072642.

[4] S. Khadke, P. Gupta, S. Rachakunta, C. Mahata, S. Dawn, M. Sharma, et al., Efficient plastic recycling and remolding circular economy using the technology of trust–blockchain, Sustainability 13 (2021) 9142. Available from https://doi.org/10.3390/su13169142.

[5] K. Bernat, Post-consumer plastic waste management: from collection and sortation to mechanical recycling, Energies 16 (2023) 3504. Available from https://doi.org/10.3390/en16083504.

[6] J. Schmidt, M. Auer, J. Moesslein, P. Wendler, S. Wiethoff, C. Lang-Koetz, et al., Challenges and solutions for plastic packaging in a circular economy, Chemie Ingenieur Technik. 93 (2021) 1751–1762.

[7] A. Pires, J. Sargedas, M. Miguel, J. Pina, G. Martinho, A case study of packaging waste collection systems in Portugal—part II: environmental and economic analysis, Waste Management. (New York, N.Y.) 61 (2017) 108–116.

[8] G. Martinho, A. Gomes, P. Santos, M. Ramos, J. Cardoso, A. Silveira, et al., A case study of packaging waste collection systems in Portugal—part I: performance and operation analysis, Waste Management (New York, N.Y.) 61 (2017) 96–107.

[9] C. Cimpan, A. Maul, M. Jansen, T. Pretz, H. Wenzel, Central sorting and recovery of MSW recyclable materials: a review of technological state-of-the-art, cases, practice and implications for materials recycling, Journal of Environmental Management. 156 (2015) 181–199.

[10] A. Gallardo, M.D. Bovea, F.J. Colomer, M. Prades, M. Carlos, Comparison of different collection systems for sorted household waste in Spain, Waste Management. (New York, N.Y.) 30 (2010) 2430–2439.

[11] J.N. Hahladakis, P. Purnell, E. Iacovidou, C.A. Velis, M. Atseyinku, Post-consumer plastic packaging waste in England: assessing the yield of multiple collection-recycling schemes, Waste Management. (New York, N.Y.) 75 (2018) 149–159.

[12] A. Chruszcz; S. Reeve Composition of Plastic Waste Collected via Kerbside, WRAP, 2018. Available from: <https://wrap.org.uk/resources/report/composition-plastic-waste-collected-kerbside>.

[13] E. Dijkgraaf, R. Gradus, Post-collection separation of plasticwaste: better for the environment and lower collection costs? Environmental and Resource Economics. 77 (2020) 127–142.

[14] K. Ragaert, L. Delva, K. van Geem, Mechanical and chemical recycling of solid plastic waste, Waste Manag. (New York, N.Y.) 69 (2017) 24–58.

[15] D. McKinnon, J. Fazakerley, R. Hultermans, Waste Sorting Plants—Extracting Value from Waste, ISWA, Vienna, Austria, 2017.

[16] A. Chidepatil, P. Bindra, D. Kulkarni, M. Qazi, M. Kshirsagar, K. Sankaran, From trash to cash: how blockchain and multi-sensor-driven artificial intelligence can transform circular economy of plastic waste? Administrative Sciences. 10 (2020) 23.
[17] H. Wilts, B.R. Garcia, R.G. Garlito, L.S. Gómez, E.G. Prieto, Artificial intelligence in the sorting of municipal waste as an enabler of the circular economy, Resources 10 (2021) 28.
[18] C. Lubongo, M.A.A. Bin Daej, P. Alexandridis, Recent developments in technology for sorting plastic for recycling: the emergence of artificial intelligence and the rise of the robots, Recycling 9 (2024) 59. Available from https://doi.org/10.3390/recycling9040059.
[19] A. Rahimi, J.M. García, Chemical recycling of waste plastics for new materials production, Nature Reviews Chemistry 1 (6) (2017) 0046.
[20] M.K. Eriksen, A. Damgaard, A. Boldrin, T.F. Astrup, Quality assessment and circularity potential of recovery systems for household plastic waste, Journal of Industrial Ecology. 23 (1) (2019) 156–168.
[21] H. Jung, G. Shin, H. Kwak, L.T. Hao, J. Jegal, H.J. Kim, et al., Review of polymer technologies for improving the recycling and upcycling efficiency of plastic waste, Chemosphere 320 (2023) 138089.
[22] B. Geueke, K. Groh, J. Muncke, Food packaging in the circular economy: overview of chemical safety aspects for commonly used materials, Journal of Cleaner Production 193 (2018) 491–505.
[23] G. Faraca, T. Astrup, Plastic waste from recycling centres: characterisation and evaluation of plastic recyclability, Waste Management (New York, N.Y.) 95 (2019) 388–398.
[24] B. Samiha, The importance of the 3R principle of municipal solid waste management for achieving sustainable development, Mediterranean Journal of Social Sciences. 4 (3) (2013) 129.
[25] N. Taneepanichskul, D. Purkiss, M. Miodownik, A review of sorting and separating technologies suitable for compostable and biodegradable plastic packaging, Frontiers in Sustainability. 3 (2022).
[26] C. Lubongo, P. Alexandridis, Assessment of performance and challenges in use of commercial automated sorting technology for plastic waste, Recycling 7 (2) (2022) 11.
[27] D. Bonello, M.A. Saliba, K.P. Camilleri, An exploratory study on the automated sorting of commingled recyclable domestic waste, Procedia Manufacturing 11 (2017) 686–694.
[28] P. Pandiaraja, S. SanthanaHari, S. Suriya, S. Karthikeyan, Convolutional neural network for solid waste segregation and management, International Journal of Advanced Science and Technology. 29 (7) (2020) 1661–1668.
[29] J.-P. Lange, Managing plastic waste—sorting, recycling, disposal, and product redesign, ACS Sustainable Chemistry & Engineering. 9 (47) (2021) 15722–15738.
[30] S.P. Gundupalli, S. Hait, A. Thakur, A review on automated sorting of source separated municipal solid waste for recycling, Waste Management (New York, N.Y.) 60 (2017) 56–74.
[31] B. Esmaeilian, B. Wang, K. Lewis, F. Duarte, C. Ratti, S. Behdad, The future of waste management in smart and sustainable cities: a review and concept paper, Waste Management (New York, N.Y.) 81 (2018) 177–195.
[32] U. Martinez-Hernandez, G. West, T. Assaf, Low-cost recognition of plastic waste using deep learning and a multi-spectral near-infrared sensor, Sensors 24 (2024) 2821. Available from https://doi.org/10.3390/s24092821.
[33] S. Vishnu, S.R.J. Ramson, M.S.S. Rukmini, A.M. Abu-Mahfouz, Sensor-based solid waste handling systems: a survey, Sensors 22 (6) (2022) 2340.
[34] A.S.Wijaya, Z. Zainuddin, M. Niswar, Design a smart waste bin for smart waste management, 2017 5th International Conference on Instrumentation, Control, and Automation (ICA), 2017, pp. 62–66.
[35] S. Lopes, S. Machado, IoT based automatic waste segregator, 2019 International Conference on Advances in Computing, Communication and Control (ICAC3), 2019, pp. 1–5.
[36] J.R. Raj, B.I.P. Rajula, R. Tamilbharathi, S. Srinivasulu, An IoT based waste segreggator for recycling biodegradable and non-biodegradable waste, 2020 6th International Conference on Advanced Computing and Communication Systems (ICACCS), 2020, pp. 928–930.
[37] N. Samion, M. Kaamin, N. Adan, A. Nadiah, A. Kadir, U. Tun, et al., The effectiveness of segregation recyclable materials by automated motorized bin, Journal of Advanced Manufacturing Technology. 12 (2018).
[38] A. Das, A. Shukla, R. Manjunatha, E.A. Lodhi, IoT based solid waste segregation using relative humidity values, 2021 Third International Conference on Intelligent Communication Technologies and Virtual Mobile Networks (ICICV), 2021, pp. 312–319.
[39] N. Kroell, X. Chen, A. Maghmoumi, M. Koenig, A. Feil, K. Greiff, Sensor-based particle mass prediction of lightweight packaging waste using machine learning algorithms, Waste Management (New York, N.Y.) 136 (2021) 253–265.

[40] H. Masoumi, S.M. Safavi, Z. Khani, Identification and classification of plastic resins using near infrared reflectance spectroscopy, International Journal of Mechanical Engineering. 6 (2012) 213–220.
[41] M. Rani, C. Marchesi, S. Federici, G. Rovelli, I. Alessandri, I. Vassalini, et al., Miniaturized near-infrared (MicroNIR) spectrometer in plastic waste sorting, Materials 12 (17) (2019) 2740.
[42] Q. Duan, J. Li, Classification of common household plastic wastes combining multiple methods based on near-infrared spectroscopy, ACS EST Eng. 1 (7) (2021) 1065–1073.
[43] X. Wu, J. Li, L. Yao, Z. Xu, Auto-sorting commonly recovered plastics from waste household appliances and electronics using near-infrared spectroscopy, Journal of Cleaner Production. 246 (2020) 118732.
[44] I. Bhati, M. Bhattacharya, An IOT-based system for classification and identification of plastic waste using near infrared spectroscopy, in: Proceedings of the 2nd International Conference on Communication, Devices and Computing: ICCDC 2019, Springer Singapore, 2020, pp. 697–703. Available from https://10.1007/978-981-15-0829-5_65.
[45] M. Abdallah, M. Abu Talib, S. Feroz, Q. Nasir, H. Abdalla, B. Mahfood, Artificial intelligence applications in solid waste management: a systematic research review, Waste Management (New York, N.Y.) 109 (2020) 231–246.
[46] W. Xia, Y. Jiang, X. Chen, R. Zhao, Application of machine learning algorithms in municipal solid waste management: a mini review, Waste Management & Research. 40 (6) (2021) 609–624.
[47] G.L. Huang, J. He, Z. Xu, G. Huang, A combination model based on transfer learning for waste classification, Concurrency and Computation: Practice and Experience. 32 (19) (2020) e5751.
[48] K. Lin, Y. Zhao, J.-H. Kuo, H. Deng, F. Cui, Z. Zhang, et al., Toward smarter management and recovery of municipal solid waste: a critical review on deep learning approaches, Journal of Cleaner Production. 346 (2022) 130943.
[49] S. Agarwal, R. Gudi, P. Saxena, One-shot learning based classification for segregation of plastic waste, 2020 Digital Image Computing: Techniques and Applications (DICTA), 2020, pp. 1–3.
[50] R. Maliks, R. Kadikis, Multispectral data classification with deep CNN for plastic bottle sorting, 2021 6th International Conference on Mechanical Engineering and Robotics Research (ICMERR), IEEE, 2021, pp. 58–65.
[51] E.R.K. Neo, J.S.C. Low, V. Goodship, K. Debattista, Deep learning for chemometric analysis of plastic spectral data from infrared and Raman databases, Resources, Conservation and Recycling. 188 (2023) 106718.
[52] J. Xia, Y. Huang, Q. Li, Y. Xiong, S. Min, Convolutional neural network with near-infrared spectroscopy for plastic discrimination, Environmental Chemistry Letters. 19 (5) (2021) 3547–3555.
[53] R. Puˇcnik, M. Dokl, Y. Van Fan, A. Vujanovic´, Z.N. Pintariˇc, K.B. Aviso, et al., A waste separation system based on sensor technology and deep learning: a simple approach applied to a case study of plastic packaging waste, Journal of Cleaner Production. 450 (2024) 141762.
[54] R.D. Kulko, A. Pletl, A. Hanus, B. Elser, Detection of plastic granules and their mixtures, Sensors 23 (2023) 3441. Available from https://doi.org/10.3390/s23073441.
[55] Z. Cai, J. Yang, H. Fang, T. Ji, Y. Hu, X. Wang, Research on waste plastics classification method based on multi-scale feature fusion, Sensors 22 (2022) 7974. Available from https://doi.org/10.3390/s22207974.
[56] G. Chowdhury, et al., Image-based plastic classification using convolutional neural networks, Journal of Waste Management (2020).
[57] B. Adebayo, et al., Hyperspectral imaging for plastic waste sorting, Journal of Environmental Science and Technology. (2019).
[58] X. Zeng, et al., Reinforcement learning for robotic waste sorting, IEEE Robotics and Automation Letters (2018).
[59] H. Yang, et al., Adaptive learning in robotic sorting systems, Journal of Intelligent Systems (2021).
[60] Y. Liu, et al., Multi-sensor fusion for plastic sorting, Sensors (2019).
[61] M. García, et al., Real-time AI processing for waste sorting, Journal of Industrial Ecology (2020).
[62] L. Zhang, et al., Data quality in AI waste sorting, Artificial Intelligence Review (2022).
[63] J. Smith, et al., Scalability of AI sorting systems, Journal of Environmental Engineering (2021).
[64] M. Artetxe, G. Lopez, Amutio, P. Jones, et al., Ethical and environmental implications of AI in waste management, Journal of Cleaner Production (2020).
[65] T. Petersen, R. Nygaard, Robotic Arms in Plastic Sorting: Efficiency and Accuracy, 2018.
[66] L. Huang, et al., Integration of Conveyor Belt Systems with Robotic Arms for Plastic Sorting, 2019.
[67] M. Kühn, et al., Speed and Efficiency of Robotic Sorting Systems, 2020.
[68] A. Jones, et al., Consistency and Accuracy in Robotic Plastic Sorting, 2021.
[69] J. Smith, et al.,Safety Improvements through Automation in Plastic Sorting, 2022.
[70] X. Li, et al., Application of CNNs in Plastic Identification, 2019.

[71] Y. Zhao, et al., NIR Spectroscopy and Machine Learning for Plastic Classification, 2020.
[72] P. Wang, et al. Predictive Maintenance Framework for Sorting Equipment, 2021.
[73] R. Kumar, et al., Optimization of Plastic Sorting Processes using Machine Learning, 2022.
[74] T. Johnson, et al., Real-time Monitoring of Sorting Processes with Machine Learning, 2021.
[75] L. Chen, et al., Continuous Improvement in Plastic Sorting Accuracy with Machine Learning, 2023.
[76] M. Garcia, et al., Case Study: AMP Robotics in Plastic Sorting, 2020.
[77] S. Miller, et al., Case Study: ZenRobotics in Plastic Sorting, 2021.
[78] J. Lee, et al., Advanced Sensors for Improved Plastic Sorting, 2024.
[79] N. Patel, et al., Improved Machine Learning Algorithms for Plastic Sorting, 2023.
[80] D. Taylor, et al., Scalability of Advanced Sorting Systems in Recycling Facilities, 2023.
[81] G. Lopez, et al., Thermochemical routes for the valorization of waste polyolefinic plastics to produce fuels and chemicals. A review, Energy & Environmental Science (2017).
[82] U. Arena, Process and technological aspects of municipal solid waste gasification. A review, Waste Management. (2012).
[83] A.M. Al-Sabagh, et al., Recent advances in the chemical recycling of PET, Green Chemistry (2016).
[84] J. Aguado, et al., Catalytic cracking of polyolefins into liquid fuels, Journal of Analytical and Applied Pyrolysis (2007).
[85] M.R. Kember, et al., Solvolysis of polyurethanes and PET for chemical recycling, Macromolecular Materials and Engineering (2019).
[86] V. Tournier, et al., An engineered PET depolymerase to break down and recycle plastic bottles, Nature (2020).
[87] A.K. Panda, et al., Thermolysis of waste plastics to liquid fuel: a suitable method for plastic waste management and manufacture of value added products—a world prospective, Renewable and Sustainable Energy Reviews (2010).
[88] K. Ragaert, et al., Mechanical and chemical recycling of solid plastic waste, Waste Management (2017).
[89] J. Hopewell, et al., Plastics recycling: challenges and opportunities, Philosophical Transactions of the Royal Society B: Biological Sciences (2009).
[90] B. Kunwar, H. Cheng, S. Chandrashekaran, B. Sharma, Plastics to fuel: a review, Renewable and Sustainable Energy Reviews. 54 (2016) 421–428.
[91] Z. Wang, J. Li, K. Burra, X. Liu, X. Li, M. Zhang, et al., Synergetic effect on CO2-assisted co-gasification of biomass and plastics, The Journal of Energy Resources Technology. 143 (2021) 031901.
[92] P. Singh, N. D´eparrois, K. Burra, S. Bhattacharya, A. Gupta, Energy recovery from cross-linked polyethylene wastes using pyrolysis and CO2 assisted gasification, Applied Energy 254 (2019) 113722.
[93] G. Elordi, M. Olazar, G. Lopez, M. Artetxe, J. Bilbao, Product yields and compositions in the continuous pyrolysis of high-density polyethylene in a conical spouted bed reactor, Ind. Eng. Chem. Res. 50 (2011) 6650–6659.
[94] T. Kruse, O. Woo, H. Wong, S. Khan, L. Broadbelt, Mechanistic modeling of polymer degradation: a comprehensive study of polystyrene, Macromolecules 35 (2002) 7830–7844.
[95] J. Zhou, Y. Qiao, W. Wang, E. Leng, J. Huang, Y. Yu, et al., Formation of styrene monomer, dimer and trimer in the primary volatiles produced from polystyrene pyrolysis in a wire-mesh reactor, Fuel 182 (2016) 333–339.
[96] T.M. Kruse, O. Sang Woo, L.J. Broadbelt, Detailed mechanistic modeling of polymer degradation: application to polystyrene, Chemical Engineering Science 56 (2001) 971–979.
[97] B. Gui, Y. Qiao, D. Wan, S. Liu, Z. Han, H. Yao, et al., Nascent tar formation during polyvinylchloride (PVC) pyrolysis, Proceedings of the Combustion Institute. 34 (2013) 2321–2329.
[98] Thermoplastic polyesters, Polym. Adv. Technol. 15 (2004) 691–700. [60] F. Samperi, C. Puglisi, R. Alicata, G. Montaudo, Thermal degradation of poly(ethylene terephthalate) at the processing temperature, Polym. Degrad. Stab. 83 (2004) 3–10.
[99] M. Artetxe, G. Lopez, M. Amutio, G. Elordi, M. Olazar, J. Bilbao, Operating conditions for the pyrolysis of poly-(ethylene terephthalate) in a conical spouted-bed reactor, Industrial & Engineering Chemistry Research. 49 (2010) 2064–2069.
[100] C. Kassargy, S. Awad, G. Burnens, K. Kahine, M. Tazerout, Experimental study of catalytic pyrolysis of polyethylene and polypropylene over USY zeolite and separation to gasoline and diesel-like fuels, Journal of Analytical and Applied Pyrolysis. 127 (2017) 31–37.
[101] D.S. Achilias, E. Antonakou, C. Roupakias, P. Megalokonomos, A. Lappas, Recycling techniques of polyolefins from plastic wastes, Global Nest Journal 10 (2008) 114e122.

[102] C. Gutierrez, M.T. Garcia, I. Gracia, A. de Lucas, J.F. Rodriguez, The selective dissolution technique as initial step for polystyrene recycling, Waste Biomass Valor. 4 (2013) 29e36.
[103] R.E. Martini, S. Barbosa, E. Brignole, Demixing of polypropylene/polystyrene blends by near-critical selective solubilization, Industrial & Engineering Chemistry Research. 45 (2006) 3393e3399.
[104] I. Devotta, R.A. Mashelkar, Role of thermodynamic and kinetic factors in polymer dissolution in mixed solvents, Chemical Engineering Communications. 156 (1996) 31e43.
[105] W.J. Cooper, P.D. Krasicky, F. Rodriguez, Dissolution rates of poly(methyl methacrylate) films in mixed-solvents, Journal of Applied Polymer Science. 31 (1986) 65e73.
[106] B.A. Miller-Chou, J.L. Koenig, Dissolution of symmetric diblock copolymers with neutral solvents, a selective solvent, a nonsolvent, and mixtures of a solvent and nonsolvent monitored by FT-IR imaging, Macromolecules 36 (2003) 4851e4861.
[107] T. Noguchi, Y. Inagaki, M. Miyashita, H. Watanabe, A new recycling system for expanded polystyrene using a natural solvent. Part 2. Development of a prototype production system, Packaging Technology and Science. 11 (1998) 29e37.
[108] J.G. Poulakis, C.D. Papaspyrides, Recycling of polypropylene by the dissolution/ reprecipitation technique.1. A model study, Resources, Conservation and Recycling. 20 (1997) 31e41.
[109] L.M. Vane, F. Rodriguez, Selected aspects of poly(ethylene-terephthalate) solution behavior - application to a selective dissolution process for the separation of mixed plastics, ACS Symposium Series. 513 (1992) 147e162.
[110] A. Arostegui, M. Sarrionandia, J. Aurrekoetxea, I. Urrutibeascoa, Effect of dissolution-based recycling on the degradation and the mechanical properties of acrylonitrile–butadiene–styrene copolymer, Polymer Degradation and Stability 91 (11) (2006) 2768–2774. Available from https://doi.org/10.1016/j.polymdegradstab.2006.03.019.
[111] Ellen MacArthur Foundation, The New Plastics Economy: Rethinking the Future of Plastics, 2016.
[112] D. Lazarevic, et al., Plastics recycling: challenges and opportunities, Philosophical Transactions of the Royal Society B: Biological Sciences (2010).
[113] European Environment Agency (EEA), The Plastic Waste Trade in the Circular Economy, 2019.
[114] Institute for Local Self-Reliance (ILSR), Recycling Means Business, 2017.
[115] D.C. Wilson, et al., Role of Informal Sector Recycling in Waste Management in Developing Countries, Habitat International, 2006.
[116] Grand View Research, Recycled Plastics Market Size, Share & Trends Analysis Report, 2021.
[117] World Economic Forum, The Global Risks Report 2020, 2020.
[118] McKinsey & Company, How Plastics-waste Recycling Could Transform the Chemical Industry, 2018.
[119] European Commission, Horizon 2020: The EU Framework Programme for Research and Innovation, 2020.
[120] OECD, Extended Producer Responsibility: Updated Guidance for Efficient Waste Management, 2016.
[121] J. Smith, Advancements in NIR technology for plastic sorting, Journal of Environmental Science (2020).
[122] A. Johnson, et al., XRF applications in recycling, Material Recycling Journal (2019).
[123] T. Lee, Machine vision systems in waste management, Autom. Recycl. (2021).
[124] R. Brown, Robotic automation in plastic recycling, The International Journal of Robotics Research (2018).
[125] D. Green, Cost analysis of recycling automation, Waste Management Review (2022).
[126] L. Zhao, Labor cost reduction through automation, Recycling Today (2019).
[127] S. Kumar, Throughput analysis of automated recycling facilities, Journal of Waste Management (2020).
[128] H. Clark, Quality improvement in recycled plastics, Polymer Recycling (2021).
[129] M. Young, Environmental benefits of automated recycling, Green Technology Journal (2019).
[130] CIEL, Plastic & Climate: The Hidden Costs of a Plastic Planet, Center for International Environmental Law, 2019.
[131] U.S. Environmental Protection Agency (EPA), Advancing Sustainable Materials Management: 2018 Fact Sheet, 2020.
[132] European Environment Agency (EEA), Plastics, the Circular Economy and Europe's Environment—A priority for Action, 2019.
[133] T. Brown, L. Jones, Workforce impact of automation in recycling industries, Industrial Technology Rev (2019).
[134] M. Green, P. Adams, Reskilling for the future: workforce development in the automated recycling sector, Journal of Vocational Education (2022).
[135] AMP Robotics, Case study: transforming recycling with AI and robotics, AMP Robotics White Paper, 2023.
[136] ZenRobotics, Educational partnerships for workforce transition, ZenRobotics Annual Report, 2021.

CHAPTER

6

Artificial intelligence for reutilizing the plastics

Hameed Khan[1], *and Kamal Kushwah*[2]

[1]Computer Science and Engineering, Jabalpur Engineering College, Jabalpur, Madhya Pradesh, India [2]Department of Applied Physics, Jabalpur Engineering College, Jabalpur, Madhya Pradesh, India

Introduction

Plastic pollution is a global environmental challenge that poses significant threats to ecosystems, human health, and the overall well-being of the planet. The exponential increase in plastic production, coupled with improper disposal and limited recycling efforts, has led to the accumulation of plastic waste in landfills, oceans, and various ecosystems. Addressing this crisis requires innovative and sustainable solutions, and artificial intelligence (AI) emerges as a powerful tool to revolutionize how we approach plastic reutilization. Rapidly developing AI is becoming increasingly common in several sectors, notably waste management. Urban waste handling plant design and operation may be revolutionized by integrating robots and AI, which will result in more sustainable methods for waste management and improved operational efficiency for solid waste management. AI technologies are already being adopted by some developed nations, such as Austria, New Zealand, the United States, Japan, Singapore, Switzerland, and Canada, to optimize the use of resources, efficiency, and recycling opportunities throughout the solid waste management cycle. AI technologies are becoming increasingly important in waste management, especially for sorting and treating solid waste.

This chapter summarizes the many forms of garbage, how they are produced, the problems that arise from them, and how AI may be used to manage waste. These applications include trash-sorting robots, intelligent bin systems, sensor-based waste surveillance, and waste-generation prediction models. The chapter also addresses how AI can facilitate the tracking and monitoring of waste materials during the recycling process, as well as the identification and mitigation of illicit dropping and disposal practices, waste composition analysis, and the optimization of recycled waste logistics and transportation. This research makes a new contribution by combining waste chemical analysis and AI to enhance waste-to-energy conversion.

Harnessing Automation and Machine Learning for Resource Recovery and Value Creation
DOI: https://doi.org/10.1016/B978-0-443-27374-2.00006-6

Plastic pollution scenario 2024

There is a plastic pollution issue throughout the globe. Over 8 billion tons of plastic have been manufactured by humanity since 1950; most of this material ended up in landfills, with just 9% being recycled. Plastic may harm the ecosystem gradually but inevitably in a number of ways, from physically strangling or killing animals who unintentionally consume it to seeping harmful chemicals into the groundwater and soil. Larger, more populous nations often create more plastic garbage overall, but the ranking drastically shifts when the data are narrowed down to highlight the top producers per capita, or person. The top ten nations in the world in terms of plastic trash production are shown in Fig. 6.1.

Machine learning in plastic composition analysis

The identification of various kinds of plastic is made possible by machine learning (ML) models, which also handle the complexity of plastic trash and facilitate effective sorting. Like many other nations, privatized trash collection businesses dispose of municipal solid waste in the country and either transport it straight to landfills or separate it at the source for

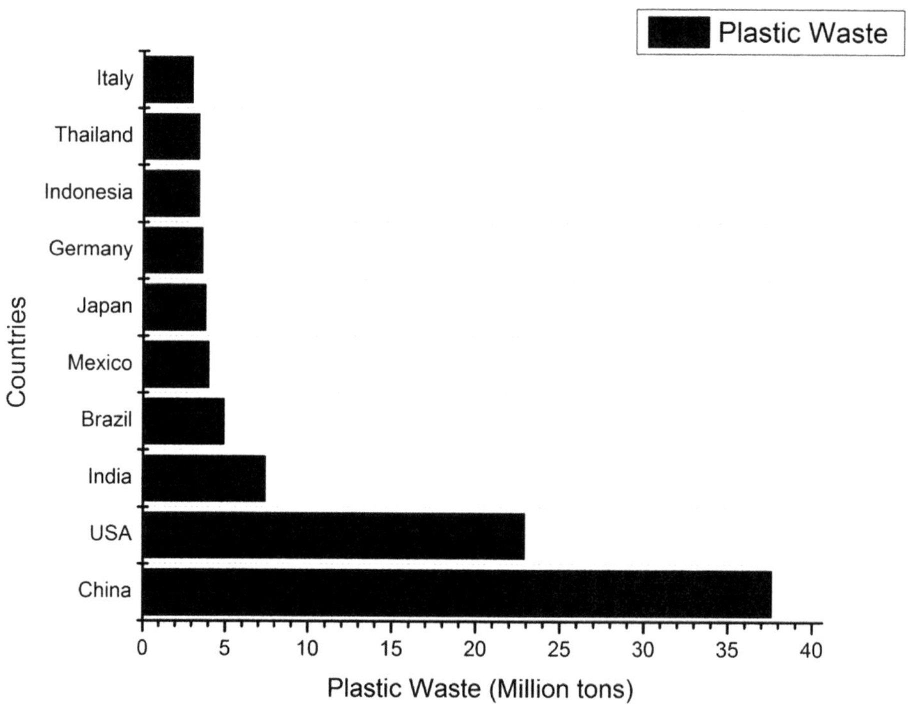

FIGURE 6.1 Top countries that produce the most plastic waste (tons).

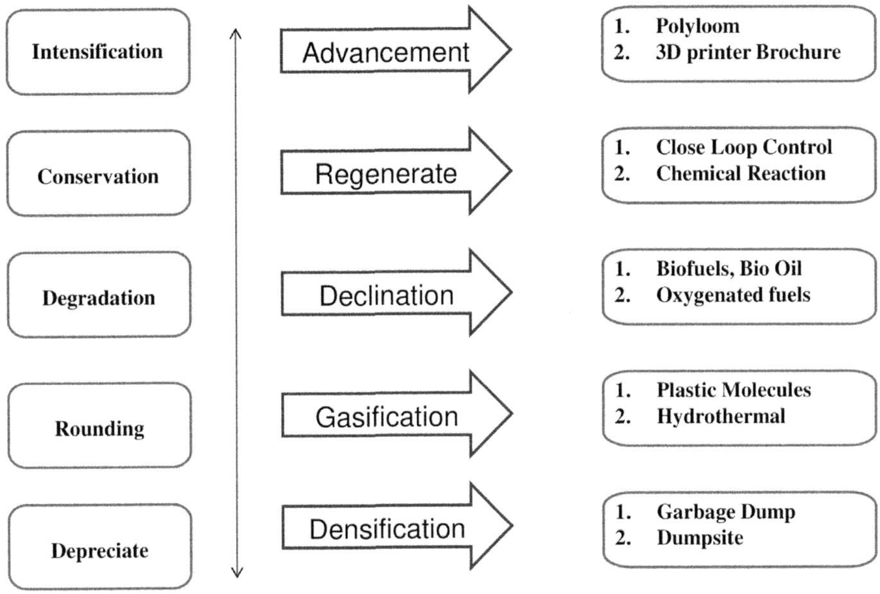

FIGURE 6.2 Revaluation of plastic waste in various stages.

recycling purposes. Revaluating plastic waste at various stages involves assessing and improving the lifecycle management of plastics, from production to disposal, and reintroducing plastic waste into the economy through sustainable practices are shown in Fig. 6.2.

ML has significantly improved plastic-type recognition, making it faster, more accurate, and scalable for waste management and recycling applications. Traditional methods of plastic sorting are often manual and prone to errors, but ML-driven systems can automate this process by identifying plastic types based on their unique properties. Table 6.1 provides a clear overview of various machine-learning techniques and how they are used for efficient plastic-type recognition.

Recycling technologies to build a sustainable future

Chemical recycling

Chemical recycling converts hard plastic back into virgin-equivalent plastic, as it was initially termed. In the waste management sector, chemical recycling generally refers to a variety of newly developed methods that convert hard plastic into recycled materials that were previously impossible or prohibitively expensive to recycle mechanically [1]. The need for secondary raw materials is growing daily due to client choice strategies, advancements in chemical recycling technologies, as shown in Fig. 6.3, and a drive toward more sustainable aims. It should be mentioned that secondary materials based on chemical recycling lower carbon emissions and substitute virgin raw materials from the production supply chain. This is how plastic trash is chemically recycled back into its constituent parts.

TABLE 6.1 Summary of machine learning techniques for plastic-type recognition.

S. no.	Factors	Descriptions
1.	Training on diverse datasets	The various datasets used to train machine learning algorithms include various kinds of plastic, colors, and shapes. The model is better equipped to detect plastics in real-world situations because this training enables it to identify patterns and characteristics linked to various polymers.
2.	Spectral analysis	Machine learning algorithms may analyze spectral data from different plastic materials. Based on their molecular makeup, distinct patterns for various polymers are provided by their spectral signatures, which enable the model to differentiate between them.
3.	Computer vision techniques	Plastics are visually identified and categorized using ML-based computer vision algorithms according to their texture, color, and physical attributes. Cameras and sensors take images of plastic objects, and the ML model uses this visual data to categorize plastics reliably.
4.	Near-infrared (NIR) spectroscopy	NIR spectroscopy and machine learning techniques examine spectra and classify plastics according to their molecular makeup. NIR technology aids in distinguishing various polymers that could have a similar appearance.
5.	Robotic sorting systems	Robotic sorting systems use ML-driven algorithms, which use sensors and cameras for real-time plastic identification. The ML model guides the robotic arms by effectively sorting plastics into the right categories, increasing sorting accuracy.
6.	Label and marking recognition	Machine learning models may be taught to identify labels and other marks on plastic objects to provide further information for precise identification. It is beneficial for labels that list the kind of plastic or provide recycling guidelines.
7.	Integration with robotic systems	Recycling facilities may automate material sorting by integrating robotic devices powered by AI. These robots recognize and sort plastics using sensors and machine learning algorithms, which increase sorting efficiency and accuracy.
8.	Contamination detection	Contaminants in recycled materials may be found and identified using AI algorithms. This covers nonplastic objects, leftovers, or anything made of incorrect polymer. The detection and removal of impurities ensures the manufacturing of high-quality recycled materials.
9.	Material property prediction	Machine learning algorithms can anticipate recycled materials' mechanical and chemical characteristics based on their composition. Manufacturers may use this information to assess if recycled materials are appropriate for a particular application.
10.	Automated sorting	Sorting various plastics is automated by robotic arms using computer vision and machine learning techniques. The ability of these devices to swiftly and precisely identify materials makes it easier to separate recyclables from nonrecyclables.
11.	Real-time decision-making	Robotic arms with integrated AI can make real-time judgments using data from cameras and sensors. This enhances the overall quality of recycled materials by guaranteeing accurate sorting and reducing mistakes.
12.	Reduced contamination	By precisely sorting items, robotic arms help to lower contamination in recycling streams. Because this process is automated, there is less chance of human mistakes, and materials are sorted precisely, producing recycled materials that are cleaner and more valuable.

Source: From B. Carrera, J. B. Mata, V. L., Piñol, and K. Kim, Environmental sustainability: a machine learning approach for cost analysis in plastic recycling classification. Resources, Conservation and Recycling 197, 107095 (2023). https://doi.org/10.1016/j.resconrec.2023.107095.

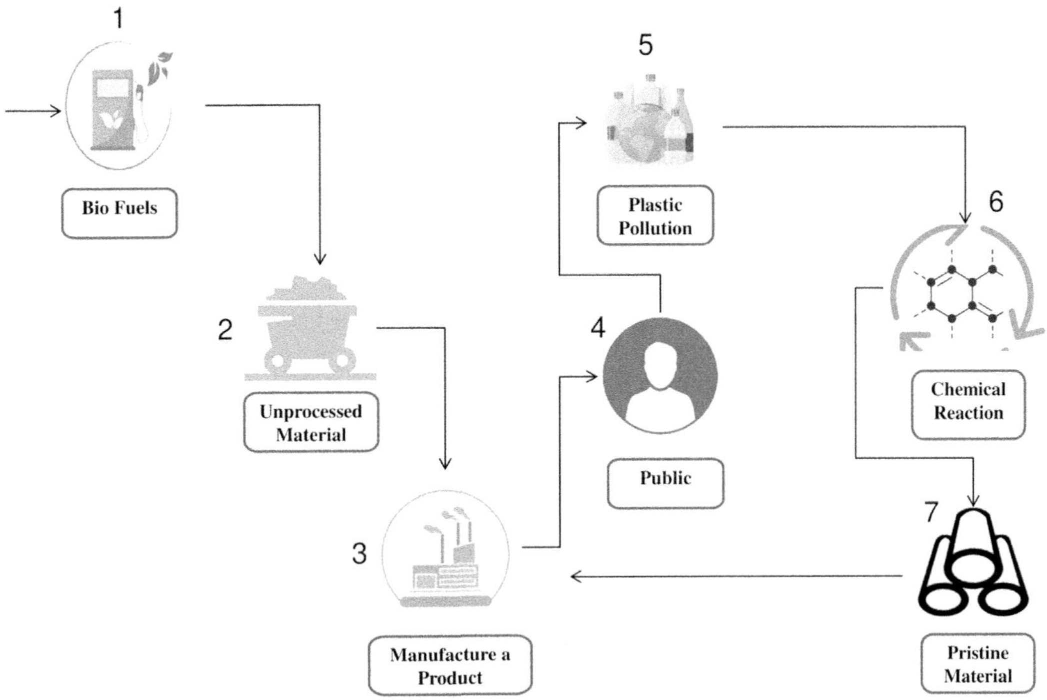

FIGURE 6.3 Chemical recycling for reducing the plastic pollution.

- **Conversion:** The conversion process is the first step in the closed-loop recycling process, called the circular economy. Here, pyrolysis and gasification may break down the long-chain hydrocarbon molecules known as polymers into smaller molecules in mixed or sorted plastics. To create new recycled plastics and other compounds, these brief molecules may be used as feedstock for novel chemical processes.
- **Decomposition:** The next step is breaking polymers into monomers to create new plastics. Solvents are used with heat or chemicals to break down polymers into monomers.
- **Purification:** The last phase is this. Unlike other chemical recycling methods, the polymer is neither altered nor broken down in this process. Solvents are instead used to extract polymers from impurities or additives.

Mechanical recycling

Mechanical recycling is a chemical method that preserves a substance's original chemical structure while transforming plastic trash into secondary plastic material [2]. The most common method for recycling plastics is mechanical recycling, which works with all kinds of thermoplastics. Specific processing procedures for mechanical recycling must be followed. These actions are enumerated below:

- **Collection:** Diverse and combined waste streams gather end-of-life plastic items.
- **Initial sorting:** The collected plastic garbage is sorted as soon as it reaches the recycling facility. Some sorting may have been completed during the collecting phase. However, further color or thickness sorting may be done at this point.
- **Shredding:** The gathered and sorted plastic trash is cut into smaller pieces before reusing.
- **Washing:** This step ensures that the plastic is clean before proceeding to the next step by cleaning the shreds of plastic trash to eliminate dust or debris. Usually, portions of food, beverages, or labels are covered with dust and debris.
- **Second sorting**: Plastics are sorted and regulated again before being transferred to extrusion.
- **Extrusion:** This process transforms plastic flakes into uniform pellets prepared to create new goods.

Recycling of dissolution

Dissolution recycling is recovering plastic waste into polymers by separating them and returning them to their original chemical structural properties. Polymers found in plastic trash are dissolved in solvents during the dissolution recycling process so they may be sorted and returned to their purest form. The following are some examples of industrial applications for the dissolving recycling process: polyvinyl chloride, polystyrene, nylon, and polypropylene. These polymers disassemble into distinct, superior monomers. Subsequently, new polymers with the same characteristics as the original materials are created using these monomers [3]. Ultimately, this raises recycling rates and lowers the environmental demand for plastic usage.

Recycling of organic materials

During organic recycling, biodegradable plastics are subjected to regulated microbiological treatment in aerobic (composting) or anaerobic (biogasification) environments. Specific polymers that microorganisms can break down into water, carbon dioxide, methane, and stabilized organic wastes may be recycled via organic recycling. Composting occurs when organic waste reaches the recycling phase in an aerobic environment [4]. Composting is mainly returned to the soil and used as biological fertilizer to support the growth of new crops. On the other hand, recycling primarily serves as energy when carried out in anaerobic environments.

Recycling of monomers

Recycling plastic in which the lengthy polymer molecules break apart into monomer building blocks is known as monomer recycling. These monomer building blocks may produce virgin plastic polymers with the same qualities [5]. The plastic's chemical components are reversed due to monomer recycling, which results in a more stable recycled plastic.

Reusing outdated online content

The recycling sector is evolving along with technology and adjusting to new developments. Recyclers think IoT-enabled waste management strategies will help them overcome the inefficiencies in their waste logistics industry, given the growth of recycling technology. Material quality assessment sensors, intelligent bins, and fill-level sensors. Fill-level sensors monitor the trash cans and notify the waste collection staff when it's time for a pickup. "Radio-frequency identification technology" is the second IoT-based innovation that trash management companies and recyclers use. It enables haulers to monitor assets while they are out in the field. It helps recyclers with trash container maintenance, work order administration, and tracking.

Robot recycling

Mechanical recycling is a specialized and well-known method for recovering the uniqueness of plastic materials. Nevertheless, companies needed help with worker shortages and contaminated garbage during mechanical recycling. Startups created recycling robots to automate and enhance sorting lines using AI-powered sorting systems to deal with this. Recycling robots help overcome human limits, increase trash-picking speed, minimize mistakes (particularly while sorting garbage), and improve picking efficiency [6]. Materials recovery facilities have the power to save operating costs, maximize waste stream quality management, and improve waste flow visibility.

Valorization of waste heat

It is the process of turning trash into resources for energy or valuable items while preserving the material's quality. Rather than being disposed of in landfills or incinerators, this technique converts organic and solid waste into chemicals or energy sources. Startups use waste valorization by providing anaerobic digesters that break down organic waste into biogas with the help of microorganisms. Furthermore, there is increased interest in waste-to-energy solutions due to the growing need for sustainable energy.

Artificial intelligence

AI is incorporated into recycling processes to increase efficiency and effectiveness without restricting human ingenuity. Instead, AI is fostering human creativity. Automation has reduced direct human exposure to hazardous waste streams, protecting worker health via material analysis, sorting, and garbage collection. By incorporating automated processes, companies might use computer vision and ML methods to identify patterns, comprehend essential data, make precise predictions, and carry out intricate activities with minimum human intervention. This led to notable advancements in the operations of different recycling businesses, enabling entrepreneurs to boost output, save expenses, and expand their operations faster.

Extension of the material life cycle

Extending the material life cycle is significantly aided by recycling technology. The firms created automated recycling solutions based on developments in recycling technology, such as chemical treatment and other tech-driven mechanical recycling. This decreased reliance on virgin raw materials by producing high-quality secondary materials and extending the life of materials.

Additional technology fixes

Similarly, various recycling technologies have been embraced or are scheduled for actual implementation, which will streamline waste management operations. One of these is combining automated route optimization software with the trash management and recycling industry. This one is among the most critical integrations. This will assist you in overcoming the time required for manual garbage collection route planning, which is in contrast to previous approaches. Moving to the most cloud-based recycling trash management software, you may cover the most stops within a deadline.

- Quickly and efficiently plan, schedule, and optimize garbage collection routes.
- Assigning different drivers and reroute or shift routes between them based on your preferences and time convenience.
- Even the amount of time your drivers spend driving is significantly decreased.
- If a driver is unavailable, reassigning the pickup routes to another driver.

- Regularly create reports detailing your trash collection, daily pickups, and team performance.
- To confirm that the service has been provided at the appointed stops, get fast proof of service (by e-sign and photos).

Some real-life implications of recycling technologies

Automated fill-level sensors: Bintel

A startup firm in Sweden creates fill-level sensors applicable to recycling stations, garbage containers, and trash cans. Depending on the needed range, Bintel's automatic fill-level sensors use narrowband IoT or low-power wide-area networking. Let us examine how it benefits the garbage collecting industry [7,8].

- Provides more insight into the state of garbage containers without requiring significant upfront costs.
- It is solid and dependable and works in all settings and applications.
- The sensors also inform recyclers about the frequency of emptying the container and trash logistics.
- Easily fitted in your current vessels;
- Maximizes each fraction's capacity by recycling station specifications.

Postconsumer textile waste chemical recycling: refibering

The US-based firm created Refiberd, which specializes in the chemical recycling of postconsumer textile waste. AI, robots, and Refiberd's proprietary green chemical recycling technology are all employed to transform discarded and old textiles into new, reusable threads. It provides recycled polyester and cellulose thread kits that enable the diversion of a substantial proportion of textile waste in landfills after turning postconsumption textile waste into reusable threads [9]. As a result, the textile manufacturing industry's requirement for virgin material decreases.

Styrene butadiene styrene thermal technologies: waste heat valuation

The technique for waste heat valorization was created by a Spanish business and is patented. High-efficiency contact technology, or HECO, produces ideal conditions for mass and energy transfer and raises waste conversion efficiency. Styrene butadiene styrene (SBS) thermal technology, you may save up to 40% on energy costs by quickly converting your waste into biofuel, clean energy, and valuable raw materials [10]. They concentrate on the primary environmental objective of enhancing and growing the procedures for recycling and recovering garbage. Their primary responsibility is to recover and repurpose the recyclable waste.

Recycling feedstock: In 2020 Neste Nordeste, a refining facility in Porvoo, Finland, finished its maiden processing run. Three thousand tons of liquefied plastic wastes were processed to produce the final product, Neste RE [11]. It serves as a raw material to make new plastic and a substitute for materials derived from fossil fuels.

Several tactics using artificial intelligence to reduce and recycle plastic waste

Waste collection that operates on its own

Numerous plastic waste fragments wind up in streams, threatening aquatic life and degrading the water's quality. Researchers in Singapore have created an AI-driven robot to address this problem. Clearbot Neo uses machine vision as it glides over the water's surface to recognize and classify garbage before picking it up. Clearbot Neo may collect one metric ton of waste each day. Along the way, it collects data on the kinds of trash and where it occurs. This information may provide insightful information on regional waste trends, which may impact management choices or legislative changes made to address specific issues in an area in the future [12]. Machine vision technology, Clearbot Neo, can steer clear of aquatic life. In this approach, the robot can clean water systems without interfering with the regular functions of the ecosystems.

Automated classification

Plastic collection is just the first step in the waste management process. Afterward, organizations must sort it out and choose how to eliminate it. However, only 29% of global e-waste finds its way to the appropriate recycling route. AI can make this process better. Machine vision-capable robots can evaluate garbage and determine the best route. Like other repetitious, data-intensive operations, AI performs this role much more quickly and accurately than humans. Businesses have found that these robots can sort materials twice as accurately as people, doubling recycled items' resale value [13]. Recycling facilities may recover as much plastic garbage as possible by automating the sorting process using AI. Algorithms are less likely to make errors that lead to recyclable plastics ending up in landfills because they remain motivated and focused.

Finding novel techniques for disposal

After sorting, AI can help find new, less harmful methods to eliminate plastic waste. Some ML algorithms can evaluate tens of millions of choices in a split second. By doing this, researchers can identify the best methods for disposing of plastic that they would not have otherwise thought of. Researchers at the University of Texas in Austin recently did this. Utilizing a powerful ML technique, they discovered an enzyme degrading plastic in hours rather than the customary years [14]. To determine which enzyme mutation would result in the best results, the software looked at a variety of them. Such discoveries can require weeks or months of laboratory work and investigation without AI. ML may accelerate the process, which can perform several precise simulations simultaneously. Enhancing the packaging, AI can help reduce plastic waste by minimizing it. Since much of this waste originates from packaging, several companies have used AI to produce less wasteful packaging.

Amazon used AI to analyze customer complaints and identify trends of overpackaging and shipping problems. The algorithms may determine how to use the least amount of material while providing sufficient protection during transportation. The resulting packaging changes resulted in a 5% reduction in each item's carbon footprint and shipping costs. Similarly, L'Oréal uses AI to forecast the likelihood of new packaging ideas succeeding. After that, the company may decide how to use less raw plastic and more recycled materials in its packaging. These algorithms could improve their forecasts with time, resulting in even more refined design modifications.

Avoiding waste in production

Predictive analytics may also reduce plastic waste by modifying production levels in response to shifting demand. Businesses may have extra inventory that might spoil before it is used because customer requirements change suddenly. AI can reduce that waste by foreseeing these developments. AI-based demand forecasts may reduce forecasting errors by 30%–50% compared to conventional approaches. Then, in reaction to incoming changes, manufacturers may scale up or down output with more reliability. They will generate less waste if they can appropriately satisfy customer requests [15]. Similarly, AI projections may help reduce supply chain breakdowns that might otherwise damage goods—even less plastic waste results from reduced packaging loss caused by product loss during transit.

Procedure for recycling plastic bottles

Recycling plastic bottles provides a way to lessen the adverse environmental effects of these thrown-away containers. Plastic bottle collecting and sorting is the first phase of recycling, followed by several cleaning and processing procedures. Let us examine each of these phases in more depth [16]. A schematic outlines collecting, classifying, processing, and cleaning plastic garbage.

Sorting and gathering bottles made of plastic

Gathering bottles from several sources, including homes, companies, and public areas, is the initial stage in recycling plastic bottles. Recycling facilities or trash management firms typically provide plastic bottle-collecting containers. These containers are positioned thoughtfully to encourage people to dispose of their bottles properly.

After they are gathered, the plastic bottles are sent to a recycling center and sorted there. The bottles are sorted at the recycling plant according to the kind of plastic resin they contain. This is important since various plastics need different processing and recycling qualities.

Plastic bottle recycling technologies

The plastic bottles go through several procedures after sorting to be made into new items. Mechanical recycling is a popular technique that involves shredding the bottles into tiny fragments and melting them to create plastic pellets. New plastic bottles and other plastic items may be made using these pellets.

Chemical recycling is another recycling method that is gaining traction. The plastic bottles are broken down into constituent molecular parts through this procedure, which may be used to make new polymers or other materials [16]. One benefit of chemical recycling is that it may recycle mixed plastics and even contaminated bottles, which are usually difficult to recycle using mechanical methods. AI plays a very important role in the process to get optimized and fast results.

Typical obstacles in recycling plastic bottles

Recycling plastic bottles is a potential option, but there are some challenges to plastic bottle recycling and it is one of the major issues. The recycling stream may get contaminated by improper disposal, such as combining bottles with nonrecyclable items or food waste. Because they are more challenging to recycle, contaminated bottles may wind up in landfills or be burned.

More infrastructure and recycling facilities are needed. Some places need more infrastructure for recycling, making it difficult for people to get rid of their plastic bottles correctly. More money should be invested in recycling facilities and better recycling education programs to tackle these issues and encourage excellent recycling rates.

Plastic bottle collecting and sorting

Recycling plastic bottles has several advantages despite the difficulties. The preservation of resources is the main advantage. We may lessen our reliance on virgin resources, such as petroleum, to manufacture plastic by recycling plastic bottles. In addition to protecting natural resources, this lowers greenhouse gas emissions from the extraction and processing of these commodities.

Another way to lessen landfill trash is to recycle plastic bottles. We can alleviate the space required for garbage disposal and discharge dangerous chemicals into the environment by keeping plastic bottles out of landfills. Recycling plastic bottles also promotes the circular economy. Plastic bottles may be recycled into new items, prolonging their useful life and lowering the need for new plastic manufacture instead of being disposed of as garbage. Reducing our dependency on limited resources and promoting sustainability are the two benefits of this transition toward a circular economy.

Plastic bottle recycling methods

It is crucial to adhere to a few recommended practices to optimize the benefits of recycling plastic bottles. Here are some pointers to assist in recycling efficiently: To reduce contamination, empty and clean plastic bottles before recycling. Taking off any lids or tops, as they might obstruct the recycling process and are often composed of a different kind of plastic. Make sure of recycling the right kinds of plastic bottles by consulting local recycling regulations. Flatten the bottles to maximize transportation efficiency and save room in the recycle container. Share knowledge about the value of recycling with others and yourself to promote broad involvement. Thus, one supports the success of plastic bottle recycling and saves the environment by adopting these habits into everyday life.

Smart waste technologies

Businesses and waste management firms currently use intelligent automation and technology in their operations all around the globe. These are the top six cutting-edge new technologies for 2024 [17,18].

Pello

Pello is a new technology designed to help companies manage their garbage collection more effectively and lessen their environmental impact. Pello offers a variety of solutions to assist businesses in achieving these objectives. First, the Pello system tracks the waste in your trash cans and gives you up-to-date information on their position and contents. It also notifies you when a collection is due and lets you know if your container has been polluted. With this knowledge, you can take charge of your company's garbage disposal and simplify waste management procedures. Additionally, it implies that trash trucks are only sent when required, which may aid in lowering traffic and greenhouse gas emissions.

Robots for recycling

AI has been in the headlines recently, not usually for the best reasons. However, the recycling field is one where AI might have a significant and positive impact. Several recycling sites currently use AI robots; they may be trained to distinguish between various materials quickly and correctly. By integrating AI robots into the process, recycling facilities can sort garbage more effectively and keep as many recyclables out of landfills as feasible. Additionally, it enables waste management businesses to work longer hours or even around the clock, significantly increasing the volume of garbage handled.

Vacuum-powered waste pipes

The practicality of conventional garbage disposal and collecting methods decreases with the growth and density of metropolitan areas. The pneumatic waste pipe is one of the most technologically advanced solutions. Instead of waiting for garbage collection, waste may be transported directly to processing plants by installing pneumatic pipes under public trash cans. The advantages of this method are twofold. First, it could significantly lower the quantity of trash trucks on the road. This may lessen the number of cars cluttering our cities and reduce hazardous pollutants. Second, to prevent overflowing dumpsters by sending debris straight from dumpsters to waste management facilities. In addition to being ugly, overflowing dumpsters may risk public health and increase local pollution.

Trash compactors powered by sunlight

A garbage truck may gather more rubbish on each round and need fewer trips if it can transport more debris. Garbage-powered garbage compactors compress garbage to expand the capacity of a dumpster as it builds up. Because of this, these intelligent containers can carry up to five times as much as conventional garbage cans. Trash compactors driven by solar energy compress rubbish and include waste level monitors. These sensors provide information about the bins' capacities, enabling users to plan pickups and expedite the collection procedure.

E-waste kiosks

Electronic garbage includes dangerous materials from outdated computers, phones, and televisions. These materials may seep into the environment and seriously endanger human health and the surrounding ecology if e-waste is disposed of inappropriately. E-waste should either be recycled or, better yet, mended and utilized rather than being dumped in a landfill. This will safeguard the surrounding environment and lessen the likelihood of dangerous chemicals infiltrating our land and water.

One startup business built specialized e-waste recycling kiosks, taking matters into their own hands. You may securely remove your unwanted devices using EcoATM kiosks. They could even pay you cash for your phone, tablet, or computer if it's still in excellent working order.

This acknowledgment is also a step toward appreciating efforts to enhance e-waste recycling. Many governments are implementing initiatives to establish effective systems for managing e-waste, as illustrated in Fig. 6.4, which shows the Indian Government's efforts to achieve the goals of the Swachh Bharat Mission.

Apps for recycling

Why not the way we recycle, given how apps have altered how we date, order cabs, and communicate? A fresh selection of applications is now available to assist companies and people in navigating the recycling industry. These applications are excellent tools for anyone who wants to keep their garbage out of the landfill. App allows you to do anything from finding a local recycling center to giving away your clothing and donating your food waste to a good cause. Furthermore, specific applications provide comprehensive details about the recyclable products in your neighborhood. Apps are a terrific method to establish a sustainable, circular economy since they often enable you to pass unwanted products on to others. To learn more about the available specialized applications app store can be visited.

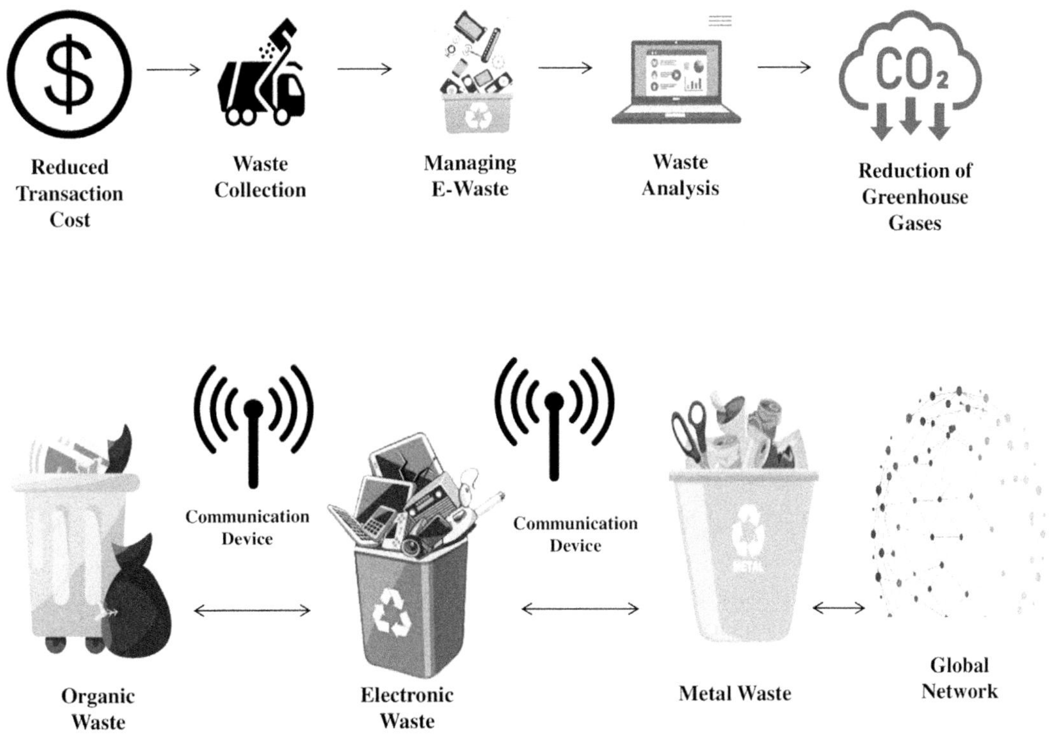

FIGURE 6.4 "Swachh Bharat Mission" to be completed by Indian government.

Robotics and automation for plastic recycling

Artificial intelligence-guided robotic sorting systems

Plastic recycling in the construction industry is revolutionized by integrating AI with robotic sorting systems. These robots can precisely recognize and classify various forms of plastic garbage according to their composition and attributes thanks to AI-powered vision capabilities. With less dependence on human labor and higher overall recycling rates, this automation greatly increases the efficiency and accuracy of plastic recycling. The robots may swiftly analyze and categorize the waste plastic, guaranteeing that it is appropriately separated for further processing and recycling. This AI-guided robotic system is playing an essential role recycling of plastic waste.

Recycled plastic 3D printing

AI-controlled 3D printing technology provides an environmentally friendly way to use recycled plastic in buildings. The 3D printers can streamline the printing process and guarantee the correct usage of recycled plastic resources by integrating AI algorithms. By analyzing the characteristics of the recycled plastic, these systems may modify the printing conditions to provide the best possible structural integrity and performance. By enabling the production of buildings and building parts out of recycled plastic, this method helps to promote the circular economy and lessens reliance on fresh plastic resources. The construction sector may adopt sustainable practices and address plastic pollution by using robots and AI in plastic recycling and 3D printing. These developments improve the precision, adaptability, and effectiveness of plastic recycling procedures, encouraging a more ecologically responsible and sustainable method of building.

Intelligent waste management

Trash tracking and monitoring, made possible by AI sensors and data analytics, is use for revolutionizing for waste management in the construction sector. Construction sites may measure and monitor the development of plastic trash in real-time by using AI technology. AI algorithms examine sensor data to find trends, patterns, and opportunities for growth. In doing so, trash reduction, recycling, and disposal procedures are optimized, and proactive waste management techniques are made possible. Construction organizations may reduce the environmental effects of plastic waste by implementing specific measures and keeping a close eye on it.

Recycle bins and collecting systems with intelligence

AI-driven smart bins and collecting systems provide a new degree of precision and efficiency in handling plastic garbage. These smart-systems automatically sort and take plastic garbage using sensors, computer vision, and ML algorithms. AI facilitates identifying and classifying diverse plastic materials, guaranteeing appropriate recycling and disposal. These intelligent solutions lower contamination and boost recycling rates by automating garbage collection and sorting. They simplify handling plastic trash, increasing its cost-effectiveness and sustainability for the building sector.

The construction sector may better control plastic waste by incorporating AI into waste management procedures. Construction sites may decrease waste output by taking proactive actions using AI-enabled tracking and monitoring systems. Smart recycling bins and collection systems may minimize environmental damage and promote a circular economy by ensuring that plastic garbage is properly sorted and processed. These creative ideas stimulate innovation and support the building sector's long-term growth.

Sustainable design predictive analytics

AI can completely transform the construction process by improving building designs to use less plastic and be more recyclable using predictive analytics. Large volumes of data, including material performance and qualities, may be analyzed by AI algorithms to provide other design options that reduce the quantity of plastic waste. By considering several criteria, including energy efficiency, environmental effect, and structural integrity, AI-driven design optimization guarantees the sustainability of building projects while reducing their dependence on plastics. By predictive analysis of AI data, a sustainable design of building can control plastic waste in systematic way.

Modeling and simulation

AI-based simulation tools are essential for assessing how building projects will affect the environment and pinpointing areas where plastic waste may be reduced. Designers and architects may evaluate a building's life cycle and run various scenarios via simulation models that use AI algorithms to identify the most sustainable choices. Examining the usage of plastic materials, their suitability for recycling, and viable substitutes are all part of this process. Using AI-driven simulations, construction experts may make well-informed judgments that result in less plastic waste and more sustainability. AI allows the construction sector to improve building designs, focusing on lowering plastic consumption and enhancing recyclability by integrating predictive analytics into sustainable design processes. AI-driven design optimization minimizes the need for plastic materials and maximizes their potential for recycling or reuse, ensuring that projects are ecologically mindful. AI-powered modeling and simulation provide essential insights into how building projects affect the environment, enabling experts to make data-driven choices and implement plans to reduce plastic waste. AI-powered sustainable design techniques open the door to more resource- and environmentally-conscious building methods.

Collaborative artificial intelligence platforms approach

Artificial intelligence-powered collaboration platforms

Cooperation and information exchange are essential in the building industry's battle against plastic pollution. Collaboration platforms driven by AI provide a forum for policymakers, academics, and construction experts to convene, exchange best practices, and pool their expertise in mitigating plastic pollution. These platforms use AI algorithms to speed up content discovery and distribution, giving users access to pertinent resources, case studies, and creative solutions. AI powered platforms support group efforts to address plastic waste in sustainable building methods by encouraging cooperation and information exchange [19].

Making decisions based on data

AI technologies may analyze and interpret large amounts of construction data, allowing for data-driven decision-making in reducing plastic waste. Using AI algorithms to analyze data, construction workers may learn much about material consumption, trash creation, recycling rates, and environmental effects. Decision-making procedures are aided by these insights, which make it possible to put into practice practical tactics to cut down on plastic waste. AI-powered data analysis enables stakeholders to prioritize sustainable practices, make well-informed decisions, and maximize resource utilization, significantly decreasing plastic pollution in the construction sector [19].

Collaborative AI platforms and data-driven decision-making facilitate communication and knowledge-sharing among construction experts, hence eliminating plastic waste. These platforms enable the adoption of new solutions by fostering stakeholder cooperation and information sharing via AI technology. Furthermore, AI-driven data analysis improves decision-making by empowering professionals in the construction industry to make well-informed decisions by drawing on insights from construction data. By integrating data-driven decision-making with collaborative platforms, the construction sector can significantly reduce plastic waste and promote sustainable building practices.

Plastic utilization

Due to the expansion of developing markets, the use of plastic has increased fourfold in the previous thirty years. Between 2000 and 2019, the world's plastics manufacturing quadrupled, reaching 460 million tons. 3.4% of greenhouse gas emissions worldwide are attributed to plastics. From 2000 to 2019, the amount of plastic garbage generated worldwide more than quadrupled to 353 million tons. Plastics with a lifespan of less than five years account for about two-thirds of plastic garbage; packaging, consumer products, and apparel and textiles account for 40%, 12%, and 11% of the waste, respectively. Just 9% of plastic garbage is recycled; the remaining 40% is disposed of as leftovers after 15% is collected for recycling. A further 19% is burnt, 50% is disposed of in landfills, and 22%, particularly in developing nations, avoid waste management systems and wind up in uncontrolled dumpsites, open pits, or terrestrial or aquatic habitats. 6.1 million tonnes (MT) of plastic debris spilled into waterways and 1.7 MT went into the ocean in 2019. An additional 109 MT of plastic debris has been collected in rivers, and there is already an estimated 30 Mt of plastic pollution in seas and oceans. The data highlight the global scale of the plastic waste problem and the need for coordinated international efforts to improve recycling, reduce plastic consumption, and develop sustainable waste management practices as shown in Fig. 6.5. The accumulation of plastics in rivers suggests that even with drastic reductions in

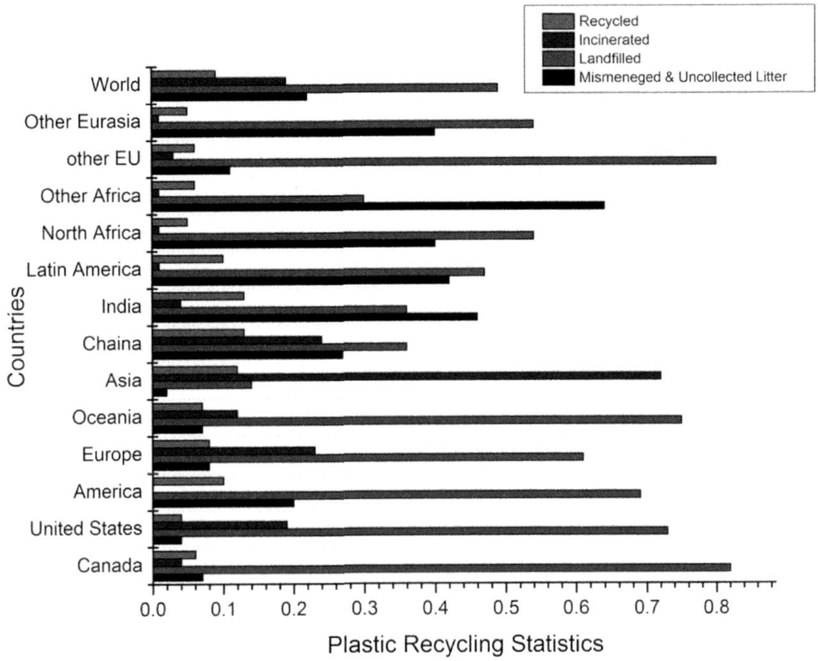

FIGURE 6.5 Displays data on plastic disposed off from different nations.

improperly handled plastic trash, leakage into the ocean may persist for decades to come. Enhancing the circularity of plastics would require harmonizing design methodologies and the regulation of chemicals, taking into account global value chains and plastics commerce. All available funding sources, including development assistance, should be tapped into as part of an international waste management strategy to assist low- and middle-income nations in covering the anticipated 25 billion euros in annual expenditures for upgrading waste management infrastructure.

Real-time examples of artificial intelligence applications in plastic reutilization include

Artificial intelligence-driven plastic sorting systems

- **Zen Robotics** uses AI-powered robotic arms for sorting plastic waste. The system utilizes ML algorithms to identify and separate various types of plastics from mixed waste streams, improving recycling efficiency.
- **Tomra Sorting Solutions** combines hyperspectral imaging with AI to sort plastic waste based on chemical composition. This technology enhances the accuracy of sorting and reduces contamination.

- **Waste Vision** uses AI algorithms to predict plastic waste generation and optimize collection routes. This system helps municipalities manage waste more effectively and reduce landfill use.

Conclusion

In summary, incorporating AI into the process of repurposing plastics is crucial in tackling the worldwide issue of plastic pollution. Because of the complexity of plastic waste and the growing need for sustainable practices, creative solutions are required, and AI proves to be a potent ally in accomplishing these objectives. AI greatly aids in the recognition and classification of various plastic kinds, assisting in overcoming the difficulties presented by the intricacy of plastic compositions. Robotic systems, computer vision, and ML models increase the effectiveness of recycling procedures, which raises recycling rates and improves the quality of recycled materials. AI influences a more sustainable and circular approach to plastics via chemical recycling, smart bin technology, and sustainable package design developments.

Furthermore, AI-powered supply chain optimization, predictive maintenance, and quality control solutions simplify the whole lifetime of plastic materials—from procurement to production. This lowers the environmental impact while improving operational effectiveness and optimizing resource use. The ethical issues, legal frameworks, and cooperative initiatives highlighted the importance of using AI responsibly when repurposing plastic. Ensuring ethical standards are followed in the sector is made possible by regulatory frameworks, which are vital in promoting transparency, fairness, and data privacy protection. Recycling plastic using AI will prove to be a boon for humanity, although adequate literacy about this issue should be spread among civilians through various means, everyone has to be aware and socially responsible for reutilization of plastic.

References

[1] T. Sayara, R. Basheer-Salimia, F. Hawamde, A. Sánchez, Recycling of organic wastes through composting: process performance and compost application in agriculture, Agronomy 10 (11) (2020) 1838. https://doi.org/10.3390/agronomy10111838.

[2] E. Brepohl, M. Paschetag, S. Scholl, Monomer recycling as complementary technology in a circular economy, Chemie Ingenieur Technik 95 (2023) 1282–1289. https://doi.org/10.1002/cite.202300052.

[3] S.Y. Lo, Y.Y. Lai, J.C. Liu, et al., Robots and sustainability: robots as persuaders to promote recycling, International Journal of Social Robotics 14 (2022) 1261–1272. https://doi.org/10.1007/s12369-021-00828-z.

[4] I. Pakere, D. Blumberga, A. Volkova, K. Lepiksaar, A. Zirne, Valorisation of waste heat in existing and future district heating systems, Energies 16 (19) (2023) 6796. https://doi.org/10.3390/en16196796.

[5] Z. Kang, Z. Huang, Q. Peng, Z. Shi, H. Xiao, R. Yin, et al., Recycling technologies, policies, prospects, and challenges for spent batteries, iScience 26 (11) (2023) 108072. https://doi.org/10.1016/j.isci.2023.108072; PMID: 37867952; PMCID: PMC10589888.

[6] N. Rathore, N.L. Panwar, Environmental impact and waste recycling technologies for modern wind turbines: an overview, Waste Management & Research 41 (4) (2023) 744–759. https://doi.org/10.1177/0734242X221135527.

[7] P. Wang, H.-R. Wei, X.-Y. Liu, R.-B. Ren, L.-Z. Wang, Identifying the long-term thermal storage stability of SBS-polymer-modified asphalt, including physical indexes, rheological properties, and micro-structures characteristics, Sustainability 13 (19) (2021) 10582. https://doi.org/10.3390/su131910582.

[8] G. Grause, A. Buekens, Y. Sakata, et al., Feedstock recycling of waste polymeric material, Journal of Material Cycles and Waste Management 13 (2011) 265–282. https://doi.org/10.1007/s10163-011-0031-z.

[9] B. Fang, J. Yu, Z. Chen, et al., Artificial intelligence for waste management in smart cities: a review, Environmental Chemistry Letters 21 (2023) 1959–1989. https://doi.org/10.1007/s10311-023-01604-3.

[10] J. Choi, B. Lim, Y. Yoo, Advancing plastic waste classification and recycling efficiency: integrating image sensors and deep learning algorithms, Applied Sciences 13 (18) (2023) 10224. https://doi.org/10.3390/app131810224.

[11] S. Narayanamoorthy, V. Annapoorani, D. Kang, D. Baleanu, J. Jeon, J.V. Kureethara, et al., A novel assessment of bio-medical waste disposal methods using integrating weighting approach and hesitant fuzzy MOOSRA, Journal of Cleaner Production 275 (2020) 122587. https://doi.org/10.1016/j.jclepro.2020.122587; ISSN 0959-6526.

[12] M.G. Kibria, N.I. Masuk, R. Safayet, et al., Plastic waste: challenges and opportunities to mitigate pollution and effective management, International Journal of Environmental Research 17 (2023) 20. https://doi.org/10.1007/s41742-023-00507-z.

[13] A.J. Graettinger, P.W. Johnson, P. Sunkari, M.C. Duke, J. Effinger, Recycling of plastic bottles for use as a lightweight geotechnical material, Management of Environmental Quality 16 (6) (2005) 658–669. https://doi.org/10.1108/14777830510623727.

[14] W. Czekała, J. Drozdowski, P. Łabiak, Modern technologies for waste management: a review, Applied Sciences 13 (15) (2023) 8847. https://doi.org/10.3390/app13158847.

[15] J. Abedin, Top 10 Smart Waste Management Technologies, Techyv.com, 2023. https://www.techyv.com/article/top-10-smart-waste-management-technologies/.

[16] F. Olan, E.O. Arakpogun, J. Suklan, F. Nakpodia, N. Damij, U. Jayawickrama, Artificial intelligence and knowledge sharing: contributing factors to organizational performance, Journal of Business Research (2022), https://doi.org/10.1016/j.jbusres.2022.03.008.

[17] H. Khan, K.K. Kushwah, S. Singh, J.S. Thakur, K.K. Sadasivuni, Machine learning in additive manufacturing, in: K. Deshmukh, S.K.K. Pasha, K.K. Sadasivuni (Eds.), Nanotechnology-Based Additive Manufacturing, 2023 https://doi.org/10.1002/9783527835478.ch21.

[18] H. Khan, K.K. Kushwah, P.K. Jhinge, G.G. Soni, R.K. Choubey, R. Kushwah, Faults Diagnosis Using AI and ML, Auerbach Publications eBooks, 2023. https://doi.org/10.1201/9781003301820-1.

[19] Global Plastics Outlook. (n.d.). https://www.oecd-ilibrary.org/environment/global-plastics-outlook_de747aef-en.

CHAPTER

7

Handling metals waste to salvage with automation

Izhar Ahmed[1], *Kondaiah Seku*[2], *Sreenivasa Babu Nidamanuri*[2], *and Krishna Kumar Koyyala*[2]

[1]Department of Engineering, College of Engineering and Technology, University of Technology and Applied Sciences, Salalah, Oman [2]Department of Engineering, College of Engineering and Technology, University of Technology and Applied Sciences, Shinas, Oman

Introduction

Scrap metal

Managing metal waste is a crucial part of environmental sustainability and resource conservation. As a finite and valuable resource, metals are extremely valuable. It is important to recycle and repurpose metal waste in order to reduce the need for new ore mining. This contributes to the preservation of natural resources and reduces the impact of mining operations on the environment. Metal waste can be polluted if it is improperly disposed of. It is important to note that metals can contaminate soil, water, and air if they are not handled properly. These negative effects on the environment can be avoided by recycling and disposing of waste properly. The processing of recyclable metals consumes considerably less energy than that of newly extracted metals. As a result, greenhouse gas emissions are reduced, and energy resources are conserved. The proper handling of metal waste is vital to the sustainability of the environment, resource conservation, economic prosperity, and public health. The circular economy strives to minimize waste and maximize resource efficiency by integrating it into its design. The Fig. 7.1 shows Europe zero waste management for the circular economy.

Scrap is a waste material that consists of pieces of metal from discarded objects, machines, or appliances, which can be transformed into useful material. The scrap also includes other types of nonmetallic materials [2]. Metals are considered one of the most valuable resources in economic terms, as their primary production involves quite high investment and operating costs. Therefore, there is a high demand for metal waste for recovery and reuse. Metals can be regenerated from scrap metal and reused several times without affecting their properties. Therefore, most of the metal production in modern days is carried out by recycling metal scrap [3].

Harnessing Automation and Machine Learning for Resource Recovery and Value Creation
DOI: https://doi.org/10.1016/B978-0-443-27374-2.00007-8

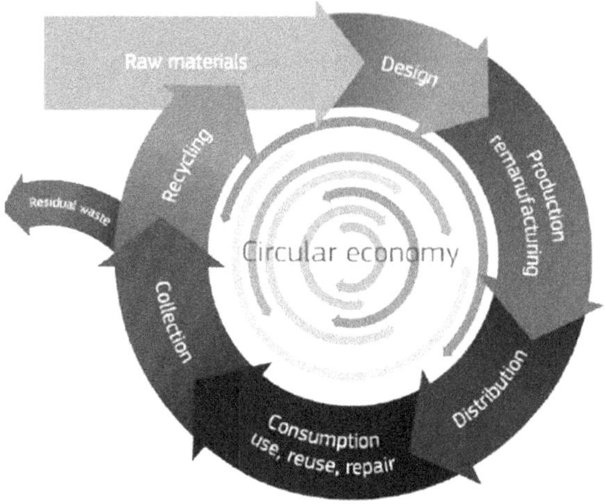

FIGURE 7.1 Circular economy: Europe Zero Waste. Europe zero waste management for circular economy [1]. *Source: From C. Hagelüken, D. Goldmann, Recycling and circular economy—towards a closed loop for metals in emerging clean technologies. Mineral Economics, 35 (3–4) (2022) 539–562, https://doi.org/10.1007/s13563-022-00319-1.*

Types of scrap metals

Types of scrap are normally classified into two main groups: ferrous scrap and nonferrous scrap.

Ferrous scrap

Ferrous scrap comes from discarded steel, several types of iron, and alloys. This scrap is called foundry scrap, as these materials are used for their strength and durability. Materials made of iron and its alloys are used on a large scale for the purpose of building automobiles, pipes, industrial containers, household appliances, etc. However, the disadvantage of ferrous scrap is that it is susceptible to corrosion under climatic conditions [4].

Ferrous-scrap is further classified into:

1. Industry scrap
   This type of scrap is generated from the manufacturing of various consumer products. It is the leftover metal after the production of various products.
2. Burr scrap
   A burr is an accumulation of metal, created on the workpiece's surface during manufacturing. It extends over the intended and actual workpiece surface and has a slightly higher volume in comparison to the workpiece [5]. These rough edges (burrs) that are formed because of the cutting process of a metal must be removed from the finished product. Generally, the burrs are removed by applying a file on the rough surfaces of the finished product. This process generates a huge quantity of burr-scrap in the industrial production line.
3. Fuel scrap
   Iron powder has high energy density and hence can be used as a fuel. Iron powder can be ignited at high temperatures in industrial furnaces, and the heat energy produced can be

used to produce electricity by steam turbines. Combustion of iron powder generates zero emissions of carbon dioxide, sulfur dioxide, and ultra-low emissions of nitrous oxide. Iron powder is produced in industrial processes or directly from iron scrap [6].

Nonferrous scrap

Nonferrous scrap metals are usually more malleable and lighter. Metals and alloys like copper, aluminum, brass, and bronze come under the nonferrous scrap category. The source for this type of scrap is household items due to their high resistance to rust and corrosion, making them the perfect choice for water pipes, roofs, and traffic signs. In addition, these metals are nonmagnetic, so they are also used as small electronic components and cables [7].

Examples of nonferrous scrap metals include:

- Copper is found in kitchenware, cables, pipes, and electric wires.
- is used to make cans and cutlery. It is found in ingots, profiles, and offcuts.
- Brass scrap is generated in the form of tinning, wires, and decorative elements.
- Bronze is used in industrial machinery, decorative items, and jewelry.

World iron scrap market (current trends)

The climate goals of different countries can be achieved using recycled ferrous scrap as a raw material. The Bureau of International Recycling (BIR) predicts an increase in the use of iron scrap to produce steel in the coming years. The three countries leading the iron scrap imports are the USA (2.5 million tons), India (5.25 million tons), and Türkiye (9.6 million tons) during 2022, shown in Table 7.1.

TABLE 7.1 Global steel scrap consumption by country.

Country	Steel scrap consumption (million tons)	Steel production in 2022 (million tons)	Dynamics
China	225	1013	–1%
European Union	78	137	–11%
USA	55	81	–7%
Turkey	30	35	–14%
Japan	30	89	–14%
Russia	28	72	–13%
Bharat	26	118	–2%
South Korea	25	66	–12%
Others	113	261	–14%

Source: From https://gmk.center/en/posts/global-scrap-market-in-the-first-half-of-the-year-current-trends/.

The European Union is considered one of the largest iron scrap consumers and exporters. The statistics for the European Union given by steel mint are 77.5 million tons in the year 2020, 88.5 million tons in the year 2021 with 14% growth, and 78 million tons in the year 2022. The Association of Rebar Producers and Exporters (IREPAS) reported that steel producers in European countries are transforming to electric arc furnace technology due to which Europe may become an iron scrap importer in the coming years. Because of high-capacity electric steel-making, Turkey consumed 30 million tons of ferrous scrap in the year 2022. In the years 2021 and 2022, Turkey imported 13 million tons and 10.7 million tons of ferrous scrap from Europe [8].

Electronic waste (e-waste)

It is a term used to describe discarded or old electronic equipment and electrical appliances used for domestic and industrial purposes. E-waste contains valuable metals that are hazardous to the environment. Recycling metals from e-waste requires special handling techniques and various recycling methods. Large household appliances generated a maximum quantity of e-waste (49%), which contains 60% of metals [9]. Diverse types of metals present in e-waste can be bulk and trace elements.

Bulk elements: Tin, copper, silicon, iron, and aluminum.

Small amounts of elements: cadmium and mercury.

Trace elements: Germanium, gallium, barium, nickel, indium, vanadium, beryllium, gold, europium, titanium, ruthenium, cobalt, palladium, manganese, silver, antimony, bismuth, selenium, platinum, arsenic, lithium, and boron [10].

According to the approximations, the toxic metal content in 500 Mt personal computers is 718,000 tons of lead, 287 tons of mercury and 1363 tons of cadmium. The extraction of metals from e-waste is a complicated process. For instance, for cobalt recycling, by using the available technology, the recovery rate of 95% is achievable, but the recovery rate is only up to 30%. The formally collected and recycled e-waste in the year 2019 was 17.4% of a total of 53.6 million tons, which was produced globally [11]. E-waste generated $57 billion in metals like gold, iron, and copper in 2019. In recent years there was a significant increase in e-waste production from 20 to 50 million metric tons. It is estimated to be doubled by the year 2050 to 111 million metric tons [11].

Future e-waste scenarios

According to UNEP, by the end of the century, metals like zinc, platinum, and copper could face a shortage in supply. Metals are crucial for clean technologies like hafnium used in electronic circuits, gallium used in electronic products, and indium used in photovoltaic cells and light-emitting diodes possibly face a shortage within 10 years. The UNEP reports that the recycling rate for fully recyclable metals like gold, bismuth nickel, palladium, antimony, etc., used in electronic products is less than 10% [10].

Recycling metals (significance)

Recycling of metal scrap is a useful component in the concept of a circular economy, aimed at conserving metal resources and reducing the impact on the environment. The energy efficiency of newly made metals is more than that of recycled ones. The extraction and refinement of metals from their ores require higher energy than that needed in recycling. It is estimated that the reclamation of gold (per unit) from electronic waste material produces 80% less release of CO_2 when compared to gold mining from natural resources. The conservation of natural resources can be achieved by recycling metals [12]. Metals have a great capacity for regeneration during the industrial recycling process, which makes it a great opportunity to raise awareness about the importance of their recovery for the care of the environment [13]. The percentage of metals in electronic circuit boards is represented in Table 7.2.

The most important benefits of scrap reuse are:

- **Reduce the use of naturally available raw materials**: The recycling of metal scrap is an effective way of reducing the use of natural resources for the manufacture of new materials and protecting the environment at the same time.
- **Control of waste generation:** During the procedures of waste management, the classification and recycling of waste metal play an important role in reclaiming the pure metal for reuse in a production line. In this way, the scrap metal can be regenerated, which would have ended up in the landfill [14].
- **Reduction in air pollution and water contamination:** The accumulation of metal scrap in landfills results in air pollution and water contamination of the water bodies around the surroundings. Therefore, proper recycling of metal waste prevents polluting the

TABLE 7.2 The percentage of metals in electronic circuit boards.

	Percentage of metals	
Metal	**Ore**	**Printed circuit board—PCB**
Copper	0.5–3.0	12.0–29.0
Zinc	1.7–6.4	0.1–2.7
Tin	0.2–0.85	1.1–4.8
Lead	0.3–7.5	1.3–3.9
Iron	30–60	0.1–11.4
Nickel	0.7–2.0	0.3–1.6
Gold	0.005	0.0029–0.112
Silver	0.005	0.01–0.52

Source: From W.A. Bizzo, R.A. Figueiredo, V.F De Andrade, Characterization of printed circuit boards for metal and energy recovery after milling and mechanical separation. Materials, 7 (6) (2014) 4555–4566. https://doi.org/10.3390/ma7064555.

environment. According to the Association of World Steel, the primary steel-making by oxygen furnace and blast furnace to produce 1000 Kg of crude steel requires approximately 800 Kg of coal, 300 Kg of limestone, along with 1400 Kg of iron ore and 120 Kg of recycled steel [15]. Whereas the electric arc furnace utilizes 16 Kg of coal and 64 Kg of limestone along with 880 Kg of recycled steel mixed with hot metal and granulated iron for the same amount of crude steel. The recovery of 1000 Kg of crude steel from scrap saves approximately 50 Kg of limestone, 580 Kg of coal, and 1.030 tons of iron ore. This results in negligible environmentally unwanted waste production.

Urban mining

The extraction and recycling of valuable metals from e-waste, old buildings, and various infrastructures is termed urban mining. The concept can reduce the traditional mining and purification of metals from their ore, hence reducing the use of natural resources. Urban mining is set to play an important part in the global move toward sustainability [16].

Urban techniques available for the effective extraction

Separating metals from solid waste is an important process and is essential for the conservation of the environment, economic development, and resource conservation. [17]. Resource recovery, economic values, energy conservation, waste reduction, environmental safety [18], prevention of contamination, compliance toward regulations, greenhouse gas emission reduction, sustainable waste management, and job creation to society are the important reasons for metals recovery from solid waste [19]. Magnetic, gravity, and electrostatic separation techniques [20]; pyrometallurgical processes; biotechnological, chemical precipitation; ion exchange; solvent extraction; chelation; flotation methods [17]; size reduction; and screening techniques are the existing traditional methods that are followed as effective extraction methods of metal from the solid waste. The latest technologies include thermal separation (TS), hydrometallurgy (leaching and metal recovery), and bioleaching technologies that have minimal environmental impact as they could reduce time and cost in extracting heavy metals (HM) from the solid waste [21,22]. These current trends are illustrated in detail in the classification of acid leaching for metal recovery, which is furnished in Fig. 7.2.

Municipal solid waste is an important rich source of recyclable components like metal, glass, paper, and plastic [13]. Automated sorting techniques such as direct sorting utilize material properties, namely magnetic susceptibility, electrical conductivity, and density, for heavy media separation by adopting external fields like magnetic, eddy current, and gravity, respectively. Indirect sorting, on the other hand, includes sensors to explore recyclables in the waste through automated machines or robots that can be utilized to sort the detected recyclable materials. Some of the current trends and methodologies are illustrated in Fig. 7.3 and in detail in the classification.

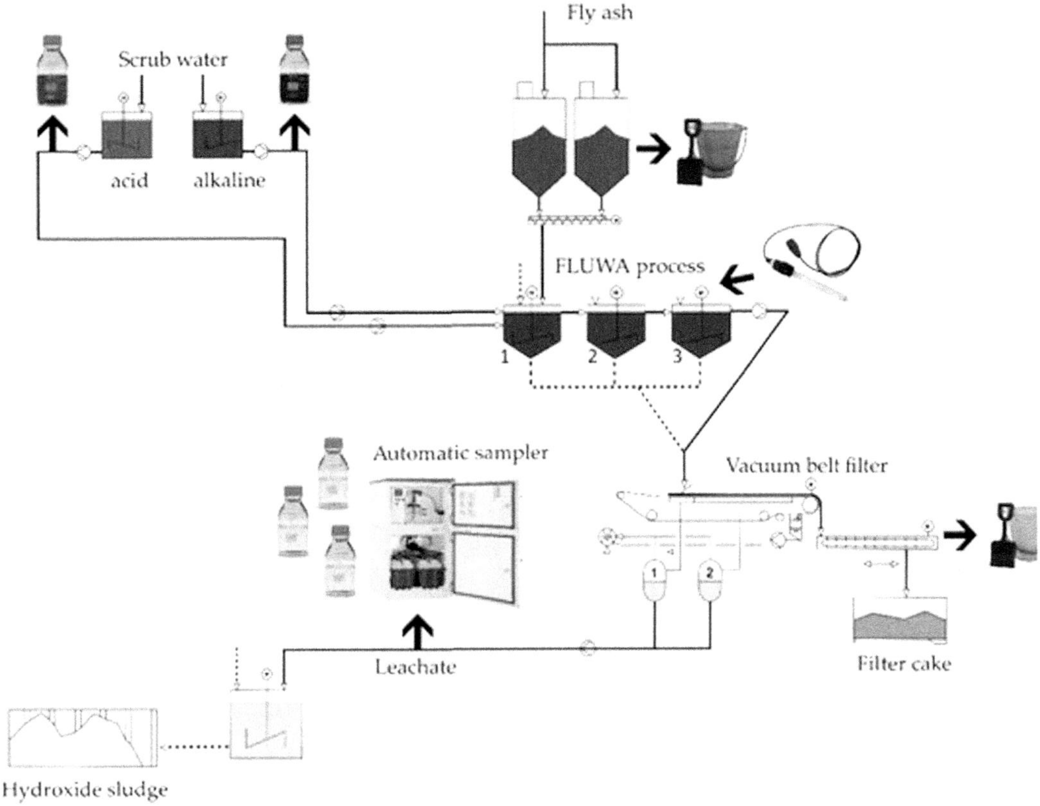

FIGURE 7.2 Acid leaching for metal recovery. *Source: From G. Weibel, A. Zappatini, M. Wolffers, S. Ringmann, Optimization of metal recovery from mswi fly ash by acid leaching: Findings from laboratory-and industrial-scale experiments. Processes, 9 (2) (2021) 1–12. https://doi.org/10.3390/pr9020352.*

Classification of solid waste extraction methods:

1. Magnetic separation
   a. Accumulated soluble nutrients are separated by employing magnetic absorption to carry the material that has magnetic properties [24].
   b. Overhead magnets, drum magnets, and magnetic pulleys are effective techniques in this category.
   c. An electromagnetic field parallel to the flow of separation with electric cables to hold the accumulated particles is deployed for effective segregation from Fig. 7.4.
   d. Advantageous for high elimination performance in the separation of ferromagnetic metals from nonferromagnetic materials.
2. Electrostatic separation
   a. This method ensures the separation of materials based on electric properties exclusively, which will exploit the variation in the electric conductivity or electrostatic charge of different materials [25].

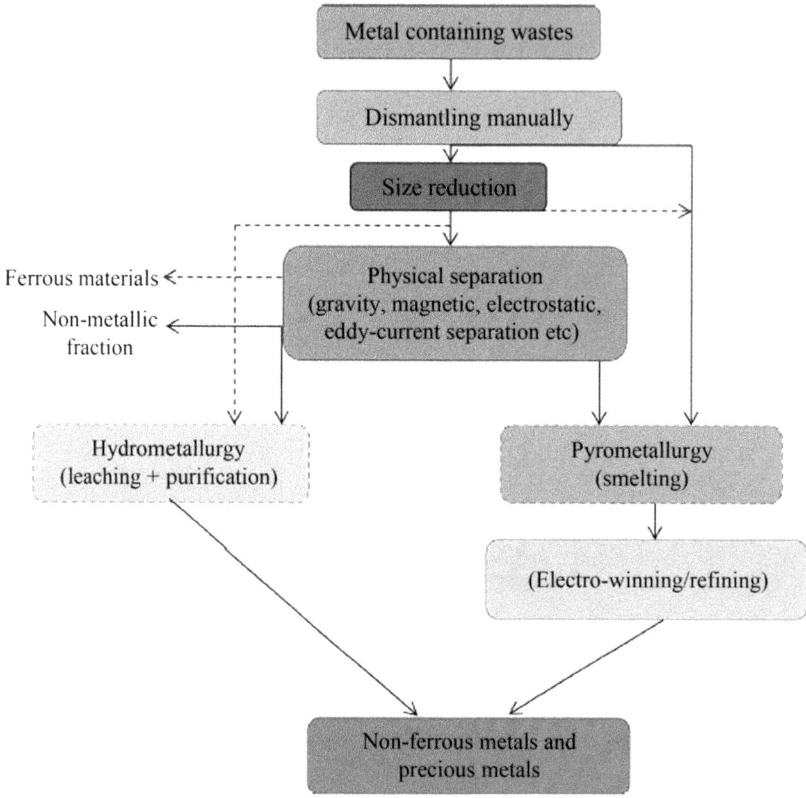

FIGURE 7.3 A diagram of several extraction processes for metal recovery from metal waste [23]. Source: *From K. Santhana, Z.N. Syahidah, K. Hesam, T.S. Mat, Din, M.F.B.M., Majid, Z. Abd, C., Sumate, K., Iwao, I., Yo, N., Mohd, C., Shreeshivadasan, O., Norzilla, Current technologies for recovery of metals from industrial wastes: an overview. Environmental Technology & Innovation, 22 (2021). https://doi.org/10.1016/j.eti.2021.101525.*

   b. This depends on the principle that if two bodies are charged alike, they repel each other, and if charged oppositely they will be attracted from Fig. 7.5.
   c. Charging, attraction–repulsion, and separation chamber are some of the conceptual techniques followed in this method.
3. Hydrometallurgical separation
   a. This method is used to separate metals from ore or waste by dissolving them using aqueous or organic solutions [14].
   b. Electric-arc furnace dust (EAFD) is one of the major activities in which the residue is produced during the processing of steel in mini-mills from Fig. 7.6.
   c. This is advantageous specifically for specific metals such as gold (Au), silver (Ag), copper (Cu), and zinc (Zn).
4. Thermal separation
   a. This method can be achieved by adopting pyrolysis, incineration, and smelting processes based on the type of metal waste [26].

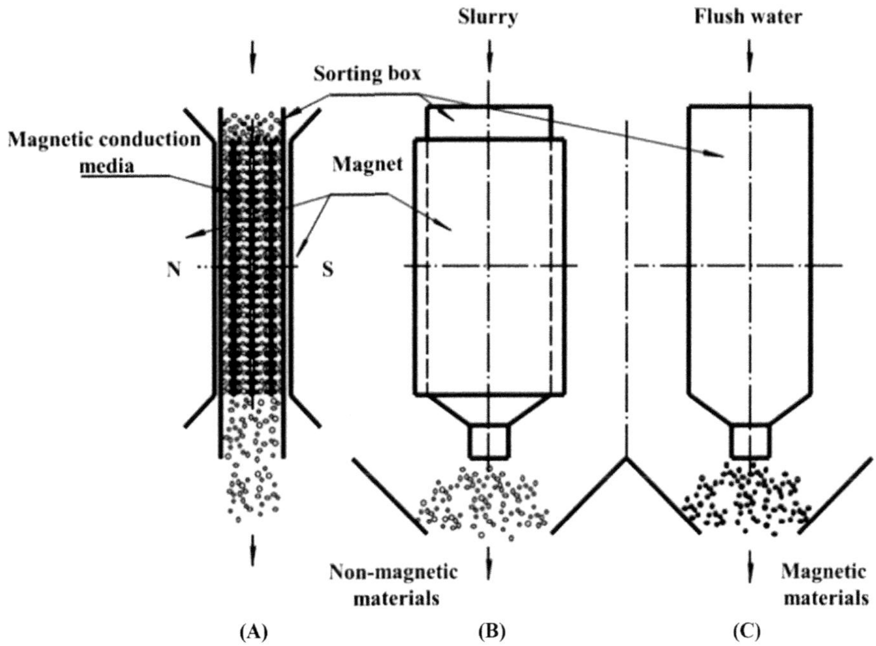

FIGURE 7.4 Magnetic separation technique [27]. Source: *From C.B. Gill, Electrostatic separation. Materials Beneficiation (1991) 141–147.*

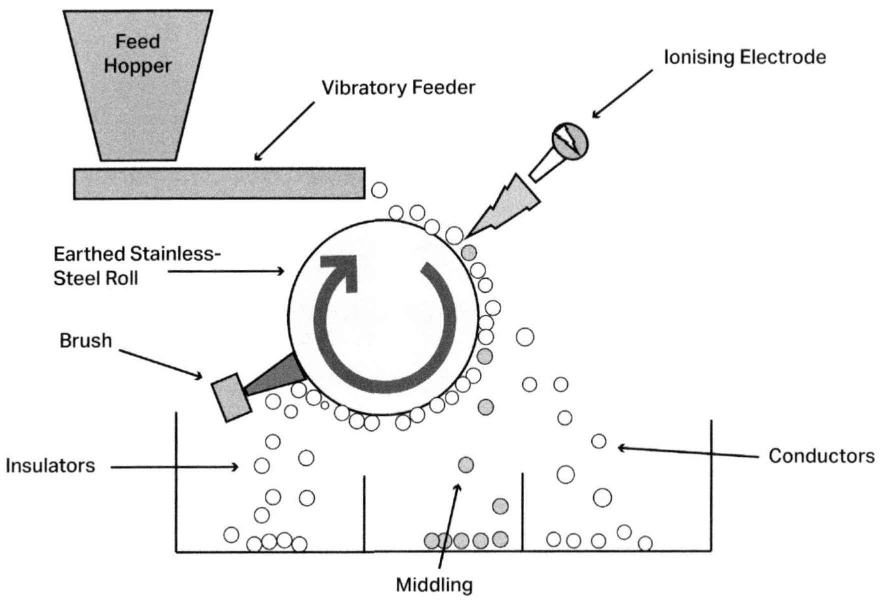

FIGURE 7.5 Electrostatic technique. Source:*https://www.electricity-magnetism.org/electrostatic-separators Bunting - Electrostatic Separator.*

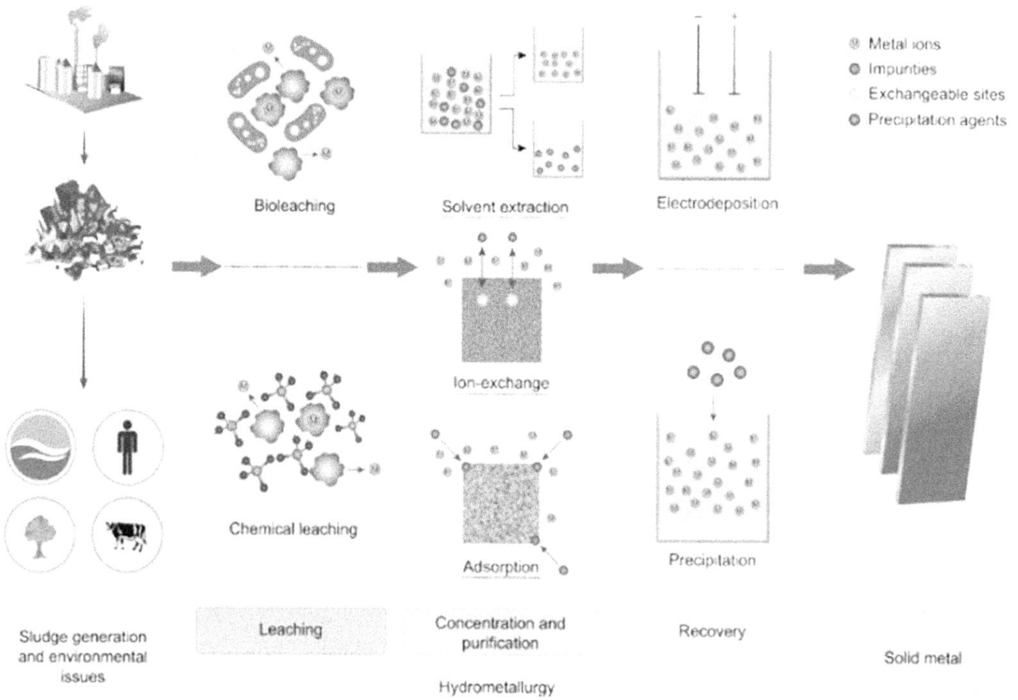

FIGURE 7.6 Hydrometallurgical separation technique [28]. Source: *From J. Joo, E.E. Kwon, J. Lee, Achievements in pyrolysis process in E-waste management sector. Environmental Pollution, (2021) 287. https://doi.org/10.1016/j.envpol.2021.117621.*

b. Pyrolysis includes heating the metal waste by isolating it from oxygen, causing it to decompose into its related compounds, gases, and solid residue.
c. Incineration is an ultra-high-temperature combustion process that oxidizes organic materials into a waste stream by leaving ash residue containing the metal waste.
d. Smelting is adopted in many aluminum and steel industries, which involves melting the metal waste at very high temperatures to separate the metal from other waste.
e. For selecting an appropriate TS methodology, various factors to be considered are the composition of the metal waste, desired metal recovery efficiency, aspects related to environmental pollution, and feasibility of economic conditions from Fig. 7.7.

Reduce and reuse of the metal waste

In today's world, the management of metal waste is of paramount importance due to its environmental impact and the depletion of natural resources. Adopting strategies to reduce and reuse metal waste is crucial for transitioning toward a circular economy model, where resources are utilized efficiently and waste is minimized. This chapter explores various techniques and approaches for reducing and reusing metal waste, thereby promoting sustainability and environmental stewardship [29].

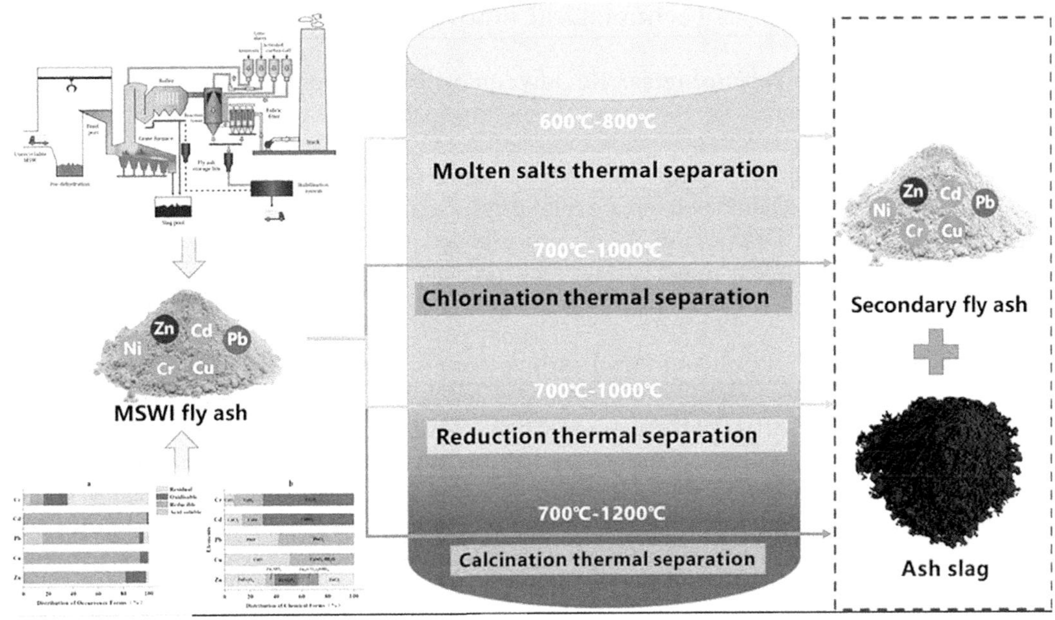

FIGURE 7.7 Thermal separation technique. Source: *From A. David, Y.D. Thangavel, R. Sankriti, Recover, recycle and reuse: an efficient way to reduce the waste. International Journal of Mechanical and Production Engineering Research and Development, 9 (3) (2019) 31–42. https://doi.org/10.24247/ijmperdjun20194.*

Reducing metal waste

Design for sustainability: Emphasize the importance of designing products with longevity, recyclability, and ease of disassembly to minimize metal waste generation.

Material efficiency: Introduce strategies for optimizing material use in manufacturing processes, such as lean manufacturing principles and light weighting [30].

Source reduction: Explore methods for reducing metal waste at the source, including process optimization, material substitution, and waste prevention initiatives [31].

Reusing metal waste

Salvage and reclamation: Discuss the practice of salvaging metal waste from end-of-life products and industrial processes for reuse in new applications [32].

Remanufacturing and refurbishment: Highlight the benefits of remanufacturing and refurbishing metal components and products to extend their lifecycle and reduce demand for virgin materials [33].

Circular supply chains: Examine the concept of circular supply chains, where metal waste is reintegrated into the production process through recycling, repurposing, and closed-loop systems [34].

Technological innovations

Advanced recycling technologies: Review emerging technologies for the efficient recycling of metal waste, such as pyrolysis, hydrometallurgy, and electrochemical processes [2].

Additive manufacturing: Explore the role of additive manufacturing (3D printing) in utilizing recycled metal feedstock to create new products with minimal waste generation [14].

Sensor-based sorting: Discuss the use of sensor-based sorting technologies to improve the efficiency and effectiveness of metal waste recycling operations [35].

Economic and policy considerations

Economic incentives: Analyze the economic benefits of reducing and reusing metal waste, including cost savings, resource conservation, and enhanced market competitiveness [36].

Regulatory frameworks: Evaluate existing policies and regulations governing metal waste management and explore potential measures to incentivize sustainable practices [37].

Public awareness and education: Emphasize the importance of raising awareness and educating stakeholders about the environmental and economic benefits of reducing and reusing metal waste [38].

Importance of automation technology in metal waste management

Automating metal waste handling can benefit organizations in several ways, including increased efficiency, reduced costs, and improved sustainability. Automating can also enhance safety and improve the environment. Automate tasks with a high volume or repetitive nature gradually. Ensure the accuracy and compatibility of data by implementing standardized data formats and protocols. Analyzing data in order to identify trends, optimize operations, and improve decision-making are some of the benefits of data analytics. Keep up to date with the latest advancements in automation technology, including robotics, artificial intelligence (AI), and machine learning (ML). In addition, it can contribute to the reduction of carbon emissions and the protection of the environment from pollution.

A significant reduction in labor costs can be achieved through automation, especially when dealing with repetitive or hazardous tasks. As well as this, it can enhance efficiency by minimizing downtime and errors, thereby reducing costs as a whole. By operating 24/7, automation increases productivity and throughput. As a result, output and revenue can be increased. Quality control can be enhanced by automating processes, which reduces human error and ensures consistent standards. By eliminating defects and reworking, costs can be reduced, and customer satisfaction can be improved. It is true that automation may create some jobs that will be lost, but it can also provide new opportunities in areas such as technology development, maintenance, and data analysis. By fostering innovation and creating new industries, automation in metal waste management can contribute to overall economic growth.

Case studies of automation technologies in industries for metal waste management

Tesla (USA): For efficient processing of end-of-life vehicles, Tesla (USA) has invested in advanced robotics and automation technologies in its recycling facilities. To recycle different materials, such as aluminum, steel, and plastic, these systems are capable of accurate separation.

Electronics manufacturing: Apple (China): Several automated recycling facilities have been invested in by Apple in order to retrieve valuable materials from discarded electronic devices. It is used in these facilities to extract metals such as copper, aluminum, and rare earth elements using advanced sorting and separation technologies.

Scrap metal recycling: Schnitzer Steel (USA): The Schnitzer Steel recycling facility has implemented an automated sorting system that uses optical sensors and AI to deliver accurate identification and separation of different types of scrap metal. A reduction in manual labor is achieved as a result of this improvement.

Construction and demolition waste: Euro Recycling (Europe): By using AI and optical sensors, Schnitzer Steel's recycling facility has implemented a system that sorts scrap metal using automated sorting. Consequently, this improvement reduces manual labor. Metals, concrete, and other materials can be efficiently separated using these systems.

There is a great deal of diversity in the applications of automation in metal waste management across different industries and geographical regions, as illustrated by these case studies. In their metal waste management operations, organizations can benefit from automation technologies by improving efficiency, reducing costs, and enhancing environmental sustainability.

Artificial intelligent technology

Metal waste management presents significant challenges, requiring innovative solutions to reduce environmental impact and promote sustainability. This topic explores the application of AI technologies in the reduction and reuse of metal waste, focusing on their potential to enhance efficiency, optimize processes, and facilitate decision-making [39]. Through a comprehensive review of existing literature, case studies, and technological advancements, this content examines AI-driven approaches for waste characterization, material recovery, process optimization, and supply chain management. Key themes addressed include ML, robotic systems, predictive analytics, and data-driven decision support tools in the context of metal waste reduction and reuse. By elucidating the capabilities and challenges of AI technologies, this paper aims to inform stakeholders, researchers, and policymakers and advance the adoption of AI-driven solutions for sustainable metal waste management practices [40].

Remediation of metal waste

Remediation strategies for metal waste involve a range of innovative and sustainable approaches to address environmental and health hazards associated with metal contaminants. Techniques such as phytoremediation and bioremediation utilize plants and microorganisms to extract or transform metals [41]. Chemical and physical methods, including chemical precipitation and soil washing, are employed to reduce metal concentrations. In situ immobilization, ion exchange, and constructed wetlands provide additional avenues for metal waste remediation. Advanced technologies such as electro-kinetic remediation and data-driven optimization contribute to efficient treatment. The integration of these strategies, considering environmental, economic, and social factors, holds promise for effective metal waste remediation, fostering a resilient and environmentally conscious society [42].

The increasing prevalence of metal waste poses significant environmental and health challenges, necessitating innovative and sustainable remediation approaches. This book provides a comprehensive detail of the various strategies employed for the remediation of metal waste in soil, water, and other ecosystems [43]. The focus is on elucidating the principles, applications, and advancements in remediation technologies that aim to mitigate the adverse effects of metal contamination on ecosystems and human health. Remediation of metal waste involves the application of strategies and technologies to address environmental and health hazards associated with the presence of metal contaminants in soil, water, or other ecosystems. The goal is to reduce or eliminate the adverse effects of metal pollution and restore affected areas to a more sustainable and safer [44].

Metal waste contamination presents complex environmental challenges that demand innovative solutions. This paper explores the application of AI technologies in the remediation of metal waste, focusing on their potential to enhance efficiency, accuracy, and sustainability [45]. The concept of artificial neural network (ANN) is represented in Fig. 7.8 with a block diagram.

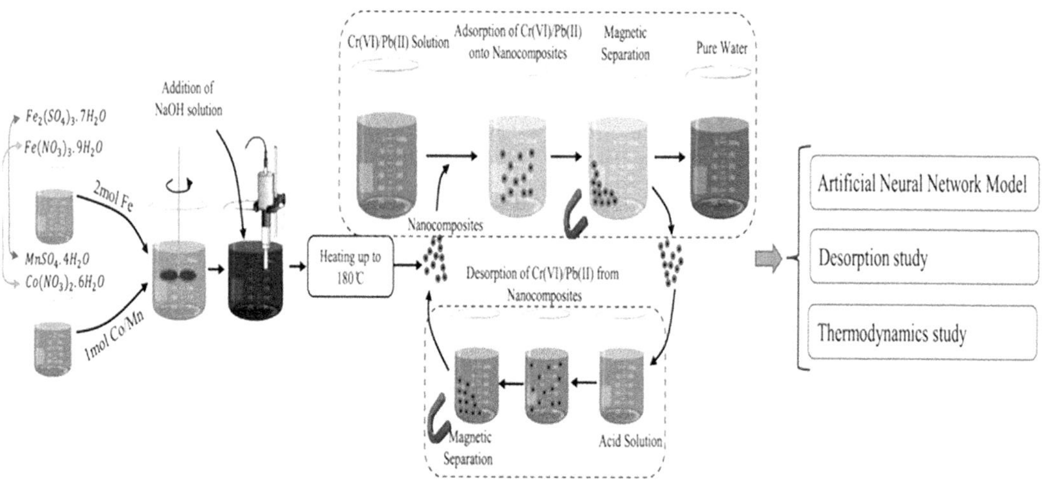

FIGURE 7.8 Artificial neural network – a block diagram [46]. Source: *From S.F. Rafie, H. Abdollahi, H. Sayahi, F.D. Ardejani, K. Aghapoor, M.H. Karimi Darvanjooghi, S. Kaur Brar, S. Magdouli, Genetic algorithm-assisted artificial neural network modelling for remediation and recovery of Pb (II) and Cr(VI) by manganese and cobalt spinel ferrite super nanoadsorbent. Chemosphere, 321 (2023). https://doi.org/10.1016/j.chemosphere.2023.138162.*

Through a comprehensive review of existing literature, case studies, and technological advancements, this paper examines AI-driven approaches for site characterization, remediation process optimization, risk assessment, and decision-making [47]. Key themes addressed include the integration of ML, remote sensing, robotics, and data analytics in metal waste remediation efforts [48]. By elucidating the capabilities and limitations of AI technologies, this paper aims to inform researchers, practitioners, and policymakers and facilitate the adoption of AI-driven solutions for sustainable remediation practice conditions [49]. Below are a few remedial techniques used regularly for the separation of metals. Fig. 7.9 represents acid mine drainage treatment.

- Phytoremediation.
- Bioremediation.
- Chemical precipitation.

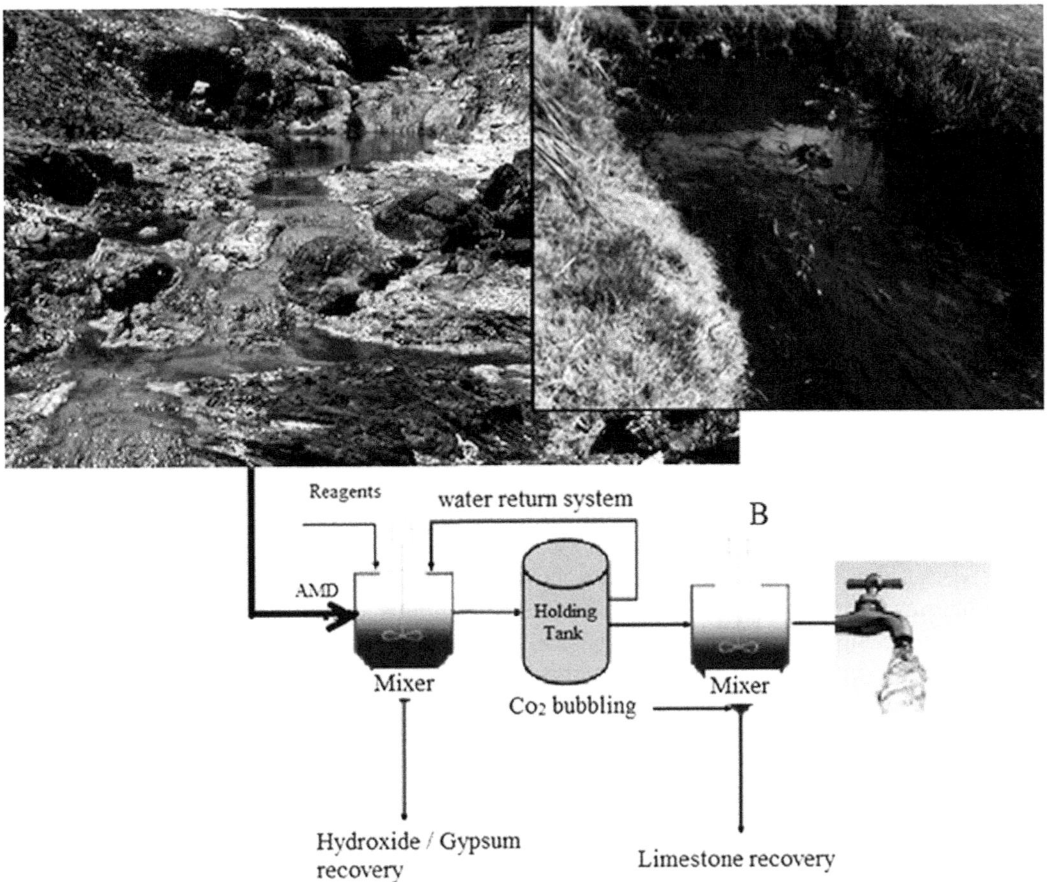

FIGURE 7.9 Treatment of acid mine drainage. Source: *From R. Shukla, S.S. Bhadauria, R.K. Shrivastava, Mining from landfills as a remediation strategy regarding open dumpsites using artificial intelligence hybrid models. Journal of Scientific and Industrial Research, 82 (4) (2023) 460–465. https://doi.org/10.56042/jsir.v82i04.72386.*

- Electrokinetic remediation.
- In situ immobilization.
- Soil washing.
- Ion exchange.
- Constructed wetlands.
- Ex situ treatment.

Metal waste contamination

Sources and impacts

1. **Sources of metal contamination:** It is believed that HMs are produced by mining, industrial production (foundries, smelters, oil refineries, petrochemical plants, pesticide production, chemical industry), untreated sewage sludge, and diffusion sources, including metal piping, traffic, and combustion byproducts from coal-fired power plants.
2. **Environmental and health impacts:** Environmental concerns related to metal mining are primarily due to the operations and waste products associated with metal extraction and processing. There are several concerns regarding the landscape, including physical disturbances. Contamination of soil and water.
3. **Need for remediation strategies**: Metal waste should be disposed of properly in order to protect the environment and human health.

Remediation of metal waste: Metal waste contamination poses significant environmental and health risks, requiring effective remediation strategies [50]. This investigation provides a comprehensive overview of remediation techniques for metal waste, focusing on both soil and water environments [51]. Through a synthesis of existing literature, case studies, and technological advancements, this research explores various remediation approaches, including physical, chemical, and biological methods [43]. Key themes addressed include the principles of remediation, site characterization, selection of appropriate techniques, and challenges associated with remediation implementation. By elucidating the complexities of metal waste remediation, this paper aims to inform stakeholders and policymakers and contribute to the development of sustainable solutions [52].

Metal waste has been recovered from a variety of sources using ML techniques. In these, two techniques are used; one is ANN, and the other one is CNN. Waqar Muhammad et al. reported recovering metal from the waste printed circuit boards (WPCBs) of mobile phones; we are reducing our dependence on natural resources while also saving energy and reducing emissions [53]. They used a ML-based model for Cu, Ni, and Pb recovery from WPCBs of mobile phones through low-temperature roasting. ML model of metal recovery is based on ANN inputs that include three variables (roasting temperature, roasting time, NH_4Cl dose).

Integration of remote sensing and robotics: Robots replace manual labor in waste sorting and collection processes, thereby increasing efficiency and reducing human effort. Currently, the following techniques are being used for the remediation of metal waste.

- Remote sensing technologies for environmental monitoring.
- Hyperspectral imaging and spectroscopy.

- LiDAR (light detection and ranging) applications.
- Robotic systems for field sampling and remediation.
- Autonomous ground vehicles (AGVs).
- Aerial drones and UAVs (unmanned aerial vehicles).

Data analytics and visualization: As part of waste management, data analytics (algorithms and methodologies) can be utilized to monitor and analyze waste bin fill levels, forecast tonnage transportation for specific materials (brown paper, scrap metal, plastic, etc.), forecast prices, and optimize routes and volumes. Generally, these methods and techniques are derived from the fields of statistics, probability, deep learning, time series modeling, and linear trend analysis. Using IoT technology and monitoring functionality, supervised control of waste levels can be achieved as well as access to historical data. In view of the abundance of proposed analytical methods, it is evident that it is possible to plan and optimize waste management practices in order to benefit the environment as well as the financial bottom line. Using the proposed technologies, metal waste can be separated and recovered rapidly. A block diagram for e-waste metal recovery is shown in Fig. 7.10.

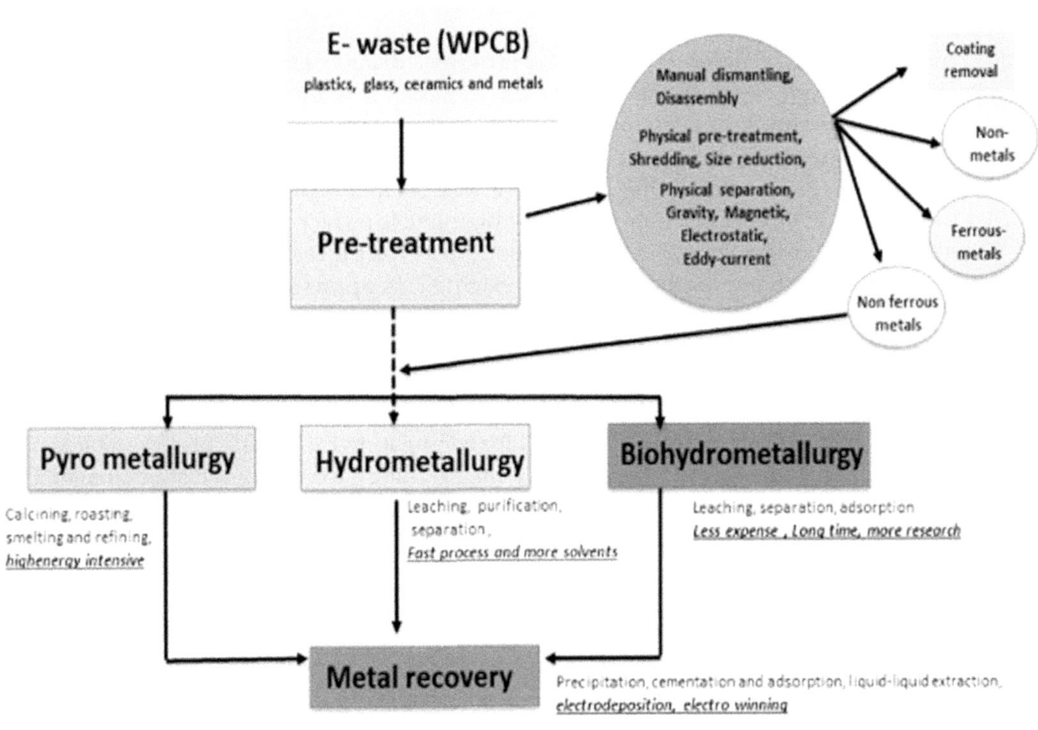

FIGURE 7.10 Block diagram for e-waste metal recovery [54]. Source: *From V. Rai, D. Liu, D. Xia, Y. Jayaraman, J.C.P. Gabriel, Electrochemical approaches for the recovery of metals from electronic waste: a critical review. Recycling, 6 (3) (2021). https://doi.org/10.3390/recycling6030053.*

Reported artificial intelligence remediation techniques for the remediation of metal wastes

AI is related to imitating human intelligence on a computer to enhance information and knowledge, which are symbolic, complex, and more output-based on the application [55]. Huang et al. described the basic characteristics of waste reduction and narrated techniques like fuzzy knowledge-based systems (hybrid intelligence systems), logic systems, and artificial neural networks to recover material and energy recovery processes of metals from waste. Fuzzy logic is a logic system close to the human brain and language over the conventional logic approach. An artificial neural network is one of the computing techniques in which a number of highly inter-connected processing elements are processed by dynamic state response to the given numerical inputs [56]. Piotr et al. proposed a vehicle model to collect e-waste from densely populated cities and factories [57]. Honggui Han et al. in 2022 reported that metallic product recognition with dual attention and multibranch residual blocks-based convolutional neural networks. It is proposed in this article that a convolutional neural network with a dual attention mechanism and multibranch residual blocks can recognize metallic products with a high level of accuracy. In the first step, a channel-spatial dual attention mechanism is introduced in order to enhance the sensitivity of the model to key features. The model can focus on key features even when there is a large amount of confusing information when analyzing metallic products. To achieve deeper and more effective feature extraction for metallic products with complex characteristic features, deep convolutional networks with multibranch residual blocks as the backbone are embedded with a dual attention module. Various datasets containing images pertaining to waste electrical and electronic equipment (WEEE) and waste household metal appliances (WHMA) are used to evaluate the proposed model. According to the results of this experiment, 94.31% and 95.88% of WEEE and WHMA samples were accurately recycled, thus achieving high levels of recycling accuracy and quality [58]. Kshirsagar et al. described the reusable robotic arm for processing and collecting useful items from the metal garbage to the maximum extent. This approach segregates the garbage into cartons and plastics from the garbage waste. Industry-oriented management techniques and exclusive robotic architecture with multifunctionalities like photo editing, motion control, grabbing, and classifications are the main studies of this article. The success rate of grasping objects like plastics (69.86%) and glass (87.78%) was achieved from robotic arm reusable technology [59]. The futuristic may be employed for segregating the metals, also possible using this methodology. Kan Hua Yu et al. narrated a methodology for calling waste management using an AI-based hybridized intelligent framework (AIHIF) for automatic recycling to optimize waste when the consumers use and discard the goods [60]. Changru Le designed a hybrid network framework based on a convolutional network for problem classification. The author used deep learning (DL), which uses neural networks with not less than two hidden layers to perform the required task. The recurring neural network (RNN) will extract and train the input data and will greatly reduce the sensitivity of DL to data samples. Then the DLRNN will be used to generate directional input and output connected into a ring. In this paper, the author tested and analyzed not only carbon emissions in the warehousing, handling, and other related activities. The results show a good understanding of the remediation of metal waste, which presents a significant challenge for environmental scientists and engineers worldwide. Conventional methods often struggle to achieve the precision and efficiency required for the effective separation and removal of metal contaminants from various environmental matrices. However, recent advancements in AI techniques offer promising solutions to address these challenges. By leveraging AI algorithms and technologies, researchers have developed innovative remediation methods for metal waste separation. This article explores the application of AI in metal waste separation, drawing insights from cutting-edge research papers to highlight the diverse approaches and advancements in this field. LRNN method and Text Retrieval Conference (TREC) dataset [61]. The recycling in mining application for gravity concentration is shown in Fig. 7.11.

The remediation of metal waste presents a significant challenge for environmental scientists and engineers worldwide. Conventional methods often struggle to achieve the precision and efficiency required for the effective separation and removal of metal contaminants from various environmental matrices. However, recent advancements in AI techniques offer promising solutions to address these challenges. By leveraging AI algorithms and technologies, researchers have developed innovative remediation methods for metal waste separation. This article explores the application of AI in metal waste separation, drawing insights from cutting-edge

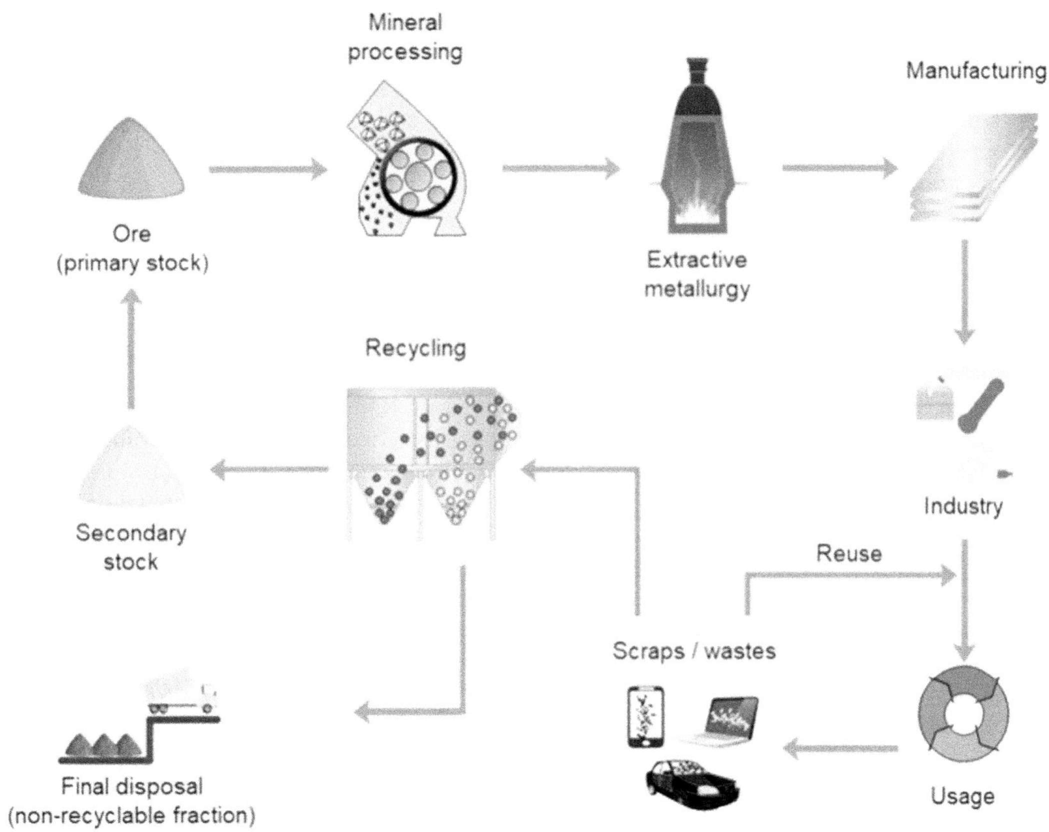

FIGURE 7.11 Recycling mining applications for gravity concentration [62]. Source: *From W.M. Ambrós, Gravity concentration in urban mining applications—a review. Recycling, 8 (6) (2023). https://doi.org/10.3390/recycling8060085.*

research papers to highlight the diverse approaches and advancements in this field. Chen et al. introduced a deep-learning approach for the identification of metal contaminants in soil samples. They trained a convolutional neural network (CNN) model on spectral data collected from soil samples using a portable spectrometer. The trained model demonstrated high accuracy in classifying different types of metal contaminants, enabling precise identification and targeted remediation strategies. The study showcases the potential of deep learning techniques in enhancing the efficiency and accuracy of metal waste separation processes [63]. Patel et al. presented the development of an autonomous drone-based system for mapping and remediation of metal waste contamination. The researchers equipped drones with AI-driven sensors and cameras capable of detecting metal contaminants from aerial images. By integrating real-time data processing and decision-making algorithms, the system autonomously identified contaminated areas and deployed remediation measures, such as soil stabilization or vegetation cover, to mitigate metal leaching and dispersion. The study highlights the potential of drone technology and AI integration for scalable and cost-effective metal waste remediation efforts [64]. Wang et al. proposed a multiagent reinforcement learning (MARL) framework for dynamic metal waste separation in complex environments. The researchers developed a decentralized MARL algorithm that enables multiple agents to collaborate and adaptively optimize separation strategies based on real-time feedback. The framework was applied to a robotic system tasked with sorting mixed metal waste streams, demonstrating superior performance compared to traditional methods. The research underscores the potential of MARL in enabling autonomous and adaptive metal waste separation processes [65]. Ninghui Li et al. proposed the use of graph neural networks (GNNs) for identifying sources of metal contamination in water bodies. They constructed a graph representation of water quality data, where nodes represent sampling locations and edges represent spatial correlations between samples. By training a GNN model on this graph data, the researchers achieved accurate identification of upstream sources of metal pollution, facilitating targeted remediation efforts. The study highlights the potential of GNNs in leveraging spatial relationships for effective metal waste source identification [66]. Zhang et al. introduced a real-time adaptive control strategy for metal waste treatment processes based on AI techniques. The researchers developed a predictive model of metal waste treatment systems using RNNs trained on historical process data. By integrating the predictive model with a model predictive control (MPC) framework, the system continuously adjusts process parameters to optimize treatment efficiency and minimize waste generation. The study demonstrates the potential of AI-driven adaptive control strategies in enhancing the performance of metal waste treatment processes [67]. Lee et al. present the design and implementation of a robotic swarm system for large-scale metal waste remediation in aquatic environments. The researchers developed a swarm intelligence algorithm that enables autonomous coordination and collaboration among robotic agents. Each robot is equipped with AI-driven sensors for detecting metal contaminants and performing remediation actions, such as sediment dredging or chemical treatment. The swarm system demonstrated efficient and scalable metal waste removal capabilities, highlighting the potential of collective intelligence in environmental remediation [68]. Kim et al. proposed a reinforcement learning (RL) approach for adaptive pump scheduling in metal waste treatment plants. The researchers developed a RL agent to dynamically optimize pump operation schedules based on real-time flow rate and contaminant concentration data. By maximizing treatment efficiency while minimizing energy consumption and operational costs, the adaptive pump scheduling algorithm improved the overall performance of metal waste treatment processes. The research demonstrates the potential of RL in enhancing the sustainability and cost-effectiveness of metal waste remediation operations [68]. Wu et al. proposed a deep RL approach for optimal sensor placement in metal waste monitoring networks. The researchers formulated the sensor placement problem as a sequential decision-making task, where an agent learns to select sensor locations to maximize information gained about metal contamination levels. By training a deep RL agent on simulated environmental scenarios, the researchers identified sensor configurations that achieve high detection accuracy while minimizing resource requirements. The study highlights the potential of AI-driven sensor placement strategies in enhancing the effectiveness of metal waste monitoring and remediation efforts [69]. Li et al. presented an adversarial training approach for robust metal waste detection in remote sensing images. The researchers developed a deep learning model trained on a combination of clean and adversarially perturbed images to improve the model's robustness to image distortions and environmental variations. The adversarially trained model demonstrated superior performance in detecting metal waste contamination in remote sensing images under challenging conditions, such as cloud cover or sensor noise. The study showcases the potential of adversarial training techniques in enhancing the reliability and accuracy of metal waste detection methods [70].

Conclusions

This chapter described the classification of ferrous scrap and nonferrous metal scrap in detail. An overview of the markets for metal waste in various countries around the world, including those in the USA, EU, Turkey, and others. A future scenario for metal and e-waste was also described. Recycling technologies play a significant role in the recovery of waste metal waste. A clear methodology for recycling metal waste is presented in this book chapter. It is important to recycle metal scrap in order to reduce the consumption of natural resources for the manufacture of new materials and protect the environment at the same time. Globalization is resulting in an increased emphasis on sustainability, and urban mining is expected to play an increasingly important role in this transition. An overview of clear extraction techniques, including magnetic separation, electrostatic separation, hydrometallurgical separation, and TS, has been discussed with detailed illustrations. Based on the reported literature, the most adaptable technique is hydrometallurgical separation over others. A significant advantage of this innovative approach is that it minimizes the need for energy-intensive physical methods while providing significant advantages in terms of purity and sustainability. Minerals with complex structures and low grades can be extracted using hydrometallurgy. However, people still have to deal with minor difficulties such as the high consumption of water, difficulties in separating solids from liquids, impurity problems, and a longer preparation time. In order to overcome the aforementioned reasons, all of these techniques are currently being implemented with AI technologies.

As a result of their rapid efficiency, effectiveness, economic viability, and feasibility, AI technologies play a significant role in the recovery of metal from metal waste. Currently, AI is being studied for its potential to reduce and reuse metal waste, particularly in terms of improving efficiency, optimizing processes, and facilitating decision-making. There are a number of key themes that are addressed throughout the AI, including ML, deep learning, robotics, predictive analytics, and data-driven decision support tools. According to the reported literature, two machine-learning techniques have been utilized to recover metal waste from a variety of sources, namely ANNs and CNNs. AI remediation techniques have been reported to include convolutional neural networks, deep convolutional networks with residual blocks with multiple branches, AIHIF, deep learning, RNNs, and deep learning recurring neural networks (DLRNN) for recovering metal from metal waste. ML models, including artificial neural networks, recurrent neural networks, radial basis function neural networks, convolutional neural networks, and long short-term memory (LSTM), have been utilized to predict and optimize performance. The RNN is capable of learning structured and detailed information, which can be used for the effective segregation of different metals from waste metals. With RNN, you can recursively apply the same set of weights to structured inputs in order to arrive at a structured prediction.

Future prospects

As of today, AI is in its infancy in terms of extracting metal from various metal waste. AI is being used to speed up the process of segmentation and extraction of metal from waste. The trained CNN and RNN models were efficiently utilized to separate metals from waste. The use of AI and data analytics can assist in identifying areas with high mineralization potential in a timely and safe manner, thereby reducing exploration costs and time. Using AI algorithms, such as deep neural networks, can enhance the recognition of images and speech, improving the quality and quantity of ore extraction. However, using ML techniques easily optimizes bioleaching processes for metal recovery. ML is a subfield of AI, with proven accuracy, reliability, and cost-effectiveness in various applications. ML is a subfield of AI that focuses on enhancing the accuracy, reliability, and cost-effectiveness of various applications. ML algorithms the capability to be trained in order to recognize various types of metals and their compositions within waste materials. By employing supervised learning techniques, it becomes possible to classify metal objects based on their specific properties and characteristics. Furthermore, unsupervised learning can be utilized to identify patterns and similarities in metal waste, greatly contributing to the efficiency of the separation process. Robots equipped with sensors and cameras are capable of identifying and extracting specific metal components from mixed waste streams. RL algorithms can enhance the efficiency of metal recovery systems through the process of learning from feedback and making real-time adjustments to sorting strategies. Robotic process automation (RPA) can be effectively employed to regulate and manage robotic arms or sorting mechanisms, with the primary objective of optimizing metal yield while reducing wastage to a minimum.

References

[1] C. Hagelüken, D. Goldmann, Recycling and circular economy—towards a closed loop for metals in emerging clean technologies, Mineral Economics 35 (3–4) (2022) 539–562. Available from https://doi.org/10.1007/s13563-022-00319-1.

[2] M. Kaya, Recovery of metals and nonmetals from electronic waste by physical and chemical recycling processes, Waste Management 57 (2016) 64–90. Available from https://doi.org/10.1016/j.wasman.2016.08.004.

[3] S. Nakamura, Y. Kondo, K. Nakajima, H. Ohno, S. Pauliuk, quantifying recycling and losses of Cr and Ni in steel throughout multiple life cycles using MaTrace-alloy, Environmental Science & Technology 51 (17) (2017) 9469–9476. Available from https://doi.org/10.1021/acs.est.7b01683.

[4] A. Prof, R.A. Anaee, L.M.H. Alzuhairi, Corrosion behavior of steel (St 37-2) by using natural products as inhibitors, in: Petroleum Medium A Thesis Submitted to the Department of Materials Engineering, University of Technology in a Partial Fulfillment of the Requirements for the Degree of Mas, 2014.

[5] P. John, R. Davis, Performance study of electrical discharge machining process in burr removal of drilled holes in Al 7075, Cogent Engineering 3 (1) (2016) 3727–3734. Available from https://doi.org/10.1080/23311916.2016.1270702.

[6] J. Yu, R. Xu, J. Zhang, A. Zheng, A review on reduction technology of air pollutant in current China's iron and steel industry, Journal of Cleaner Production 414 (2023) 137659. Available from https://doi.org/10.1016/j.jclepro.2023.137659.

[7] M.D. O'Toole, N. Karimian, A.J. Peyton, Classification of nonferrous metals using magnetic induction spectroscopy, IEEE Transactions on Industrial Informatics 14 (8) (2018) 3477–3485. Available from https://doi.org/10.1109/TII.2017.2786778.

[8] A. Kumar, M. Holuszko, D.C.R. Espinosa, E-waste: an overview on generation, collection, legislation and recycling practices, Resources, Conservation and Recycling 122 (2017) 32–42. Available from https://doi.org/10.1016/j.resconrec.2017.01.018.

[9] K. Parajuly et al., Future e-waste scenarios, StEP (Bonn), UNU ViE-SCYCLE UNEP IETC, 2019, p. 19.

[10] V. Forti, C.P. Baldé, R. Kuehr, G. Bel, The global e-waste monitor 2020: quantities, flows and the circular economy potential. United Nations University (UNU)/United Nations Institute for Training and Research (UNITAR) – co-hosted SCYCLE Programme, International Telecommunication Union (ITU) & International Solid Waste Association (ISWA), Bonn/Geneva/Rotterdam, 2020.

[11] S.C. Chakraborty, et al., Metals in e-waste: occurrence, fate, impacts and remediation technologies, Process Safety and Environmental Protection 162 (2022) 230–252. Available from https://doi.org/10.1016/j.psep.2022.04.011.

[12] W.A. Bizzo, R.A. Figueiredo, V.F. De Andrade, Characterization of printed circuit boards for metal and energy recovery after milling and mechanical separation, Materials (Basel) 7 (6) (2014) 4555–4566. Available from https://doi.org/10.3390/ma7064555.

[13] S. Krishnan, et al., Current technologies for recovery of metals from industrial wastes: an overview, Environmental Technology & Innovation 22 (2021) 101525. Available from https://doi.org/10.1016/j.eti.2021.101525.

[14] J.G. Backes, J. Suer, N. Pauliks, S. Neugebauer, M. Traverso, Life cycle assessment of an integrated steel mill using primary manufacturing data: actual environmental profile, Sustainability 13 (6) (2021) 1–18. Available from https://doi.org/10.3390/su13063443.

[15] L.H. Xavier, M. Ottoni, L.P.P. Abreu, A comprehensive review of urban mining and the value recovery from e-waste materials, Resources, Conservation and Recycling 190 (2023) 106840. Available from https://doi.org/10.1016/j.resconrec.2022.106840.

[16] N.K. Soliman, A.F. Moustafa, Industrial solid waste for heavy metals adsorption features and challenges; a review, Journal of Materials Research and Technology 9 (5) (2020) 10235–10253. Available from https://doi.org/10.1016/j.jmrt.2020.07.045.

[17] H.B. Sharma, K.R. Vanapalli, B. Samal, V.R.S. Cheela, B.K. Dubey, J. Bhattacharya, Circular economy approach in solid waste management system to achieve UN-SDGs: solutions for post-COVID recovery, Science of the Total Environment 800 (2021) 149605. Available from https://doi.org/10.1016/j.scitotenv.2021.149605.

[18] M.K. Awasthi, et al., Sustainable Management of Solid Waste, Elsevier B.V, 2019. Available from 10.1016/B978-0-444-64200-4.00006-2.

[19] R. Kumari, S.R. Samadder, A critical review of the pre-processing and metals recovery methods from e-wastes, Journal of Environmental Management 320 (2022) 115887. Available from https://doi.org/10.1016/j.jenvman.2022.115887.

[20] H. Wang, F. Zhu, X. Liu, M. Han, R. Zhang, A mini-review of heavy metal recycling technologies for municipal solid waste incineration fly ash, Waste Management & Research 39 (9) (2021) 1135–1148. Available from https://doi.org/10.1177/0734242X211003968.

[21] S. Gulliani, M. Volpe, A. Messineo, R. Volpe, Recovery of metals and valuable chemicals from waste electric and electronic materials: a critical review of existing technologies, RSC Sustainability 1 (5) (2023) 1085–1108. Available from https://doi.org/10.1039/d3su00034f.

[22] G. Weibel, A. Zappatini, M. Wolffers, S. Ringmann, Optimization of metal recovery from mswi fly ash by acid leaching: findings from laboratory-and industrial-scale experiments, Processes 9 (2) (2021) 1–12. Available from https://doi.org/10.3390/pr9020352.

[23] C.M. Mehta, W.O. Khunjar, V. Nguyen, S. Tait, D.J. Batstone, Technologies to recover nutrients from waste streams: a critical review, Critical Reviews in Environmental Science and Technology 45 (4) (2015) 385–427. Available from https://doi.org/10.1080/10643389.2013.866621.

[24] J. Ku, K. Wang, Q. Wang, Z. Lei, application of magnetic separation technology in resource utilization and environmental treatment, Separations 11 (5) (2024). Available from https://doi.org/10.3390/separations11050130.

[25] V. Gunarathne, et al., Hydrometallurgical processes for heavy metals recovery from industrial sludges, Critical Reviews in Environmental Science and Technology 52 (6) (2022) 1022–1062. Available from https://doi.org/10.1080/10643389.2020.1847949.

[26] D. He, et al., Thermal separation of heavy metals from municipal solid waste incineration fly ash: a review, Chemical Engineering Journal 467 (March) (2023). Available from https://doi.org/10.1016/j.cej.2023.143344.

[27] C.B. Gill, Electrostatic separation, Materials Beneficiation (1991) 141–147.

[28] J. Joo, E.E. Kwon, J. Lee, Achievements in pyrolysis process in E-waste management sector, Environmental Pollution 287 (April) (2021) 117621. Available from https://doi.org/10.1016/j.envpol.2021.117621.

[29] A. David, Y.D. Thangavel, R. Sankriti, Recover, recycle and reuse: an efficient way to reduce the waste, International Journal of Mechanical and Production Engineering Research and Development 9 (3) (2019) 31–42. Available from https://doi.org/10.24247/ijmperdjun20194.

[30] C. Geng, et al., Recovery of metals from municipal solid waste incineration fly ash and red mud via a co-reduction process, Resources, Conservation and Recycling 154 (2020) 104600. Available from https://doi.org/10.1016/j.resconrec.2019.104600 November 2019.

[31] K. Kabirifar, M. Mojtahedi, C. Wang, V.W.Y. Tam, Construction and demolition waste management contributing factors coupled with reduce, reuse, and recycle strategies for effective waste management: a review, Journal of Cleaner Production 263 (2020) 121265. Available from https://doi.org/10.1016/j.jclepro.2020.121265.

[32] A. Islam, et al., Advances in sustainable approaches to recover metals from e-waste–a review, Journal of Cleaner Production 244 (2020). Available from https://doi.org/10.1016/j.jclepro.2019.118815.

[33] G.D. Corder, A. Golev, D. Giurco, 'Wealth from metal waste': translating global knowledge on industrial ecology to metals recycling in Australia, Minerals Engineering 76 (2015) 2–9. Available from https://doi.org/10.1016/j.mineng.2014.11.004.

[34] Y. Ruslinda, S. Raharjo, Y. Dewilda, Hidayatullah, R. Aziz, Minimization of household hazardous solid waste (HHSW) with 4R concepts (reduce, reuse, recycle and recovery) in Padang City, Indonesia, IOP Conference Series: Materials Science and Engineering 602 (1) (2019). Available from https://doi.org/10.1088/1757-899X/602/1/012055.

[35] W. Gumley, An analysis of regulatory strategies for recycling and re-use of metals in Australia, Resources 3 (2) (2014) 395–415. Available from https://doi.org/10.3390/resources3020395.

[36] L. Zhang, Z. Xu, A review of current progress of recycling technologies for metals from waste electrical and electronic equipment, Journal of Cleaner Production 127 (2016) 19–36. Available from https://doi.org/10.1016/j.jclepro.2016.04.004.

[37] Y. Xuan, Q. Yue, Forecast of steel demand and the availability of depreciated steel scrap in China, Resources, Conservation and Recycling 109 (2016) 1–12. Available from https://doi.org/10.1016/j.resconrec.2016.02.003.

[38] H. Ohno, et al., Optimal recycling of steel scrap and alloying elements: input-output based linear programming method with its application to end-of-life vehicles in Japan, Environmental Science & Technology Journal 51 (22) (2017) 13086–13094. Available from https://doi.org/10.1021/acs.est.7b04477.

[39] A. Rajesh, M. Asaad, Artificial intelligence and machine learning in surgery, The American Journal of Surgery 89 (1) (2023) 9–10. Available from https://doi.org/10.1177/00031348221117024.

[40] M. Koi, et al., Sa1670 prediagnostic regular use of aspirin and/or non-steroidal anti-inflammatory drugs (NSAIDS) is associated with colon cancer survival in an age- and race-dependent manner, Gastroenterology 158 (6) (2020) S-376. Available from https://doi.org/10.1016/s0016-5085(20)31634-6.

[41] L. Liu, W. Li, W. Song, M. Guo, Remediation techniques for heavy metal-contaminated soils: Principles and applicability, Science of the Total Environment 633 (2018) 206–219. Available from https://doi.org/10.1016/j.scitotenv.2018.03.161.

[42] R. van Herwijnen, T.R. Hutchings, A. Al-Tabbaa, A.J. Moffat, M.L. Johns, S.K. Ouki, Remediation of metal contaminated soil with mineral-amended composts, Environmental Pollution 150 (3) (2007) 347–354. Available from https://doi.org/10.1016/j.envpol.2007.01.023.

[43] Y.W. Jeon, Optimization of ultrasonification of slaughter blood for protein solubilization, Environmental Engineering Research 20 (2) (2015) 163–169. Available from https://doi.org/10.4491/eer.2014.018.

[44] S. Khalid, M. Shahid, N.K. Niazi, B. Murtaza, I. Bibi, C. Dumat, A comparison of technologies for remediation of heavy metal contaminated soils, Journal of Geochemical Exploration 182 (2017) 247–268. Available from https://doi.org/10.1016/j.gexplo.2016.11.021.

[45] R. Shukla, S.S. Bhadauria, R.K. Shrivastava, Mining from landfills as a remediation strategy regarding open dumpsites using artificial intelligence hybrid models, Journal of Scientific & Industrial Research (India) 82 (4) (2023) 460–465. Available from https://doi.org/10.56042/jsir.v82i04.72386.

[46] S.F. Rafie, et al., Genetic algorithm-assisted artificial neural network modelling for remediation and recovery of Pb (II) and Cr(VI) by manganese and cobalt spinel ferrite super nanoadsorbent, Chemosphere 321 (2023). Available from https://doi.org/10.1016/j.chemosphere.2023.138162 December 2022.

[47] M. Abbasi, A. El Hanandeh, Forecasting municipal solid waste generation using artificial intelligence modelling approaches, Waste Management 56 (2016) 13–22. Available from https://doi.org/10.1016/j.wasman.2016.05.018.

[48] R. Gholami, A. Kamkar-Rouhani, F. Doulati Ardejani, S. Maleki, Prediction of toxic metals concentration using artificial intelligence techniques, Applied Water Science 1 (3–4) (2011) 125–134. Available from https://doi.org/10.1007/s13201-011-0016-z.

[49] M. Zafar, A. Aggarwal, E.R. Rene, K. Barbusiński, B. Mahanty, S.K. Behera, Data-driven machine learning intelligent tools for predicting chromium removal in an adsorption system, Processes 10 (3) (2022). Available from https://doi.org/10.3390/pr10030447.

[50] C.R. Evanko, D.A. Dzombak, October, Remediation of Metals-Contaminated Soils and Groundwater 01 Ground-Water Remediation Technologies Analysis Center, Pittsburgh, PA, 1997, pp. 1–61.

[51] A. Bhandari, R.Y. Surampalli, P. Champagne, S.K. Ong, R.D. Tyagi, I.M.C. Lo, Remediation technologies for soils and groundwater, Remediation Technologies for Soils and Groundwater 60 (2007) 1–449. Available from https://doi.org/10.1061/9780784408940.

[52] M.A. Hashim, S. Mukhopadhyay, J.N. Sahu, B. Sengupta, Remediation technologies for heavy metal contaminated groundwater, Journal of Environmental Management 92 (10) (2011) 2355–2388. Available from https://doi.org/10.1016/j.jenvman.2011.06.009.

[53] W.M. Ashraf, R. Panda, P.R. Jadhao, K.K. Pant, V. Dua, Machine learning based metal recovery from the waste printed circuit boards of mobile phones for circular economy and sustainable environment, Energy Procedia 37 (2023). Available from https://doi.org/10.46855/energy-proceedings-10780.

[54] V. Rai, D. Liu, D. Xia, Y. Jayaraman, J.C.P. Gabriel, Electrochemical approaches for the recovery of metals from electronic waste: a critical review, Recycling 6 (3) (2021). Available from https://doi.org/10.3390/recycling6030053.

[55] Y.L. Huang, L.T. Fan, Artificial intelligence for waste minimization in the process industry, Computers in Industry 22 (2) (1993) 117–128. Available from https://doi.org/10.1016/0166-3615(93)90059-A.

[56] M.M. Gupta, Fuzzy information and decision processes, IFAC (1982). Available from https://doi.org/10.1016/s1474-6670(17)63380-9.

[57] P. Nowakowski, K. Szwarc, U. Boryczka, Combining an artificial intelligence algorithm and a novel vehicle for sustainable e-waste collection, Science of the Total Environment 730 (2020) 138726. Available from https://doi.org/10.1016/j.scitotenv.2020.138726.

[58] H. Han, Q. Zhang, F. Li, Y. Du, Y. Gu, Y. Wu, Metallic product recognition with dual attention and multi-branch residual blocks-based convolutional neural networks, Circular Economy 1 (2) (2022) 100014. Available from https://doi.org/10.1016/j.cec.2022.100014.

[59] P.R. Kshirsagar, et al., Artificial intelligence-based robotic technique for reusable waste materials, Computational Intelligence and Neuroscience 2022 (2022). Available from https://doi.org/10.1155/2022/2073482.

[60] K.H. Yu, Y. Zhang, D. Li, C.E. Montenegro-Marin, P.M. Kumar, Environmental planning based on reduce, reuse, recycle and recover using artificial intelligence, Environmental Impact Assessment Review 86 (2021) 106492. Available from https://doi.org/10.1016/j.eiar.2020.106492.
[61] C. Li, Construction of the reverse resource recovery system of e-waste based on DLRNN, Computational Intelligence and Neuroscience 2021 (2021). Available from https://doi.org/10.1155/2021/2143235.
[62] W.M. Ambrós, Gravity concentration in urban mining applications—a review, Recycling 8 (6) (2023). Available from https://doi.org/10.3390/recycling8060085.
[63] S. Lin, et al., Remediation of emerging contaminated sites due to uncontrolled e-waste recycling, Chemical Engineering Journal 430 (2022) 133169. Available from https://doi.org/10.1016/j.cej.2021.133169.
[64] P. Galgali, S. Palimkar, A. Adhikari, R. Patel, J. Routh, Remediation of potentially toxic elements -containing wastewaters using water hyacinth – a review, International Journal of Phytoremediation 25 (2) (2023) 172–186. Available from https://doi.org/10.1080/15226514.2022.2068501.
[65] C.C. dos Santos Accioly Lins, F.M. de Moraes Ramos-Perez, A. dos Anjos Pontual, M.L. dos Anjos Pontual, E.H.L. do Nascimento, Digital oral radiography, Digitization in Dentistry (2021) 65–88. Available from https://doi.org/10.1007/978-3-030-65169-5_3.
[66] N. Li, Y. Chen, Municipal solid waste classification and real-time detection using deep learning methods, Urban Climate 49 (2023) 101462. Available from https://doi.org/10.1016/j.uclim.2023.101462.
[67] A.N. Matheri, F. Ntuli, J.C. Ngila, T. Seodigeng, C. Zvinowanda, Performance prediction of trace metals and cod in wastewater treatment using artificial neural network, Computers & Chemical Engineering 149 (2021) 107308. Available from https://doi.org/10.1016/j.compchemeng.2021.107308.
[68] N. Kumar, et al., Efficient computational stochastic framework for performance optimization of E-waste management plant, Journal of King Saud University Computer and Information Sciences 34 (8 Part A) (2022) 4712–4728. Available from https://doi.org/10.1016/j.jksuci.2022.05.018.
[69] M. Mowbray, et al., Machine learning for biochemical engineering: a review, Biochemical Engineering Journal 172 (2021) 108054. Available from https://doi.org/10.1016/j.bej.2021.108054.
[70] C. Zhao, et al., Hyperspectral image classification with multi-attention transformer and adaptive superpixel segmentation-based active learning, , IEEE Transactions on Image Processing 32 (2023) 3606–3621. Available from https://doi.org/10.1109/TIP.2023.3287738.

Further reading

S.P. Gundupalli, S. Hait, A. Thakur, A review on automated sorting of source-separated municipal solid waste for recycling, Waste Management 60 (2017) 56–74. Available from https://doi.org/10.1016/j.wasman.2016.09.015.
M. Ji, B. Li, A. Majdi, T. Alkhalifah, F. Alturise, H.E. Ali, Application of nano remediation of mine polluted in acid mine drainage water using machine learning model, Chemosphere 311 (2023). Available from https://doi.org/10.1016/j.chemosphere.2022.136926 October 2022.
W. Zhenwu, L. Jingshe, X. Kuangdi, Iron and steel scrap, The ECPH Encyclopedia of Mining and Metallurgy 1 (703) (2023) 1–2. Available from https://doi.org/10.1007/978-981-19-0740-1_969-1.

CHAPTER

8

Machine learning: a better means for metal waste to reprocess

Aya Nabil Sayed, Md. Mosarrof Hossen, Tamim Al-Hasan, Mohammad Noorizadeh, and Faycal Bensaali

Department of Electrical Engineering, College of Engineering, Qatar University, Doha, Qatar

Introduction

The escalating growth of the global population, coupled with the depletion of natural resources, presents a myriad of challenges for our world. While some nations exert immense pressure on the earth's resources, others grapple with the harsh realities of food and water scarcity. Consequently, there is a heightened focus on minimizing waste and implementing recycling strategies, particularly for end-of-life products (ELPs), reflecting the prevalent concerns of current generations [1]. Industrial activities contribute substantially to solid and liquid waste production, generating millions of tons annually with untapped reservoirs of valuable metals. Nonferrous metals like aluminum, copper, chromium, nickel, manganese, lead, titanium, and zinc, and precious metals such as silver, gold, and platinum remain available for recycling as scrap materials. As we approach 2025, global municipal solid waste production is anticipated to surge by approximately 2.2 billion tonnes annually [2], necessitating more advanced recovery technologies. Alarming figures from 2019 reveal that out of the 53.6 million metric tons (Mt) of electronic waste generated worldwide, a mere 17.4% was officially recorded and documented as being appropriately collected and recycled [3]. The improper disposal of solid wastes around workplaces or households, resulting in the eventual deposition of waste in landfills, is frequently problematic. Such negligence not only jeopardizes human lives but also poses a significant threat to ecosystems on a large scale. Addressing these challenges is crucial for the sustainable future of our planet. In the face of escalating environmental concerns and the ever-growing demand for metals in various industries, the imperative to revolutionize traditional methods of metal waste reprocessing has become increasingly evident [4]. The advent of machine learning (ML) presents a transformative avenue to address the challenges associated with the inefficient utilization of metal scrap and the environmental ramifications of traditional recycling practices [5,6]. While serving as foundational pillars of recycling practices, traditional

Harnessing Automation and Machine Learning for Resource Recovery and Value Creation
DOI: https://doi.org/10.1016/B978-0-443-27374-2.00008-X

methods of metal waste reprocessing face challenges in optimizing efficiency, minimizing environmental impact, and meeting the escalating demand for metals across diverse industries. Recognizing the pressing need for sustainable alternatives, this chapter explores the potential of ML as a dynamic and efficient means for metal waste reprocessing. ML, with its capacity to analyze vast datasets and adapt to complex patterns, has the potential to enhance every facet of the metal recycling process, from sorting and separation to extraction and alloy composition.

Increased waste volumes and a recycling trend mark the evolution of waste management from logistics to manufacturing. Despite this shift, logistics remains pivotal, presenting opportunities for further optimization through digitalization. Key emerging approaches include the integration of smart bins, autonomous vehicles, advanced routing systems, and the utilization of near-field communication (NFC)/radio-frequency identification (RFID) tags in both products and waste. Numerous peer-reviewed papers delve into these innovations, focusing on waste collection and logistics in "smart cities" [7–9]. The Chair of Automation at Montanuniversitaet Leoben is actively researching the integration of imaging and deep learning (DL) for automated waste classification in processing plants. This involves capturing in situ waste images with industrial cameras and employing DL, particularly deep convolutional neural networks (CNNs), for automatic classification based on type, contents, and state. Transfer learning is utilized in lower layers to detect primitive features, allowing for quicker training and successful network convergence. Noteworthy studies on digital image analysis methods have been conducted [10–12]. Waste sorting using robotic technologies faces challenges related to waste heterogeneity, surface contamination, and dynamic positioning of objects on conveyor belts. Innovative solutions incorporate material recognition algorithms, artificial intelligence (AI), and robotic systems capable of multitasking. Robotic technologies find applications in construction site waste sorting, automatic quality assurance, and enhancements in areas like construction and demolition waste. Manufacturers envision a combined approach, incorporating robotic technologies and optical sensor sorting systems, as an effective solution for municipal solid waste, particularly packaging waste. Research projects like RadioRoSo and RoMaNS focus on modernizing robotic systems for sorting nuclear waste. The Danish Institute of Technology explores AI-based robotic systems for sorting hazardous waste, focusing on battery detection in electrical devices. Concrete examples include the ERO robot for concrete recycling and Apple's recycling robots, LIAM and DAISY, designed for disassembling iPhones and recovering selected materials through advanced automation techniques, including DL technology [13,14].

This chapter delves into the current landscape of metal waste reprocessing, highlighting the limitations of existing methodologies and underscoring the potential advantages ML offers. By leveraging advanced algorithms and data-driven decision-making, ML models can significantly enhance the precision and efficiency of metal recycling operations, ultimately contributing to the conservation of natural resources and the reduction of environmental impact. As we explore the synergy between ML and metal waste reprocessing, we aim to unravel the untapped potential of artificial intelligence in fostering a more sustainable and circular approach to metal utilization. Through comprehensive analysis and case studies, we will unveil ML's promising prospects for revolutionizing the metallurgical industry and propelling us toward a greener, more resource-efficient future.

The subsequent sections of this chapter are organized as follows: "Fundamentals of metal waste reprocessing" explores the foundational principles of metal waste reprocessing. "Role of machine learning in environmental sciences" highlights the contributions of ML to environmental sciences. "Machine learning applications in metal waste reprocessing" provides a detailed discussion of ML's applications in managing metal waste. "Challenges, opportunities, and future directions" examines key challenges, opportunities, and potential future research directions. Finally, "Conclusion" summarizes the chapter's main insights and takeaways.

Fundamentals of metal waste reprocessing

Improper treatment of metal waste can hinder our ecosystem. Landfills filled with waste metals serve as breeding grounds for environmental degradation. As these metals disintegrate, toxic compounds and heavy metals such as lead, mercury, and chromium can leak into the soil and water [15]. This pollution disturbs sensitive ecosystems, damaging plant and animal life, and can even make its way into the food chain, creating possible human health hazards. Furthermore, our reliance on raw metals for manufacturing needs massive mining operations. This practice has a substantial environmental cost, resulting in deforestation, habitat degradation, and air and water pollution [16]. Massive landscapes are cleared to make space for mining, displacing species and damaging vital ecosystems. Mining also emits toxic chemicals into the atmosphere and water, affecting nearby populations and contributing to overall environmental deterioration [17]. Furthermore, typical metal extraction and processing are highly energy-intensive, consuming large amounts of fossil fuels and contributing to greenhouse gas emissions and climate change.

Fortunately, there are several methods available to mitigate these concerns and encourage sustainable metal waste procedures. Metal recycling offers a viable alternative to relying solely on virgin metals [18]. Recycling discarded metal significantly reduces the demand for new mining activities, preserves precious natural resources, and mitigates the environmental impact of extraction. Recycling also consumes far less energy than virgin metal fabrication, reducing our carbon footprint and contributing to climate change mitigation efforts. Metal waste must be collected, sorted, and processed correctly to avoid inappropriate disposal and negative environmental consequences. This involves building efficient metal waste collection systems, employing effective sorting techniques to segregate various metal types, and applying environmentally friendly processing technology. Investing in creative methods to save energy and reduce environmental impact during metal processing is crucial. This includes investigating technologies like bioleaching, which utilizes microorganisms to extract metals from waste, and closed-loop systems, which limit waste output while enhancing resource recovery [19].

This section will delve into the traditional approach to metal waste reprocessing, exploring the sequential steps of collection, sorting, processing, and melting that have long been foundational in recycling practices [20]. Understanding the traditional approach to metal waste reprocessing is paramount as it forms the foundation upon which emerging approaches can be integrated into the regulatory cycle of metal waste management. Examining the established methods allows us to identify critical stages, such as collection, sorting, processing, and melting, where innovations and sustainable practices can be seamlessly incorporated. This provides valuable insights into the challenges and opportunities inherent in metal waste management, informing the development and implementation of strategies that contribute to a more sustainable and environmentally conscious recycling process. Following that, we will address the environmental impact associated with these traditional methods, acknowledging the energy-intensive processes and potential emissions involved in metal extraction and purification.

Traditional approach

Metal waste recycling is crucial in mitigating environmental impact and conserving valuable resources. The traditional methods of metal waste reprocessing follow a sequential series of steps, each contributing to the overall efficiency and effectiveness of the recycling process. The journey begins with collecting metal waste and concludes with transporting the recycled material to end-users. Below is a detailed description of each step:

Collection

The collection phase aggregates metal waste from diverse sources, from industrial manufacturing processes and construction sites to end-of-life consumer products. This initial step is critical in determining the composition and volume of the metal waste stream. Collection strategies may vary, encompassing municipal recycling programs, scrap yards, and dedicated collection centers. Efficient collection systems ensure a robust supply of metal scrap for subsequent processing stages.

Sorting

Upon collection, metal waste undergoes an initial sorting process where workers manually classify materials based on visual inspection and basic physical properties. Manual sorting helps separate ferrous from nonferrous metals and distinguish different types of alloys. While this step is essential, it is labor-intensive, time-consuming, and subject to human error. Innovations in automated sorting technologies, including advanced sensors and artificial intelligence, aim to enhance accuracy and reduce the reliance on manual labor.

Compacting

Compacting involves compressing metal waste into more manageable forms, reducing its overall volume. This step is crucial for optimizing transportation logistics and facilitating downstream processing stages. Mechanical presses or hydraulic compactors are commonly used to achieve compaction. By minimizing the volume of metal waste, compacting enhances the efficiency of subsequent processing steps, including shredding and melting.

Shredding

Shredding is a pivotal stage where metal waste, now in a compacted form, undergoes further size reduction. Industrial shredders with rotating blades or hammers break the material into smaller, uniform pieces. This step is essential for facilitating subsequent processes, such as melting. However, traditional shredding methods may generate fine particles, posing challenges in material recovery. Advanced shredding technologies, coupled with innovations in material separation, aim to optimize the efficiency of this stage.

Melting

Melting is the transformative process where shredded metal scrap is subjected to high temperatures until it reaches a molten state. This step is integral to metals recycling as it allows for the purification and separation of different materials. Traditional melting methods may involve furnaces powered by fossil fuels, contributing to the energy-intensive nature of metal recycling. Innovations, such as induction heating and plasma technology, offer more energy-efficient alternatives, addressing environmental concerns associated with traditional melting processes.

Purification

Following melting, the molten metal undergoes purification to refine its composition. This step involves the removal of impurities, oxides, and other elements that may have entered the metal during the recycling process. Purification ensures that the recycled metal meets specified quality standards and can be used to produce new materials without compromising performance or structural integrity.

Solidifying

Once the purification process is complete, the molten metal is solidified into the desired form. Casting or molding techniques shape the metal into ingots, billets, or other configurations suitable for subsequent manufacturing processes. Solidification ensures that the recycled metal can be seamlessly integrated into various industrial applications, ranging from construction and automotive industries to electronics and consumer goods.

Transportation

The final step in the traditional method involves transporting the recycled metal to end-users, such as manufacturers or construction companies. Efficient transportation logistics play a crucial role in minimizing the overall environmental impact of metal recycling. This step completes the cycle, as recycled metal is reintegrated into the production chain, contributing to the circular economy and reducing the demand for primary metal extraction.

This method has proven effective in reducing the need for virgin raw materials and minimizing the ecological footprint associated with metal production. Fig. 8.1 illustrates a simplified representation of the traditional metal waste recycling process, highlighting the sequential stages from collection to creating reusable metal forms. While conventional methods have been successful, ongoing advancements in recycling technologies continue to

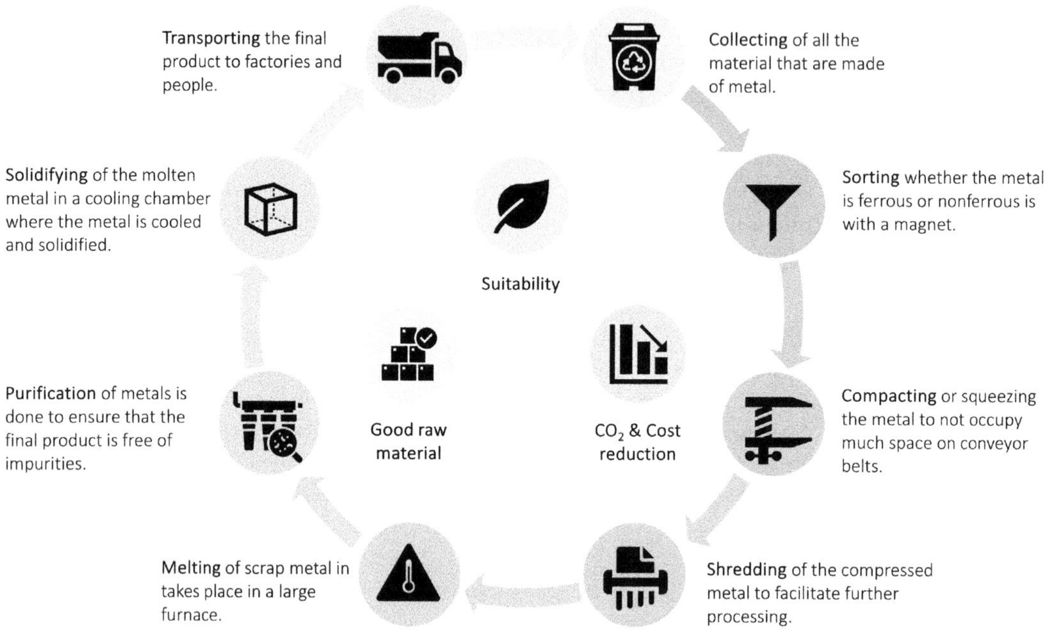

FIGURE 8.1 Metal waste recycling. Traditional approach to metal waste recycling.

enhance efficiency and sustainability, contributing to the broader effort of creating a more circular and eco-friendly economy.

Environmental impact

The environmental effect of the conventional metal waste reprocessing approach is a significant problem in the current sustainability debate. Analyzing the ecological footprint of these activities is critical for establishing methods that support environmental conservation aims. The energy-intensive nature of conventional metal recycling, particularly during the melting stage, presents substantial ecological predicaments. The significant energy inputs necessary to heat metal waste to molten states increase greenhouse gas emissions, exacerbating the recycling industry's carbon impact. As worldwide efforts to minimize dependency on fossil fuels and combat climate change increase, addressing the energy consumption of metal recycling becomes critical.

Beyond energy constraints, job risks related to manual sorting raise additional environmental and social problems. Workers involved in manual separation operations face possible health concerns due to dangerous compounds and sharp items in metal trash. The safety and well-being of workers are critical to building ethical and sustainable metal recycling procedures. Moreover, contamination concerns with recyclables persist for conventional systems. Although they can be somewhat successful, hand-sorting processes have the potential to inadvertently mix incompatible elements, which lowers the grade of recovered metals. The integrity of recycled materials is jeopardized by this contamination, which necessitates extra processing steps to remove it and increases energy and resource consumption. The need to switch to more environmentally friendly methods of recycling metal waste is growing in response to these problems.

Role of machine learning in environmental sciences

The role of ML in environmental sciences and waste management has been a topic of increasing interest and research in recent years. ML techniques have been applied in various aspects of ecological sustainability, including waste classification, sorting, recycling, and prediction of waste generation rates [21–27]. This section provides a detailed account of the application of ML in environmental sciences and waste management, explores specific ML applications in waste classification and sorting, and discusses notable case studies and successful implementations of ML in improving environmental sustainability.

ML has been widely applied in environmental sciences and waste management to address challenges related to waste classification, sorting, recycling, and predicting waste generation rates. Various studies have demonstrated the effectiveness of ML techniques in these areas. For instance, the study by Wu et al. [28] discussed using support-vector networks in ML, highlighting the construction of different ML with arbitrary types of decision surfaces. This approach has implications for waste classification and sorting, as it allows for developing tailored decision surfaces for different waste materials. Furthermore, ML models have been used for waste classification and sorting, leveraging techniques such as image processing and intelligent classification systems [21]. These applications have significantly improved the efficiency and accuracy of waste sorting processes, contributing to enhanced recycling and waste management practices. Furthermore, ML has been instrumental in predicting organic solid waste treatment and recycling processes, thereby facilitating the optimization of waste management strategies [29].

Moreover, the study by Mohammed et al. [25], conducted a study on automated waste-sorting and recycling classification using artificial neural networks (ANNs) and feature fusion. Their work exemplifies the application of ML in waste classification and recycling, demonstrating the potential for advanced predictive modeling in waste management. In addition, Eshbayev et al. [23] proposed a natural language processing (NLP) and ML-based approach to increase the efficiency of environmental management operations and engineering practices. Their contributions underscore the potential of ML in guiding environmental preservation and addressing growing challenges in sustainable development. This approach aligns with the need for more effective opportunities in waste management and environmental sustainability. Furthermore, Imran et al. [30] highlighted the potential of data mining and ML-based hybrid mechanisms in waste management. Their study emphasized using ML to produce detailed and timely waste generation information, enabling waste management authorities to effectively manage waste amounts in specific areas. This application of ML demonstrates its potential to provide actionable insights for waste management and environmental sustainability.

Moreover, Hussain et al. [31] explored waste management and prediction of air pollutants using an internet of things (IoT) and ML approach. Their study emphasized the significance of addressing unmanaged garbage as a primary cause of environmental pollution. By leveraging ML techniques, their approach aimed to mitigate the environmental impact of waste through predictive modeling and IoT-based interventions. Besides, Haji and Sallow [32] conducted a review of IoT for smart environment monitoring based on Python, emphasizing the role of IoT technologies and ML in environmental monitoring systems. The review highlighted the integration of sensors, ML, and classification methods in smart environment monitoring, showcasing the potential of IoT-based solutions in enhancing environmental monitoring practices. The study underscored the importance of leveraging IoT and ML for efficient and comprehensive environmental monitoring. Additionally, a study by Al-Duhayyim et al. [33] presented an intelligent hazardous waste detection and classification model using ensemble learning techniques. Their work focused on smart waste material categorization using ML tools, such as CNN models and support vector machines (SVM), among others, to enhance waste classification and management. This study exemplifies the potential of ML in addressing hazardous waste management challenges.

Furthermore, Ochoa-Barragán et al. [34] proposed a hybrid ML-mathematical programming optimization approach for municipal solid waste management. Their case study of New York City demonstrated the use of extensive socioeconomic datasets to train ML models, highlighting the potential of ML in optimizing waste management strategies during the pandemic. Notable success stories in the application of ML in environmental sustainability include the characterization of groundwater potential using GIS-based ML models, which have enabled the spatial mapping and assessment of groundwater resources [27]. Additionally, ML has been employed in environmental monitoring applications, utilizing techniques such as SVMs and ANNs to enhance the accuracy and effectiveness of monitoring systems [24]. Moreover, the use of explainable artificial intelligence (XAI) methods has shown promise in predicting wildfire ignition causes and addressing critical environmental concerns such as wildfire occurrence and size [22].

The integration of ML in waste management has also led to significant advancements in circular economy concepts for sustainable electronic waste management [26]. By harnessing ML pipelines, researchers have developed automated tools for optimizing data science processes, contributing to the efficient management of environmental data and resources [35]. Furthermore, the application of DL architectures has enabled comprehensive waste prediction and estimation, particularly in the context of construction materials and hazardous waste [36]. In addition to waste classification and recycling, ML has been instrumental in developing appropriate technology for waste recycling, emphasizing the importance of learning assistance and technological tools in waste management practices [37]. Moreover, AI-based e-waste management has emerged as a critical component of environmental planning, addressing the challenges associated with hazardous waste and extended producer responsibility [38].

The application of ML in environmental sciences extends beyond waste management, encompassing diverse areas such as water level forecasting, digital soil mapping, and climate prediction. For instance, ML techniques have been employed in the forecasting of water levels in Dongting Lake and its relationship with the Three Gorges Dam [39], demonstrating the potential of long short-term memory (LSTM) networks in environmental predictions. Furthermore, ML has been utilized for digital soil mapping, addressing challenges and suggesting solutions for accurate soil mapping [39]. Additionally, ML has been applied to predict long-term precipitation variability in the Poyang Lake Basin, China, showcasing the diverse applications of ML in climate-related predictions [27]. Moreover, the use of ML in environmental monitoring applications, such as the calibration of airborne particulate sensors, has demonstrated the potential of ML in enhancing sensor calibration processes [40].

In summary, the integration of ML in environmental sciences and waste management has yielded significant advancements, ranging from waste classification and sorting to ecological monitoring and sustainable waste management practices. The summary of the earlier studies can be found in Table 8.1. These applications have demonstrated notable success stories, showcasing the potential of ML in addressing complex environmental challenges and promoting environmental sustainability.

Machine learning applications in metal waste reprocessing

Reprocessing metal waste using ML is bringing cutting-edge innovation into an age-old art. ML redefines recycling by combining data-driven insights with the practicality of industrial practice. Its algorithms turn every scrap metal piece into a canvas, waiting for the brushstrokes of predictive analysis to reveal its hidden value. In this study, we explore the revolutionary possibilities of ML, shedding light on the way forward for a more environmentally friendly and productive metal waste management industry. Fig. 8.2 demonstrates a unique idea for a waste collection planning system, which includes identifying and categorizing waste appliances. Individual images of the equipment slated for disposal are taken and sent to a server running image recognition software to identify them. Following the identification and classification of e-waste, the collecting company established a strategy for effective retrieval depending on the size and type of the waste items. Artificial intelligence-based systems might help plan and optimize collection vehicle routes. The core image identification model is based on a CNN architecture.

Similarly, a system was suggested in [42] for predicting mass, with and without feature extraction and selection of scrap metal. This system evaluated different feature extraction methods, such as principal component analysis (PCA), using a combined dataset of cast (C), wrought (W), and stainless steel (SS) images. The evaluation was done using advanced ML and DL algorithms. Subsequently, the most effective mass prediction framework was combined with a DenseNet classifier, resulting in multiple outputs for both object classification and mass prediction. The proposed architecture includes a DenseNet neural network for classification and a backpropagation neural network (BPNN) for mass prediction, which utilizes up to 24 features extracted from depth images. The flowchart of the suggested integrated architecture is shown in Fig. 8.3.

Data acquisition and preprocessing

The data collection and preprocessing phase establishes the foundation for further analysis and modeling for metal waste reprocessing. Below, we explore the complex

TABLE 8.1 Summary of the studies found using machine learning in environmental sciences and applications.

Method or algorithm used	Application area	Key points	References
ANNs	Waste classification and recycling	Study on automated waste-sorting and recycling classification.	Mohammed et al. [25]
NLP, ML	Environmental preservation, sustainable development	NLP and ML-based approach for environmental management.	Eshbayev et al. [23]
Data mining, ML	Waste management, environmental sustainability	– Data mining and ML-based hybrid mechanisms are used in waste management. – Emphasis on producing detailed and timely waste generation information.	Imran et al. [30]
IoT, ML	Waste management, environmental pollution	Exploration of waste management and prediction of air pollutants using IoT and ML.	Hussain et al. [31]
Hybrid ML, mathematical programming optimization	Municipal solid waste management	– Hybrid ML-mathematical programming optimization approach for municipal solid waste management. – Case study of New York City using extensive socioeconomic datasets.	Ochoa-Barragán et al. [34]
Ensemble Learning	Hazardous waste management and detection	– Intelligent hazardous waste detection using ensemble learning. – The focus was on smart waste material classification using CNNs for feature extraction and ensemble ML for classification.	Al-Duhayyim et al. [33]
ML, IoT	Smart environment monitoring, environmental monitoring practices	– Review of IoT for smart environment monitoring based on Python. – Emphasis on IoT technologies and ML in environmental monitoring systems. – Integration of sensors, ML, and classification methods in smart environment monitoring. – Importance of leveraging IoT and ML for efficient environmental monitoring.	Haji and Sallow [32]
CNN	Waste management	Waste classification and sorting using image processing and intelligent classification systems.	Abu-Qdais et al. [21]
–	Organic solid waste treatment and recycling	Prediction of organic solid waste treatment and recycling processes for optimizing waste management strategies.	Guo et al. [29]
–	Hazardous building material management	Emphasizing the ML's potential in hazardous material management and the need for an environmental inventory for circular construction.	Wu et al. [28]
LR, DNN, and RF	Environmental sustainability	GIS-based ML models map and assess groundwater potential.	Wang et al. [27]

–	Environmental monitoring	Exploring WSN architectures for environmental monitoring applications, emphasizing their integration with ML to enhance intelligence, autonomy, and decision-making.	Adu-Manu et al. [24]
–	Circular economy	Integration of ML in circular economy concepts for sustainable electronic waste management.	Srivastav et al. [26]
ML Pipelines	Environmental data optimization	Development of automated tools for optimizing data science processes in environmental data and resource management.	Olson and Moore [35]
LSTM	Waste prediction	Applying DL to predict construction and hazardous waste.	Utku and Kaya [36]
–	Waste recycling technology	ML's role in developing appropriate technology for waste recycling.	Ayu Arwati et al. [37]
AI	E-waste management	Artificial intelligence-based e-waste management addresses hazardous waste challenges and extends producer responsibility.	Chen et al. [38]
LSTM	Water level forecasting	Forecasting water levels in Dongting Lake using ML techniques with LSTM.	Liang et al. [39]
XAI	Wildfire prediction	XAI methods for predicting wildfire ignition causes in Southern France.	Bountzouklis et al. [22]
RF	Sensor calibration, climatological study	Calibration of airborne particulate sensors using ML in environmental monitoring applications.	Wijeratne et al. [40]

ANNs, Artificial neural network; *RF*, random forest; *NLP*, natural language processing; *ML*, machine learning; *IoT*, Internet of Things; *CNN*, convolutional neural network; *LR*, logistic regression; *DNN*, deep neural networks; *SVM*, support vector machine; *DL*, deep learning; *LSTM*, long short-term memory; *XAI*, explainable artificial intelligence.

procedures involved in gathering, sanitizing, and readying data for ML uses, particularly in reprocessing metal waste.

Gathering information from various sources

- **Metal waste streams:** The first step in the data acquisition process is gathering various data streams from metal waste sources, such as manual input, industrial sensors, and historical records. These resources include information on weight, composition, and quality measurements, among other characteristics of metal trash [43].
- **Sensor networks:** Industrial sensor networks are essential for gathering real-time data on the temperature, density, and chemical makeup of metal waste. These streams of data make it possible to continuously monitor and analyze recycling operations, which facilitates data-driven decision-making [44].

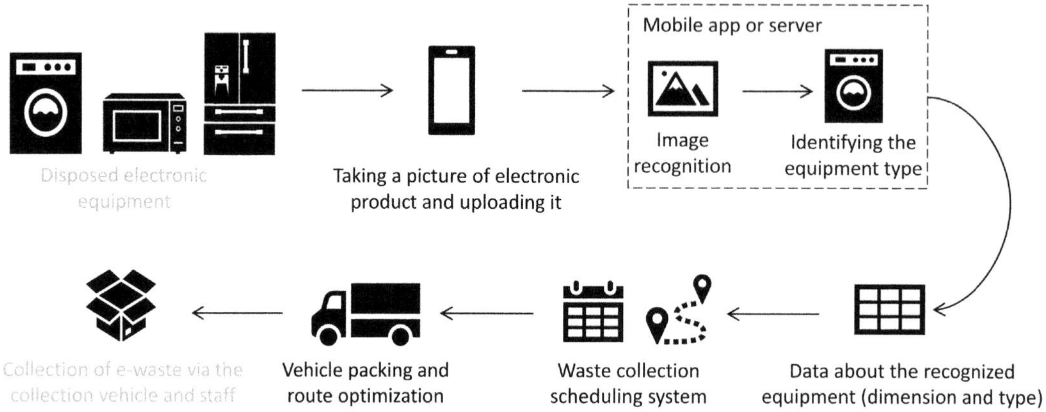

FIGURE 8.2 Metal waste classification. A technique that employs picture classification to detect waste items and optimize the collection schedule [41].

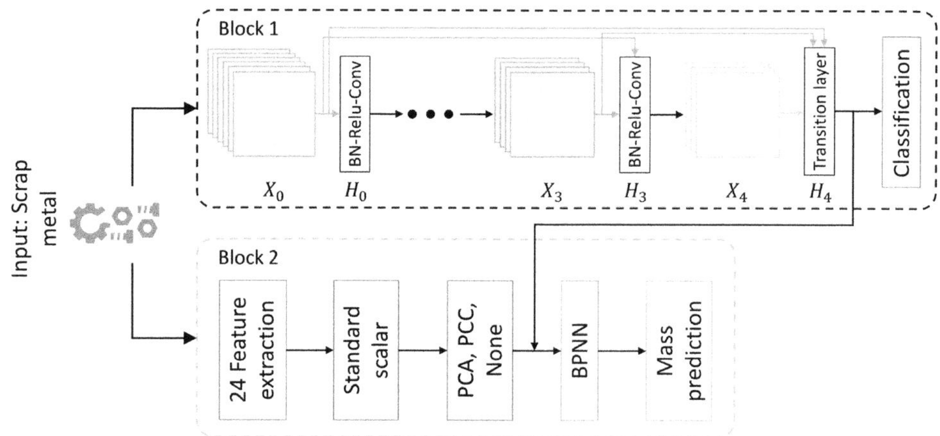

FIGURE 8.3 Mass estimation and classification of scrap metal. An approcah for scrap metal type classifiaction in Block 1 and mass prediction in Block 2 [42].

- **Historical records and databases:** Historical records and databases offer important insights into material movement, quality standards, and recycling procedures from the past. By examining historical data, researchers can find trends, patterns, and opportunities for improvement in the recycling of metal waste [45].

Techniques for preparing data for quality assurance

- **Data cleaning and filtering:** Inaccuracies, outliers, and missing values are common in raw data obtained from metal waste sources, and they can have a negative impact on the performance of ML models. To guarantee data quality and consistency, data cleaning and filtering techniques, including noise reduction, imputation, and outlier identification, are used [46].

- **Feature engineering:** This process converts unstructured data into meaningful features that encapsulate the essential characteristics of metal waste. Reducing computational complexity and extracting pertinent information are made possible by methods like feature selection, normalization, and dimensionality reduction.
- **Heterogeneous data source integration:** There may be differences in the format, organization, and granularity of metal waste data gathered from various sources. In order to facilitate smooth analysis and modeling, integration techniques are used to harmonize and combine diverse data sources [47].

Evaluation and validation of data quality

- **Data quality metrics:** Ensuring the dependability and precision of ML models requires evaluating the quality of the data that has been gathered. Measures like accuracy, consistency, and completeness are used to assess the quality of data and spot any possible problems or biases [47].
- **Methods of validation:** ML models are evaluated on unobserved data using validation techniques like holdout validation and cross-validation. Researchers can assess the model's robustness and generalization by dividing the dataset into training and validation sets [48].
- **Moral aspects to take into account:** When collecting and preparing data, ethical issues, including data privacy, permission, and responsible data usage, are crucial. Ensuring adherence to moral standards and legal requirements safeguards people's right to privacy and encourages prudent data management [49].

Machine learning algorithms and models

The subsection dedicated to exploring algorithms and models is a cornerstone in understanding the intricate processes involved. Here, we explore a wide range of ML approaches designed to address the challenges of recycling metal waste. These algorithms enable the way toward more environmentally friendly recycling methods by utilizing optimization, anomaly detection, and predictive modeling.

- **Supervised learning for predictive modeling:** Regression and classification are two supervised learning approaches fundamental to predictive modeling in recycling metal waste. These algorithms facilitate tasks such as material categorization and quality evaluation by enabling precise predictions of material qualities through the study of labeled data [50].
- **Unsupervised learning for anomaly detection and process optimization:** Unsupervised learning methods, including clustering and dimensionality reduction, play a vital role in uncovering hidden patterns and structures within metal waste data. By identifying anomalies and groupings among data points, these algorithms aid in process optimization and anomaly detection, enhancing overall efficiency [51].

- **Deep learning for image analysis and sequential data processing:** DL, characterized by neural networks with multiple layers, offers unparalleled capabilities in tasks such as image analysis and sequential data processing. CNNs excel in tasks involving image recognition, while recurrent neural networks (RNNs) are well-suited for processing sequential data like time-series sensor readings [52].
- **Ensemble techniques to enhance predictive capabilities:** To improve prediction performance, ensemble learning techniques such as gradient boosting and random forests utilize the combined knowledge of several base learners. Ensemble approaches reduce overfitting and increase the resilience of metal waste reprocessing systems by combining the predictions of several models [53].
- **Hybrid methods to increase model effectiveness:** Hybrid ML techniques use several algorithms or modalities to take advantage of their complementary advantages. For example, DL architectures combined with conventional regression models can provide better predictions by utilizing both organized and unstructured data sources [46].

Applications and limitations

Incorporating ML poses both difficult hurdles and intriguing applications in the dynamic field of recycling metal waste. ML presents revolutionary approaches to improve resource recovery and streamline recycling procedures by utilizing sophisticated algorithms and insights derived from data. Nevertheless, despite the potential advantages, industry participants need to overcome several challenges and complexities related to using ML technology. The many applications of ML techniques in reprocessing metal waste are examined in this subsection, along with the main obstacles and constraints that must be removed for effective use.

- **Automated systems for sorting:** ML algorithms are revolutionizing the field of recycling metal trash by facilitating the creation of automated sorting systems. By applying cutting-edge ML algorithms, these systems quickly classify various metal waste streams according to their distinct qualities, such as composition, form, and magnetic properties. These automated sorting systems increase recycling efficiency by utilizing ML to quickly and accurately separate metal debris, which maximizes resource recovery rates and reduces landfill waste. Large amounts of metal trash may also be processed using these systems in a scalable and economical manner, significantly improving recycling operations in various industrial sectors [54]. Nowakowski and Pamula [55] proposed a novel waste management system that combined two key elements: Waste classification with deep learning: (1) A CNN was used to automatically classify waste from images captured by cameras within brightly designed trash boxes. This real-time classification enabled the identification of recyclable and reusable materials, minimizing misusage, and (2) an IoT-based trash box monitoring: The trash boxes had multiple sensors that collected and transmitted data for real-time monitoring. This data and the waste classification results could be used to optimize waste management strategies. The authors acknowledged limitations due to insufficient training data and highlighted using pretrained (fine-tuned) models for waste classification. Fig. 8.4 illustrates the overall system architecture with a block diagram.

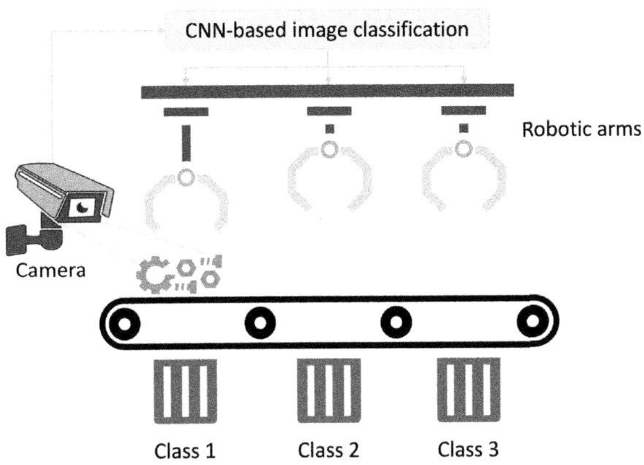

FIGURE 8.4 Automated sorting. Smart metal sorting method based on image classification and robotic arms automation [41,60].

- **Quality assessment:** Reprocessing metal waste requires careful consideration of quality evaluation, and ML approaches play a key role in enabling predictive quality control procedures. The mechanisms in question utilize advanced ML algorithms to evaluate the quality of goods made from recycled metal, taking into account various input factors such as mechanical qualities, surface flaws, and material composition. ML algorithms make it possible to accurately anticipate the quality of a product, guaranteeing that it complies with client requirements and strict quality standards by evaluating large datasets and finding complex patterns within them. Additionally, by locating inefficiencies and bottlenecks in the production process, ML improves reprocessing operations, improving overall process output and quality. ML-driven quality evaluation systems allow enterprises to reach higher efficiency and product excellence in metal waste recycling operations via ongoing refinement and optimization [56,57].
- **Practical limitations:** Even though ML has several advantages for recycling metal waste, some obstacles and restrictions must be overcome before it can reach its full potential. The availability and quality of training data for ML models is one of the main obstacles. Problems including sparsity, inconsistency, and bias are common in metal waste data, and they can negatively impact ML algorithms' performance and capacity for generalization. Furthermore, ML models' interpretability and transparency present serious difficulties, especially in sectors where stakeholder confidence and regulatory compliance are crucial [58]. Model complexity and interpretability must be balanced to guarantee that ML solutions are visible, comprehensible, and consistent with legal standards. Furthermore, there are logistical obstacles to overcome when implementing ML solutions on a wide scale in the industry, such as resource limitations, computational complexity, and infrastructure integration. Researchers, industry players, and policymakers must work together to create ML solutions that are reliable, scalable, and easy to understand in order to overcome these obstacles and promote long-term improvements in the reprocessing of metal waste [59].

Challenges, opportunities, and future directions

Since raw materials are always in demand, coupled with the increasing global awareness of the need to adopt sustainable practices, efficient and reliable metal waste management and processing becomes even more critical. Deploying ML can improve metal waste recycling to optimize resource utilization, reduce environmental impact, and enhance efficiency. However, implementing ML solutions in the complex and often challenging environment of metal waste management presents various challenges that need to be addressed alongside the exciting opportunities and promising future directions [61].

Challenges

Data availability and quality

Data availability and quality are critical pillars for effective ML applications in metal waste management. However, building a solid and resilient database poses substantial issues that demand careful attention. The following are the primary challenges:

- **Complexity of metal waste types:** Metal waste streams may be quite varied, comprising a wide range of alloys, impurities, and textures (e.g., shredded bits, wires, and castings). This complex nature makes it challenging to acquire representative data that covers the whole potential range of metal waste materials.
- **Hazardous environment:** Metal manufacturing facilities frequently operate in dangerous settings with dust, heat, and other potential risks. Sensor deployment and data collection under these conditions necessitate using specialist equipment and safety procedures.
- **Data security and privacy:** In today's connected world, ensuring the security and confidentiality of data obtained from several sources, such as operating procedures and sensors, is critical. This is particularly true when the information in question can give an unfair advantage or be considered sensitive because of its nature or commercial worth. For any organization, regardless of sector, safeguarding this data from unauthorized access, use, disclosure, interruption, alteration, or destruction is essential.
- **Ethical considerations in data collection:** To collect data in an ethical manner, it is imperative to establish transparency and trust among all parties involved. This entails giving clear information about the intended use of the data, obtaining the informed permission of data providers, and following responsible data management procedures at all times. This promotes collaboration in a setting where all participants are valued and aware of how their contributions are being utilized.

Data preprocessing and labeling

Data preprocessing and labeling are critical initial steps toward effectively applying ML models for metal waste management. However, those phases provide substantial obstacles that need careful consideration. Below are the main obstacles that may be encountered:

- **Time-consuming and labor-intensive:** Manually labeling massive databases with precise characteristics such as metal type and processing stage information can prove time-consuming and costly. It requires specialist personnel with experience in metal processing and data tagging, resulting in high prices and longer project schedules. Furthermore, the sheer amount of data involved in large-scale installations might render manual tagging almost impossible [62].

- **Expertise gap:** Precise data labeling frequently necessitates a unique combination of ML and metal processing knowledge. People possessing this particular skill set are difficult to come by and often need specialized training. This may make it more difficult for organizations, particularly smaller ones with limited budgets and fewer resources, to implement ML solutions.
- **Standardization and consistency:** Reliable and impartial ML models depend on maintaining consistent labeling methods across many facilities and processes. Labeling errors result in inaccurate data, affecting the model's performance and possibly introducing bias. This might have serious repercussions, such as producing inaccurate forecasts.

Integration with existing infrastructure

As the metal processing sector embraces the promise of ML technology, a considerable issue emerges: integrating these new systems with current infrastructure. This sophisticated process smoothly integrates ML technologies into the processing plant's control systems, equipment, and data sources. Here's a deeper look at the challenges involved:

- **Legacy systems:** Many metal processing facilities employ older control systems that may need to be redesigned to work with more modern ML models. These systems frequently rely on out-of-date communication protocols and proprietary data formats, posing substantial compatibility issues with modern ML software. Integrating these diverse systems is similar to connecting two gadgets with incompatible connectors; communication needs extra effort or specialized adapters.
- **Real-time integration requirements:** To integrate ML, operational data from sensors, equipment, and control systems must be available in real-time. This continual flow of information is critical for the ML model to work correctly, necessitating smooth communication between new and old systems. Maintaining this real-time feature is analogous to supporting a continuous conversation—any delays or disturbances in data flow can undermine the effectiveness of communication and decision-making processes.

Opportunities

ML provides a promising toolkit for improving metal waste management. A few intriguing opportunities are listed below:

- **Enhanced sorting and segregation:** ML techniques, particularly CNNs, may be trained to detect metals in waste streams through picture recognition. Sorting plants may automate the separation of metals from other recyclables, increasing the efficiency and quality of the recovered metals [63]. A trained CNN may be used to instruct robotic arms or sorting instruments to accurately pick and separate metals based on visual recognition. This dramatically decreases human mistakes and contamination in the recycling system. Furthermore, more precise separation results in higher purity of the recovered metals, translating into better quality for reuse in manufacturing. This consequently eliminates the demand for virgin metal extraction while minimizing environmental effects.

- **Real-time bin monitoring:** ML could be combined with IoT sensors mounted on recycling containers. These sensors can monitor fill levels and metal content, allowing for more efficient collection routes and avoiding overflowing bins. Waste management companies may use this data in conjunction with ML algorithms to determine when bins are reaching capacity. They can subsequently devise efficient collection routes to ensure timely emptying and avoid overflowing bins. To elaborate, ML may be trained to examine data from other sensors that identify the presence and kind of metals in the bin. This enables tailored collection tactics, which prioritize containers with high metal content for effective recycling.
- **Predictive maintenance:** Metal processing facilities use machinery to sift, shred, and separate metals. Sensors on these devices measure vibrations, temperature, and power usage. ML algorithms can use sensor data to discover trends predicting future equipment problems. Early warning signals can be generated for preventive maintenance and avoid costly breakdowns. Processing plants may maximize productivity by proactively identifying and resolving maintenance needs, reducing downtime, and ensuring a consistent supply of recyclable metals.

Future directions

Metal waste management is an essential part of environmental sustainability and resource conservation. As we move forward, we may examine numerous potential avenues to improve metal waste management:

- **Digitalization and data analytics:** Using digitalization and data analytics technologies may improve metal waste management procedures by allowing for real-time monitoring, predictive maintenance, and resource allocation. The integration of sensors, IoT devices, and ML algorithms can increase metal recycling facility efficiency and decision-making skills.
- **Edge AI and decentralized processing:** ML models frequently demand a lot of data processing capacity on the cloud, which can be restricted, particularly in real-time applications. Edge AI may be used in the future, with processing taking place directly on devices such as cameras or sensors on recycling bins. This would allow for speedier decision-making and less dependency on cloud infrastructure.
- **Integration with robotics and automation:** As ML and robotics advance, we should expect more advanced sorting robots. These robots might be equipped with enhanced grippers and manipulation capabilities to handle a broader range of metal forms and sizes, boosting more sorting accuracy and efficiency.
- **Public awareness and education:** Raising public awareness and education on the necessity of metal waste management and the advantages of recycling can help to drive behavior change and promote responsible use and disposal. Educational campaigns, outreach programs, and community participation may help develop a recycling and waste reduction culture.

Conclusion

This chapter has looked at the remarkable capacity of ML to change metal waste recycling. We investigated how ML may optimize and modify several phases of the conventional process, from initial sorting and separation to extraction and distribution. By exploiting ML's capabilities, the sector may advance beyond old methodologies, attaining more efficiency while using less energy and relying less on human resources. The capacity of ML to assess large datasets and adapt to complex patterns offers enormous potential for improving precision and efficiency in metal recycling. Ultimately, this results in a more sustainable and circular metal-use approach. We provided persuasive evidence for this transformational technology by reviewing a broad set of case studies, including ML uses in environmental research and metal waste recycling. Furthermore, a thorough discussion was held on the main issues and barriers, such as data availability and quality and integration with existing infrastructures, that needed to be addressed to fully utilize ML in the context of metal waste recycling.

References

[1] S. Uniyal, R. Paliwal, B. Kaphaliya, R.K. Sharma, Human Overpopulation, IGI Global, 2020, pp. 20–30. Available from 10.4018/978-1-5225-9276-1.ch002.

[2] D. Moya, C. Aldás, G. López, P. Kaparaju, Municipal solid waste as a valuable renewable energy resource: a worldwide opportunity of energy recovery by using waste-to-energy technologies, Energy Procedia 134 (2017) 286–295. Available from https://doi.org/10.1016/j.egypro.2017.09.618, http://www.sciencedirect.com/science/journal/18766102.

[3] V. Forti, C.P. Balde, R. Kuehr, G. Bel, The global e-waste monitor 2020: quantities, flows and the circular economy potential, (2020). https://ewastemonitor.info/wp-content/uploads/2020/11/GEM_2020_def_july1_low.pdf.

[4] G. Bellenfant, A.-G. Guezennec, F. Bodenan, Hugues, D. Cassard, Reprocessing of mining waste: combining environmental management and metal recovery? Mine Closure 2013: Proceedings of the Eighth International Seminar on Mine Closure 2013 (2013) 571–582.

[5] K.N. Sami, Z.M.A. Amin, R. Hassan, Waste management using machine learning and deep learning algorithms, International Journal on Perceptive and Cognitive Computing 6 (2) (2020) 97–106. Available from https://doi.org/10.31436/ijpcc.v6i2.165.

[6] J. Teng, Y. Shi, Z.H. Liu, H.Q. Li, M.X. He, Z.H. Li, et al., Rapid identification of hazardous heavy metal-containing waste by combining EDXRF with machine learning: taking zinc smelting waste as an example, Resources, Conservation and Recycling 198 (2023). Available from https://doi.org/10.1016/j.resconrec.2023.107155, http://www.elsevier.com/locate/resconrec.

[7] B. Esmaeilian, B. Wang, K. Lewis, F. Duarte, C. Ratti, S. Behdad, The future of waste management in smart and sustainable cities: a review and concept paper, Waste Management 81 (2018) 177–195. Available from https://doi.org/10.1016/j.wasman.2018.09.047, http://www.elsevier.com/locate/wasman.

[8] M.M. Hossen, M.E. Majid, S.B.A. Kashem, A. Khandakar, M. Nashbat, A. Ashraf, et al., A reliable and robust deep learning model for effective recyclable waste classification, IEEE Access 12 (2024) 13809–13821. Available from https://doi.org/10.1109/ACCESS.2024.3354774, http://ieeexplore.ieee.org/xpl/RecentIssue.jsp?punumber=6287639.

[9] T. Anagnostopoulos, A. Zaslavsky, K. Kolomvatsos, A. Medvedev, P. Amirian, J. Morley, et al., Challenges and opportunities of waste management in IoT-enabled smart cities: a survey, IEEE Transactions on Sustainable Computing 2 (3) (2017) 275–289. Available from https://doi.org/10.1109/TSUSC.2017.2691049, http://ieeexplore.ieee.org/xpl/RecentIssue.jsp?punumber=7274860.

[10] S.P. Gundupalli, S. Hait, A. Thakur, Multi-material classification of dry recyclables from municipal solid waste based on thermal imaging, Waste Management 70 (2017) 13–21. Available from https://doi.org/10.1016/j.wasman.2017.09.019, http://www.elsevier.com/locate/wasman.

[11] M.M. Hossen, A. Ashraf, M. Hasan, M.E. Majid, M. Nashbat, S.B.A. Kashem, et al., GCDN-Net: garbage classifier deep neural network for recyclable urban waste management, Waste Management 174 (2024) 439–450. Available from https://doi.org/10.1016/j.wasman.2023.12.014, https://www.sciencedirect.com/science/journal/0956053X.

[12] G. Bonifazi, G. Capobianco, S. Serranti, Asbestos containing materials detection and classification by the use of hyperspectral imaging, Journal of Hazardous Materials 344 (2018) 981–993. Available from https://doi.org/10.1016/j.jhazmat.2017.11.056, http://www.elsevier.com/locate/jhazmat.

[13] B. Heater, Apple has a new iphone recycling robot named 'daisy, (2018). https://techcrunch.com/2018/04/19/apple-has-a-new-iphone-recycling-robot-named-daisy/.

[14] R. Sarc, A. Curtis, L. Kandlbauer, K. Khodier, K.E. Lorber, R. Pomberger, Digitalisation and intelligent robotics in value chain of circular economy oriented waste management – a review, Waste Management 95 (2019) 476–492. Available from https://doi.org/10.1016/j.wasman.2019.06.035.

[15] N. Vongdala, H.D. Tran, T.D. Xuan, R. Teschke, T.D. Khanh, Heavy metal accumulation in water, soil, and plants of municipal solid waste landfill in Vientiane, Laos, International Journal of Environmental Research and Public Health 16 (1) (2019). Available from https://doi.org/10.3390/ijerph16010022, http://www.mdpi.com/journal/ijerph.

[16] A.J.P. Smolders, R.A.C. Lock, G. Van der Velde, R.I. Medina Hoyos, J.G.M. Roelofs, Effects of mining activities on heavy metal concentrations in water, sediment, and macroinvertebrates in different reaches of the Pilcomayo River, South America, Archives of Environmental Contamination and Toxicology 44 (3) (2003) 314–323. Available from https://doi.org/10.1007/s00244-002-2042-1.

[17] S. Luckeneder, S. Giljum, A. Schaffartzik, V. Maus, M. Tost, Surge in global metal mining threatens vulnerable ecosystems, Global Environmental Change 69 (2021) 102303Available from https://doi.org/10.1016/j.gloenvcha.2021.102303.

[18] L. Zhang, Z. Xu, A review of current progress of recycling technologies for metals from waste electrical and electronic equipment, Journal of Cleaner Production 127 (2016) 19–36. Available from https://doi.org/10.1016/j.jclepro.2016.04.004.

[19] T. Rohwerder, T. Gehrke, K. Kinzler, W. Sand, Bioleaching review part A: progress in bioleaching: fundamentals and mechanisms of bacterial metal sulfide oxidation, Applied Microbiology and Biotechnology 63 (3) (2003) 239–248. Available from https://doi.org/10.1007/s00253-003-1448-7.

[20] M. Kaya, Recovery of metals and nonmetals from electronic waste by physical and chemical recycling processes, Waste Management 57 (2016) 64–90. Available from https://doi.org/10.1016/j.wasman.2016.08.004, http://www.elsevier.com/locate/wasman.

[21] H. Abu-Qdais, N. Shatnawi, E. AL-Alamie, Intelligent solid waste classification system using combination of image processing and machine learning models, Waste Management, Research Square, Jordan, 2023Available from https://www.researchsquare.com/browse, 10.21203/rs.3.rs-2573812/v1.

[22] C. Bountzouklis, D.M. Fox, E.D. Bernardino, Predicting wildfire ignition causes in Southern France using eXplainable Artificial Intelligence (XAI) methods, Environmental Research Letters 18 (4) (2023). Available from https://doi.org/10.1088/1748-9326/acc8ee, http://iopscience.iop.org/journal/1748-9326.

[23] O.A. Eshbayev, S.M. Mirzaliev, R.U. Rozikov, D.M. Kuzikulova, G.A. Shakirova, NLP and ML based approach of increasing the efficiency of environmental management operations and engineering practices, IOP Conference Series: Earth and Environmental Science 1045 (1) (2022) 012058. Available from https://doi.org/10.1088/1755-1315/1045/1/012058.

[24] K.S. Adu-Manu, J.-D. Abdulai, F. Engmann, M. Akazue, J.K. Appati, G.E. Baiden, et al., WSN architectures for environmental monitoring applications, Journal of Sensors 2022 (2022) 1–18. Available from https://doi.org/10.1155/2022/7823481.

[25] M.A. Mohammed, M.J. Abdulhasan, N.M. Kumar, K.H. Abdulkareem, S.A. Mostafa, M.S. Maashi, et al., Automated waste-sorting and recycling classification using artificial neural network and features fusion: a digital-enabled circular economy vision for smart cities, Multimedia Tools and Applications 82 (25) (2023) 39617–39632. Available from https://doi.org/10.1007/s11042-021-11537-0, https://www.springer.com/journal/11042.

[26] A.L. Srivastav, Markandeya, N. Patel, M. Pandey, A.K. Pandey, A.K. Dubey, et al., Concepts of circular economy for sustainable management of electronic wastes: challenges and management options, Environmental Science and Pollution Research 30 (17) (2023) 48654–48675. Available from https://doi.org/10.1007/s11356-023-26052-y, https://www.springer.com/journal/11356.

[27] D. Wang, J. Qian, L. Ma, W. Zhao, D. Gao, X. Hou, et al., Characterizing Groundwater Potential Using GIS-Based Machine Learning Model in Chihe River Basin, China, Research Square, China, 2021Available from https://www.researchsquare.com/browse, 10.21203/rs.3.rs-1044219/v1.

[28] P.-Y. Wu, K. Mjörnell, C. Sandels, M. Mangold, Machine learning in hazardous building material management: research status and applications, Recent Progress in Materials 03 (02) (2021) 1. Available from https://doi.org/10.21926/rpm.2102017-1.

[29] H.N. Guo, S.B. Wu, Y.J. Tian, J. Zhang, H.T. Liu, Application of machine learning methods for the prediction of organic solid waste treatment and recycling processes: A review, Bioresource Technology 319 (2021). Available from https://doi.org/10.1016/j.biortech.2020.124114, http://www.elsevier.com/locate/biortech.

[30] Imran, S. Ahmad, D.H. Kim, Quantum GIS based descriptive and predictive data analysis for effective planning of waste management, IEEE Access 8 (2020) 46193–46205. Available from https://doi.org/10.1109/ACCESS.2020.2979015, http://ieeexplore.ieee.org/xpl/RecentIssue.jsp?punumber=6287639.

[31] A. Hussain, U. Draz, T. Ali, S. Tariq, M. Irfan, A. Glowacz, et al., Waste management and prediction of air pollutants Using IoT and machine learning approach, Energies 13 (15) (2020) 3930. Available from https://doi.org/10.3390/en13153930.

[32] S.H. Haji, A.B. Sallow, IoT for smart environment monitoring based on python: a review, Asian Journal of Research in Computer Science (2021) 57–70. Available from https://doi.org/10.9734/ajrcos/2021/v9i130215.

[33] M. Al-Duhayyim, S.S. Alotaibi, S. Al-Otaibi, F.N. Al-Wesabi, M. Othman, I. Yaseen, et al., An intelligent hazardous waste detection and classification model using ensemble learning techniques, Computers, Materials & Continua 74 (2) (2023) 3315–3332. Available from https://doi.org/10.32604/cmc.2023.033250.

[34] R. Ochoa-Barragán, A.D.C. Munguía-López, J.M. Ponce-Ortega, A hybrid machine learning-mathematical programming optimization approach for municipal solid waste management during the pandemic, Environment, Development and Sustainability, Springer Science and Business Media B.V., Mexico, 2023Available from https://www.springer.com/journal/10668, 10.1007/s10668-023-03354-2.

[35] R.S. Olson, J.H. Moore, Identifying and Harnessing the Building Blocks of Machine Learning Pipelines for Sensible Initialization of a Data Science Automation Tool, Springer Science and Business Media LLC, 2018, pp. 211–223. Available from 10.1007/978-3-319-97088-2_14.

[36] A. Utku, S.K. Kaya, Deep learning based a comprehensive analysis for waste prediction, Operational Research in Engineering Sciences: Theory and Applications 5 (2) (2022) 176–189. Available from https://doi.org/10.31181/oresta190822135u, https://oresta.rabek.org/index.php/oresta/article/view/274/101.

[37] I.G. Ayu Arwati, E. Nina Saparina Yuliani, N. Endah Retno Wuryandari, D. Lutfiana, Development and application of appropriate technology to recycle waste performed, Dinasti International Journal of Management Science 2 (4) (2021) 561–568. Available from https://doi.org/10.31933/dijms.v2i4.192.

[38] J. Chen, S. Huang, S. BalaMurugan, G.S. Tamizharasi, Artificial intelligence based e-waste management for environmental planning, Environmental Impact Assessment Review 87 (2021) 106498Available from https://doi.org/10.1016/j.eiar.2020.106498.

[39] C. Liang, H. Li, M. Lei, Q. Du, Dongting lake water level forecast and its relationship with the Three Gorges Dam based on a long short-term memory network, Water 10 (10) (2018) 1389. Available from https://doi.org/10.3390/w10101389.

[40] L.O.H. Wijeratne, D.R. Kiv, A.R. Aker, S. Talebi, D.J. Lary, Using machine learning for the calibration of airborne particulate sensors, Sensors (Switzerland) 20 (1) (2020). Available from https://doi.org/10.3390/s20010099, https://www.mdpi.com/1424-8220/20/1/99/pdf.

[41] M.W. Rahman, R. Islam, A. Hasan, N.I. Bithi, M.M. Hasan, M.M. Rahman, Intelligent waste management system using deep learning with IoT, Journal of King Saud University – Computer and Information Sciences 34 (5) (2022) 2072–2087. Available from https://doi.org/10.1016/j.jksuci.2020.08.016, http://www.journals.elsevier.com/journal-of-king-saud-university-computer-and-information-sciences/.

[42] D.J. Díaz-Romero, S. Van den Eynde, W. Sterkens, B. Engelen, I. Zaplana, W. Dewulf, et al., Simultaneous mass estimation and class classification of scrap metals using deep learning, Resources, Conservation and Recycling 181 (2022). Available from https://doi.org/10.1016/j.resconrec.2022.106272.

[43] H. Saleem, M. Khosravi, S. Maroufi, V. Sahajwalla, A.P. O'Mullane, Repurposing metal containing wastes and mass-produced materials as electrocatalysts for water electrolysis, Sustainable Energy & Fuels 6 (21) (2022) 4829–4844. Available from https://doi.org/10.1039/d2se01068b.

[44] L.R. Adrianto, S. Pfister, Prospective environmental assessment of reprocessing and valorization alternatives for sulfidic copper tailings, Resources, Conservation and Recycling 186 (2022) 106567Available from https://doi.org/10.1016/j.resconrec.2022.106567.

[45] K. Hara, H. Yabar, Historical evolution and development of waste management and recycling systems-analysis of Japan's experiences, Journal of Environmental Studies and Sciences 2 (4) (2012) 296–307. Available from https://doi.org/10.1007/s13412-012-0094-8, http://www.springer.com/environment/journal/13412.

[46] R. Kumari, S.R. Samadder, A critical review of the pre-processing and metals recovery methods from e-wastes, Journal of Environmental Management 320 (2022). Available from https://doi.org/10.1016/j.jenvman.2022.115887, https://www.sciencedirect.com/journal/journal-of-environmental-management.

[47] F. Wang, J. Huisman, A. Stevels, C.P. Baldé, Enhancing e-waste estimates: improving data quality by multivariate input-output analysis, Waste Management 33 (11) (2013) 2397–2407. Available from https://doi.org/10.1016/j.wasman.2013.07.005.

[48] H.L. Vu, K.T.W. Ng, A. Richter, C. An, Analysis of input set characteristics and variances on k-fold cross validation for a recurrent neural network model on waste disposal rate estimation, Journal of Environmental Management 311 (2022). Available from https://doi.org/10.1016/j.jenvman.2022.114869, https://www.sciencedirect.com/journal/journal-of-environmental-management.

[49] L. Barth, L. Schweiger, R. Benedech, M. Ehrat, From data to value in smart waste management: optimizing solid waste collection with a digital twin-based decision support system, Decision Analytics Journal 9 (2023) 100347Available from https://doi.org/10.1016/j.dajour.2023.100347.

[50] X. Luo, Y. Li, Q. Wang, J. Mu, Y. Liu, Machine learning based modeling for predicting the compressive strength of solid waste material-incorporated magnesium phosphate cement, Journal of Cleaner Production 442 (2024) 141172Available from https://doi.org/10.1016/j.jclepro.2024.141172.

[51] G. Hu, W. Pfingsten, Data-driven machine learning for disposal of high-level nuclear waste: a review, Annals of Nuclear Energy 180 (2023) 109452Available from https://doi.org/10.1016/j.anucene.2022.109452.

[52] S. Mishra, R. Yaduvanshi, P. Rajpoot, S. Verma, A.K. Pandey, D. Pandey, An integrated deep-learning model for smart waste classification, Environmental Monitoring and Assessment 196 (3) (2024). Available from https://doi.org/10.1007/s10661-024-12410-x, https://www.springer.com/journal/10661.

[53] B. Fang, J. Yu, Z. Chen, A.I. Osman, M. Farghali, I. Ihara, et al., Artificial intelligence for waste management in smart cities: a review, Environmental Chemistry Letters 21 (4) (2023) 1959–1989. Available from https://doi.org/10.1007/s10311-023-01604-3, https://www.springer.com/journal/10311.

[54] A.J. Whitworth, J. Vaughan, G. Southam, A. van der Ent, P.N. Nkrumah, X. Ma, et al., Review on metal extraction technologies suitable for critical metal recovery from mining and processing wastes, Minerals Engineering 182 (2022). Available from https://doi.org/10.1016/j.mineng.2022.107537, http://www.journals.elsevier.com/minerals-engineering/.

[55] P. Nowakowski, T. Pamuła, Application of deep learning object classifier to improve e-waste collection planning, Waste Management 109 (2020) 1–9. Available from https://doi.org/10.1016/j.wasman.2020.04.041, http://www.elsevier.com/locate/wasman.

[56] M. Erkinay Ozdemir, Z. Ali, B. Subeshan, E. Asmatulu, Applying machine learning approach in recycling, Journal of Material Cycles and Waste Management 23 (3) (2021) 855–871. Available from https://doi.org/10.1007/s10163-021-01182-y, http://link.springer.de/link/service/journals/10163/index.htm.

[57] N.H. Zulkernain, N. Basant, C.C. Ng, Kriti, M. Salari, S. Mallick, Recovery of precious metals from e-wastes through conventional and phytoremediation treatment methods: a review and prediction, Journal of Material Cycles and Waste Management 25 (5) (2023) 2726–2752. Available from https://doi.org/10.1007/s10163-023-01717-5, https://www.springer.com/journal/10163.

[58] B.K. Reck, T.E. Graedel, Challenges in metal recycling, Science 337 (6095) (2012) 690–695. Available from https://doi.org/10.1126/science.1217501, http://www.sciencemag.org/content/337/6095/690.full.pdf.

[59] S. Krishnan, N.S. Zulkapli, H. Kamyab, S.M. Taib, M.F.B.M. Din, Z.A. Majid, et al., Current technologies for recovery of metals from industrial wastes: an overview, Environmental Technology and Innovation 22 (2021). Available from https://doi.org/10.1016/j.eti.2021.101525, http://www.journals.elsevier.com/environmental-technology-and-innovation/.

[60] A.U. Gondal, M.I. Sadiq, T. Ali, M. Irfan, A. Shaf, M. Aamir, et al., Real time multipurpose smart waste classification model for efficient recycling in smart cities using multilayer convolutional neural network and

perceptron, Sensors 21 (14) (2021). Available from https://doi.org/10.3390/s21144916, https://www.mdpi.com/1424-8220/21/14/4916/pdf.

[61] M.T. Munir, B. Li, M. Naqvi, Revolutionizing municipal solid waste management (MSWM) with machine learning as a clean resource: opportunities, challenges and solutions, Fuel 348 (2023) 128548Available from https://doi.org/10.1016/j.fuel.2023.128548.

[62] T. Fredriksson, D.I. Mattos, J. Bosch, H.H. Olsson, Data labeling: an empirical investigation into industrial challenges and mitigation strategies. Lecture notes in computer science (including subseries lecture notes in artificial intelligence and lecture notes in bioinformatics), 12564 (2020) 202–216. https://www.springer.com/series/558 10.1007/978-3-030-64148-1_13.

[63] S. Majchrowska, A. Mikołajczyk, M. Ferlin, Z. Klawikowska, M.A. Plantykow, A. Kwasigroch, et al., Deep learning-based waste detection in natural and urban environments, Waste Management 138 Elsevier Ltd, Poland, 2022, pp. 274–284. Available from https://www.sciencedirect.com/science/journal/0956053X, 10.1016/j.wasman.2021.12.001.

CHAPTER

9

Waste classification using convolutional neural networks tuned by modified metaheuristics algorithm

Ana Tasic, Luka Jovanovic, Miroslav Popovic, Nebojsa Bacanin, Miodrag Zivkovic, and Milos Dobrojevic

Department of Informatics and Computing, Singidunum University, Belgrade, Serbia

Introduction

With the population rising to an outstanding number of 7.9 billion in 2021 [1], it is not a surprise that waste rates are increasing worldwide, and that is truly concerning [2]. Facing 2050, waste is estimated to grow by 70%, thus adequate waste management will be required [2]. One way to fight climate change is carbon neutrality, whose purpose is to balance the carbon dioxide (CO_2) emissions with the amount of CO_2 absorbed from the atmosphere so that the increase of CO_2 would be minimal [3]. There are a lot of approaches for achieving that goal; one that interests us is the reuse of solid wastes [3]. Two more concepts that intertwine with the reuse of solid waste are the upcycling and zero waste concepts [4–6]. Although there is no single definition for zero waste, the main concept is recycling and reusing all products and by-products in one society [7,8]. On the other hand, the upcycling concept is about "upgrading" waste into more valuable products [9].

Waste classification is significant for numerous reasons, as it helps in controlling and mitigating the environmental and health repercussions related to waste. Different types of waste can have diverse impacts on the environment, as hazardous waste may contaminate soil, water, and air, and pose a serious threat to both ecosystems and human society [10–13]. Precise classification of waste may facilitate the application of appropriate disposal options that can avert the impact on the environment [14]. Moreover, it helps in ensuring health and safety, as inappropriate management of hazardous waste can deteriorate human health [15,16]. Accurate identification and classification of hazardous substances guarantees safe handling, transport, and disposal procedures, therefore assuring safe conditions for both waste disposal workers and the general public and preventing pollution. It also helps in resource recovery, as it can identify specific types of waste that may be recycled and promote a sustainable society [17]. Finally, it helps in

Harnessing Automation and Machine Learning for Resource Recovery and Value Creation
DOI: https://doi.org/10.1016/B978-0-443-27374-2.00009-1

adherence to the regulatory standards, as many countries worldwide have procedures, regulations, and guidelines that govern the proper care and disposal of different types of waste. Accurate categorization ensures compliance with these standards, thus averting legal implications and penalties [18–21].

Therefore, we can see the importance of classifying, recycling, and reusing solid waste for a brighter future. A lot of factors, such as urbanization, industrialization, and linear approaches to the consumption of products that are consumed and then discarded, are impacting the various ways of waste management. Furthermore, performance in waste management in different countries varies because of different economic, social, and cultural situations [22]. For instance, Singapore has the lowest priority for landfilling since it is a small island country with a lot of people living there; therefore, that puts waste minimalization and solid waste incineration at the top of the waste management hierarchy [23]. An important aspect of waste behavior is how convenient it is for someone to dispose of their household waste.

Analysis of one study on students from Bulgaria, Plovdiv, and Sweden, Kalmar, showed that although students live in European Union countries with the same waste management legislation, there is a significant difference in waste behavior. The Sweden group survey showed more positive results in waste behavior, even though both groups showed the same attitude toward waste behavior. Understandably, the Bulgarian group had more dissatisfaction with local facilities for waste management [22]. Waste classification, when done manually, is one of the most intensive and time-consuming parts of the recycling process. Segregation and classification tasks are performed in harsh working environments, and people are working in direct contact with waste, which can be harmful. Also, workers sorting effectiveness decays over the period of working hours [24,25]. Nevertheless, understanding the classification of waste is crucial in assessing its possible dangers and the impact it may have on the environment [26,27].

Computer vision is one of the fields of artificial intelligence that is capable of mimicking human vision, a good potential for waste classification [28]. Since the accumulation of waste is too fast for people to take care of manually, a more productive way is needed. Deep learning algorithms are ones that can automatize the waste classification process and accelerate the time-consuming process of sorting, as well as keep people away from harmful environments or hazardous materials. Many algorithms are used to differentiate images and objects, yet the most used one is the convolutional neural network (CNN). Combined with sorting equipment, it can lead to faster, more standardized, and safer detection of waste class [24,29]. Additionally, compared to human workers, robotic equipment, owing to deep algorithms, can sort different types of waste at the same time, while workers can only focus on one type [25].

The problem with neural networks in general is reflected in the fact that each model has a set of hyperparameters that needs to be tuned for each particular problem that needs to be solved, backed up by the no free lunch theorem [30]. It is a critical aspect that directly affects the model's performance and includes adjusting different parameters to allow the network to learn and generalize from the training dataset [31,32]. The process of fine-tuning hyperparameters of the neural network is recognized as an NP-hard optimization problem. This categorization signifies that determining the optimal configuration of hyperparameters within the vast domain of possible combinations has a computational complexity that is escalating exponentially with the increase of the search realm. The exploration of hyperparameter settings for neural networks poses challenges due to the extremely complex interplay of nonlinearity and high dimensionality in the parameter space, leading to impossible computational demands and resource intensiveness. Consequently, researchers frequently turn to stochastic methods and heuristic algorithms to navigate this intricate terrain effectively and pinpoint near-optimal hyperparameter configurations for neural networks [33].

Recently, one of the most promising families of stochastic methods that are regularly used to tune the hyperparameters of neural networks is metaheuristic algorithms. Metaheuristic algorithms are known for their ability to determine high-quality solutions in challenging optimization problems, and they are capable of efficiently exploring the vast and complex search spaces of the model's hyperparameters [34–37]. Utilizing these metaheuristic algorithms for hyperparameter tuning facilitates a more streamlined and powerful exploration of the intricate and multidimensional space, aiding in the determination of optimal or nearly optimal configurations for neural networks. This approach proves especially beneficial in addressing the above-mentioned computational complexities with respect to the NP-hard characteristics of hyperparameter optimization. Convolutional neural networks have also been tuned by metaheuristic algorithms with significant success, as it is evident from numerous recent publications [34,38–41].

This research explores the potential of convolutional neural networks optimized by an improved version of a novel crayfish optimization algorithm (COA) [42] for the task of waste classification utilizing image data. The proposed method is evaluated on a real-world dataset. Therefore, the scientific contributions of this work may be summarized as follows:

- Usage of CNN algorithms and their impact on waste classification.
- Utilization of a modified crayfish algorithm that was designed to enhance the original algorithm for improving the CNN's performance in the waste classification task.
- Proposal of a unified framework for waste classification with the application of computer vision algorithms.

The remainder of this document is organized in the following manner. Section "Related works" puts forward a brief literature overview of the waste management problem, convolutional neural networks, and metaheuristic algorithms. The utilized metaheuristics algorithm and the proposed modifications to the basic version are given in Section "Methods". The simulation setup is explained in Section "Simulation setup", followed by the simulation outcomes, which are shown in Section "Experimental outcomes". Section "Conclusions" winds up the manuscript and indicates promising directions of the research that lies ahead.

Related works

There is a big potential in solid waste management with the help of artificial intelligence. So far, numerous artificial intelligence models have been applied in this industry for better waste classification and management. A few of them are artificial neural networks (ANN) and different types of ANN approaches like CNN, feedforward networks, linear regression, and multilayer perceptrons (MLP) [43–45]. Other models like adaptive neuro-fuzzy inference system (ANFIS) [46,47], K-means [48,49], random forest (RF) [50,51], etc. have also been employed to some extent in this industry.

Furthermore, artificial intelligence is applied for forecasting in other parts of ecology. Some studies focus on predicting volatile organic compounds (VOCs) and their interaction with the environment and the other way around. Algorithms that are used for this matter are extreme gradient boosting (XG boosting) and ANFIS trained with metaheuristic algorithms like GA and particle swarm optimization (PSO) [52,53]. As mentioned, the recognition of images with computing power is changing the perspective on solutions for waste management and waste classification. From a variety of deep learning methods, CNN is one of the few that stands out. Since outstanding performance in image classification, different architectures of CNN have been proposed for the utilization of waste datasets for image classification. Some of them are VGGNet, AlexNet, ResNet-50, MobileNet-v2, EfficientNet, and DenseNet121. The accuracy of these architectures on TreshNet, self-composed ones, and other datasets varies, yet some ranges of accuracy from recent studies are from 83% to an astonishing 98% [54–59].

Reviewed work suggests that researchers have yet to explore the full potential of optimization metaheuristics, specifically the most recent algorithms such as the COA [42], for tackling waste classification in an optimized way. This work aims to bridge the observed gap in the literature.

Convolutional neural network

ANN, from which CNN is derived, is a computational system that is forged in a way that neural networks in the human brain work. The system is composed of multilayers in which each layer is comprised of nodes (representing neurons in biology). Some of the basic and most commonly used architectures are feedforward networks, which, as their name tells, information through them flows forward from the input layer to the output layer without forming loops. Also, all the nodes from the previous layer are connected with every node forward. One more important concept is the backpropagation mechanism, which allows these feedforward networks to learn from mistakes. Provided that there are differences in actual output and generated output of networks, which is called error, information is propagated backward so weights can be adjusted for better results in future performances [60–63].

Although ANN can process image data, there are a lot of difficulties it will get through processing it depending on the image size, like overfitting, and more computational power since it will need more nodes in layers for more complex images. As more suitable for images, the algorithm stands out: CNN, since before fully connected layers, there were convolutional layers that were sparsely connected, and therefore it will need less computational power. Furthermore, input for them can be three-dimensional which is more suitable when speaking of image input, and in ANN it is one or two-dimensional.

The representative layer of CNN is the convolutional layer. In this layer, through image input, the kernel strides like some filter and searches for patterns in its data. Both the image and kernel are comprised of matrices that multiply, and after the equation, the product goes in the feature map or activation map. The equation for feature map values calculation:

$$F_{i,j} = (I^*K)[i, j] \tag{9.1}$$

where $F_{i,j}$– denotes i, j-th entry in a feature map, I is input data, and K represents kernel.

Since dimensionality is a problem for image detection, there is also a pooling layer in CNN that helps reduce image dimensionality and as a consequence, we got less computation power needed, fewer parameters, and less overfitting. The equation used in the pooling layer:

$$Y_{vug} = \max_{(p,q)\in R_{ug}} \alpha_{vpg} \tag{9.2}$$

where Y_{vug} is the max pooling operation value, max(p,q) $x \in A$ R_{ug} represents a maximum value over the region defined by $x \in A$ R_{ug}, and a_{vpq} denotes input elements at location (p,q).

Metaheuristic optimization

Tuning hyperparameters is one of the biggest challenges in waste classification with CNN. The approach to finding optimally tuned hyperparameters for each dataset or problem is a polynomial hard (NP-hard) challenge [52]. These challenges cannot be realized with standard methods, or if done manually, they are extremely time-consuming. On the other hand, metaheuristics are good for handling complex tasks (NP-hard). Furthermore, the No Free Lunch (NFL) theorem [30] states that there is no perfect algorithm for all tasks or problems; thus, there are a lot of different algorithms for optimization.

Since there are a lot of different metaheuristic algorithms, based on the distinct behaviors or patterns of nature, metaheuristic algorithms can be distinguished roughly into four types: swarm intelligence, physics-based, evolutionary, and human-inspired algorithms. Numerous algorithms have gained recognition in the field of optimization research due to their consistently strong overall performance. Noteworthy instances include the particle swarm optimization (PSO) [64] and the genetic algorithm (GA) [65]. Additional examples encompass algorithms inspired by collaborative behaviors observed in nature, such as the artificial bee colony (ABC) [66] algorithm, firefly algorithm (FA) [67], and the relatively recent reptile search algorithm (RSA) [68]. Nevertheless, more abstract concepts have also served as sources of inspiration for algorithms, as evidenced by the self-adaptive step-size search (SSAS) [69] and COLSHADE [70] algorithms. The utilization of optimization metaheuristics has proven beneficial in addressing numerous complex problems. These metaheuristics, often associated with the optimization of the hyperparameters of other algorithms, have demonstrated successful applications in diverse domains, including environmental sciences [71,72], green energy [73,74], finance and banking [75,76], and medicine [77,78].

Hybridization stands out as a widely embraced strategy in the realm of optimization, aimed at overcoming limitations associated with existing optimizers. A significant number of instances in the literature have illustrated that a combined approach can yield outcomes that surpass the performance of the original baseline methods. This approach finds applications in various domains, including computer security [79,80], intrusion detection

[81,82], cloud computing [83,84], and spam detection [85,86]. Notably, the field of medical applications [87–90] shows promise, witnessing significant enhancements through the adoption of hybrid algorithms.

CNNs have also been heavily hybridized with metaheuristic algorithms, as shown in [78,91–93]. Some scientific papers made comparisons with and without applied metaheuristics on CNNs, and numerous papers showed increased accuracy after optimization. For example, Wei-Lung Mao et al.'s research paper applied GA to optimize DenseNet121, and the results were 95% for nonoptimized CNN and 99.6% for the optimized one [94].

Methods

This section starts by providing a brief overview of the basic COA algorithm, followed by listing noted deficiencies of the elementary algorithm and discussing room for improvement. Next, a modified variant of COA metaheuristics is suggested, which is later used in the conducted experiments.

Basic variant of crayfish optimization algorithm

This paper utilizes a COA [42], inspired by the crayfish behavior through different environmental conditions. COA consists of two stages: exploitation and exploration. In the exploration stage, the algorithm is based on crayfish intolerance for higher temperatures (>30), so they seek shelter in caves. On the other hand, the exploitation stage is built on their behavior to compete over the caves if there is a rival nearby; otherwise, if temperatures are below 30 degrees they go seeking food.

Through the exploration stage in the algorithm, the cave is presented with an equation:

$$X_{\text{shade}} = \frac{X_G + X_L}{2} \tag{9.3}$$

where X_G stands for the optimal position by the iterations and X_L for the ideal position of the current population. While there are no other crayfish to compete and that would be rand below 0.5, crayfish will join the cave through the equation:

$$X_{i,j}^{t+1} = X_{i,j}^{t} + C_2 \times rand \times \left(X_{\text{shade}} - X_{i,j}^{t}\right) \tag{9.4}$$

where t acts as the current iteration number, $t+1$ represents the iteration number of the next generation, and C_2 is represented with the equation below:

$$C_2 = 2 - \frac{t}{T} \tag{9.5}$$

where T represents maximum iterations.

Presuming there are rival crayfish which in this case means rand is above 0.5, the exploitation stage starts and crayfish compete through two equations:

$$X_{i,j}^{t+1} = X_{i,j}^{t} = X_{i,j}^{t} - X_{z,j}^{t} + X_{\text{shade}} \tag{9.6}$$

where z stands for random crayfish, presented with the equation:

$$z = \text{round}(rand \times (N - 1)) + 1 \tag{9.7}$$

When there are ideal temperature conditions for crayfish (< 30 degrees), therefore feeding phase initializes. Food location is represented as $X_{food}=X_G$. The equation for the foraging phase:

$$X_{(<30),t+1_{I,j}} = X_{t_{I,j}} + X_{food} \times p \times (\cos(2 \times \pi \times rand) - \sin(2 \times \pi \times rand)) \tag{9.8}$$

If the food is too big for crayfish, it will chop it and make it more suitable for eating. The criterion that is used to determine if food is large or not is $Q>(C_3 + 1)/2$; if the criterion is met, then crayfish chop food, and that is represented through the equation:

$$X_{food} = \exp\left(, \frac{1}{Q}\right) \times X_{food} \tag{9.9}$$

where Q represents the size of the food with the equation:

$$Q = C_3 \times rand \times \frac{\text{fitness}_i}{\text{fitness}_{food}} \tag{9.10}$$

where C_3 represents the largest food, and the value is constant 3, fitness_i represents the fitness value of the ith crayfish, and fitness food represents the fitness value of the food location.

On the contrary, if food is not that large, and we determine that through the formula $Q \le \frac{C3+1}{2}$, crayfish will eat directly, and that is represented through the equation:

$$X_{i,j}^{t+1} = \left(X_{i,j}^{t} - X_{food}\right) \times p + p \times rand \times X_{i,j}^{t} \tag{9.11}$$

Proposed improved crayfish optimization algorithm

The COA algorithm exhibited very promising results on the benchmark functions; however, as it is one of the latest metaheuristic algorithms, its potential and possible usages are not fully examined yet. Moreover, there is still potential room for modifications and improvements to obtain an even better level of performance. For instance, the exploration of the elementary COA may be enhanced, as well as the converging speed of the algorithm. Keeping all this in mind, several modifications of the basic algorithm were proposed in this research.

The proposed method utilizes the standard initialization procedure defined by the following equation:

$$X_{i,j} = lb_j + \psi \cdot \left(ub_j - lb_j\right) \tag{9.12}$$

where $X_{i,j}$ depicts the ith solution's j-th parameter, with lb_j and ub_j denoting the lower and upper search limits of component j. Pseudo-random value ψ is taken from the normal distribution in the range [0,1]. Nevertheless, as shown by [95], the incorporation of the quasireflection-based learning (QRL) procedure into Eq. (9.12) increases the coverage of the search realm. For each particular component j (X_j), a quasireflexive-opposite element (X_j^{qr}) is generated in this manner.

$$X_j^{qr} = rnd\left(, \frac{lb_j - ub_j}{2}, x_j \right) \tag{9.13}$$

where the *rnd* denotes a random number in the enclosed section $\left[, \frac{lb_j - ub_j}{2}, x_j \right]$. The proposed initialization method commences with the initialization of $NP/2$ solutions by the QRL procedure, which is not increasing the complexity of the algorithm with respect to the *FFEs*. *FFEs* (fitness function evaluations) are commonly used to denote and assess the complexity of the metaheuristic algorithms, as the most resource-intensive operation in the entire algorithm is the evaluation of the fitness function. Algorithm 9.1 showcases the suggested initialization method.

ALGORITHM 9.1

Initialization procedure according to the QRL.

Stage 1: Produce population P^{init} of $NP/2$ solutions by traditional initialization shown by Eq. (9.12).

Stage 2: Produce QRL population P^{qr} based on the P^{init} through employing Eq. (9.13).

Stage 3: Create overall initial population P by combining both P^{init} and P^{qr} ($P \cup P^{qr}$).

Stage 4: Calculate the fitness of every solution belonging to the P.

Stage 5: Sort all individuals in P according to the fitness.

In order to maintain the population diversity and consequently improve the exploration capabilities of the algorithm, in the first *round* ($T/2$) iterations, the worst solution (that has the worst fitness) in each iteration is simply replaced by its QRL opposite solution without any additional evaluation (not increasing the complexity of the algorithm in terms of *FFEs*). This approach assumes that if the solution has poor fitness, there is a considerable probability that its opposite may fall in more promising parts of the search realm.

Another simple modification that aims to improve the exploitation of the COA algorithm is inspired by the genetic algorithm (GA) [65] and the uniform crossover operator. During the second half of the algorithm execution (last $T/2$ iterations), where it is assumed that the search has converged to the promising part of the search domain, the two worst solutions are replaced in each iteration in the following way. The two best solutions (with the best fitness value) are blended in a uniform way, where the degree of the crossover is guided by the parameter named *pc*. This parameter defines the crossover probability for each gene (parameter), and the optimal value has been empirically established as pc=0.1. This procedure will result in producing two novel solutions, which will simply replace the two weakest solutions, without additional evaluation (again, not increasing the complexity of the modified algorithm in terms of *FFEs*). These two novel solutions will be evaluated in the next round. This procedure helps in fine-tuning the search for the best solutions so far.

The novel algorithm was simply named modified COA–MCOA and the pseudocode is depicted in Algorithm 9.2. It is worth mentioning again that the modifications are not introducing additional evaluations of the solutions (*FFEs*); therefore, the complexity of the MCOA is equal to the complexity of the elementary variant of the algorithm.

ALGORITHM 9.2

Pseudocode of the suggested MCOA.

Initialize population of solutions P according to the QRL procedure shown in Algorithm 9.1

while t is less than T **do**

Rank solutions in P by calculating their fitness

for Each solution X in P do

Applying the COA search to perform optimization

if $t < T/2$ then

Generate a new solution as QRL opposite of the worst solution

Replace the worst solution in P with this novel solution

else

Produce two novel solutions by applying a crossover operator over the two best solutions in the population

Replace two worst solutions by these novel solutions **end if**

end for

end while

return The best-determined individual in P

Simulation setup

Dataset

The dataset that was employed in the simulations is a publicly available waste classification dataset, accessible on https://www.kaggle.com/datasets/techsash/waste-classification-data. This dataset is comprised of 22,500 photos of organic and recyclable items. It was created to help with waste management and to aid in solving several problems that occur when waste ends up filling landfills. These problems include eutrophication, poisoning of the animals that consume toxic waste indirectly, and finally, land, water, and air pollution.

As mentioned before, photos are separated into two classes: organic and recycled. The dataset has been divided into two subsets—train data (85%, consisting of 22,564 photos) and test data (the remaining 15%, consisting of 2513 photos). Dataset balance between recyclable and organic material alongside sample input images can be seen in Fig. 9.1.

Experimental configuration

The goal of the experiments was to develop lightweight CNNs (as simple as possible), with a maximum of two CNN layers with the appropriate max pooling layers and a maximum of two dense layers. The proposed MCOA was utilized to determine the optimal

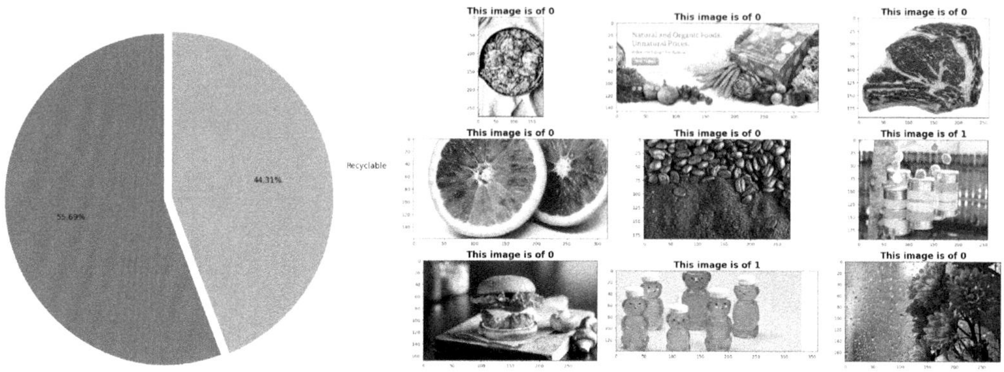

FIGURE 9.1 Dataset class distribution and sample inputs.

values of CNN parameters for this particular task, as the choice of parameters has a significant impact on the model's performance level.

The parameters that are being optimized in the following simulations were chosen based on the significance of their impact on the model's behavior. These parameters, their respective search limits, and types are as follows:

- learning rate, limits [0.0001,0.003], continuous value;
- dropout rate, limits [0.05,0.2], continuous value;
- number of epochs, limits [2,10], integer value;
- count of convolutional layers, limits [1,2], integer value
- count of dense layers, limits [1,2], integer value
- number of neurons per dense layer - number of filters per convolutional layer, limit [32,128], integer value.

The performance of the introduced MCOA algorithm was compared to several cutting-edge metaheuristic algorithms that were utilized in the same setup to optimize the identical set of CNN parameters. The competitor algorithms included the elementary variant of COA, genetic algorithm (GA) [65], particle swarm optimization (PSO) [96], artificial bee colony (ABC) [66], Harris Hawk optimization (HHO) algorithm [97], and COLSHADE algorithm [70]. These algorithms were developed independently by the authors of this research, with the control parameter values as recommended by their authors. Each metaheuristic algorithm was executed with 5 solutions in the population (N=5) and five iterations per run, over the course of 30 independent runs. A small number of solutions and iterations were used as the conducted optimization experiments were extremely resource-demanding. Images were also resized to 64×64 for the same reason.

The framework where the simulations were executed was implemented in the Python programming language, making use of the adequate machine learning and deep learning library collection, which includes Keras, sci-kit-learn, and seaborn for visualizing the experimental outcomes. The experiments were conducted on the following hardware: Intel i9 13900K processor and GPU Nvidia 1080TI with 12GB memory. While reporting the results in Section "Experimental outcomes", the following naming convention was used:

CNN tuned by MCOA metaheuristics was named CNN-MCOA, CNN tuned by GA algorithm was labeled CNN-GA, and so on.

The dataset was preprocessed as follows. Keraas ImageDataGenerator was used, and each pixel was fitted in the range [0,1]. Images were resized to 64 × 64 to mitigate the extremely resource-demanding nature of the simulations. The batch size for the generator was set to 256. Since the images are in color, three channels were used; therefore, CNN input size is 256×64×64×3 (batch×height×width×channels).

Metrics

The algorithms were evaluated by a traditional collection of metrics. To analyze the classification abilities of the model, accuracy is given by Eq. (9.14), precision is given by Eq. (9.15), recall is calculated as Eq. (9.16), and *f*1-score is defined by Eq. (9.17).

$$\text{Accuracy} = \frac{TP + TN}{TP + TN + FP + FN} \tag{9.14}$$

$$\text{Precission} = \frac{TP}{FP + TP} \tag{9.15}$$

$$\text{Recall} = \frac{TP}{FN + TP} \tag{9.16}$$

$$F1\_\text{score} = 2 \times \frac{\text{Precision} \times \text{Recall}}{\text{Precission} + \text{Recall}} \tag{9.17}$$

where TP and TN indicate true positives and negatives, while FP and FN mark false positives and negatives. A supplementary metric used in this research is Cohen's Kappa defined by Eq. (9.18) as it is able to deal with imbalanced data. Cohen's Kappa is a common choice where interrater reliability is important, as it can evaluate the consistency between several observers.

$$K = \frac{P_{\text{obs}} - P_{\text{exp}}}{1 - P_{\text{exp}}} \tag{9.18}$$

where the observed agreement values (TP+TN) are denoted as P_{obs}, while the expected agreement values $[(TP + FN) \times (TP + FP) + (FP + TN) \times (FN + TN)]$ are denoted as P_{exp}. Cohen's coefficient K is within the limits [0,1], where a value closer to 0 indicates less agreement.

In the conducted experiments, the classification error rate (1−accuracy) was set as the objective function that is required to be minimized, while Cohen's kappa coefficient was utilized as the indicator.

Flowcharts of the proposed framework and the convolutional network practical application are provided in figure and figure respectively (Figs. 9.2 and 9.3).

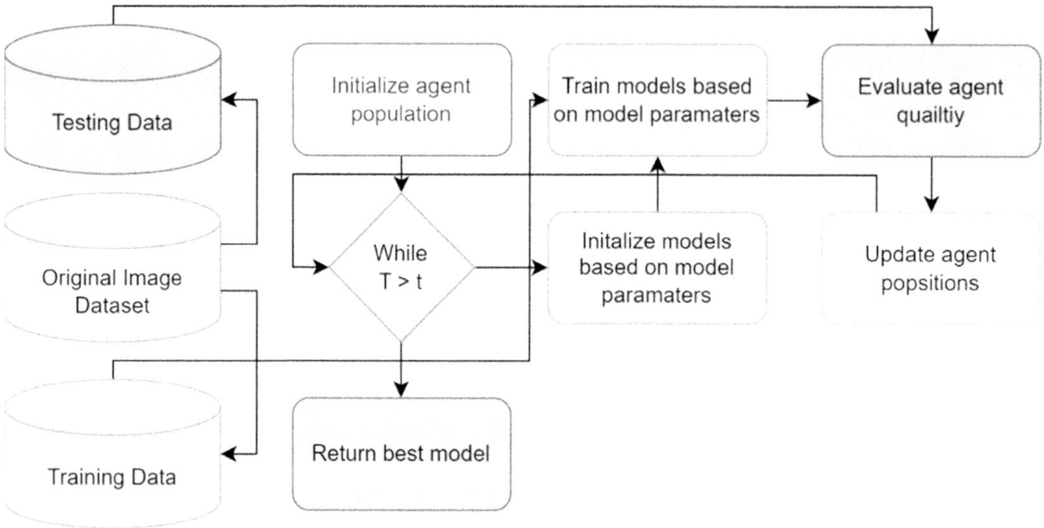

FIGURE 9.2 Optimization framework flowchart.

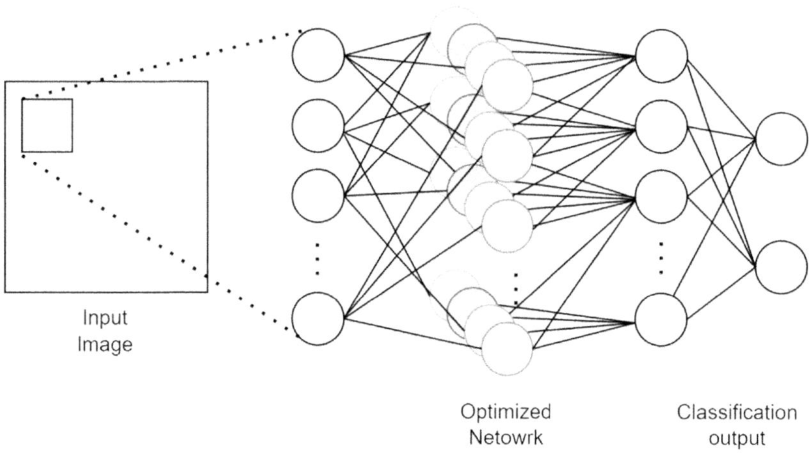

FIGURE 9.3 Convolutional network classification architecture.

Experimental outcomes

Comparisons in the best and worst mean and median executions for each of the evaluated optimizers in terms of objective and indicator functions are provided in Tables 9.1 and 9.2, respectively, where the highest rating in each category is highlighted in bold letters. The introduced optimizer attained the best outcomes for the best-constructed optimizer in both cases, attaining an error rating of 0.120175 and a Cohen's kappa coefficient of 0.752806, while the

TABLE 9.1 Objective function comparisons between evaluated optimizes.

Method	Best	Worst	Mean	Median	Std	Var
CNN-MCOA	**0.120175**	0.130123	**0.124818**	0.124154	0.004088	1.67E-05
CNN-COA	0.125746	0.137684	0.132644	0.134501	0.005047	2.55E-05
CNN-GA	0.123756	0.134103	0.129991	0.132113	0.004482	2.01E-05
CNN-PSO	0.122165	0.140072	0.129327	0.125746	0.007737	5.99E-05
CNN-ABC	0.121767	0.129725	0.125083	**0.123756**	0.003382	1.14E-05
CNN-HHO	0.126940	0.133705	0.130787	0.131715	0.002839	8.06E-06
CNN-COLSHADE	0.122563	**0.126940**	0.125348	0.126542	**0.001976**	**3.91E-06**

TABLE 9.2 Indicator function comparisons between evaluated optimizes.

Method	Best	Worst	Mean	Median	Std	Var
CNN-MCOA	**0.752806**	0.732318	0.743104	0.744187	0.008399	7.05E-05
CNN-COA	0.745468	0.716524	0.729372	0.726123	0.012037	1.45E-04
CNN-GA	0.746132	0.724235	0.732770	0.727943	0.009569	9.16E-05
CNN-PSO	0.750708	0.718452	0.737207	0.742462	0.013682	1.87E-04
CNN-ABC	0.751729	0.734461	**0.745329**	**0.749796**	0.007725	5.97E-05
CNN-HHO	0.739455	0.726623	0.731928	0.729705	0.005469	2.99E-05
CNN-COLSHADE	0.748982	**0.740187**	0.743736	0.742041	**0.003785**	**1.43E-05**

models that showcase the highest performance in the worst-case executions are constructed by the COLSHADE optimizer. This in turn resulted in the best results in terms of stability being attained by the COLSHADE algorithm. The best median results are demonstrated by the introduced algorithm in terms of objective function outcomes, while the ABC algorithm demonstrated the best outcomes in terms of indicator function for mean values. Finally, in terms of median outcomes, the ABC demonstrated the best outcomes across both metrics.

Distribution comparisons between the optimizers can be observed visually in Fig. 9.4. Distributions attest to the stability of the introduced modified version of the algorithm that showcases a definite improvement in comparison with the original algorithm. Furthermore, compared to other optimizers that might showcase higher stability, the introduced algorithm showcases the ability to locate more promising regions in the search space.

Further improvements are evident in terms of convergence shown in Fig. 9.5. The improvement of the modified version of the algorithm is evident as the introduced version converges faster toward a better outcome compared to the original algorithm. Additionally, as the introduced algorithm avoided local optima traps and converged toward a higher quality solution, the introduced modifications are deemed beneficial.

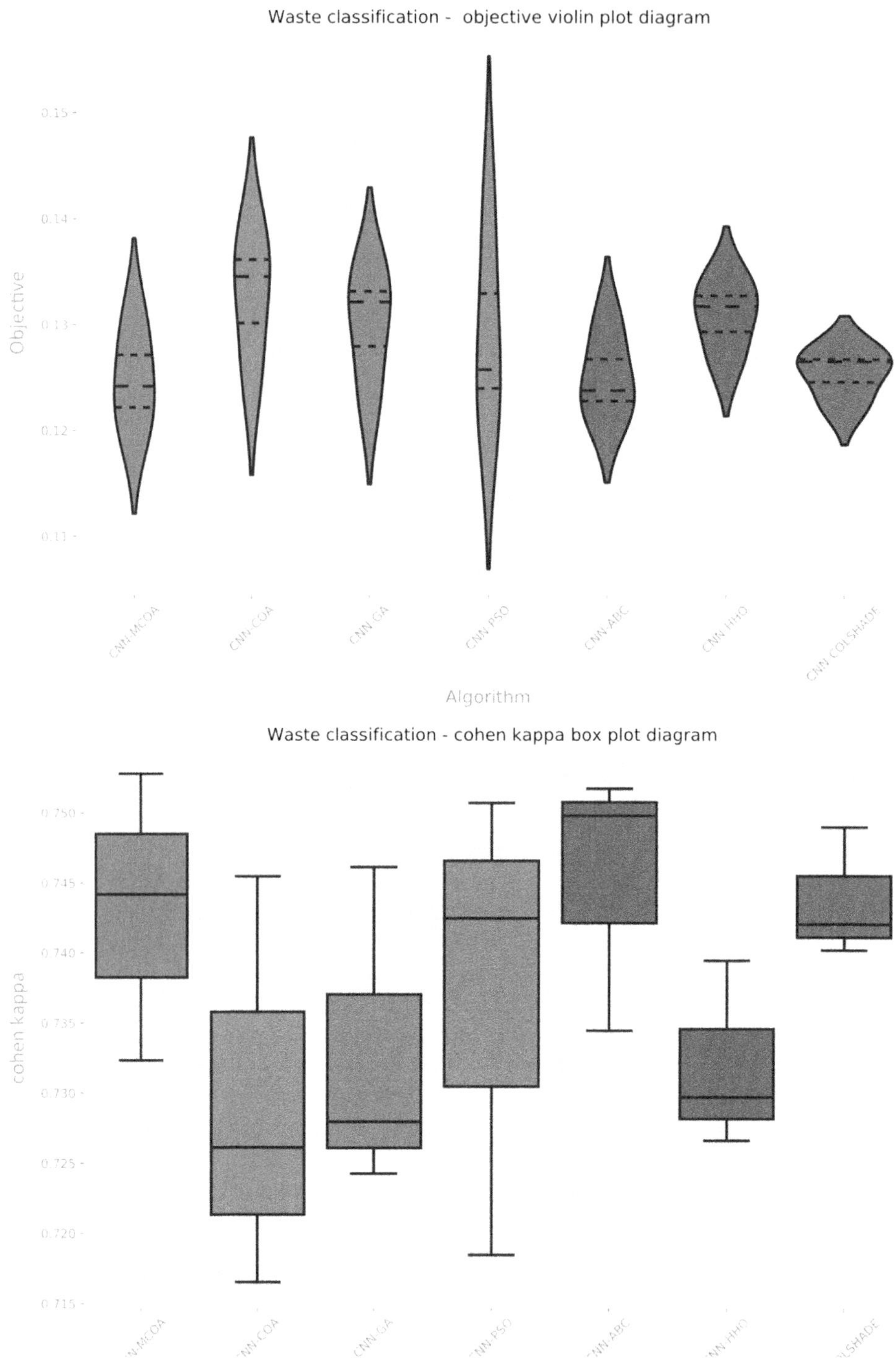

FIGURE 9.4 Distribution graphs for each optimizer in terms of objective and indicator functions.

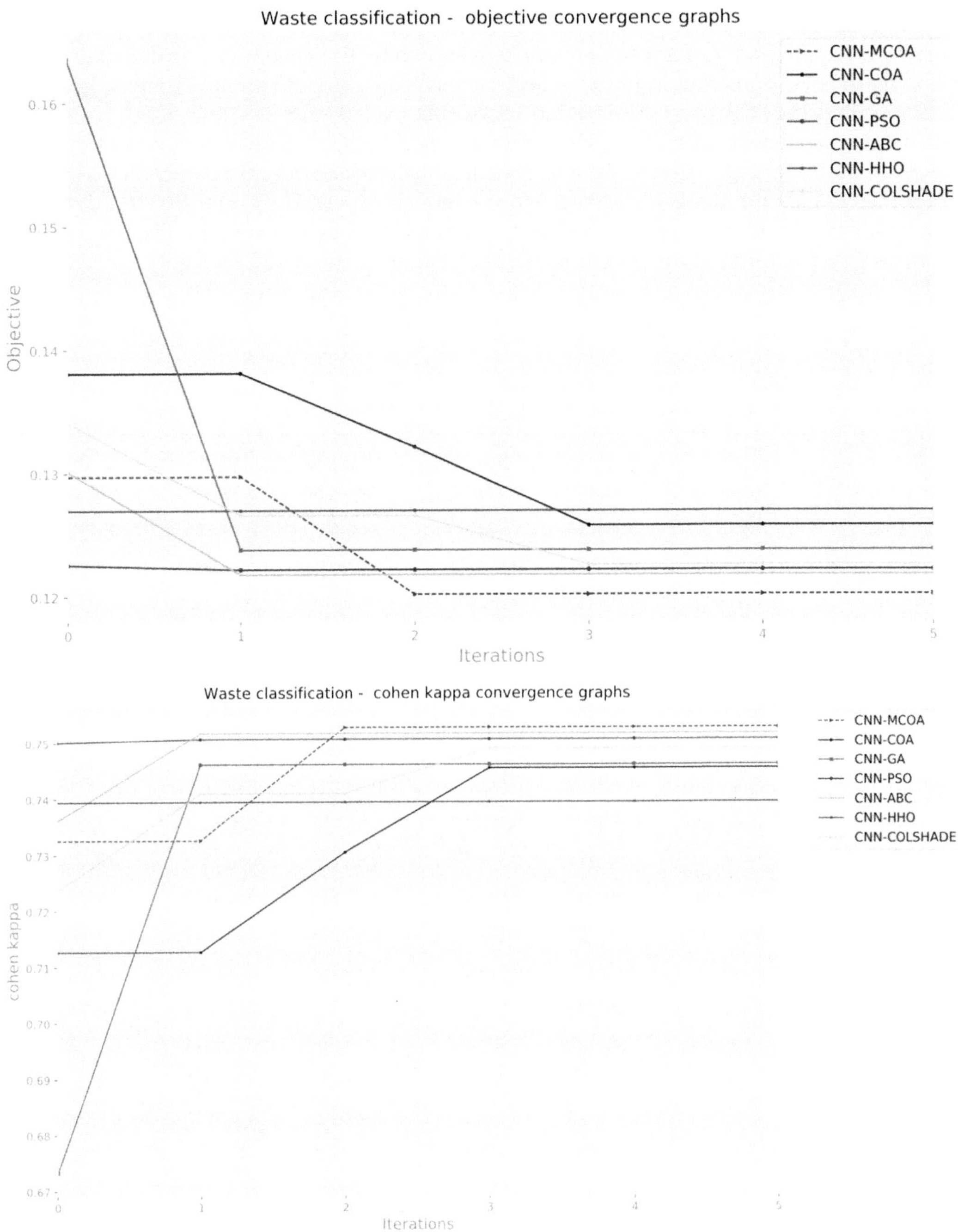

FIGURE 9.5 Convergence graphs for each optimizer in terms of objective and indicator functions.

Comparisons between the best-performing models generated by each optimizer are presented in Table 9.3. A definite dominance is shown by the introduced modified algorithm attaining the best outcomes in terms of accuracy as well as macro and micro averages across multiple metrics. However, the good performance of the original COA cannot be understated, as this version of the algorithm attained better outcomes in terms of inorganic detection and attained better recall and *f*1-scores in terms of organic detection.

Further details for the best-performing model are provided graphically through precision-recall (PR) and receiver operating characteristic (ROC) curves in Fig. 9.6, and the confusion matrix for said model is provided in Fig. 9.7.

The architecture of the best-constructed model can be seen in Fig. 9.8.

TABLE 9.3 Detailed metric comparison between the best performing models.

Method	Metric	Organic	Recyclable	Accuracy	Macro avg	Weighted avg
CNN-MCOA	precision	0.852470	**0.924528**	**0.879825**	**0.888499**	**0.884355**
	recall	**0.948608**	0.793165	**0.879825**	0.870887	**0.879825**
	f1-score	0.897973	0.853824	**0.879825**	**0.875898**	**0.878437**
CNN-COA	precision	**0.891132**	0.853464	0.874254	0.872298	0.874464
	recall	0.882227	**0.864209**	0.874254	**0.873218**	0.874254
	f1-score	0.886657	**0.858803**	0.874254	0.872730	0.874331
CNN-GA	precision	0.856209	0.907426	0.876244	0.881818	0.878873
	recall	0.935046	0.802158	0.876244	0.868602	0.876244
	f1-score	0.893893	0.851551	0.876244	0.872722	0.875157
CNN-PSO	precision	0.871099	0.887392	0.877835	0.879245	0.878309
	recall	0.916488	0.829137	0.877835	0.872812	0.877835
	f1-score	0.893217	0.857276	0.877835	0.875247	0.877313
CNN-ABC	precision	0.871102	0.887392	0.878233	0.881818	0.878873
	recall	0.916477	0.829134	0.878233	0.868602	0.876244
	f1-score	0.893218	0.857255	0.878233	0.872722	0.875157
CNN-HHO	precision	0.856102	0.907464	0.869213	0.879245	0.878309
	recall	0.935025	0.802166	0.869213	0.872812	0.877835
	f1-score	0.893888	0.851501	0.869213	0.875247	0.877313
CNN-COLSHADE	precision	0.861203	0.902000	0.877437	0.881601	0.879256
	recall	0.930050	0.811151	0.877437	0.870601	0.877437
	f1-score	0.894303	0.854167	0.877437	0.874235	0.876543
	support	1401	1112			

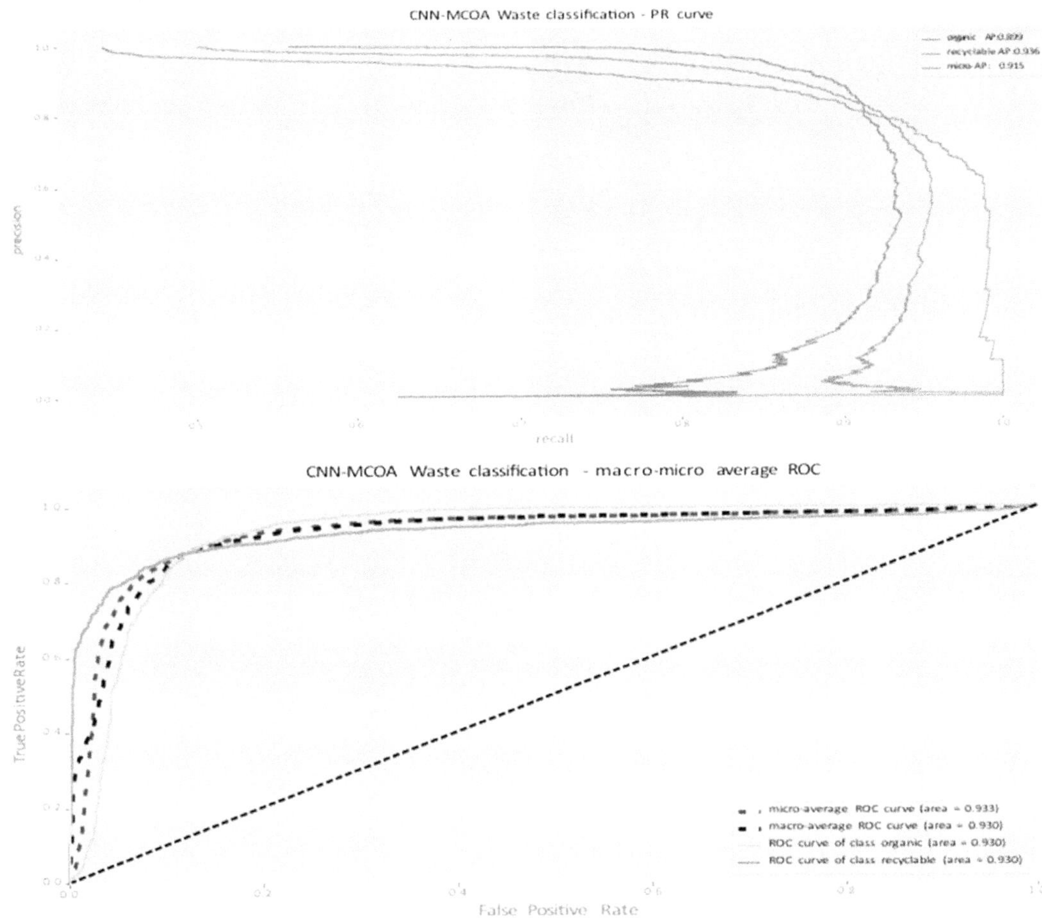

FIGURE 9.6 PR and ROC graphs for the best-performing model.

Finally, to facilitate experimental repeatability, the parameters selected for each algorithm are provided in Table 9.4.

Conclusion

Waste classification stands as a pivotal process essential for effectively managing and alleviating the environmental and health repercussions associated with waste. This process is integral in safeguarding the environment, as the mismanagement of toxic waste poses a significant threat to soil, water, and air, thereby causing profound damage to ecosystems and human well-being. The implementation of accurate waste classification techniques plays a vital role in devising appropriate disposal methods, thereby preventing environmental

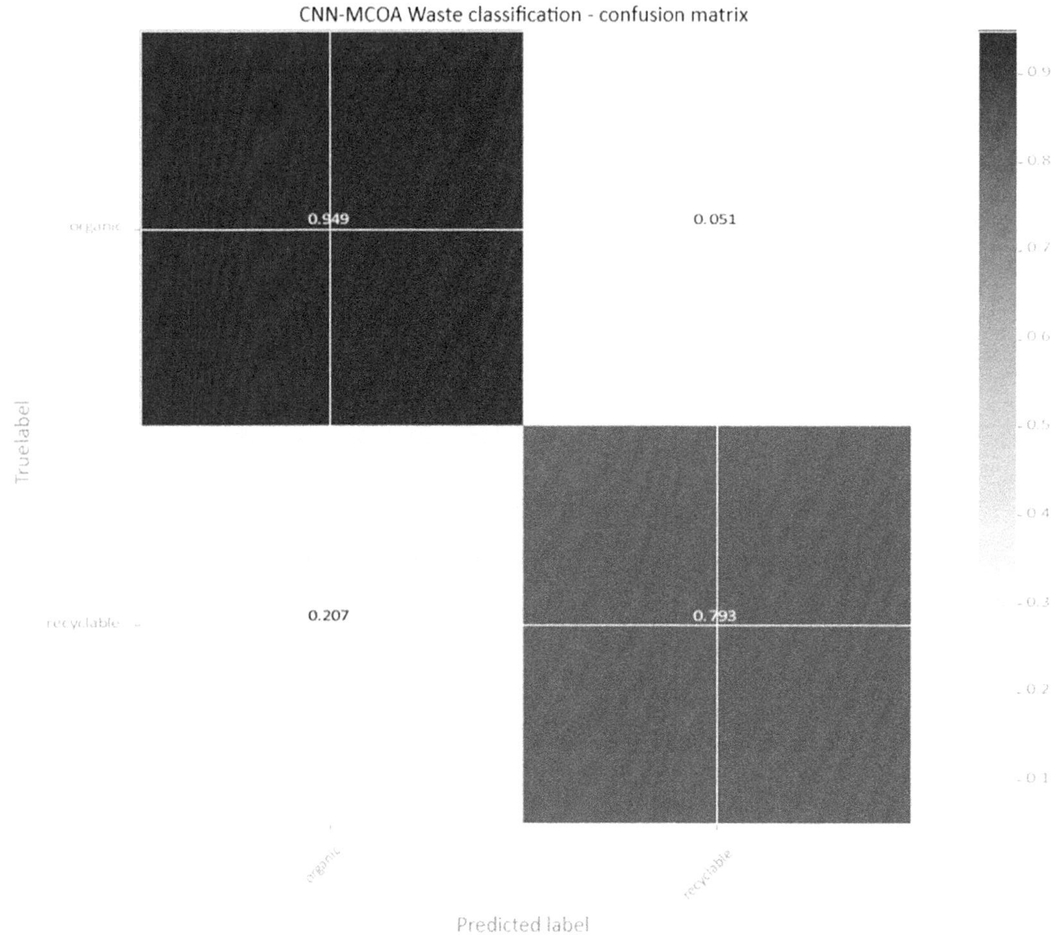

FIGURE 9.7 Best performing model confusion matrix.

degradation. This study delves into the exploration of AI's potential to enhance waste classification in real-world scenarios. Employing a computer vision approach, we base our classification on CNN. Recognizing the paramount importance of proper hyperparameter selection, we evaluate various contemporary optimizers to identify their effectiveness in optimizing models through an iterative selection process. Furthermore, we introduce a modified version of the COA tailored specifically for the requirements of this study. Upon evaluating the proposed approach on real-world data, the best-performing models are consistently those configured by the introduced classifier. These models exhibit an accuracy surpassing 87.98%, highlighting the significant potential for addressing this urgent and vital task in real-world waste management scenarios.

TABLE 9.4 Parameter selections made by each optimizer for the respective best-performing models.

Parameter	CNN-MCOA	CNN-COA	CNN-GA	CNN-PSO	CNN-ABC	CNN-HHO	CNN-COLSHADE
Learning rate	0.003000	0.003000	0.003000	0.000798	0.000930	0.000880	0.000611
Dropout	0.130158	0.200000	0.200000	0.050000	0.096698	0.063579	0.050000
Epochs	10	8	6	10	9	4	10
Layers CNN	2	1	2	2	2	1	2
Layers dense	2	1	2	2	2	1	2
CNN L1 neurons	106	45	46	32	64	95	111
CNN L2 neurons	128	76	60	49	38	108	40
Dense L1 neurons	128	128	86	52	68	99	104
Dense L2 neurons	32	67	78	128	117	107	52

Like any study, this work has certain limitations that should be acknowledged. One notable constraint is the significant computational costs associated with optimizing CNN. Our evaluation has focused on a limited set of optimizers, with only a fraction of popular optimizers being considered. Furthermore, the dataset employed in this study exhibits some imbalance, necessitating the exploration of objective metrics designed to handle imbalanced data, such as Cohen's kappa, in future works.

In terms of real-world uses, mounted cameras could theoretically be used in combination with well-performing classification models to help separate waste for better and easier recycling. Conveyor belts with blowers could be used to make the process of separation less tedious and more efficient.

Looking ahead, future research endeavors aim to delve deeper into refining the introduced methodology. Additionally, there is an aspiration to discover further applications for the modified algorithm introduced in this study. These avenues of exploration are envisioned to contribute to the ongoing development and enhancement of the proposed approach.

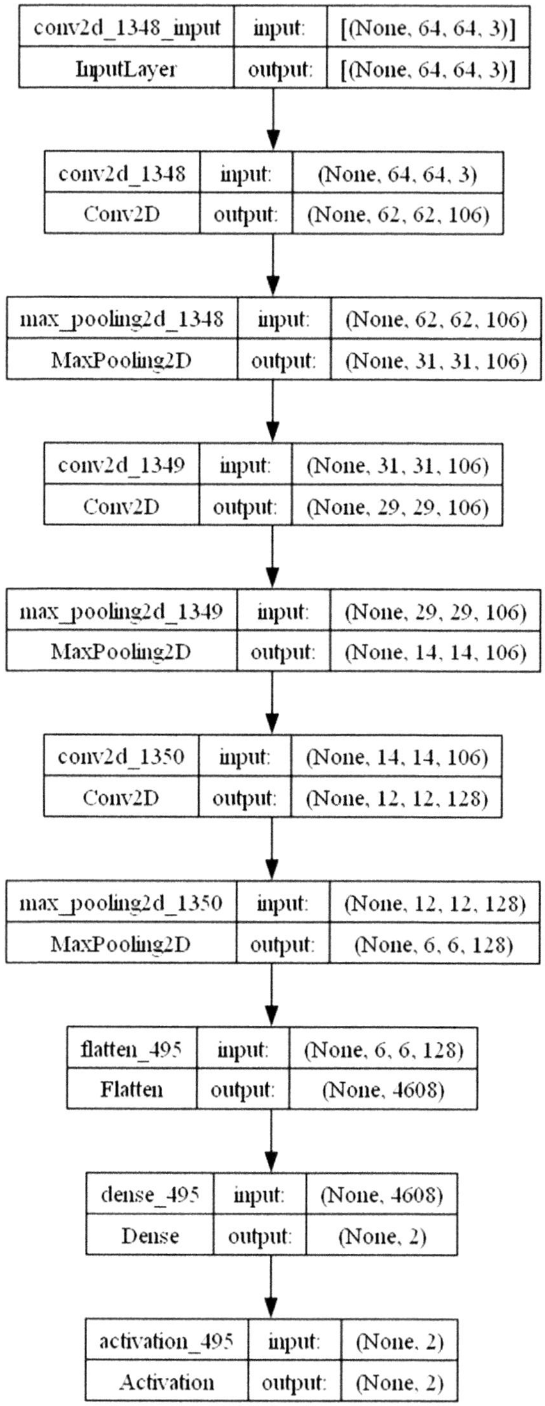

FIGURE 9.8 Best constructed model architecture.

Acknowledgment

This research was supported by the Science Fund of the Republic of Serbia, Grant No. 7502, Intelligent MultiAgent Control and Optimization applied to Green Buildings and Environmental Monitoring Drone Swarms—ECOSwarm. This research was supported by the Science Fund of the Republic of Serbia, Grant No. 7373, Characterizing crises-caused air pollution alternations using an artificial intelligence-based framework—crAIRsis.

References

[1] B. Koul, M. Yakoob, M.P. Shah, Agricultural waste management strategies for environmental sustainability, Environmental Research 206 (112285) (2022) 112285.

[2] R.G. Anuardo, M. Espuny, A.C.F. Costa, O.J. Oliveira, Toward a cleaner and more sustainable world: a framework to develop and improve waste management through organizations, governments and academia, Heliyon 8 (4) (2022) e09225.

[3] J.M. Chen, Carbon neutrality: toward a sustainable future, Innovation (Cambridge) 2 (3) (2021) 100127.

[4] Q. Song, J. Li, X. Zeng, Minimizing the increasing solid waste through zero waste strategy, Journal of Cleaner Production 104 (2015) 199–210.

[5] A.U. Zaman, A comprehensive review of the development of zero waste management: lessons learned and guidelines, Journal of Cleaner Production 91 (2015) 12–25.

[6] X. Zhao, M. Korey, K. Li, K. Copenhaver, H. Tekinalp, S. Celik, et al., Plastic waste upcycling toward a circular economy, Chemical Engineering Journal 428 (2022) 131928.

[7] A.U. Zaman, S. Lehmann, Challenges and opportunities in transforming a city into a "zero waste city", Challenges 2 (4) (2011) 73–93. Available from https://doi.org/10.3390/challe2040073, https://www.mdpi.com/2078-1547/2/4/73.

[8] A.K. Awasthi, V.R.S. Cheela, I. D'Adamo, E. Iacovidou, M.R. Islam, M. Johnson, et al., Zero waste approach towards a sustainable waste management, Resources, Environment and Sustainability 3 (100014) (2021) 100014.

[9] C. Wegener, Upcycling, Creativity – A New Vocabulary, Palgrave Macmillan UK, London, 2016, pp. 181–188.

[10] M. El-Fadel, A.N. Findikakis, J.O. Leckie, Environmental impacts of solid waste landfilling, Journal of Environmental Management 50 (1) (1997) 1–25.

[11] L. Rushton, Health hazards and waste management, British Medical Bulletin 68 (1) (2003) 183–197.

[12] P. Alam, K. Ahmade, Impact of solid waste on health and the environment, International Journal of Sustainable Development and Green Economics (IJSDGE) 2 (1) (2013) 165–168.

[13] V.K. Gaur, P. Sharma, R. Sirohi, M.K. Awasthi, C.-G. Dussap, A. Pandey, Assessing the impact of industrial waste on environment and mitigation strategies: a comprehensive review, Journal of Hazardous Materials 398 (2020) 123019.

[14] W.-j Wang, X.-y You, Benefits analysis of classification of municipal solid waste based on system dynamics, Journal of Cleaner Production 16279 (2021) 123686.

[15] J.K. Seadon, Sustainable waste management systems, Journal of Cleaner Production 18 (16–17) (2010) 1639–1651.

[16] J. Reno, Waste and waste management, Annual Review of Anthropology 44 (2015) 557–572.

[17] Z. Usmani, M. Sharma, A.K. Awasthi, N. Sivakumar, T. Lukk, L. Pecoraro, et al., Bioprocessing of waste biomass for sustainable product development and minimizing environmental impact, Bioresource Technology 322 (2021) 124548.

[18] X. Wen, Q. Luo, H. Hu, N. Wang, Y. Chen, J. Jin, et al., Comparison research on waste classification between china and the EU, Japan, and the USA, Journal of Material Cycles and Waste Management 16 (2014) 321–334.

[19] A. Liu, F. Ren, W.Y. Lin, J.-Y. Wang, A review of municipal solid waste environmental standards with a focus on incinerator residues, International Journal of Sustainable Built Environment 4 (2) (2015) 165–188.

[20] E.S. Windfeld, M.S.-L. Brooks, Medical waste management–a review, Journal of Environmental Management 163 (2015) 98–108.

[21] H.B. Sharma, K.R. Vanapalli, V.S. Cheela, V.P. Ranjan, A.K. Jaglan, B. Dubey, et al., Challenges, opportunities, and innovations for effective solid waste management during and post covid-19 pandemic, Resources, Conservation and Recycling 162 (2020) 105052.

[22] K. Stoeva, S. Alriksson, Influence of recycling programmes on waste separation behaviour, Waste Management (New York, N.Y.) 68 (2017) 732–741.
[23] R. Bai, M. Sutanto, The practice and challenges of solid waste management in singapore, Waste Management (New York, N.Y.) 22 (5) (2002) 557–567.
[24] F.A. Azis, H. Suhaimi, E. Abas, Waste classification using convolutional neural network, Proceedings of the 2020 2nd International Conference on Information Technology and Computer Communications, ACM, New York, NY, USA, 2020.
[25] H. Wilts, B.R. Garcia, R.G. Garlito, L.S. Gomez, E.G. Prieto, Artificial intelligence in the sorting of municipal waste as an enabler of the circular economy, Resources 10 (4) (2021) 28.
[26] S. Nanda, F. Berruti, Municipal solid waste management and landfilling technologies: a review, Environmental Chemistry Letters 19 (2021) 1433–1456.
[27] W. Ma, M. de Jong, F. Zisopoulos, T. Hoppe, Introducing a classification framework to urban waste policy: analysis of sixteen zero-waste cities in china, Waste Management 165 (2023) 94–107.
[28] X. Li, Y. Shi, Computer vision imaging based on artificial intelligence, 2018 International Conference on Virtual Reality and Intelligent Systems (ICVRIS), IEEE, Hunan, 2018.
[29] X. Cai, F. Shuang, X. Sun, Y. Duan, G. Cheng, Towards lightweight neural networks for garbage object detection, Sensors (Basel) 22 (19) (2022) 7455.
[30] D.H. Wolpert, W.G. Macready, No free lunch theorems for optimization, IEEE Transactions on Evolutionary Computation 1 (1) (1997) 67–82.
[31] M. Feurer, F. Hutter, Hyperparameter optimization, Automated Machine Learning: Methods, Systems, Challenges (2019) 3–33.
[32] L. Yang, A. Shami, On hyperparameter optimization of machine learning algorithms: theory and practice, Neurocomputing 415 (2020) 295–316.
[33] J. Bergstra, R. Bardenet, Y. Bengio, B. Kegl, Algorithms for hyper-parameter optimization, Advances in Neural Information Processing Systems 24 (2011) 2546–2554.
[34] B. Akay, D. Karaboga, R. Akay, A comprehensive survey on optimizing deep learning models by metaheuristics, Artificial Intelligence Review (2022) 1–66.
[35] S. Nematzadeh, F. Kiani, M. Torkamanian-Afshar, N. Aydin, Tuning hyperparameters of machine learning algorithms and deep neural networks using metaheuristics: a bioinformatics study on biomedical and biological cases, Computational Biology and Chemistry 97 (2022) 107619.
[36] N. Bacanin, C. Stoean, M. Zivkovic, M. Rakic, R. Strulak-Wojcikiewicz, R. Stoean, On the benefits of using metaheuristics in the hyperparameter tuning of deep learning models for energy load forecasting, Energies 16 (3) (2023) 1434.
[37] M. Mousapour Mamoudan, A. Ostadi, N. Pourkhodabakhsh, A.M. Fathollahi-Fard, F. Soleimani, Hybrid neural network-based metaheuristics for prediction of financial markets: a case study on global gold market, Journal of Computational Design and Engineering 10 (3) (2023) 1110–1125.
[38] I. Strumberger, E. Tuba, N. Bacanin, M. Zivkovic, M. Beko, M. Tuba, Designing convolutional neural network architecture by the firefly algorithm, 2019 International Young Engineers Forum (YEF-ECE), IEEE, Costa da Caparica, 2019, pp. 59–65.
[39] A. Gaspar, D. Oliva, E. Cuevas, D. Zald´ıvar, M. Perez, G. Pajares, Hyperparameter optimization in a convolutional neural network using metaheuristic algorithms, Metaheuristics in Machine Learning: Theory and Applications, Springer, Cham, 2021, pp. 37–59.
[40] O.N. Oyelade, A.E. Ezugwu, Characterization of abnormalities in breast cancer images using nature-inspired metaheuristic optimized convolutional neural networks model, Concurrency and Computation: Practice and Experience 34 (4) (2022) e6629.
[41] N. Bacanin, M. Zivkovic, M. Salb, I. Strumberger, A. Chhabra, Convolutional neural networks hyperparameters optimization using sine cosine algorithm, Sentimental Analysis and Deep Learning: Proceedings of ICSADL 2021, Songkhla, Springer, 2021, pp. 863–878.
[42] H. Jia, H. Rao, C. Wen, S. Mirjalili, Crayfish optimization algorithm, Artificial Intelligence Review 56 (Suppl 2) (2023) 1919–1979.
[43] M. Abdallah, M.A. Talib, S. Feroz, Q. Nasir, H. Abdalla, B. Mahfood, Artificial intelligence applications in solid waste management: a systematic research review, Waste Management (New York, N.Y.) 109 (2020) 231–246.

[44] B. Fang, J. Yu, Z. Chen, A.I. Osman, M. Farghali, I. Ihara, et al., Artificial intelligence for waste management in smart cities: a review, Environmental Chemistry Letters (2023) 1–31.

[45] R. Naveenkumar, J. Iyyappan, R. Pravin, S. Kadry, J. Han, R. Sindhu, et al., A strategic review on sustainable approaches in municipal solid waste management and energy recovery: role of artificial intelligence, economic stability and life cycle assessment, Bioresource Technology (2023) 129044.

[46] M.K. Younes, Z. Nopiah, N.A. Basri, H. Basri, M.F. Abushammala, M. KNA, Solid waste forecasting using modified anfis modeling, Journal of the Air & Waste Management Association 65 (10) (2015) 1229–1238.

[47] O. Adeleke, S.A. Akinlabi, T.-C. Jen, I. Dunmade, Prediction of municipal solid waste generation: an investigation of the effect of clustering techniques and parameters on anfis model performance, Environmental Technology 43 (11) (2022) 1634–1647.

[48] N. Sharma, R. Litoriya, A. Sharma, Application and analysis of k-means algorithms on a decision support framework for municipal solid waste management, Advanced Machine Learning Technologies and Applications: Proceedings of AMLTA 2020, Jaipur, Springer, 2021, pp. 267–276.

[49] X. Du, D. Niu, Y. Chen, X. Wang, Z. Bi, City classification for municipal solid waste prediction in mainland china based on k-means clustering, Waste Management 144 (2022) 445–453.

[50] G.-W. Cha, H.J. Moon, Y.-M. Kim, W.-H. Hong, J.-H. Hwang, W.-J. Park, et al., Development of a prediction model for demolition waste generation using a random forest algorithm based on small datasets, International Journal of Environmental Research and Public Health 17 (19) (2020) 6997.

[51] G.-W. Cha, H.-J. Moon, Y.-C. Kim, Comparison of random forest and gradient boosting machine models for predicting demolition waste based on small datasets and categorical variables, International Journal of Environmental Research and Public Health 18 (16) (2021) 8530.

[52] L. Jovanovic, G. Jovanovic, M. Perisic, F. Alimpic, S. Stanisic, N. Bacanin, et al., The explainable potential of coupling metaheuristics-optimized-xgboost and shap in revealing vocs' environmental fate, Atmosphere 14 (1) (2023). Available from https://doi.org/10.3390/atmos14010109, https://www.mdpi.com/2073-4433/14/1/109.

[53] A. Yonar, H. Yonar, Modeling air pollution by integrating ANFIS and metaheuristic algorithms, Modeling Earth Systems and Environment 9 (2) (2023) 1621–1631.

[54] N. Nnamoko, J. Barrowclough, J. Procter, Solid waste image classification using deep convolutional neural network, Infrastructures 7 (4). https://doi.org/10.3390/infrastructures7040047. https://www.mdpi.com/2412-3811/7/4/47.

[55] A. Masand, S. Chauhan, M. Jangid, R. Kumar, S. Roy, Scrapnet: an efficient approach to trash classification, IEEE Access 9 (2021) 130947–130958. Available from https://doi.org/10.1109/ACCESS.2021.3111230.

[56] T. Gupta, R. Joshi, D. Mukhopadhyay, K. Sachdeva, N. Jain, D. Virmani, et al., A deep learning approach based hardware solution to categorise garbage in environment, Complex & Intelligent Systems 8 (2022) 1129–1152.

[57] N. Li, Y. Chen, Municipal solid waste classification and real-time detection using deep learning methods, Urban Climate 49 (2023) 101462.

[58] H. Zhang, H. Cao, Y. Zhou, C. Gu, D. Li, Hybrid deep learning model for accurate classification of solid waste in the society, Urban Climate 49 (2023) 101485.

[59] R. Tripathi, H. Shetty, K. Patil, P. Ingawale, M. Trivedi, Intelligent waste material classification using efficientnet-b3 convolutional neural network for enhanced waste management, International Conference on Self Sustainable Artificial Intelligence Systems (ICSSAS) 2023 IEEE, Erode, 2023, pp. 132–137.

[60] S. Albawi, T.A. Mohammed, S. Al-Zawi, Understanding of a convolutional neural network, International Conference on Engineering and Technology (ICET) 2017 IEEE, Antalyia, 2017, pp. 1–6.

[61] J. Gu, Z. Wang, J. Kuen, L. Ma, A. Shahroudy, B. Shuai, et al., Recent advances in convolutional neural networks, Pattern Recognition 77 (2018) 354–377.

[62] W.H.L. Pinaya, S. Vieira, R. Garcia-Dias, A. Mechelli, Convolutional neural networks, Machine Learning, Elsevier, Academic Press, Cambridge, MA, 2020, pp. 173–191.

[63] M. Krichen, Convolutional neural networks: a survey, Computers 12 (8) (2023) 151.

[64] D. Wang, D. Tan, L. Liu, Particle swarm optimization algorithm: an overview, Soft computing 22 (2018) 387–408.

[65] S. Mirjalili, S. Mirjalili, Genetic algorithm, Evolutionary Algorithms and Neural Networks: Theory and Applications (2019) 43–55.

[66] D. Karaboga, B. Basturk, On the performance of artificial bee colony (abc) algorithm, Applied Soft Computing 8 (1) (2008) 687–697.

[67] X.-S. Yang, Firefly algorithm, stochastic test functions and design optimisation, International Journal of Bio-Inspired Computation 2 (2) (2010) 78–84.

[68] L. Abualigah, M.A. Elaziz, P. Sumari, Z.W. Geem, A.H. Gandomi, Reptile search algorithm (rsa): a nature-inspired meta-heuristic optimizer, Expert Systems with Applications 191 (2022) 116158.

[69] S.K. Azad, O. Hasancebi, An elitist self-adaptive step-size search for structural design optimization, Applied Soft Computing 19 (2014) 226–235.

[70] J. Gurrola-Ramos, A. Hernandez-Aguirre, O. Dalmau-Cedeño, Colshade for real-world single-objective constrained optimization problems, 2020 IEEE congress on evolutionary computation (CEC), IEEE, Glasgow, 2020, pp. 1–8.

[71] L. Jovanovic, D. Jovanovic, N. Bacanin, A. Jovancai Stakic, M. Antonijevic, H. Magd, et al., Multi-step crude oil price prediction based on lstm approach tuned by salp swarm algorithm with disputation operator, Sustainability 14 (21) (2022) 14616.

[72] L. Jovanovic, N. Bacanin, V. Simic, J. Mani, M. Zivkovic, M. Sarac, Optimizing machine learning for space weather forecasting and event classification using modified metaheuristics, Soft Computing (2023) 1–20.

[73] N. Bacanin, L. Jovanovic, M. Zivkovic, V. Kandasamy, M. Antonijevic, M. Deveci, et al., Multivariate energy forecasting via metaheuristic tuned long-short term memory and gated recurrent unit neural networks, Information Sciences 642 (2023) 119122.

[74] R. Damaseviˇ cius, L. Jovanovic, A. Petrovic, M. Zivkovic, N. Bacanin, D. Jovanovic, et al., Decomposition aided attention-based recurrent neural networks for multistep ahead time-series forecasting of renewable power generation, PeerJ Computer Science 10,2,e1795. https://doi.org/10.7717/peerj-cs.1795.

[75] N. Bacanin, M. Zivkovic, L. Jovanovic, M. Ivanovic, T.A. Rashid, Training a multilayer perception for modeling stock price index predictions using modified whale optimization algorithm, Computational Vision and Bio-Inspired Computing: Proceedings of ICCVBIC 2021, Springer, Coimbatore, 2022, pp. 415–430.

[76] M. Todorovic, N. Stanisic, M. Zivkovic, N. Bacanin, V. Simic, E.B. Tirkolaee, Improving audit opinion prediction accuracy using metaheuristics-tuned xgboost algorithm with interpretable results through shap value analysis, Applied Soft Computing 149 (2023) 110955.

[77] L. Jovanovic, N. Bacanin, M. Zivkovic, M. Antonijevic, B. Jovanovic, M.B. Sretenovic, et al., Machine learning tuning by diversity oriented firefly metaheuristics for industry 4.0, Expert Systems (2023) e13293.

[78] M. Zivkovic, N. Bacanin, M. Antonijevic, B. Nikolic, G. Kvascev, M. Marjanovic, et al., Hybrid cnn and xgboost model tuned by modified arithmetic optimization algorithm for covid-19 early diagnostics from x-ray images, Electronics 11 (22) (2022) 3798.

[79] M. Salb, L. Jovanovic, N. Bacanin, M. Antonijevic, M. Zivkovic, N. Budimirovic, et al., Enhancing internet of things network security using hybrid cnn and xgboost model tuned via modified reptile search algorithm, Applied Sciences 13 (23) (2023) 12687.

[80] M. Dobrojevic, M. Zivkovic, A. Chhabra, N.S. Sani, N. Bacanin, M.M. Amin, Addressing internet of things security by enhanced sine cosine metaheuristics tuned hybrid machine learning model and results interpretation based on shap approach, PeerJ Computer Science 9 (2023) e1405.

[81] N. AlHosni, L. Jovanovic, M. Antonijevic, M. Bukumira, M. Zivkovic, I. Strumberger, et al., The xgboost model for network intrusion detection boosted by enhanced sine cosine algorithm, International Conference on Image Processing and Capsule Networks, Springer, Bangkok, 2022, pp. 213–228.

[82] M. Zivkovic, L. Jovanovic, M. Ivanovic, N. Bacanin, I. Strumberger, P.M. Joseph, Xgboost hyperparameters tuning by fitness-dependent optimizer for network intrusion detection, Communication and Intelligent Systems: Proceedings of ICCIS 2021, Springer, Delhi, 2022, pp. 947–962.

[83] N. Bacanin, M. Zivkovic, T. Bezdan, K. Venkatachalam, M. Abouhawwash, Modified firefly algorithm for workflow scheduling in cloud-edge environment, Neural Computing and Applications 34 (11) (2022) 9043–9068.

[84] B. Predic, L. Jovanovic, V. Simic, N. Bacanin, M. Zivkovic, P. Spalevic, et al., Cloud-load forecasting via´ decomposition-aided attention recurrent neural network tuned by modified particle swarm optimization, Complex & Intelligent Systems (2023) 1–21.

[85] M. Salb, L. Jovanovic, M. Zivkovic, E. Tuba, A. Elsadai, N. Bacanin, Training logistic regression model by enhanced moth flame optimizer for spam email classification, Computer Networks and Inventive Communication Technologies: Proceedings of Fifth ICCNCT 2022, Springer, Coimbatore, 2022, pp. 753–768.

[86] N. Bacanin, M. Zivkovic, C. Stoean, M. Antonijevic, S. Janicijevic, M. Sarac, et al., Application of natural language processing and machine learning boosted with swarm intelligence for spam email filtering, Mathematics 10 (22) (2022) 4173.

[87] M. Stankovic, L. Jovanovic, V. Marevic, A. Balghouni, M. Zivkovic, N. Bacanin, Predicting credit card churn using support vector machine tuned by modified reptile search algorithm, International Conference on Worldwide Computing and Its Applications, Springer, Udaipur, 2023, pp. 63–77.

[88] L. Jovanovic, M. Kljajic, V. Mizdrakovic, V. Marevic, M. Zivkovic, N. Bacanin, Predicting credit card churn: application of xgboost tuned by modified sine cosine algorithm, 2023 3rd International Conference on Smart Data Intelligence (ICSMDI), IEEE, Trichy, 2023, pp. 55–62.

[89] A. Minic, L. Jovanovic, N. Bacanin, C. Stoean, M. Zivkovic, P. Spalevic, et al., Applying recurrent neural networks for anomaly detection in electrocardiogram sensor data, Sensors 23 (24) (2023) 9878.

[90] D. Pilcevic, M.D. Jovicic, M. Antonijevic, N. Bacanin, L. Jovanovic, M. Zivkovic, et al., Performance evaluation of metaheuristics-tuned recurrent neural networks for electroencephalography anomaly detection, Frontiers in Physiology 14. 1267011.

[91] F.E.F. Junior, G.G. Yen, Particle swarm optimization of deep neural networks architectures for image classification, Swarm and Evolutionary Computation 49 (2019) 62–74.

[92] A. Darwish, D. Ezzat, A.E. Hassanien, An optimized model based on convolutional neural networks and orthogonal learning particle swarm optimization algorithm for plant diseases diagnosis, Swarm and Evolutionary Computation 52 (2020) 100616.

[93] N. Bacanin, M. Zivkovic, F. Al-Turjman, K. Venkatachalam, P. Trojovsky, I. Strumberger, et al., Hybridized sine cosine algorithm with' convolutional neural networks dropout regularization application, Scientific Reports 12 (1) (2022) 6302.

[94] W.-L. Mao, W.-C. Chen, C.-T. Wang, Y.-H. Lin, Recycling waste classification using optimized convolutional neural network, Resources, Conservation and Recycling 164 (105132) (2021) 105132.

[95] S. Rahnamayan, H.R. Tizhoosh, M.M.A. Salama, Quasi-oppositional differential evolution, 2007 IEEE Congress on Evolutionary Computation, 2007, IEEE, Singapore, 2229–2236. https://doi.org/10.1109/CEC.2007.4424748.

[96] J. Kennedy, R. Eberhart, Particle swarm optimization, Proceedings of ICNN'95-International Conference on Neural Networks Vol. 4 IEEE, Perth, WA, 1995, pp. 1942–1948.

[97] A.A. Heidari, S. Mirjalili, H. Faris, I. Aljarah, M. Mafarja, H. Chen, Harris hawks optimization: algorithm and applications, Future Generation Computer Systems 97 (2019) 849–872.

CHAPTER

10

Exploring the potential of lightweight computer vision YOLOv8 models for effective waste classification and management

Ana Tasic[1], *Luka Jovanovic*[1], *Nebojsa Bacanin*[1], *Miodrag Zivkovic*[1], *Stanislava Kozakijevic*[2], *Milos Antonijevic*[1], *and Miroslav Popovic*[1]

[1]Department of Informatics and Computing, Singidunum University, Belgrade, Serbia
[2]Department of Psychology, Faculty of Media and Communication, Singidunum University, Belgrade, Serbia

Introduction

Waste recycling has become the standard of human life in the 21st century. They contribute in many aspects to the progress of humanity and the preservation of nature, people destroys and pollutes the earth like no other animal on planet earth. Thus the production of plastics, metals and other raw materials pollutes the ozone layer, as well as the nature from which necessary raw materials are extracted. Waste recycling to some extent reduces the process that has a harmful effect on the environment, so its development is necessary for the sake of preserving the planet earth [1]. The reuse of materials is key in mitigating climate change, such as carbon neutrality, reduction of emission gases, and thus reduction of the greenhouse effect. As we know, waste recycling is one of the prerequisites for the survival of planet earth.

Waste recycling creates new jobs, but also helps in waste management, which nowadays reflects the progress of a society, in addition to the fact that it can also affect environmental pollution. In the wake of the current world inflation, it is increasingly difficult for underdeveloped countries to follow these trends, so this is another of the world's challenges. Where we encounter extremes such as ecologically clean Sweden and the river of garbage in Bangladesh. Certainly, the most difficult part of the job is to classify the waste, and

Harnessing Automation and Machine Learning for Resource Recovery and Value Creation
DOI: https://doi.org/10.1016/B978-0-443-27374-2.00010-8

sometimes it is completely impossible. Waste classification could be a vital part of securing the apparatus utilized within the reusing prepare. Metal pieces can harm plastic and elastic reusing apparatus [2,3]. Because of all this, it is necessary to work on sensitizing people, and highlighting the importance of waste recycling, but also the recycling process and waste classification. This is where new technologies can come to the rescue, to replace human carelessness and improve the waste recycling process. One of them is certainly computer vision, which can play a significant role in the waste classification process itself.

This paper analyzes the results of Computer Vision for the purposes of waste recycling. Computer vision is one of the AI-based data science tools that enables computers to infer important data from digital images, recordings, and other visual input—and to take action or make suggestions based on that data. AI enables computer vision to acquire, recognize and analyze all kinds of visual information. AI models, reasoning, and models can rapidly consume, assimilate, and learn from vast amounts of labeled and unlabeled visual information. It also contributes to the fact that people are not directly exposed to waste and harmful substances (risk to health) and in the long term reduces the costs of waste recycling, and also gives better results due to faster and more accurate processing [4]. It should be added that computer vision is cheap and effective. An alternative to human work, but it can also contribute to the improvement of the system, and leave time for people to redirect to some other tasks that can also contribute to the improvement of the quality of waste management.

The development of YOLO (You Only Look Once) algorithms is an important contribution to the field of computer vision. New versions appear from year to year and each one brings a greater degree of precision as well as ease of use. In this work, the latest version of YOLOV8 was used, which gave excellent results. Three YOLOV8 architectures of the nano, small and medium model were used. Where the medium architecture achieved the highest degree of accuracy of 99.9% on a dataset of 7155 images. It should be noted that the model was trained only on 15 epochs with a batch size of 16 and an image size of 512. This all goes in favor of the application and efficiency achieved by the YOLO algorithms. YOLO models are famous for their impressive speed and accuracy and their ability to detect objects in images quickly and reliably YOLOv8 models are cheap, easy to train, and can run on limited hardware. However, there are challenges associated with using YOLO models, such as different architectures, training, and appropriate dataset labeling.

Related works

AI classification algorithms are part of machine learning (ML), their primary role is to classify data, to tell which category something belongs to. They are used in different areas, from image detection, spam classification, diabetes detection, and many other fields. We will list the most famous ones when talking about ML: K-Nearest Neighbors, Naive Bayes, Decision Tree and Neural Networks. Each algorithm gives different results on certain topics, so their application differs from case to case.

What artificial intelligence (AI) provides is imitating the human brain in making decisions and easily traversing a large set of data. AI models have proven to be better than physical and statistical models, that's why today a large number of developed countries want to apply them. The very integration of computer vision and AI in various fields of science has been the subject of extensive research in the past. The contribution of numerous works in these areas gives hope for the further progress of science, but opens new chapters for scientific work. This section makes a parallel and presents a number of works where through the relevant literature you can see the application of AI algorithms both through ecology and the classifications themselves through computer vision. In addition, it should be mentioned that there is a gap in the literature itself when it comes to the application of the new YOLOV8 models on the topic of waste classification. The use of AI in ecology is quite widespread, it has its application in monitoring air pollution. The study of Wang and Kong developed air quality predictive modeling based on improved decision tree algorithms to enhance the prediction accuracy and time performance [5].

Classification is a category or division in a system that categorizes or organizes objects into groups or types. The choice of classification algorithm depending on the model is presented in Ref. [6]. Performance Analysis and Comparison of Machine and Deep Learning Algorithms for IoT Data Classification is analyzed in Ref. [7]. A set of prediction models with distinct ML algorithms in order to monitor students' academic progress can be seen in Ref. [8]. Several related AI approaches, such as ML, have been applied to research on landscape patterns and ecological processes [9]. AI Technologies also used for Forecasting Air Pollution and Human Health [10]. An Example of predicting Bark Beetle Outbreaks harnessing Deep Learning is given in Ref. [11].

Image Classification is a fundamental task in vision recognition that aims to understand and categorize an image as a whole under a specific label. Facial recognition is a technology that enables the matching of faces in digital images or video frames to a face database, which is now widely used for mobile phone face lock, smart door locking, etc. [11]. In Ref. [12], an open-source visual analysis tool, Uni-Evaluator, is developed to support a unified model evaluation for classification, object detection, and instance segmentation in computer vision. Deep neural networks (DNNs) play an important role in everyday lives, from aiding in menial tasks to solving world issues such as cancer cell detection. However, few pieces of research have been conducted using DNNs and deep learning models as a medium to help classify and detect trash, in efforts to solve global waste crisis.

To address this issue, this work focuses on YOLO-Green, a novel real-time object detection model designed specifically for trash detection [13]. Object detection of flying objects remains challenging due to large variance object spatial sizes/aspect ratios, rate of speed, occlusion, and clustered backgrounds. To address some of the presented challenges while simultaneously maximizing performance, it is utilized in the current state of the art single-shot detector, YOLOv8, in an attempt to find the best trade-off between inference speed and mAP [14].

Conventional waste separation relies heavily on manual separation of objects by humans, which is inefficient, expensive, time consuming, and prone to subjective errors caused by limited knowledge of waste classification. However, advances in AI research has led to the adoption of ML algorithms to improve the accuracy of waste classification from images. In Ref. [15], a waste classification dataset is used to evaluate the performance of a bespoke five-layer convolutional neural network (CNN) when trained with two different image resolutions.

Methods

Computer vision

Computer vision is the field of computer science that focuses on replicating parts of the complexity of the human vision system and enabling computers to identify and process objects in images and videos in the same way that humans do. It is a field of AI, it is based on data from digital records, images and other visual inputs. Computer vision has wide applications in industry, because a computer can perceive things faster than the pupil of the human eye. It is currently represented in many economic fields and its role will grow more and more over the years. Computer vision is a combination of image processing and pattern recognition [16,17]. Among the primary tools in computer vision are CNNs, which work according to the principle of grouping pixels in layers, thus creating a representation of data. CNNs translate pixels into lines, followed by their combination, thus forming characteristics such as eyes, where later combinations create complex objects such as a face. The primary purpose of Computer Vision is to create models and data extracts and information from images, while Image Processing is about implementing computational transformations for images, such as sharpening, contrast, among others [18]. Functionally, computer vision and human vision are the same [19], with the aim of interpreting spatial data, that is, data indexed by more than one dimension. One of the main elements that aim to contribute to the functionality of CNN itself are kernels, filters and pooling layers. Kernels, that is, small matrix filters, have the role of performing convolution over the input image, thus highlighting features such as edges and texture. Pooling layers, on the other hand, have the role of reducing the output data of the convolution, without leaving out important information. Max pooling is the most common form, its purpose is to present the maximum value within a region as the entire region. All this contributes to reducing the amount of data, so the CNN itself is resistant to minimal changes in the position of the features of the images themselves. The development of computer vision started in the 60 s of the last century, and it only experienced its flowering in the last decades, given the current flowering of this branch, its contribution and application will certainly be greater and greater.

Computer vision works by using an algorithm and optical sensors to stimulate human visualization to automatically extract valuable information from an object [20]. It also combined with lighting systems to facilitate image acquisition continued with image analysis. In more detail, the stages of image analysis are: (1) image formation, in which image of object is captured and stored in computer, (2) image preprocessing, whereby quality of image is improved to enhance the image detail, (3) image segmentation, in which the object image is identified and separated from the background, (4) image measurement, where several significant features are quantized, and (5) image interpretation, where the extracted images are then interpreted.

Research on image processing has been directed to join ML and computing processes that can recognize patterns of increasingly diverse objects. ML is closely related to computational statistics which consisted of spam filtering, optical character recognition, search engines and computer vision. There are extensive arrays of studies observing algorithms to reduce noise effect such as linear filtering of the Gaussian-based algorithm.

In the last five years Deep Nets have enabled enormous advances in computer vision and the study of biological visual systems. Evolutionary computation approaches have been widely used for image analysis with significant achievement [21]. In light of the combinatorial complexity of natural images, that the standard methods for benchmarking vision algorithms (Deep Nets and others) are inadequate and can yield misleading impressions about their strengths and so tougher and more challenging performance measures should be developed [22]. The problem of perpetual view generation—long range generation of novel views corresponding to an arbitrarily long camera trajectory given a single image is discussed in Ref. [23].

Modeling in computer vision has long been dominated by CNNs. A Transformer is a model architecture that eschews recurrence and instead relies entirely on an attention mechanism to draw global dependencies between input and output. On the other hand, the evolution of network architectures in natural language processing (NLP) has taken a different path, where the prevalent architecture today is instead the Transformer [24]. Ref. [25] explains how to expand the applicability of Transformer such that it can serve as a general-purpose backbone for computer vision, as it does for NLP and as CNNs do in vision.

In recent years, the computer vision community has made great strides towards highly realistic image generation. In particular, Generative Adversarial Networks (GANs) [26] emerged as a powerful class of generative models. In Ref. [27], it is introduced GIRAFFE, a novel method for generating scenes in a controllable and photorealistic manner while training from raw unstructured image collections.

Text-to-image generation has traditionally focused on finding better modeling assumptions for training on a fixed dataset. These assumptions might involve complex architectures, auxiliary losses, or side information such as object part labels or segmentation masks supplied during training. A simple approach for this task based on a transformer that autoregressively models the text and image tokens as a single stream of data can be seen in Ref. [28].

YOLO models

The detection of objects in real time is increasingly prevalent, from video surveillance, robotics, through autonomous vehicles. The application of the YOLO model is extremely important in areas such as autonomous vehicle systems, where they can detect various obstacles, such as cyclists and pedestrians, but also in sports analysis. In the field of agriculture, detection and classification of crops, pesticides and diseases. In the field of forensics, facial recognition. In the field of medicine, cancer detection, pill identification, all with the aim of improving diagnosis and therefore more effective treatment of patients. In security systems, YOLO models can detect suspicious activities, but also social distancing, as well as detecting whether someone is wearing a mask or not. In traffic, they can play a role in recognizing traffic signs as well as recognizing traffic signs. As well as in the monitoring of wild animals in nature, but also in the possible identification of endangered wild species as well as changes in nature and therefore ecosystem management. YOLO stands out for its extraordinary speed and accuracy, from the first version to the eighth, YOLO has grown and progressed. YOLO uses a pretrained architecture which speeds up the learning process on the dataset [29].

The evolution of the YOLO model has continuously progressed, the following text shows their development on a scaled basis, with an emphasis on key developments and innovations. The evolution of the YOLO model has continuously progressed, the following text shows their development on a scaled basis, with an emphasis on key developments and innovations. YOLOv1 is the first version of this series of models appeared in 2015, presented by Joseph Redmon, Santosh Divvala, Ross Girshick, Ali Farhadi in Ref. [30]. At that time, it was a step forward due to its single stage approach to object detection, vastly faster than multistage RCNN models and DPM. The unique and fast architecture of the YOLOv1 model processed 45 frames per second while the smaller version of the Fast YOLO network processed 155 frames per second, while at the time achieving twice the mAP of other real-time detectors. Compared to other detection systems, YOLO makes more localization errors, but it is unlikely to predict false detections where none exist. The next version is YOLOv2 introduced in 2016, also known as YOLO9000, due to the fact that it has the ability to detect 9000 object categories, while working in real time. This enhanced version performs standard detection tasks such as PASCAL VOC and Common Object in Context (COCO). At 67 FPS, YOLOv2 gets 76.8 mAP on VOC 2007. At 40 FPS, YOLOv2 gets 78.6 mAP [31]. This version certainly shows significant improvements in speed and accuracy. Improved the next version of YOLOv3, released in 2018, based on the Darknet53 architecture compared to YOLOv2 using Darknet-19, it followed a design change. The results are as follows for image resolution 320 × 320 YOLOv3 runs in 22 Ms at 28.2 mAP, however, three times faster [32]. YOLOv3 has the ability to predict bounding boxes on three different scales, and based on that it detects frames of different sizes, also YOLOv3 relies on independent logistic classifiers instead of softmax function, an advantage when the frame contains multiple classes. The continuation of the development of the YOLOv3 model, headed by Joseph Redmon and his team, continues Alexey Bochkovskiy with collaborators through the YOLOv4 model, in 2020 [33]. They introduce concepts such as "bag of freebies" and "bag of specials," which contribute to improving the performance of the model, but also new techniques such as mosaic data augmentation and self-adversarial training (SAT), through which the model learns and generalizes better. The YOLOv5 model was also released in 2020, its implementation was done through PyTorch YOLOv3, it was developed by the company Ultralytics, and its presentation was published on the GitHub platform. Compared to YOLOv4, it primarily contributes to more efficient implementation. Controversy abounds between which of the two models produces better results. However, in the scientific paper, they point out that the YOLOv5l algorithm surpasses the results of YOLOv4 and YOLOv3, this refers to the accuracy of detection, but the speed of its conclusion is slower [34]. YOLOX was introduced in 2021, developed by a group of researchers from the company Megvii Technology, based on an approach without anchor boxes. Although YOLOX-L with a slightly larger number of parameters (54.2 M) achieves an AP of 50.0%, which is 1.8% more than YOLOv5-L's number of parameters (47.1 M), where the AP is 48.2% [35]. YOLOv6 developed in 2022 by Meituan Inc, YOLOv6-L achieves AP 52.5%, presents numerous innovations to improve accuracy and speed, focuses on industrial applications [36]. YOLOv7 introduced in 2022 is enhanced with E-ELAN efficient learning, model scaling and a "bag-of-freebies" approach. YOLOv7 introduced in 2022, improved with E-ELAN efficient learning, model scaling and "bag-of-freebies" approach. One of the authors is Alexey Bochkovskiy, who also worked on YOLOv4. The authors point out that they achieved the best results

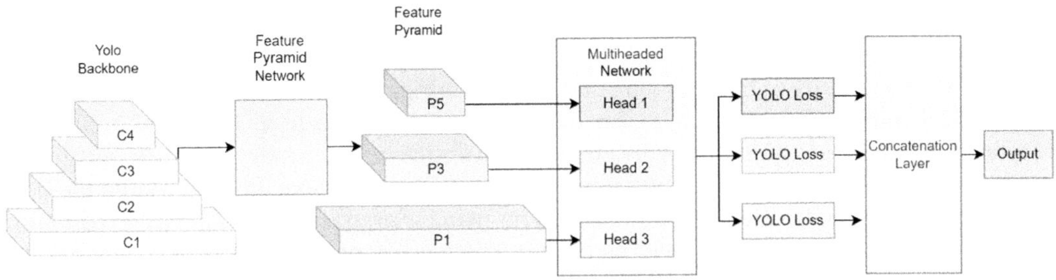

FIGURE 10.1 YOLO model general architectural layout.

compared to all previous real-time detectors achieving an AP of 56.8% [37]. YOLOv8 was released by Ultraltics in January 2023, YOLOv8 is anchor-free, reducing the number of box predictions and speeding up the Nonmaximum impression (NMS). It is built on the concepts of computer vision and deep learning. YOLOv8 include detection, segmentation, pose estimation, tracking, and classification. What sets it apart is easy installation using the pip package and even easier to run with just one line of code. The ultralitics company that developed YOLOv5 presented it through the Github platform, hence no formal scientific paper was presented that would provide specific research comparing it to other models and this paper is deprived of that type of detail.

In Ref. [38], they compared the three main image classification algorithms Single Shot Detection, Faster Region based Convolutional Neural Networks, and YOLO. The Microsoft COCO dataset was used. The results indicate that YOLO achieved the best results. Compared to other players on the market, YOLO has an advantage in real time due to its speed, the others cannot process images as quickly. As stated in the paper, CNN has the ability to remove highlights and identify objects in any image. They have an advantage in solving cases related to the diagnosis of deformities. YOLO has many advantages, especially in real time, it can train the entire model in parallel (unique detection of model objects). The best balance of speed and accuracy. A general architectural layout of YOLO models is provided in Fig. 10.1.

Transfer learning

Transfer learning in the field of AI plays an important role, namely, it allows models to use the acquired knowledge in one field and transfer it to another, often similar tasks. This has a great contribution especially if the data for training the task is limited. This possibility to apply the knowledge gained on one task to another, provides numerous opportunities that facilitate and speed up the processes themselves. Its application is widely discussed, we will list only some areas such as computer vision, NLP, but also for medical purposes. In computer vision, ImageNet should be highlighted, which are used for scene or object recognition. While in NLP, tools such as BERT and GPT, which can be used to perform classification or language understanding tasks. By using transfer learning, it is possible for the model to start learning from scratch, instead using prior knowledge obtained from related tasks. It contributes to fast convergence and it is noticeable that it contributes to the generalization of the model on new tasks. In addition, with transfer learning, we can use available resources more efficiently, for example, with a smaller amount of specific data, and thus applications develop faster.

One of the methods of transfer learning is fine-tuning, that is, using a model trained on a large and general dataset, such as ImageNet, for image detection. It mainly replaces or retrains the last layers of the network, which contributes to the model retaining the general characteristics of a large dataset, but also to adapting to a new specific task. While the other would be Feature Extraction, which is based on using a pretrained model as a fixed feature extractor. This means that the weights are not changed in most layers of the network, but only an output layer or layers are added at the end of the new model that are adapted to the new task.

The advantage of transfer learning is its efficiency to quickly and efficiently train the model, especially when there is a limited amount of data. Also, performance is better by using already acquired knowledge, which contributes to achieving better results, based on transfer learning. As for the danger, there is the risk that the model becomes overly specialized for the original task, it can be the danger of generalization to the new task.

In this paper, transfer learning is used to improve the accuracy and efficiency of the YOLOv8 model for waste calcification. It starts with the pretrained YOLOv8 model, which has been trained on a broad dataset where it was trained to recognize objects, after which fine-tuning is approached on a dataset based on images of different types of waste. Fine-tuning is based on the fact that the model learns some specific characteristics that relate to the waste classification itself, such as some materials such as paper, plastic, metal, etc. This contributes to this by reducing the need for large data sets, which are specialized for waste classification, and this is especially useful, especially since it is a big job to collect such a large amount of data. It also leads to faster and more efficient training of the model, that has the knowledge necessary for the task. And in the end, transfer learning plays a significant role in improving the accuracy of classification, because the model itself has previously acquired knowledge and diversity from more extensive data sets. In addition to all that, transfer learning is a ubiquitous technique in ML that facilitates and improves the effectiveness of training, but also the performance of the model itself. Although there is a risk of adaptability, this does not prevent transfer learning from being one of the key tools for all researchers in the field of ML.

Simulation configuration

To evaluate the proposed approach a publicly available dataset of labeled images of waste is utilized(https://universe.roboflow.com/material-identification/ garbage-classification-3/dataset/2) [39]. The dataset encompasses a total of seven classes including biodegradable items, cloth, cardboard, metal, glass, paper, and plastic. The dataset is separated into training and testing with 7324 images in the training and 2098 images in the validation portion of the dataset.

A comparison is carried out between various YOLO architectures including the nano, small and medium models in an attempt to find a stable solution that yields suitable outcomes while maintaining computational demands as low as possible. Models are trained under identical conditions with 15 training epochs and a batch size of 32 samples. Each model is evaluated using the following metrics:

$$Precision = \frac{TruePositives}{TruePositives + FalsePositives} \tag{10.1}$$

$$Recall = \frac{TruePositives}{TruePositives + FalseNegatives} \tag{10.2}$$

$$mAP@50 = \frac{1}{n}\sum_{i=1}^{n} AP_i^{(50)} \tag{10.3}$$

$$mAP@50 - 95 = \frac{1}{n}\sum_{i=1}^{n} AP_i^{50-95} \tag{10.4}$$

where n denotes the number of classes, mAP@50 is the Average Precision for class i at a threshold of 0.5, and mAP@50–95 is the average precision for class i averaged over thresholds from 0.5 to 0.95. A real-time usage example is provided in Fig. 10.2.

A flowchart of the training process of the utilized model is provided in Fig. 10.3.

Simulation outcomes

The following section showcases the outcomes of the described simulations. Each models outcomes are presented separately with examples of classification and respective metrics during training and testing. Finally, the overall simulation outcomes are discusses and a suitable solution proposed.

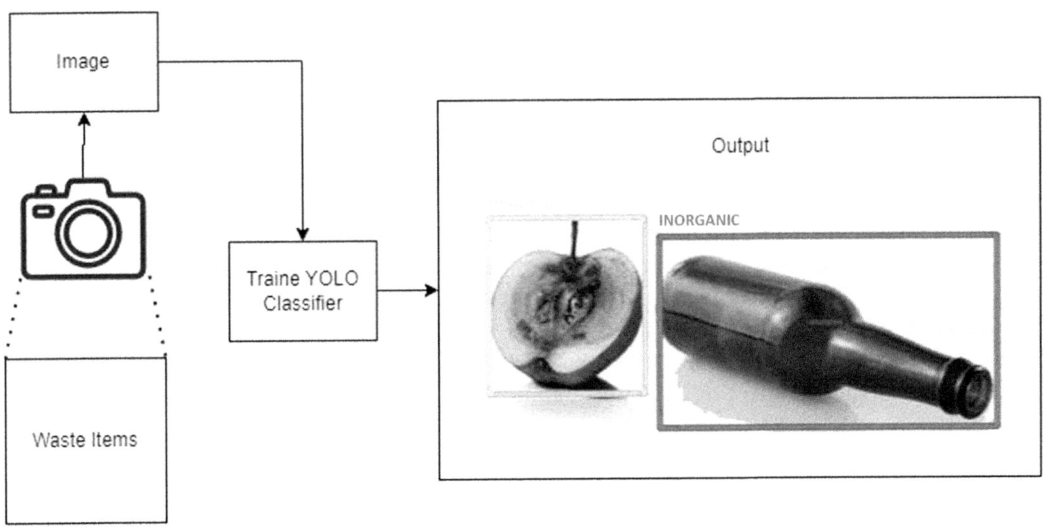

FIGURE 10.2 YOLO model training procedures.

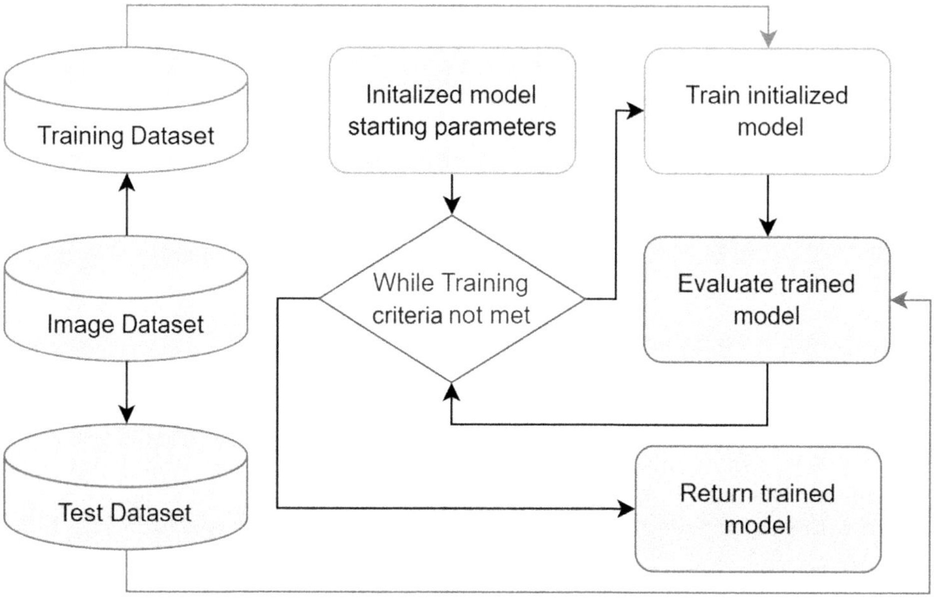

FIGURE 10.3 YOLO model training procedures.

YOLOv8 nano model outcomes

Metrics tracked during 15 training epochs in terms of precision, recall, mAP50 and mAP50–95 for the nano YOLOv8 model are provided in Table 10.1. The models precision peaks in the final epoch with a score of 0.53028. Recall peaks at 0.43117 and mAP50 of 0.47479 is reached. The mAP50–95 score is somewhat lower at 0.32117, however, this is to be expected.

Convergence graphs during training as well as validation are provided in Fig. 10.4. A spike in convergence rates is noticeable around the 5 training epochs. Spikes can indicate a lower stability of models in the long term. With a simple architecture, prone to larger changes, the nano model has a lower stability to models tested in the later stages. As a trade of the convergence is faster, and computational demands are significantly lower.

A confusion matrix for the final model is provided in Table 10.2. The model shows moderate success at identifying several materials, with the best results shown on glass and metal items, likely due to the distinct texture and color patterns present in objects made with these materials. The model struggles with identifying paper and often confuses paper for background. However, this is likely due to the lack of distinct features on paper items. A sample of model predictions as opposed to true labels is provided on a sample training batch used for the nano model in Figs. 10.5 and 10.6.

YOLOv8 small model outcomes

Metrics tracked during 15 training epochs in terms of precision, recall, mAP50 and mAP50–95 for the small YOLOv8 model are provided in Table 10.2. The models precision

TABLE 10.1 Nano model metrics during 15 training epochs.

Epoch	precision	recall	mAP50	mAP50–95
1	0.27629	0.25533	0.19645	0.11650
2	0.27240	0.25808	0.20183	0.10735
3	0.33123	0.27434	0.24286	0.14131
4	0.45443	0.30513	0.31065	0.18827
5	0.34561	0.27987	0.28883	0.18080
6	0.38647	0.33107	0.31882	0.19620
7	0.42187	0.36021	0.36212	0.23134
8	0.48694	0.36767	0.39614	0.25941
9	0.46638	0.35548	0.38500	0.24432
10	0.47033	0.36236	0.39860	0.25927
11	0.48417	0.38165	0.40679	0.26834
12	0.48826	0.40753	0.42571	0.28632
13	0.50576	0.41530	0.45020	0.30516
14	0.52867	0.40482	0.45627	0.30780
15	0.53028	0.43117	0.47479	0.32117

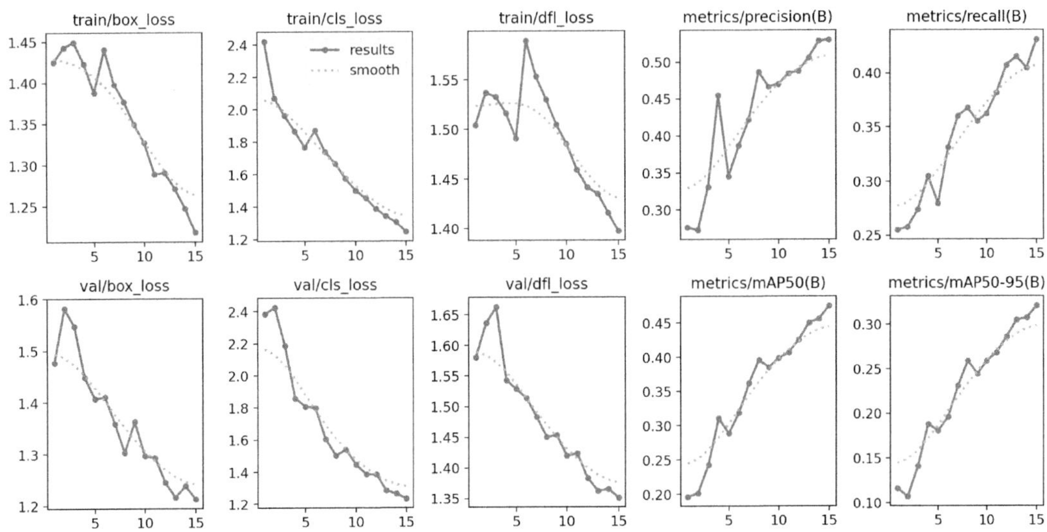

FIGURE 10.4 Nano model training and validation metrics during traning.

TABLE 10.2 Small model metrics during 15 training epochs.

Epoch	precision	recall	mAP50	mAP50–95
1	0.34358	0.24529	0.22540	0.12490
2	0.28867	0.23367	0.19881	0.10197
3	0.32765	0.32927	0.26769	0.15491
4	0.42662	0.30133	0.32558	0.20063
5	0.34369	0.32711	0.31175	0.19399
6	0.39909	0.33697	0.32910	0.20385
7	0.42811	0.33985	0.34083	0.21734
8	0.46746	0.39512	0.41051	0.26930
9	0.49339	0.38487	0.41243	0.26156
10	0.49324	0.39057	0.43189	0.28802
11	0.53935	0.37803	0.42937	0.28950
12	0.55014	0.42206	0.46520	0.31624
13	0.54693	0.42034	0.47863	0.33153
14	0.54890	0.43028	0.49311	0.33599
15	0.56049	0.43643	0.49993	0.34380

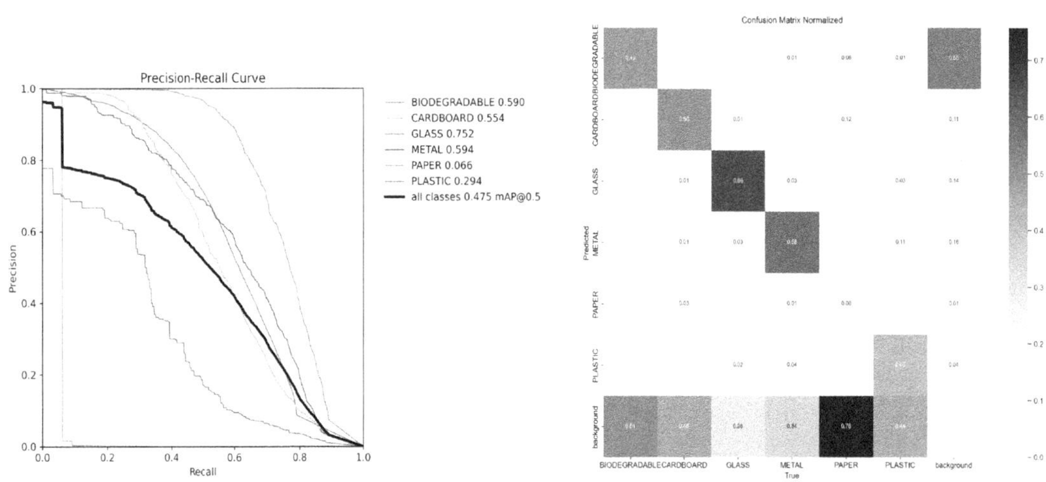

FIGURE 10.5 Nano model PR curve and confusion matrix.

peaks in the final epoch with a score of 0.56049 Recall peaks at 0.43643 and mAP50 of 0.49993 is reached. The mAP50–95 score is somewhat lower at 0.34380, however, this is to be expected. These outcomes suggest that the increased complexity of the small architecture over the nano architecture can yield improved overall outcomes across all metrics. Depending on the

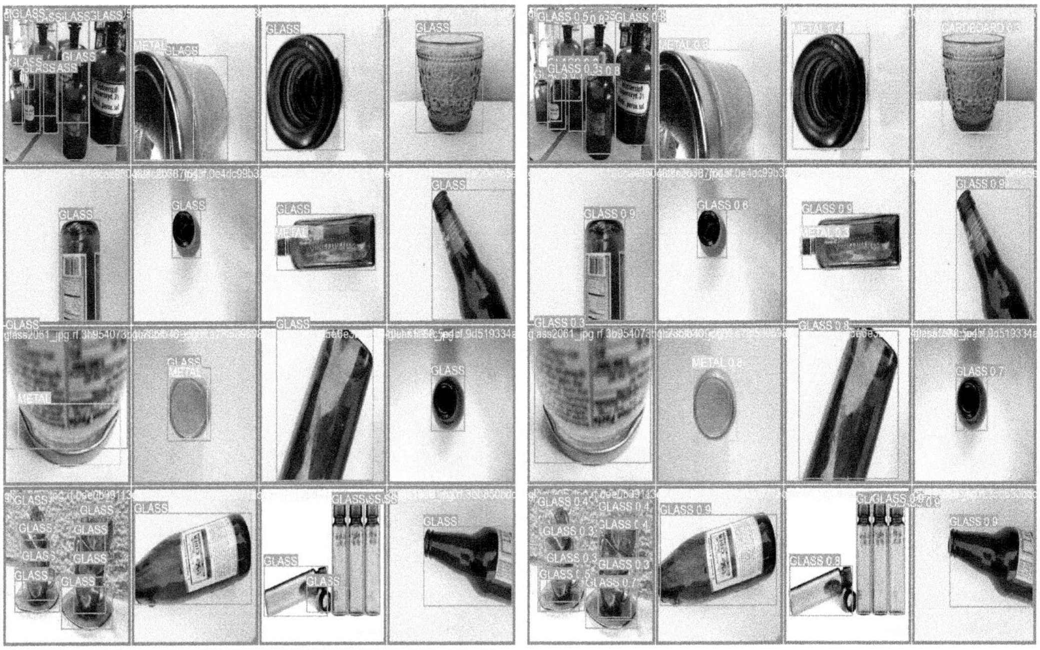

FIGURE 10.6 Dataset image samples with model predicted labels.

demands of the model this improvement may justify the increased computational demeans of training and deploying the small architecture instead of the nano model.

Convergence graphs during training as well as validation are provided in Fig. 10.7. While several smaller spikes are present in the convergence graphs these are significantly less severe than those observed in the simpler nano model. The more complex architecture has the ability to absorb some of the more drastic changes make the model more stable are resilient to sharp changes in performance. Nevertheless, this comes at a cost of a somewhat slower convergence and need for longer training times as well as a need for more training data.

A confusion matrix for the final model is provided in Table 10.2. Similarly to the nano model, the small model struggles with paper items. However, a slight improvement can be noted when identifying biodegradable items. Both models shown similarly good performance for glass and metal items. A sample of model predictions as opposed to true labels is provided on a sample training batch used for the small model in Figs. 10.8 and 10.9.

YOLOv8 medium model outcomes

Metrics tracked during 15 training epochs in terms of precision, recall, mAP50 and mAP50–95 for the medium YOLOv8 model are provided in Table 10.3. The models precision peaks in the final epoch with a score of 0.54981 Recall peaks at 0.49302 and mAP50 of 0.33738 is reached. The mAP50–95 score is lower at 0.33738. The model marginally outperforms the small model at a significantly higher computational demand. Additionally, the modest outcomes suggest that further training is needed for the medium model to reach a full

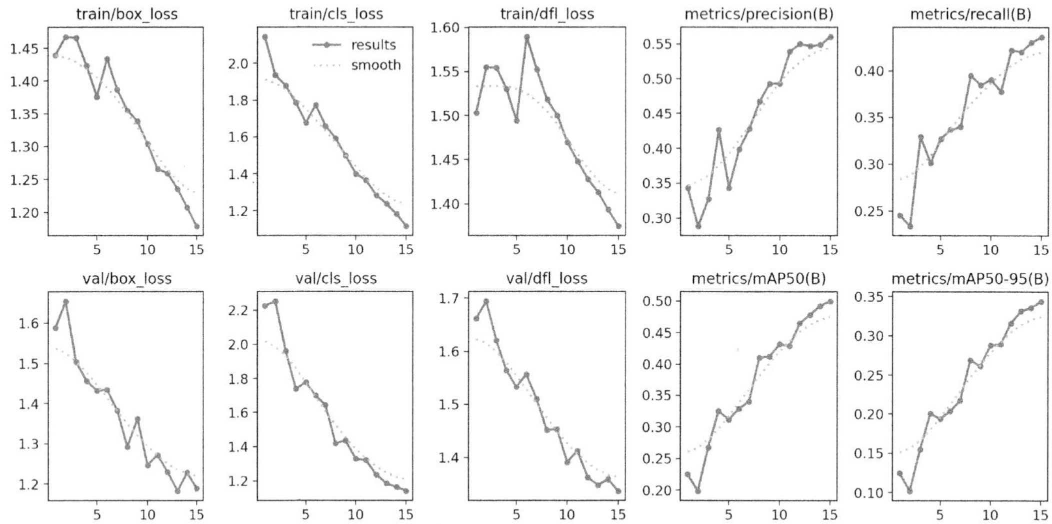

FIGURE 10.7 Small model training and validation metrics during traning.

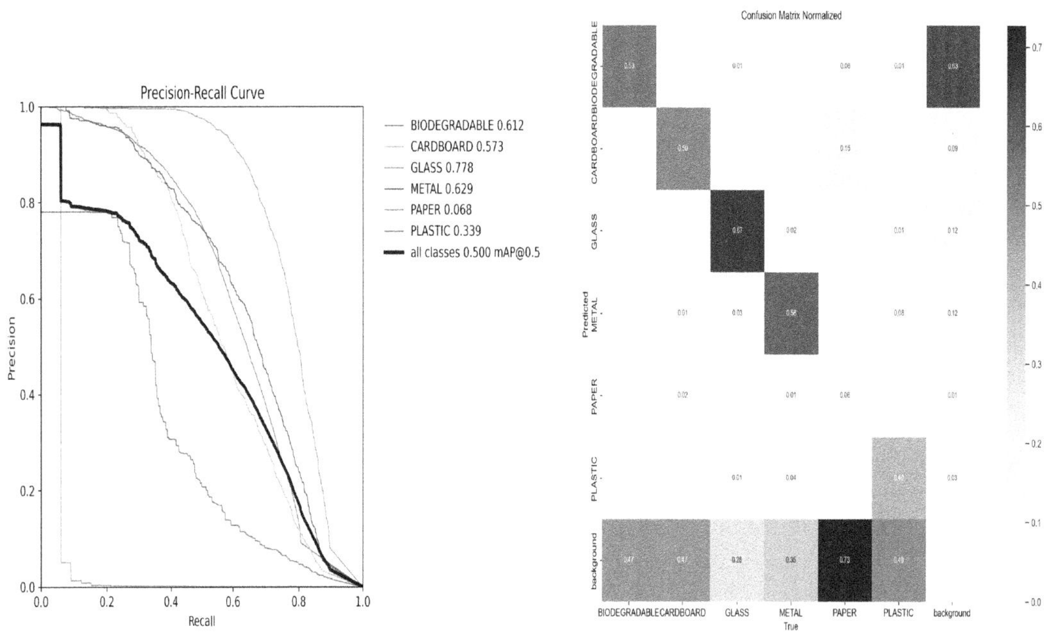

FIGURE 10.8 Small model PR curve and confusion matrix.

potential. Convergence graphs during training as well as validation are provided in Fig. 10.10. Several smaller spikes can be seen in terms of loss function, however, the complex architecture is able to absorb larger changes in behavior leading to an overall stable if slower convergence.

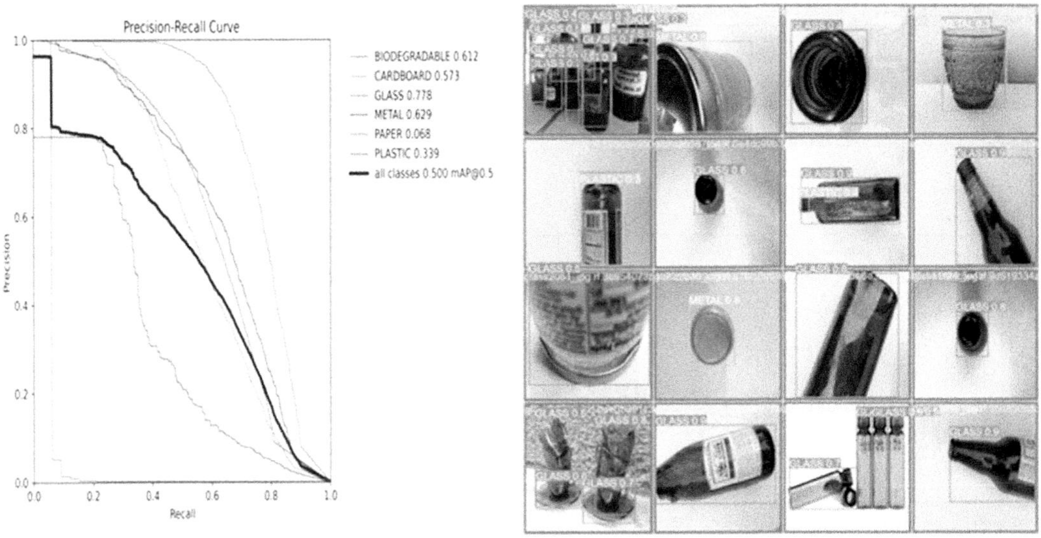

FIGURE 10.9 Small model labels versus predictions sample classifications.

TABLE 10.3 Medium model metrics during 15 training epochs.

Epoch	precision	recall	mAP50	mAP50–95
1	0.31090	0.22641	0.21044	0.11519
2	0.29128	0.25179	0.22509	0.12051
3	0.31489	0.31278	0.22594	0.12355
4	0.36493	0.30628	0.29449	0.17433
5	0.36500	0.31685	0.31089	0.19618
6	0.37035	0.33759	0.31391	0.19569
7	0.40738	0.34077	0.33287	0.20853
8	0.43603	0.37569	0.37583	0.24750
9	0.47406	0.37271	0.39254	0.24608
10	0.46055	0.38889	0.39968	0.26248
11	0.48237	0.38655	0.40561	0.27592
12	0.52167	0.42297	0.45675	0.31115
13	0.51987	0.41664	0.45406	0.31121
14	0.54981	0.42265	0.47783	0.32555
15	0.53502	0.44625	0.49302	0.33738

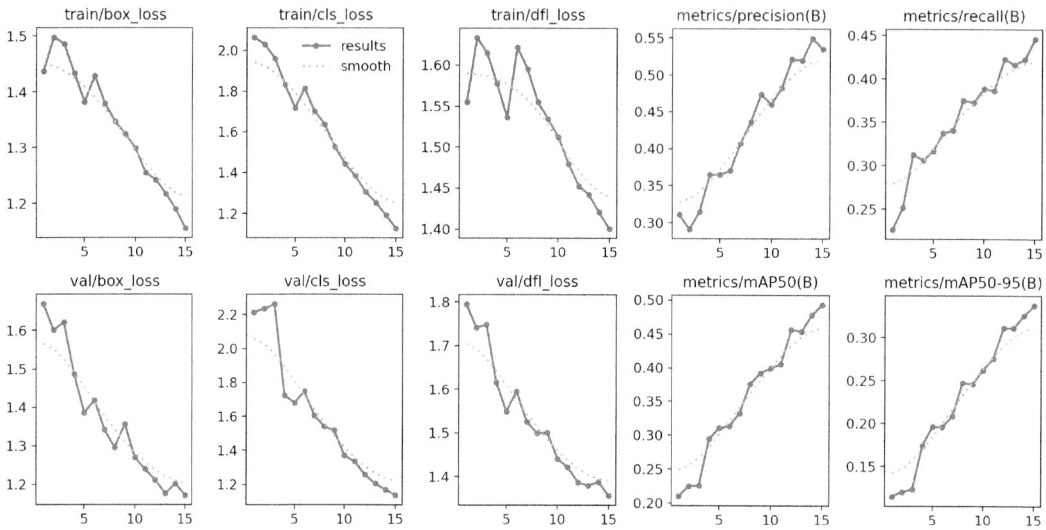

FIGURE 10.10 Medium model training and validation metrics during training.

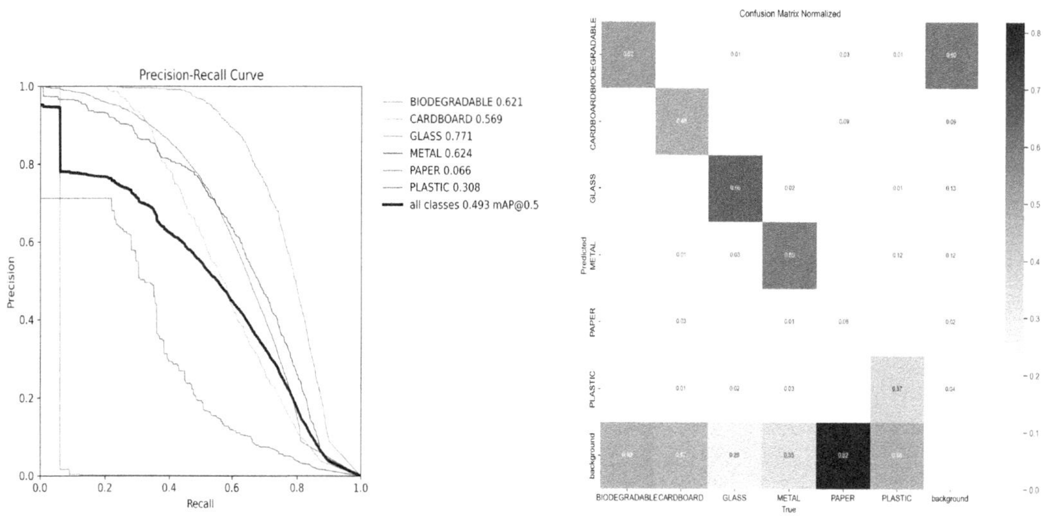

FIGURE 10.11 Medium model PR curve and confusion matrix.

Finally the confusion matrix and PR curve for the medium model are provided in. When considering the confusion matrix of the medium model an overall drop in performance is seen in the medium model with lower precision on all classes. The more complex architecture requires additional training in order to meet the performance of simpler models. A sample of model predictions as opposed to true labels is provided on a sample training batch used for the medium model in Figs. 10.11 and 10.12.

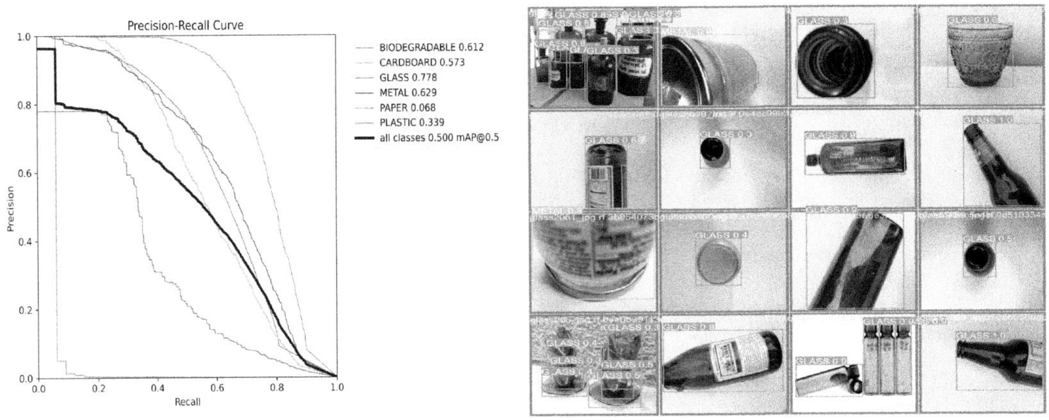

FIGURE 10.12 Medium model labels versus predictions sample classifications.

Conclusion

This work explores the applicative potential of a relatively recently introduced YOLOv8 models on waste item classifications using computer vision. Waste classification is a major societal issue and with ever increasing amounts of waste generated by people around the world, effective techniques for managing and reducing waste are needed. Recycling is a major way of reducing and reusing waste. However, recycling faces certain challenges in order to be effective. One major challenge is waste separation. Waste items are often comprised of several materials and the processes for recycling these materials are not often compatible. While separating could be handled manually, this is a tedious and potentially dangerous process for workers. This work explores the use of computer vision as a step towards the amortization of trash separation.

As computer vision can be a computationally expensive procedure, this work utilized lightweight architecture in order to explore an affordable alternative to more expensive processes. Several YOLOv8 architecture are explored including nano, medium and small models. While all trained models showed potential with moderate outcomes, the small model shows the optimal trade-off between computational complexity and attained outcomes. Models showcase good results on metal and glass items which is especially important since these materials are among the most highly receivable. However, models do struggle in identifying paper, often unable to tell paper apart from backgrounds.

It is important to note that this work faces certain limitations. Namely, limited computational power constrains the number of training epochs, batch sizes and training epochs allocated to each model. Future works hope to address this by further expanding on the training process, incorporating more data and larger batches seeking higher model generalization. Additionally, optimization metaheuristics hope to be combined with YOLO models to help optimize model parameters for the given problem improving classification outcomes further.

References

[1] O.A. Lamma, The impact of recycling in preserving the environment, International Journal of Advanced Research 7 (11) (2021) 297–302.
[2] G. Davis, Counting (green) jobs in Queensland's waste and recycling sector, Waste Management & Research 31 (9) (2013) 902–909.
[3] N.J. Nipun, Sustainable Innovation for Reducing Negative Environmental Impact in the Textile Industry: A Comparative Case Study between Sweden and Bangladesh, (2023).
[4] S.Y. Islam, M.G.R. Alam, Computer vision-based waste detection and classification for garbage management and recycling, The Fourth Industrial Revolution and Beyond: Select Proceedings of IC4IR+, Springer, 2023, pp. 389–411.
[5] Y. Wang, T. Kong, Air quality predictive modeling based on an improved decision tree in a weather-smart grid, IEEE Access 7 (2019) 172892–172901.
[6] Z. Wu, J. Zhang, S. Hu, Review on classification algorithm and evaluation system of machine learning, 2020 13th International Conference on Intelligent Computation Technology and Automation (ICICTA). IEEE (2020).
[7] Vakili, M., Ghamsari, M. and Rezaei, M. (2020) Performance Analysis and Comparison of Machine and Deep Learning Algorithms for IoT Data Classification, arXiv:2001.09636. arXiv.org e-Print Archive.
[8] I. Khan, A.R. Ahmad, N. Jabeur, M.N. Mahdi, An artificial intelligence approach to monitor student performance and devise preventive measures, Smart Learning Environments 8 (1) (2021).
[9] M.-S. Stupariu, S.A. Cushman, A.-I. Ple¸soianu, I. P˘atru-Stupariu, C. Fu¨rst, Machine learning in landscape ecological analysis: a review of recent approaches, Landscape Ecology 37 (5) (2022) 1227–1250.
[10] S. Subramaniam, N. Raju, A. Ganesan, N. Rajavel, M. Chenniappan, C. Prakash, et al., Artificial intelligence technologies for forecasting air pollution and human health: a narrative review, Sustainability 14 (16) (2022) 9951.
[11] W. Rammer, R. Seidl, Harnessing deep learning in ecology: an example predicting bark beetle outbreaks, Frontiers in Plant Science 10 (2019) 1327.
[12] C. Chen, Y. Guo, F. Tian, S. Liu, W. Yang, Z. Wang, et al., A Unified Interactive Model Evaluation for Classification, Object Detection, and Instance Segmentation in Computer Vision, (2023) arXiv:2308.05168. arXiv.org e-Print Archive.
[13] W. Lin, YOLO-green: A real-time classification and object detection model optimized for waste management, 2021 IEEE International Conference on Big Data (Big Data) (2021).
[14] D. Reis, J. Kupec, J. Hong, A. Daoudi, Real-Time Flying Object Detection with YOLOv8, (2023) arXiv:2305.09972. arXiv.org e-Print Archive.
[15] N. Nnamoko, J. Barrowclough, J. Procter, Solid waste image classification using deep convolutional neural network, Infrastructures 7 (4) (2022) 47.
[16] O. Cosido, A. Iglesias, A. Galvez, R. Catuogno, M. Campi, L. Teran, et al., Hybridization of convergent photogrammetry, computer vision, and artificial intelligence for digital documentation of cultural heritage: a case study of the Magdalena Palace, 2014 International Conference on Cyberworlds. IEEE (2014).
[17] J. Long, E. Shelhamer, T. Darrell, Fully convolutional networks for semantic segmentation, IEEE Conference on Computer Vision and Pattern Recognition (CVPR). IEEE (2015).
[18] O.H. Babatunde, L. Armstrong, J. Leng, D. Diepeveen, A survey of computer-based vision systems for automatic identification of plant species, Journal of Agricultural Informatics 6 (1) (2015).
[19] S. Ullman, L. Assif, E. Fetaya, D. Harari, Atoms of recognition in human and computer vision, Proceedings of the National Academy of Sciences of the United States of America 113 (10) (2016) 2744–2749.
[20] S. Matiacevich, D. Celis Cofré, P. Silva, J. Enrione, F. Osorio, Quality parameters of six cultivars of blueberry using computer vision, International Journal of Food Science 2013 (2013) 419535.
[21] B. Bi, P. Xue, S. Mesejo, M. Cagnoni, A Survey on Evolutionary Computation for Computer Vision and Image Analysis: Past, Present, and Future Trends, (2022) IEEE.
[22] A.L. Yuille, C. Liu, Deep Nets: What Have They Ever Done for Vision? (2018) arXiv.org e-Print Archive.
[23] A. Liu, R. Tucker, V. Jampani, A. Makadia, N. Snavely, A. Kanazawa, Infinite Nature: Perpetual View Generation of Natural Scenes from a Single Image, (2020) arXiv.org e-Print Archive.
[24] A. Vaswani, N. Shazeer, N. Parmar, J. Uszkoreit, L. Jones, A.N. Gomez, et al., Attention Is All You Need, (2017) arXiv.org e-Print Archive.

[25] Z. Liu, Y. Lin, Y. Cao, H. Hu, Y. Wei, Z. Zhang, et al., Swin Transformer: Hierarchical Vision Transformer Using Shifted Windows, (2021) IEEE.

[26] I.J. Goodfellow, J. Pouget-Abadie, M. Mirza, B. Xu, D. Warde-Farley, S. Ozair, et al., Generative Adversarial Networks, (2014) arXiv:1406.2661. arXiv.org e-Print Archive.

[27] M. Niemeyer, A. Geiger, GIRAFFE: Representing Scenes as Compositional Generative Neural Feature Fields, (2020) arXiv.org e-Print Archive.

[28] A. Ramesh, M. Pavlov, G. Goh, S. Gray, C. Voss, A. Radford, et al., Zero-Shot Text-to-Image Generation, (2021) arXiv.org e-Print Archive.

[29] J. Terven, D. Cordova-Esparza, A Comprehensive Review of YOLO: From YOLOv1 and Beyond, (2023) arXiv:2304.00501.

[30] J. Redmon, S. Divvala, R. Girshick, A. Farhadi, You only look once: unified, real-time object detection, Proceedings of the IEEE Conference on Computer Vision and Pattern Recognition (2016) 779–788.

[31] J. Redmon, A. Farhadi, YOLO9000: better, faster, stronger. In Proceedings of the IEEE conference on computer vision and pattern recognition (pp. 7263–7271), 2017.

[32] J. Redmon, A. Farhadi, YOLOv3: an Incremental Improvement, (2018) arXiv:1804.02767. arXiv.org e-Print Archive.

[33] A. Bochkovskiy, C.-Y. Wang, H.-Y.M. Liao, YOLOv4: Optimal Speed and Accuracy of Object Detection, (2020) arXiv:2004.10934. arXiv.org e-Print Archive.

[34] Nepal, U. and Eslamiat, H. (2022) Comparing YOLOv3, YOLOv4, and YOLOv5 for autonomous landing spot detection in faulty UAVs, Sensors, 22(2). Available at: https://doi.org/10.3390/s22020464, https://www.mdpi.com/1424-8220/22/2/464.

[35] Z. Ge, S. Liu, F. Wang, Z. Li, J. Sun, YOLOX: Exceeding YOLO Series in 2021, (2021) arXiv:2107.08430. arXiv.org e-Print Archive.

[36] C. Li, L. Li, H. Jiang, K. Weng, Y. Geng, L. Li, et al., YOLOv6: A Single-Stage Object Detection Framework for Industrial Applications, (2022) arXiv:2209.02976. arXiv.org e-Print Archive.

[37] C.-Y. Wang, A. Bochkovskiy, H.-Y.M. Liao, YOLOv7: Trainable Bag-of-Freebies Sets New State-of-the-Art for Real-Time Object Detectors, (2022) arXiv:2207.02696. arXiv.org e-Print Archive.

[38] S. Srivastava, A.V. Divekar, C. Anilkumar, I. Naik, V. Kulkarni, V. Pattabiraman, Comparative analysis of deep learning image detection algorithms, Journal of Big Data 8 (1) (2021).

[39] Roboflow (2022) Garbage. Available at: <https://universe.roboflow.com/material-identification/garbage-classification-3> (accessed 26.04.24).

CHAPTER

11

Waste to value added: role of automation in organic waste

Mithra Geetha[1], *Nihal Poovadichalil*[2], *Md Nawshad*[1,2], *Muni Raj Maurya*[1], *Asad Ullah*[1,2], *Syed Faizan Haider*[2], *and Anwarul Hasan*[2]

[1]Center for Advanced Materials, Qatar University, Doha, Qatar [2]Mechanical Engineering Department, Qatar University, Doha, Qatar

Introduction

Overview of organic waste

Organic waste (OW) plays a vital role within our ecosystem and daily routines. The diversity of its manifestations, including remnants from meals, garden debris, and agricultural byproducts, underscores the necessity of comprehending its characteristics and control for the promotion of sustainable lifestyles. By delving into the types and characteristics of OW, one can appreciate its significance in ecological balance and environmental stewardship. OW's ability to naturally decompose through microbial action sets it apart from inorganic waste, highlighting the need for specialized strategies to handle it effectively.

Here we will discuss the definition and types of OW, offering a detailed exploration of its categories, including food scraps, yard waste, and agricultural waste. Food scraps, commonly found in household kitchens, present unique management challenges due to their perishable nature. Yard waste, generated from garden maintenance activities, offers opportunities for sustainable practices like composting and mulching. Agricultural waste, produced through farming, encompasses crop residues and manure, requiring innovative approaches for responsible disposal and utilization. This comprehensive review intends to provide researchers and academics with a coherent comprehension of the generation and handling of OW, facilitating well-informed decision-making and efficient waste management strategies. Furthermore, the environmental consequences of OW, including emissions of greenhouse gasses from decomposition processes and the possibility of soil pollution, highlight the pressing need for the implementation of effective waste management measures.

Harnessing Automation and Machine Learning for Resource Recovery and Value Creation
DOI: https://doi.org/10.1016/B978-0-443-27374-2.00011-X

OW, originating from unused or discarded materials generated by animal and human actions, has become a notable worldwide issue associated with the swift expansion of population and urban development, circumstances that have exacerbated challenges in waste handling [1]. In the past, OW was commonly repurposed or recycled for purposes like fodder or fertilizer; however, contemporary approaches have evolved, leading to heightened financial burdens on municipalities and environmental apprehensions [1]. The recycling of OW, encompassing food scraps, vegetable remnants, and fruit discards, holds the potential to generate valuable resources such as fertilizers through techniques like composting and fermentation, processes that bolster plant development and mitigate unpleasant odors [2]. Novel methodologies for managing OW, including those integrating solar heating and wastewater treatment, present cost-effective, efficient remedies suitable for rural areas and agricultural environments [3]. Governments across the globe are implementing diverse economic and administrative strategies to divert organic municipal solid waste (MSW) away from landfills toward treatment facilities, thereby encouraging the utilization of organic fertilizers and realizing substantial economic and environmental advantages [4]. In countries like Hungary, the byproduct of pálinka distillation yields considerable quantities of fruit waste, which can be combined with mineral supplements during composting to regulate pH levels and yield valuable compost or fertilizer [5]. This composting procedure not only promotes the breakdown and synthesis reactions but also facilitates the adsorption of heavy metals, underscoring the potential of sustainable waste management practices [5]. In essence, the valorization of OW through a medley of innovative and conventional approaches is imperative in tackling the escalating waste management hurdles and advancing global environmental sustainability.

Importance of waste management

Waste management is a local issue that has far-reaching global impacts [6]. It is an extremely complex issue that must be addressed in every aspect of its operations [7]. Proper waste management has become a global concern because of its economic, environmental, social, and sustainability implications [8]. The waste management process involves the collection, transportation, and use of various bins for waste disposal [9]. Several waste management and treatment methods are employed, including composting, bio-refining, incineration pyrolysis, recycling, and biogas production [10]. A more sustainable waste management approach emphasizes reducing waste production, classifying waste, reusing, recycling, and recovering energy, rather than relying on traditional methods like landfilling, open dumps, and open incineration [11]. Monitoring is a crucial function in waste management, essential for addressing challenges related to waste generation, collection, transportation, treatment, and disposal processes [12]. Effective waste management can enhance environmental quality and public health, while poor waste management can contribute to water, soil, and air pollution [13]. Inefficient waste management practices, including the use of unsecured landfills, insufficient technical knowledge, limited scientific and economic resources, and the absence of waste emergency policies, can result in serious consequences for both the community and workers [14–16]. Improper waste management poses significant risks to human health and the environment. Beyond being unsightly, it leads to air pollution, contaminates water bodies when dumped, and depletes the ozone layer when burned, exacerbating climate change. Conventional methods of waste disposal, such as burning, ocean dumping, and roadside dumping, are often inadequate. Poorly managed waste harms humans, animals, plants, and the environment, underscoring the need for better practices [17].

The effects of ineffective household solid waste management (SWM) on public health can be categorized into several areas: physical (chemical and radiation injuries), biological (diseases such as dysentery, diarrhea, gastrointestinal issues, food poisoning, worm infections, dengue, cholera, fever, leptospirosis, and bacterial infections), noncommunicable diseases (including cancers of the liver, kidney, pancreas, larynx, and non-Hodgkin lymphoma), psychosocial (issues like disturbing odors, unsightly waste, and cognitive and stress-related problems), and ergonomic health risks (such as repetitive movements, poor body posture, and excessive force) [18]. Waste management offers several economic advantages, such as financial stability, job creation, and strengthened community bonds [19]. SWM is a significant environmental issue with direct impacts on air, water, and soil quality, as well as public health. The rising volume of waste globally makes effective waste management increasingly challenging [20]. Unregulated disposal of solid waste can have negative effects on both public health and the environment [21]. Effective SWM is essential for sustainable urban development. It involves the collection, transportation, and recycling of solid waste to reduce its adverse impacts on the environment and human health [17]. However, SWM is causing significant issues, leading to a decline in air, land, and water quality, which negatively affects natural ecosystems and public health [22]. Improper management of MSW at landfill sites often attracts animals, rodents, mosquitoes, vultures, and scavengers, which can pose health risks and even lead to fatalities among frontline workers and waste pickers [23].

Concept of waste to value added

The conversion of waste into valuable products, commonly known as the "waste-to-value" concept, represents a fundamental strategy in contemporary resource management. This strategy involves converting discarded materials into useful commodities through various methods like recycling, upcycling, and bioconversion. The goal is not only to address the mounting issue of waste but also to extract economic, environmental, and social benefits from materials that would otherwise contribute to landfill volumes. By integrating sustainability principles, waste-to-value projects aim to fulfill present requirements without compromising future resources, thus closely harmonizing with the objectives of sustainable development.

One prominent illustration of waste-to-value projects involves converting food waste (FW) into biogas using anaerobic digestion (AD), a method that not only cuts down on greenhouse gas emissions but also generates renewable energy for various uses. The combination of FW with animal manure in anaerobic codigestion has been proven to significantly increase biogas production. For example, when FW was codigested with pig manure, it resulted in a peak cumulative biogas production of 418 mL/g volatile solids (VS), surpassing chicken, and goat manure, which yielded 408 mL/g VS and 319 mL/g VS, respectively [24].

Converting waste into valuable products is crucial for several reasons, encompassing environmental, economic, and social dimensions. Environmentally, the transformation of waste mitigates pollution and reduces the harmful effects of waste decomposition, such as methane (CH_4) emissions from FW, which contribute to global warming [25,26]. From an economic perspective, the recovery and utilization of waste materials have the potential to substantially reduce production expenses and establish fresh sources of income. An illustration of this is how MSW possesses valuable metals and other materials that can be transformed into novel goods, thereby promoting a circular economy and diminishing the demand for untouched raw materials [27]. From a societal perspective, the transformation of waste materials into valuable goods has the potential to generate employment opportunities and facilitate the advancement of expertise within the realms of recycling and upcycling, thereby cultivating a more sustainable and equitable economic environment. Additionally, through the redirection of waste away from landfills and incineration sites, the recycling and upcycling sectors play a significant role in mitigating environmental contamination and preserving the finite resources of the planet for forthcoming generations.

Transforming waste into valuable resources requires a holistic strategy that integrates various methods and technologies to improve resource recovery and minimize environmental consequences. The first stages involve the sorting and separation of waste materials, followed by shredding and dewatering to simplify handling. Advanced techniques like pyrolysis and gasification are highly effective for treating refuse-derived fuels, providing better energy recovery efficiency and lower pollutant emissions compared to traditional incineration, while also generating valuable by-products such as syngas and chemical platforms [27]. Biological processes, like composting, AD, and vermicomposting, are essential in converting OW into valuable products like compost and biogas. AD, for example, can repurpose food and agricultural waste into sustainable electricity, heat, and nutrient-dense fertilizers, promoting a circular economy in agricultural settings [28]. Innovative technologies like hydrothermal carbonization and plasma arc gasification are promising solutions for waste conversion, improving resource recovery efficiency and scope. Recycling and upcycling processes further reduce production costs and environmental impact by extracting and reprocessing materials for reuse [29]. Policy measures and economic incentives such as extended producer responsibility, subsidies, and carbon credits are crucial for supporting integrated processes that promote a circular economy by converting waste into valuable resources. Success stories from different countries demonstrate the effectiveness of these strategies in achieving the transformation of waste into wealth, emphasizing the significance of implementing comprehensive waste management policies [29].

In conclusion, the notion of waste-to-value introduces a revolutionary approach to sustainable resource management through the transformation of discarded materials into valuable products. This approach not only tackles the urgent issue of waste accumulation but also unlocks economic, environmental, and social advantages. Through the integration of cutting-edge technologies and sustainable principles, waste-to-value initiatives contribute to a circular economy, mitigate environmental consequences, and establish novel revenue streams. Policy backing and economic incentives play a vital role in nurturing the prosperity of these projects. Ultimately, the conversion of waste into valuable resources harbors vast potential in advancing a more sustainable and all-encompassing economy while preserving natural resources for future generations.

Role of automation in waste management

Artificial intelligence (AI) is a rapidly evolving technology that is increasingly being adopted across various industries, with waste management being a notable example [30]. The advent of AI is driving significant changes in the waste management industry, leading to greater sustainability and profitability [31]. Several developed countries, such as Austria, Germany, New Zealand, the USA, the UK, Japan, Singapore, Switzerland, South Korea, and Canada, have started adopting AI technologies to enhance resource utilization, improve efficiency, and increase recycling opportunities throughout the SWM cycle [32]. AI has the potential to revolutionize municipal waste management by improving the efficiency of waste collection, processing, and classification. AI technologies such as smart garbage bins, sorting robots, predictive models, and wireless sensors can monitor waste levels, forecast collection needs, and optimize the operation of processing facilities. By integrating AI, municipalities can lower costs, enhance safety, and minimize the environmental impact of waste management (Fig. 11.1) [31].

Smart bins

Traditional garbage bins only serve the purpose of collecting waste, and sanitation workers must perform manual checks to determine the trash level. This method is inefficient for regular waste inspections. Additionally, frequent filling of these bins can lead to the breeding of disease-causing organisms and insects [33]. In contrast, smart garbage bins are designed to automatically monitor the level of waste and send timely notifications Fig. 11.2. These bins use sensors to collect data, which is transmitted via a network. Intelligent bin systems can improve garbage collection efficiency, reduce disease spread, and enhance the overall urban environment. However, the high cost of implementing these systems poses a challenge for widespread adoption. Additionally, environmental factors like temperature and humidity can affect their performance, necessitating regular maintenance by dedicated personnel. Despite these challenges, it is essential to continue developing and promoting smart garbage bins for future use [31].

Robotic technology in waste sorting

AI enhances waste management through several key applications. Real-time monitoring of garbage bins helps optimize waste collection routes and prevent overflow. Intelligent sorting systems improve recycling efficiency and reduce contamination, while robotic waste sorting utilizes robotic arms to speed up and precision-sort waste, minimizing manual labor. AI can also predict maintenance needs for waste-sorting equipment, reducing downtime and extending equipment lifespans. Additionally, AI-driven optimization considers factors like traffic, weather, and population density to boost the efficiency of waste collection and processing [31]. Employing robots in SWM can significantly improve the efficiency of garbage classification. However, these robots need advanced visual and operational capabilities to effectively operate in complex, diverse, and unpredictable environments typical of waste processing facilities [34]. Researchers are exploring ways to integrate waste-sorting robots into existing waste management systems, such as using them to sort waste before it reaches landfills. Additionally, a waste-sorting robot has been developed employing deep learning techniques

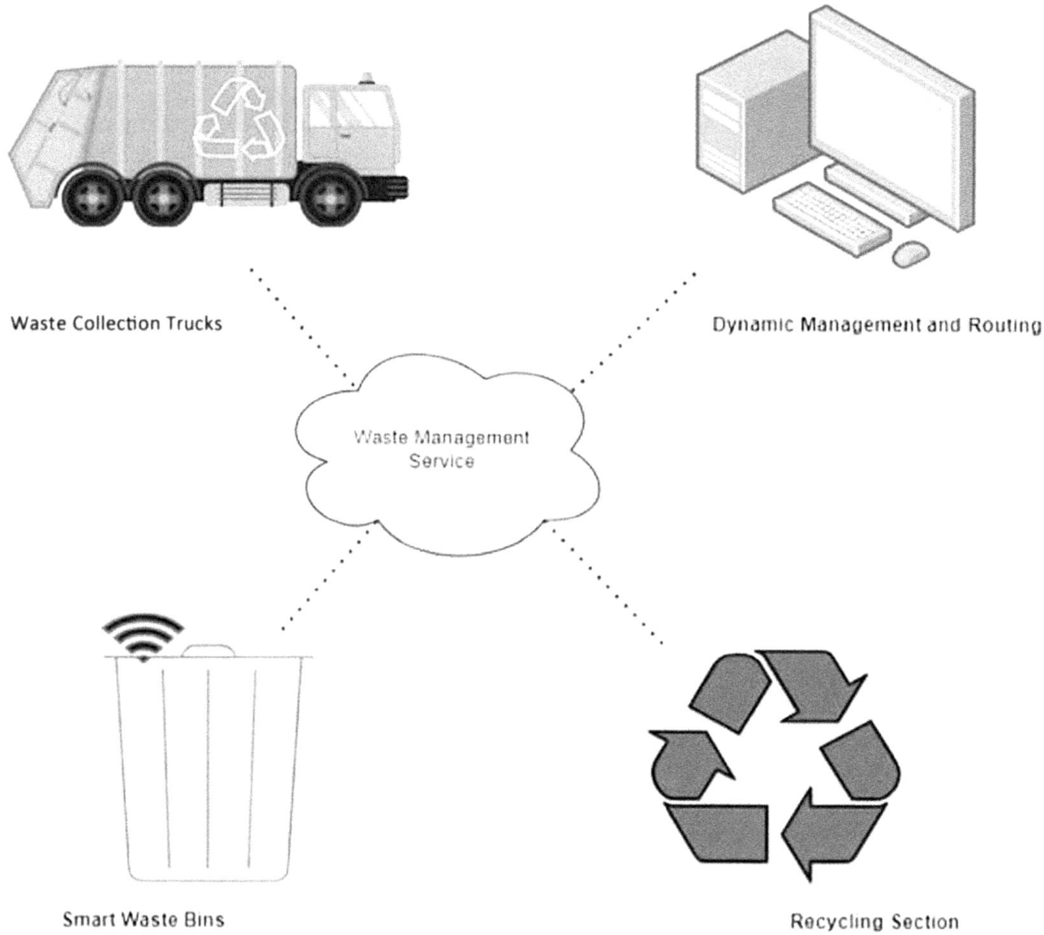

FIGURE 11.1 Artificial intelligence in waste management. *Source: From D.K. Gude, H. Bandari, A.K.R. Challa, S. Tasneem, Z. Tasneem, S.B. Bhattacharjee, et al., Transforming urban sanitation: enhancing sustainability through machine learning-driven waste processing, Sustainability 16 (17) (2024) 7626.*

and optical sensors, enabling it to accurately identify and classify various types of waste [35]. In conclusion, trash-sorting robots have the potential to greatly enhance waste management efficiency, reduce labor costs, and improve sorting accuracy. However, some critics argue that the high installation and maintenance costs of these robots make them less practical compared to traditional waste-sorting methods [31].

Sensor-based waste monitoring

A sensor installed on the garbage bin can gather data such as smell, weight, and humidity to classify the trash. Additionally, it monitors the bin's environment and tracks its filling

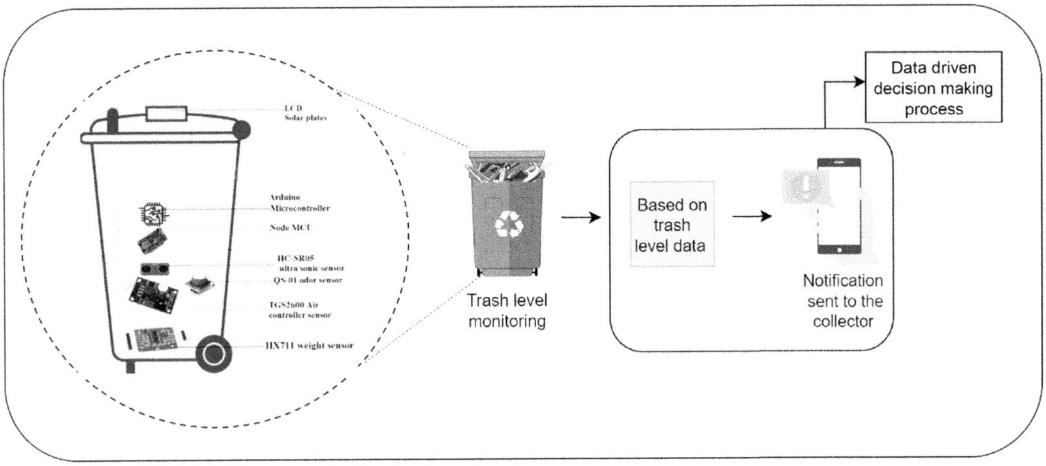

FIGURE 11.2 Smart bin used in SGCS (smart garbage collection system). *Source: From A. Hussain, U. Draz, T. Ali, S. Tariq, M. Irfan, A. Glowacz, et al., Waste management and prediction of air pollutants using IoT and machine learning approach, Energies 13 (15) (2020) p. 3930. https://doi.org/10.3390/en13153930.*

level. Users can view the real-time status of the bins on a platform, as the information is transmitted via the internet [31]. Sensor-based waste monitoring technology uses sensors to track waste generation, identify waste sources, and assess the effectiveness of waste management strategies in a particular area [36]. A typical wireless sensor network architecture for SWM systems include a range of sensors, such as temperature, humidity, odor, infrared, gas, and sound sensors. These sensors provide real-time monitoring of various parameters, enabling more effective control of the waste treatment process [31]. The development of wireless sensor networks has advanced rapidly, with numerous studies focusing on their application in waste monitoring. These studies primarily involve tracking garbage levels and using the network to notify users [37].

Types of organic waste

The categorization of OW plays a pivotal role in optimizing waste management and resource recovery, enabling the application of customized processing techniques to enhance the efficiency and efficacy of conversion technologies. As an illustration, AD can be employed to convert FW into biogas, whereas composting is more appropriate for yard waste, yielding soil amendments rich in nutrients. Accurate classification of waste minimizes contamination, thereby enhancing the quality of final products like compost or biogas, while also boosting recycling rates, decreasing reliance on landfills, and mitigating greenhouse gas emissions, thus fostering environmental sustainability and bolstering the circular economy. Through the categorization of OW, communities can establish more efficient waste management approaches, ultimately resulting in improved resource utilization and decreased environmental footprint.

Effective waste management commences with the accurate categorization of OW into various classifications: agricultural waste, FW, yard waste, and industrial OW. Each classification possesses distinct attributes and necessitates particular procedures for handling and processing in order to enhance resource recuperation and reduce environmental repercussions.

Agricultural waste

Agricultural waste encompasses all the residual materials produced as a result of a wide range of agricultural practices, such as cultivating crops, raising livestock, and carrying out industrial activities related to agriculture. Agricultural waste can be mainly categorized into four main categories including crop residue, livestock waste, agro-industrial waste and aqua-cultural waste (Fig. 11.3). These materials comprise a diverse array of items like leftover plant matter from harvesting, animal excrement, discarded byproducts from food processing, as well as any other biodegradable substances of organic origin.

Improper management and disposal of agricultural waste can lead to significant environmental pollution, including soil degradation, water contamination, and air pollution. This necessitates the development of efficient waste management strategies to mitigate these adverse effects.

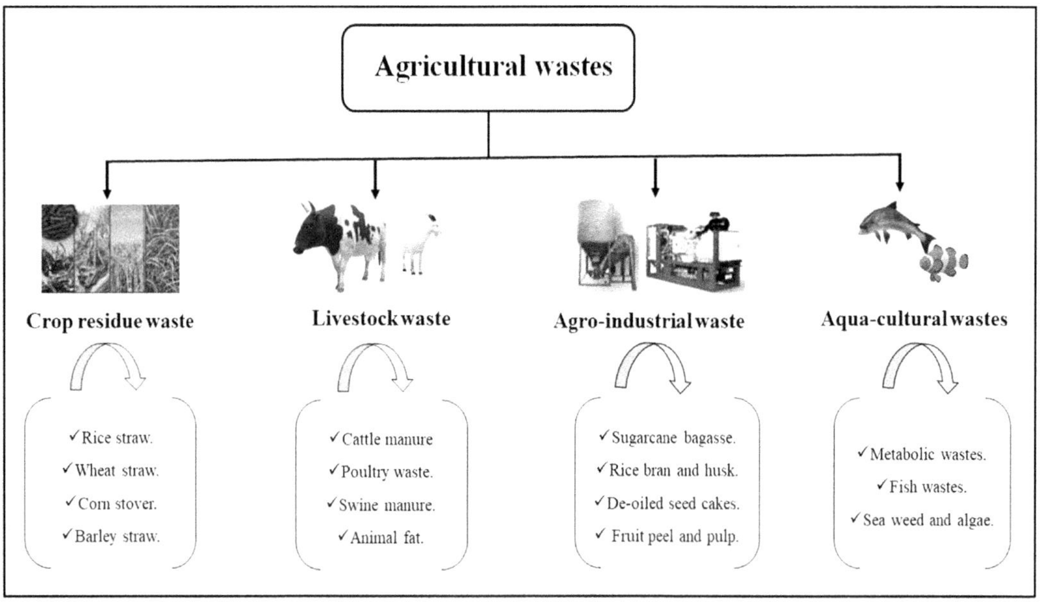

FIGURE 11.3 Different sources of agricultural waste categories. *Source: From S. Muhammad, H.P.S. Abdul Khalil, S. Abd Hamid, Y.M. Albadn, A.B. Suriani, S. Kamaruzzaman, A. Mohamed, A.A. Allaq, E.B. Yahya, Insights into agricultural-waste-based nano-activated carbon fabrication and modifications for wastewater treatment application, Agriculture 12 (10) (2022) 1737. https://doi.org/10.3390/agriculture12101737.*

Traditionally, agricultural production was circular, with minimal waste as residues were reused or recycled. However, modern agricultural practices have shifted toward a linear model, generating large quantities of waste that pose economic, environmental, and health challenges.

Agricultural waste is rich in nutrients such as nitrogen, phosphorus, and potassium, which can be valorized through various processes like composting, AD, and fermentation. These processes not only reduce waste but also produce valuable by-products like biogas, biofertilizers, and bioalcohols.

Current waste management models emphasize circular and bioeconomic approaches that focus on waste reduction, reuse, and recycling. These models aim to transform agricultural waste into raw materials for various industrial applications, thereby promoting sustainability and economic benefits [38].

Advanced technologies such as simultaneous saccharification and fermentation, and consolidated bioprocessing are being explored to enhance the efficiency of agricultural waste conversion into valuable products. These technologies help in reducing process time, costs, and environmental impact [39]. Composting is one of the most effective methods for managing organic agricultural waste. It improves soil fertility, enhances soil structure, and reduces reliance on synthetic fertilizers. However, it is crucial to use compost in moderation to prevent potential negative environmental impacts [40].

Sustainable agricultural waste management is essential for achieving environmental, economic, and social goals. Continued research and innovation in waste management technologies and practices will be key to transforming agricultural waste into valuable resources and achieving a zero-waste agricultural system.

Food waste

FW is defined as any food that is still edible but is thrown away or not consumed for various reasons, contributing to the accumulation of unnecessary waste. The sources of this types of FW includes home based, agriculture based, restaurant based and wedding party based FW (Fig. 11.4). This type of waste plays a crucial role in the composition of OW found in landfills, adding to the environmental challenges associated with waste management. Moreover, globally, FW stands out as a prominent contributor to the emission of greenhouse gases (GHGs), further exacerbating the issue of climate change.

When FW decomposes in landfills, it generates CH_4, which is a powerful greenhouse gas with a high potential to contribute to global warming and climate change. The process of composting FW, on the other hand, has the ability to effectively diminish the release of these harmful emissions into the atmosphere, thereby playing a crucial role in promoting environmentally friendly practices and offering a more sustainable solution for managing waste and its environmental impact. Composting FW not only mitigates greenhouse gas emissions but also recycles valuable nutrients back into the soil. This process enhances soil fertility and structure, reducing the need for synthetic fertilizers [41].

Proper management of FW through composting can lead to economic benefits by reducing waste disposal costs and creating compost that can be sold or used to improve agricultural productivity. It also has social benefits, such as reducing the environmental footprint of communities and promoting sustainable practices. One of the main challenges in

FIGURE 11.4 Sources of food waste generation. *Source: From A. Sarker, M.K. Ghosh, T. Islam, M. Bilal, R. Nandi, M.L. Raihan, M.N. Hossain, J. Rana, S.K. Barman, J.E. Kim, Sustainable food waste recycling for the circular economy in developing countries, with special reference to Bangladesh. Sustainability (Switzerland) 14 (19) (2022) 12035. https://doi.org/10.3390/su141912035.*

FW composting is managing emissions of CH_4 and N_2O. Research suggests that increasing the turning frequency during the thermophilic phase and reducing watering can help lower these emissions [42].

Numerous global and local efforts have been established to effectively address the issue of FW in a more environmentally sustainable manner. An example of this is the legislation implemented in California, which has set ambitious targets to efficiently manage millions of metric tons of OW by the year 2025. A considerable proportion of this waste includes FW, and the primary goal is to bring about significant reductions in greenhouse gas emissions.

Yard waste

Yard waste, which is commonly found in outdoor spaces around residential areas, generally encompasses a variety of organic materials that are biodegradable in nature, including but not limited to grass clippings, fallen leaves, twigs, and other remnants of garden maintenance and landscaping tasks that are routinely carried out. These materials are typically a byproduct of the regular upkeep and beautification efforts in gardens and outdoor spaces, stemming from activities such as mowing lawns, pruning plants, and clearing out debris to maintain a tidy and visually appealing environment.

When improperly managed, yard waste can contribute to significant environmental issues, including greenhouse gas emissions. Composting yard waste, however, can mitigate these emissions by reducing CH_4 and N_2O production compared to landfilling. Composting yard waste not only helps in reducing greenhouse gas emissions but also produces valuable compost that can enhance soil health and fertility. This process recycles nutrients back into the soil, promoting sustainable agricultural practices.

Various strategies can be employed to optimize the composting process and further reduce emissions. For instance, adjusting the C/N ratio and managing moisture levels can significantly impact the efficiency and environmental footprint of composting operations [43]. Recent studies have explored the addition of different organic materials, such as lignite and agro-FW, to yard waste composting. These additives can enhance nitrogen retention and reduce ammonia (NH_3) and greenhouse gas emissions, making the composting process more efficient and environmentally friendly [44].

Incorporating yard waste into composting systems is in accordance with a wider range of sustainable agricultural practices that aim to promote environmental and economic sustainability. This practice aids in the movement toward a bioeconomy framework by diminishing the need for artificial fertilizers and encouraging the utilization of natural fertilizers, ultimately leading to enhancements in soil composition, structure, and the activity of beneficial microorganisms.

Industrial organic waste

Industrial OW is defined as waste materials that are biodegradable in nature and are produced as a byproduct of various industrial activities such as food processing, agriculture, and other industries that are based on organic materials. It is of utmost importance to implement effective strategies for the management of this type of waste in order to ensure the long-term sustainability of the environment and to facilitate the recovery and utilization of valuable resources.

Industrial OW refers to waste materials that are biodegradable and are generated as a result of industrial processes like food processing, agriculture, and other industries that work with organic materials. Proper management of this waste is crucial for maintaining environmental sustainability and for extracting valuable resources. Implementing effective strategies for handling industrial OW is essential to reduce its impact on the environment and to promote resource recovery and utilization. By managing industrial OW efficiently, we can minimize pollution, conserve resources, and contribute to a healthier and more sustainable ecosystem.

By integrating various additives into the composting process, such as biochar derived from organic matter and co-substrates like FW, it is possible to enhance the efficiency and effectiveness of the decomposition of organic materials. These additional components play a crucial role in accelerating the breakdown of organic matter, resulting in a more efficient and faster composting process. Furthermore, the incorporation of such materials can lead to a significant reduction in nitrogen losses during the composting process, ultimately contributing to a more sustainable and environmentally friendly approach. Additionally, the presence of these additives can also have a positive impact on the overall quality of the final compost product, ensuring that it is rich in nutrients and beneficial microorganisms essential for soil health and plant growth [45].

Improper disposal of industrial OW can have severe consequences on the environment, such as contamination of soil and water, increased greenhouse gas emissions, and the creation of unpleasant odors. To address these issues, it is crucial to implement effective waste management practices. Composting stands out as a highly sustainable method for handling industrial OW. By converting waste into nutrient-rich compost, composting not only helps in reducing the environmental impact but also improves soil fertility and structure, supporting sustainable agricultural practices.

Challenges in organic waste management

SWM is required for achieving sustainable metropolitan development by collecting, transporting, and recycling the solid waste in advance for minimizing the harmful effects on the environment and humans. The waste management process comprises the collection, transportation, and deployment of different bins for disposing of waste and recycling (Fig. 11.5) challenging stages in SWM [9].

Collection and transportation

Waste collection is a critical component of waste management, though its prevalence varies significantly based on economic status. Upper-middle- and high-income countries typically provide nearly universal waste collection services [6]. According to the World Bank in 2018, low-income countries collect approximately 48% of waste in urban areas, but this rate falls significantly to 26% in rural regions [46]. Today, a variety of MSW collection systems are being adopted in modern cities [47]. Several key steps are crucial in designing an effective and sustainable MSW collection system. These include defining the waste fractions to be sorted and collected, selecting appropriate storage levels, locating disposal points and determining their volume, choosing suitable waste containers based on factors like the number of residents per disposal point, daily waste generation rate, sorting rate, collection frequency, and waste density, and mapping the distribution of disposal points for each waste fraction. Additionally, indicators such as transport distances and the number of vehicles needed are traditionally used to assess

FIGURE 11.5 Key stages in solid waste management (SWM). *Source: From S.V. Akram, R. Singh, A. Gehlot, M. Rashid, A.S. AlGhamdi, S.S. Alshamrani, & D. Prashar, Role of wireless aided technologies in the solid waste management: a comprehensive review. Sustainability 13 (23) (2021) 13104. https://doi.org/10.3390/su132313104.*

the efficiency of the MSW collection system [48]. The door-to-door collection system has been found to boost user participation in sorting and collecting solid waste. This approach has particularly led to an increase in the amount of recyclable waste collected, which in turn has helped reduce its environmental impact, including human toxicity, global warming, photochemical oxidation, acidification, and eutrophication [49]. Costs related to waste collection and transport, including fuel consumption and vehicle expenses, are crucial factors in designing an efficient and sustainable MSW management (MSWM) system. Typically, default parameters used for cost estimation and greenhouse gas emissions include fuel consumption (liters per ton of waste) and the number of collection vehicles. Three levels for each parameter—low, middle, and high—are defined, with scenarios suggesting that reducing the number of vehicles and fuel consumption leads to lower waste collection and transport costs per ton and enhances overall system sustainability [48]. A sustainable solid waste collection (SWC) model must also consider public health principles, legislative requirements, social impacts, economic factors, and other environmental considerations [50]. The SWC model encompasses various functional elements, including collection, sorting, separation, transfer, and disposal. It aims to address concerns related to esthetics, health, economic factors, energy conservation, and other environmental impacts [51]. Among all the steps, waste collection is considered the most crucial component for achieving sustainable SWC. However, SWC is a complex service that involves various operational and logistical activities, including equipment maintenance, preparation crews, health and safety measures, significant expenditures, and environmental concerns. It also requires consumer awareness and proper management of destination facilities for different types of waste [52]. The most common optimization objectives in waste management include optimizing collection routes to reduce distance and time, managing waste content, minimizing costs, and addressing environmental impacts. To achieve these goals in real-world scenarios, various constraints are considered. Common constraints include vehicle capacity limitations, time restrictions, characteristics of collection components, environmental factors, and regulatory requirements [53]. Globally, research on developing SWC models is advancing rapidly. Various methods are being employed to establish these models, including mixed-integer linear programming, deterministic, stochastic, probabilistic, and fuzzy approaches. Additionally, numerous meta-heuristic approaches are gaining popularity in waste collection optimization due to their simplicity, flexibility, and robustness [54]. Recently, the development of intelligent technologies has significantly advanced SWC systems. Technologies such as geographical information systems, global positioning systems (GPS), remote sensing, radio-frequency identification (RFID), ZigBee, image processing, and smart bins equipped with various sensors have been introduced and applied to enhance the effectiveness of SWC systems [53]. Advanced technologies not only lower maintenance costs and travel distances but also save significant amounts of time and reduce carbon emissions [55]. SWC operations can differ across countries due to various socio-economic factors, including population size, financial resources, road congestion, air pollution, availability of human labor, and the number of collection vehicles. Additionally, the waste collection process may vary based on the type of building, such as residential or commercial structures [56]. Typically, waste collection vehicles gather waste from pickup points and transport it to intermediate facilities or collection stations. As the volume of waste increases and travel distances to these facilities lengthen, transportation costs rise. Therefore, optimizing the transfer of waste to collection stations is crucial for minimizing expenses. One study proposed a robust multi-objective function for SWM that aims to minimize both costs and emissions. This approach involves optimally determining the locations of recycling plants, landfills, and waste transport vehicles, as well as selecting appropriate technologies, while considering the three dimensions of sustainability: social, economic, and environmental [57]. Designing an effective SWC system that takes into account the municipality's economic conditions is challenging and involves numerous factors. The cost of SWC is influenced by factors such as the municipality's area, geographic conditions (e.g., hilly or flat terrain), population size, land use (e.g., residential, industrial, or commercial), transportation modes, and the number of personnel involved. A significant portion of operational costs is attributed to fuel consumption, which depends on driving distance, road gradient, vehicle condition, and waste load. Environmental impacts can be categorized into three main areas: noise control, traffic congestion, and air pollution. Traffic congestion can exacerbate issues such as extended collection hours, increased fuel consumption, higher expenses, and more air pollution. Air pollution is a major concern, as waste collection vehicles can emit considerable amounts of pollutants, including GHGs [53]. SWC can ensure environmentally responsible waste disposal, prevent open dumping, and protect water and soil from chemically toxic waste. Additionally, SWC has the potential to generate renewable energy from OW [58,59]. Moreover, SWC can significantly reduce carbon emissions and prevent waste, especially plastics, from entering the oceans. Additionally, SWC technology enhances air quality in cities, supports urban development, and aids in slum upgrading by promoting efficient recycling and SWM [53]. An optimized SWC system also positively impacts human health by reducing the incidence of diseases such as Hepatitis A–C, HIV/AIDS, and malaria [60]. SWC can serve as a sustainable model to foster economic growth and create potential job opportunities [61,62].

Segregation and sorting

Segregation involves separating waste into distinct categories at the source, including recyclables, OW, hazardous materials, and general refuse. Effective segregation is crucial for efficient waste management, as it minimizes contamination, supports recycling efforts, and ensures the safe handling of hazardous materials. This process usually requires educating and motivating individuals and businesses to sort their waste properly before disposal.

Sorting takes place after segregation, often at a waste processing facility. During sorting, waste is further categorized using methods such as manual labor, conveyor belts, and advanced technologies like optical sensors and robotic arms. The goal of sorting is to isolate materials—such as plastics, metals, glass, and paper—for appropriate recycling or disposal. This process improves recycling efficiency, reduces landfill use, and aids in recovering valuable materials, which in turn lowers environmental impact and conserves resources. Proper collection and sorting from the source can effectively reduce the costs of subsequent waste management steps, optimize yield and profitability, and enhance material reuse efficiency while minimizing environmental burdens [63].

Treatment and processing

A study identifies five treatment technologies for managing FW in developing countries: animal feeding, composting, AD, incineration, and landfills. However, animal feeding and incineration are rarely used. The predominant methods for waste disposal are illegal open dumping and landfilling (90%), followed by composting (1%), AD (0.6%), and other methods (8.4%). Although incineration can reduce waste volume and extend landfill lifespan, it is not widely used for FW due to its high moisture content and the need for costly pretreatment processes [64]. The treatment of OW has several disadvantages, one of which is the production of bioaerosols. Bioaerosols are tiny airborne particles, ranging from 0.001 to 100 μm, that originate from biological sources. They can contain pathogenic or nonpathogenic microorganisms, either dead or alive, as well as fragments of living organisms. Bioaerosols may include bacteria, fungi, viruses, and other particles such as endotoxins, β-glucans, and spores [65]. Exposure to bioaerosols can pose a risk for respiratory diseases, including aspergillosis, asthma, rhinitis, and chronic obstructive pulmonary disease (COPD). Additionally, such exposure may lead to gastrointestinal disorders, as well as skin and eye inflammation [66].

Environmental and health impacts

Effective SWM helps reduce adverse health and environmental effects, conserves resources, and enhances urban livability. However, unsustainable SWM practices, which are worsened by rapid urbanization and financial and institutional constraints, can harm public health and environmental sustainability. Common disposal methods include uncontrolled dumping, open-air incineration, and landfilling. These practices result in negative consequences such as air and water pollution, land degradation, CH_4 emissions, hazardous leachate, and climate change. The environmental and health impacts of these methods impose significant costs, disproportionately affecting marginalized social groups [11].

Greenhouse gas emissions

Decomposing OW releases GHGs, such as CH_4 and carbon dioxide (CO_2), which contribute to climate change. CH_4, in particular, is a powerful greenhouse gas emitted from landfills and AD processes.

Land and water pollution

Mismanagement of OW, including illegal dumping or poor composting practices, can lead to contamination of soil and water. Leachate from landfills may taint groundwater, while runoff from composting sites can pollute surrounding water bodies.

Resource consumption

Inefficient waste management can result in excessive use of natural resources. For instance, using chemical fertilizers during composting can deplete soil nutrients and negatively impact soil health.

Biodiversity loss

Landfills and inadequately managed composting operations can disturb local ecosystems and wildlife habitats, leading to habitat degradation and a decline in biodiversity.

Health impacts

Respiratory issues

The breakdown of OW generates bioaerosols with microorganisms and particulate matter, which can cause respiratory conditions such as asthma, COPD, and other infections.

Gastrointestinal disorders

Contact with pathogens in decomposing OW can result in gastrointestinal problems, including infections and foodborne illnesses.

Skin and eye irritation

Handling decomposing OW can cause skin rashes and irritation. Additionally, exposure to bioaerosols can lead to eye inflammation and discomfort.

Vector-borne diseases

Poorly managed OW can attract vectors like flies, rodents, and mosquitoes, which can transmit diseases and spread infections (Fig. 11.6).

To address these issues, effective OW management practices such as proper segregation, composting, and treatment are crucial.

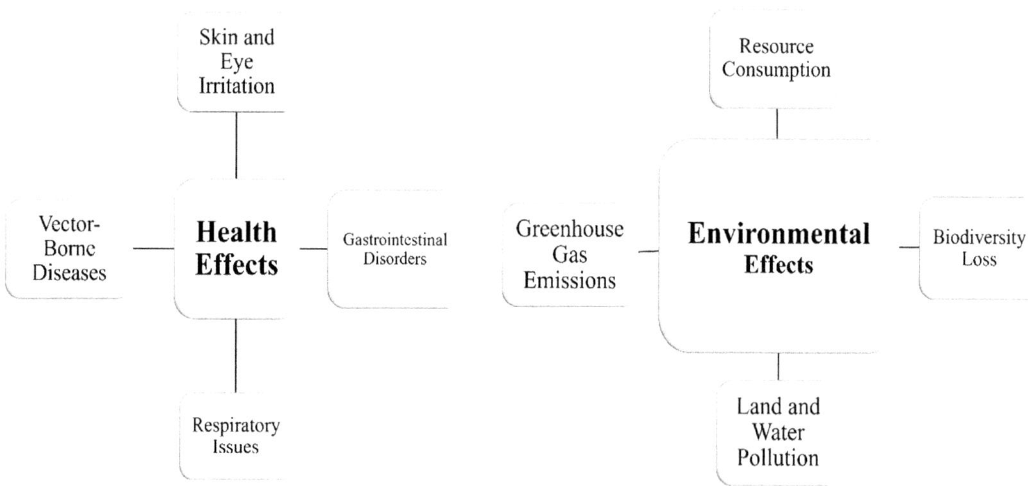

FIGURE 11.6 Environmental and health impacts on waste management.

Automation technologies in organic waste management

Sensors and Internet of Things devices

Sensors and Internet of Things (IoT) devices play a crucial role in modernizing OW management by providing real-time data, enhancing process efficiency, and enabling better decision-making.

Various types of sensors are employed in managing OW effectively. Firstly, ultrasonic level sensors are commonly used to measure the waste levels inside bins and trigger alarms when the maximum capacity is reached. In addition to level sensors, automatic hazardous gas sensors are also utilized. These sensors detect the emission of stinky and hazardous gases from OW, which can negatively impact the environment. Early detection enables authorities to take appropriate actions promptly [67]. Moreover, the transformation of OW into inert and sterilized material is increasingly important. To manage this process, Arduino-based electronic systems are used to control process parameters. These systems are integrated with user recognition and product traceability through RFID technology [68].

Automated sorting systems

Automated waste sorting techniques can be broadly classified into direct and indirect methods. Direct sorting techniques leverage the inherent physical properties of materials, such as magnetic susceptibility, electrical conductivity, and density, to separate different types of waste [69]. For instance, magnetic separation uses magnets to attract ferrous metals, while eddy current separation exploits induced currents to separate nonferrous metals, and gravity separation relies on density differences to separate materials using a heavy media suspension. Numerous direct technique processes such as Screw press [70], disk screen [71], Magnetic drum [72], magnetic head pulley, Eddy current [73], have been developed for automatic segregation of OW.

On the other hand, indirect sorting techniques depend on advanced sensor technologies to identify and locate recyclable materials within the waste stream [74]. These techniques utilize various sensors, such as optical, near-infrared (NIR), and X-ray sensors, to detect material characteristics and guide automated machines or robots to perform the sorting. Optical sorting, for example, uses cameras and light to differentiate materials based on color and shape [75], while NIR sorting employs spectroscopy to identify different types of plastics [76]. Indirect sorting is increasingly favored for its precision and ability to handle complex waste streams.

Robotics in waste handling

Robotics in waste management has emerged as a crucial technology, addressing inefficiencies and safety concerns in traditional waste storage and sorting processes. These systems use advanced sensors, such as optical, infrared, and X-ray sensors, combined with AI and machine learning (ML) algorithms, to accurately store, identify and sort various types of waste materials.

Improper collection and transportation of MSW in open containers pose significant environmental and health risks, contributing to pollution and disease spread [77]. This problem persists despite the high costs associated with its transportation and storage (https://doi.org/10.1007/s12008-023-01320-w). Advanced robots are becoming popular in addressing shortcomings of traditional waste storage and transportation systems. Any municipality can use advanced development to replace the traditional collection bins and their system because underground bins have only inlets above ground surface for garbage collection. For example, researchers developed a new sensor-based e-smart device to enhance the storage and transport of solid waste, utilizing technology for filling, picking, dropping, and weighing waste, along with GPS module tracking for energy conversion. Multi-compartment bins and underground containers, with capacities ranging from 4.5 to 8 m^3, can efficiently manage waste volumes. The e-smart system, featuring filling alarms, route tracking, and ultrasonic sensor controls, offered a cost-effective solution, reducing the impact of atmospheric conditions, health hazards, and odors. It can reduce open dumping and operational costs by up to 40% [77].

Beside storage and transportations, Robotic sorting systems are also becoming a widely adopted technology in waste management [78]. Robotic sorting systems typically feature robotic arms and conveyor belts that enable high-speed processing and high-volume handling of waste. These systems can distinguish between different materials, such as plastics, metals, and organics, thereby reducing contamination in recyclable streams and improving the quality of recycled materials [79]. The integration of machine vision and AI allows robots to learn and adapt to different sorting tasks, enhancing their efficiency and effectiveness over time. The primary advantages of robotic waste sorting include increased operational efficiency, reduced labor costs, and improved worker safety, as robots can handle hazardous materials and operate in dangerous environments [80]. However, these systems inherit some challenges such as the high initial investment costs, the complexity of integrating robotic systems with existing waste management infrastructure, and the need for continuous updates to keep pace with technological advancements [81]. Despite these challenges, the potential for robotics to revolutionize waste sorting and contribute to more sustainable waste management practices is significant. The ongoing development of AI and ML will likely enhance the capabilities of robotic sorters, making them even more precise and adaptable in the future.

Data analytics and machine learning

Data analytics and ML are transforming waste management by optimizing collection, transportation, storage, sorting, and transformation processes.

In waste collection, predictive analytics can forecast waste generation patterns, allowing for more efficient route planning and scheduling, thereby reducing fuel consumption and operational costs. Numerous mathematical and ML models have been deployed in forecasting of solid waste generation. For example, time series analysis, using statistical models like auto regressive integrated moving average (ARIMA) and seasonal ARIMA (SARIMA), is extensively employed in waste forecasting to analyze temporal data and predict future trends based on historical patterns. These models have proven effective in capturing seasonal variations and long-term trends in waste generation, as evidenced by various studies in the literature [82]. In addition, Regression models, both linear and nonlinear, are vital for waste forecasting. They analyze historical data to predict future waste volumes and elucidate relationships between variables such as population growth, economic factors, and waste generation, offering insights into waste production dynamics [83]. Similarly, classification models such as SVM, decision tree algorithms are being utilized for effectively categorizing waste types and predicting the likelihood of specific waste streams based on input variables [84].

ML algorithms can also analyze vast amounts of data from sensors and IoT devices to predict fill levels in waste bins, triggering timely pickups and preventing overflow. This real-time monitoring capability enables immediate responses to emerging issues, improving the efficiency and responsiveness of waste management systems. IoT sensors in waste bins, collection vehicles, and processing facilities provide real-time data on fill levels, temperature, and operational statuses [85]. This data is transmitted to centralized platforms and analyzed in real-time using advanced analytics techniques [86].

For transportation, data analytics helps in route optimization, minimizing travel time and emissions. Real-time monitoring also enhances operational efficiency by identifying inefficiencies or delays in collection routes promptly. Adjustments can be made dynamically, such as rerouting collection vehicles to avoid traffic congestion or optimizing routes based on current fill levels. This proactive approach minimizes fuel consumption, reduces operational costs, and improves service delivery to residents [87]. In terms of storage, smart systems can monitor environmental conditions and ensure waste is stored under optimal conditions to prevent hazards.

Furthermore, ML/DL also enhances sorting by improving the accuracy and speed of automated systems, identifying and segregating different types of waste materials with greater precision than manual methods [88]. Additionally, data-driven insights facilitate the transformation of waste into valuable resources, such as energy or compost, by optimizing processes like AD or composting [89]. Overall, the integration of data analytics and ML in waste management enhances operational efficiency, reduces environmental impact, and contributes to sustainable urban development.

Blockchain for waste tracking and transparency

Blockchain technology can revolutionize waste tracking and transparency by providing a decentralized and immutable ledger for all waste management activities. For instance, IBM and Veolia's collaboration on a blockchain platform for waste management demonstrates how blockchain can track the entire lifecycle of waste, from generation to disposal, ensuring that data remains transparent and tamper-proof. In this system, each transaction—such as waste collection, sorting, and recycling—is recorded on the blockchain, which helps in reducing fraud, enhancing data integrity, and improving regulatory compliance [90]. This technology not only streamlines operations through smart contracts but also enables real-time tracking, allowing stakeholders, including waste generators and recyclers, to access and verify data effortlessly. The increased transparency and efficiency achieved through blockchain can significantly enhance public trust and facilitate better waste management practices [91].

Composting systems

An automated composting system is a mechanized setup that has been specifically engineered and developed to efficiently oversee and regulate the entire composting procedure with minimal involvement from human operators. The primary objective of this system is to enhance and streamline the breakdown of OW materials into a highly beneficial and nutrient-rich compost product, all while ensuring maximum effectiveness and productivity in the composting process.

Automated composting systems produce high-quality compost that can improve soil fertility and crop yield, benefiting agricultural applications. These systems efficiently manage OW like livestock manure, FW, and green waste, reducing landfill use and encouraging recycling.

Overview of composting process

Initial Preparation involves gathering materials, such as organic matter like livestock manure, FW, green waste, and bulking agents such as sawdust, for the purpose of composting. Subsequently, these gathered materials are combined in precise proportions to ensure a harmonious balance of carbon and nitrogen content, a factor pivotal for facilitating microbial activity and optimizing the decomposition process.

After the initial preparation, the composting process goes through several distinct phases. The first phase is known as the Mesophilic Phase, characterized by moderate temperatures between 20°C and 40°C. During this phase, mesophilic microorganisms thrive and aid in breaking down easily degradable compounds. These microorganisms play a crucial role in the decomposition process. Following the mesophilic phase, the compost enters the thermophilic phase where temperatures rise significantly to a range of 40°C–70°C. This increase in temperature creates an ideal environment for thermophilic microorganisms to flourish. These microorganisms are responsible for breaking down more complex organic matter and pathogens present in the compost. Finally, the compost progresses into the cooling and maturation phase. In this phase, the temperature gradually decreases, allowing mesophilic microorganisms to reestablish themselves. This stage is essential for further stabilizing the compost and enhancing humification, which is the process of organic matter transforming into humus, a vital component of healthy soil.

Automated aeration and turning

Automated aeration and turning, which involve the use of mechanical equipment, are essential processes employed in the composting method to guarantee the adequate supply of oxygen and thorough mixing of materials. These procedures play a crucial role in promoting efficient microbial activity and facilitating the decomposition process, ultimately contributing to the overall success of the composting operation.

Automated aeration and turning in composting ensure proper oxygen supply and material mixing for efficient microbial activity and decomposition. They prevent anaerobic conditions, regulate compost temperature for pathogen destruction, and reduce manual labor, making the process more efficient. Automated turning in composting provides several advantages such as uniform mixing of compost materials, even distribution of moisture, nutrients, and microorganisms, breaking up clumps, increasing surface area for microbial activity, and controlling odors by preventing anaerobic pockets.

Modern automated composting systems are designed to incorporate cutting-edge technology in the form of sensors and controls. These innovative components are responsible for continuously monitoring crucial factors such as temperature, moisture levels, and oxygen concentrations within the composting environment. The main purpose of this meticulous monitoring is to facilitate precise adjustments that are essential for optimizing the overall composting process. By utilizing programmable systems, operators have the ability to schedule turning and aeration activities at specific intervals. This strategic approach ensures that consistent and optimal conditions are maintained to support the necessary microbial activity. Ultimately, the integration of such advanced technology significantly boosts the efficiency and effectiveness of composting operations, leading to more successful outcomes [92].

Monitoring and control systems

Automated composting systems frequently make use of IoT technology in order to establish a connection between compost bins and a network, which in turn facilitates the capability for real-time monitoring and control over a multitude of parameters including temperature, moisture, and pH levels. Through this seamless integration, users are empowered with the ability to remotely manage the composting system, thereby elevating both the efficiency and efficacy of the composting process to new heights.

These systems employ a range of sensors to monitor environmental conditions within the compost bin. Actuators are used to make real-time adjustments based on sensor data, ensuring optimal conditions for composting. For example, moisture sensors can trigger irrigation systems to maintain the desired moisture level. Advanced control systems in automated composting use algorithms like proportional-integral-derivative (PID) to regulate environmental conditions. This ensures that the system can maintain ideal conditions for composting without the need for constant human intervention.

Monitoring systems in automated composting setups are furnished with sensors that gage essential parameters like temperature, moisture, oxygen levels, and pH. The role of temperature sensors is to guarantee that the compost heap remains within the thermophilic range, a crucial factor for effective decomposition and elimination of pathogens. Moisture sensors are responsible for identifying and sustaining the optimal moisture levels, thus preventing the compost from either excessively drying out or becoming overly saturated. Oxygen sensors are tasked with overseeing the air circulation within the compost heap to ensure proper aeration and to avoid anaerobic conditions that may result in undesirable odors and decelerated decomposition. pH sensors play a pivotal role in monitoring the acidity or alkalinity of the compost, variables that can influence microbial activity and the pace of decomposition.

Control systems utilize the data obtained from monitoring sensors to automatically make adjustments and uphold optimal conditions for composting. For instance, in cases where temperature sensors detect excessive heat within the compost, the system is capable of activating cooling mechanisms or modifying aeration to reduce the temperature. Likewise, if the moisture levels deviate from the ideal range, the control system can introduce water or improve drainage to rectify the situation. The operation of aeration fans or turning mechanisms is determined by oxygen and moisture measurements to ensure consistent blending and appropriate air dispersion across the compost heap.

Advanced monitoring and control systems for composting include data analysis, reporting features, and real-time reporting to track progress, identify trends, and predict issues. Alerts and notifications can be set up to inform operators of any deviations from optimal conditions for prompt intervention. Integration with other automated systems, such as automated feeders or temperature-controlled environments, enhances the overall efficiency of composting operations. By automating the control of multiple parameters, these systems minimize manual intervention and ensure consistent composting conditions. Programmable settings allow operators to customize the composting process according to specific requirements or types of OW.

Benefits of automated composting

Enhanced efficiency and effectiveness

Automated composting systems utilize IoT technology to monitor and control various parameters such as temperature, moisture, and pH levels in real-time. This ensures optimal conditions for composting, making the process more efficient and effective.

Reduced labor and time

By automating the composting process, these systems significantly reduce the need for human labor and time. Sensors and actuators manage the composting conditions, allowing for minimal manual intervention.

Real-time monitoring and control

The integration of IoT allows for real-time monitoring and control of the composting process. Users can remotely manage the system through user-friendly interfaces, often accessible via mobile applications or web platforms.

Data logging and analysis

Automated composting systems often include data logging capabilities, which allow for the collection and analysis of historical data. This helps in optimizing system performance and identifying any issues that may arise over time.

Environmental sustainability

These systems contribute to environmental sustainability by reducing the need for chemical fertilizers and minimizing waste. By optimizing composting conditions, they help produce high-quality compost that can enhance soil fertility and plant growth.

Advanced control algorithms

The use of advanced control algorithms, such as PID, ensures that the system maintains ideal conditions for composting without constant human intervention. This leads to a more consistent and reliable composting process.

Deep learning and artificial intelligence integration

Some automated composting systems incorporate deep learning algorithms, such as convolutional neural networks (CNNs), to enhance their monitoring and control capabilities. These algorithms can be used for tasks like waste segregation, ensuring that only compostable materials are processed.

User-friendly interfaces

Modern automated composting systems are designed with user-friendly interfaces, making them accessible to a wider range of users, including those who are new to composting. This ease of use encourages more people to adopt composting practices.

Reduced environmental impact

By optimizing conditions and reducing the need for manual intervention, automated composting systems can significantly reduce their environmental impact. They help in minimizing waste and reducing the emission of GHGs, contributing to more sustainable practices.

Versatile applications

Automated composting systems can be applied in various fields, including kitchen waste management and vermicomposting. Intelligent systems can handle tasks from waste segregation to compost soil transportation, making the entire process more efficient and less labor-intensive.

By integrating these advanced technologies and methodologies, automated composting systems offer numerous benefits that enhance efficiency, sustainability, and ease of use.

Biogas production from organic waste

Introduction to biogas production

Biogas is a gaseous biofuel produced through the fermentation of biomass in an oxygen-free environment and at a controlled temperature by microorganisms. It primarily consists of CH_4 and CO_2, with smaller amounts of water vapor (H_2O), oxygen (O2), nitrogen (N2) and hydrogen sulfide (H2S) [93]. Biogas is derived from the biochemical process of AD, where microorganisms break down biodegradable biological materials in the absence of oxygen. The composition and properties of biogas are influenced by the type of substrate being digested, the digestion plant's design, and the process's conditions. Additionally, its chemical composition affects how biogas production operations are optimized to enhance its utilization [94]. Various waste materials are utilized in the AD process to sustainably produce biogas. These include organic components of MSW, industrial waste, wastewater sludge, and solid waste from agricultural activities. Additionally, process water and agricultural waste from livestock and crop production are also used in the AD process [95]. The efficiency of the AD process and the quantity of biogas produced are influenced by several factors, including total solids (TS), total volatile solids (TVS), moisture content (MC), hydraulic retention time (HRT), the type of substrate used, pH, the presence of toxic metals in the substrate and the C/N ratio of the substrate [96]. The production of biogas from OW involves four distinct stages, each facilitated by different microorganisms (Fig. 11.7).

1. **Hydrolysis:** Complex organic molecules such as proteins, carbohydrates, and fats are broken down into simpler molecules like amino acids, sugars, and fatty acids through the action of enzymes and anaerobic bacteria.
2. **Acidogenesis:** The simpler organic molecules generated during hydrolysis are converted into volatile fatty acids (VFAs) in this stage.
3. **Acetogenesis:** In this stage, the VFAs produced are transformed into acetic acid, hydrogen, and CO_2.
4. **Methanogenesis:** The final stage of the AD process, where acetic acid and hydrogen are converted into CH_4 and CO_2 [97].

The conversion of OWs to biogas involves several challenges at different stages of the process:

1. Pretreatment: This stage often requires the removal of contaminants and the preparation of waste materials to make them suitable for AD. Challenges include ensuring uniformity in the waste material and effectively managing any toxic substances.
2. Hydrolysis: During hydrolysis, complex organic compounds are broken down into simpler molecules. The challenge lies in efficiently managing this breakdown, as it can be slow and may require optimal conditions to ensure complete conversion.

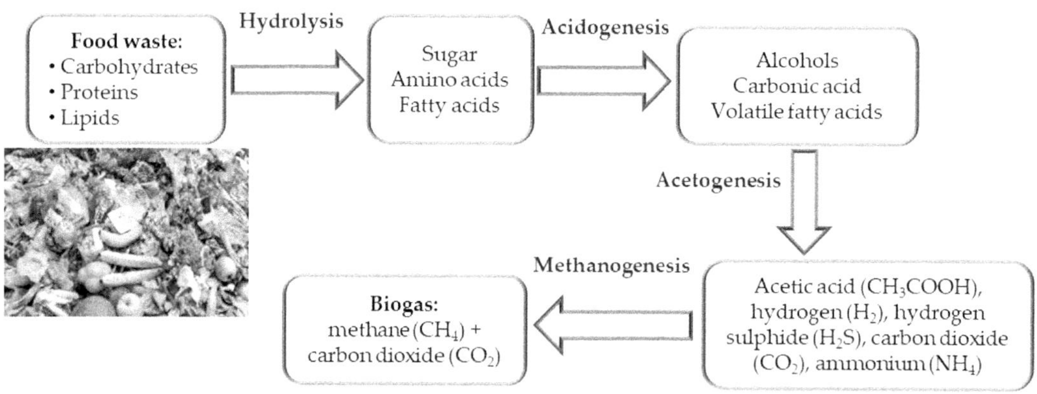

FIGURE 11.7 Steps of anaerobic digestion process. *Source: From M. Ferdeș, B.Ș. Zăbavă, G. Paraschiv, M. Ionescu, M.N. Dincă, G. Moiceanu, Food waste management for biogas production in the context of sustainable development. Energies 15 (17) (2022) 6268. https://doi.org/10.3390/en15176268.*

3. Bioconversion: In the bioconversion phase, the simpler molecules are further processed into biogas. Challenges here include maintaining the right microbial balance and optimizing conditions to maximize biogas production.
4. Fuel separation: Finally, separating biogas from the digestate and removing impurities can be challenging. Effective separation is crucial for ensuring high-quality biogas and managing residual waste.

Addressing these challenges is essential for improving the efficiency and effectiveness of biogas production from OW [98,99].

Biogas has numerous applications, primarily for energy generation. Using biogas and biomethane as energy sources offers environmental advantages, including reducing greenhouse gas emissions, substituting fossil fuels, and improving energy production efficiency [93].

Automated anaerobic digesters

AD is commonly used to process various OWs while simultaneously producing renewable energy and nutrient-rich digestate. The effectiveness of AD systems is influenced by multiple factors, such as growth conditions, operational parameters, system design, and digester type. However, AD processes can suffer from instability, which negatively impacts biogas production. The complexity of AD, driven by factors like substrate composition, temperature, pH levels, hydraulic retention time, and microbial dynamics, makes monitoring and optimization challenging (Fig. 11.8). To address these issues, significant efforts have been made to develop strategies for controlling and predicting AD performance. Recently, ML has become a valuable tool in optimizing the AD process, predicting uncertain parameters, detecting disturbances, and enabling real-time monitoring. ML is especially useful for understanding the nonlinear relationships within AD systems, improving process control, operational safety, and performance forecasting [100–103]. ML involves a three-step cycle:

Training: The algorithm is provided with a training dataset to learn patterns and relationships within the data.
Validation: A separate dataset is used to refine and optimize the model by adjusting the classifier's hyperparameters.
Testing: A different dataset is employed to evaluate the final accuracy and performance of the model [104].

Various ML algorithms, including artificial neural networks (ANN), adaptive neuro-fuzzy inference systems (ANFIS), k-nearest neighbors (KNN), random forests (RF), and support vector machines (SVM), have been used to model the nonlinear and intricate relationships within the AD process [105]. Various models have been created to represent the different processes involved in AD. However, the main challenges with these modeling approaches are their complexity in development and the need for prior knowledge [106]. Therefore, these models are commonly utilized as valuable tools for planning and designing AD systems. Examples include ADM1, AQUASIM, ANESSA, and GRAINIT BIOGAS [107]. Among these models, AD Model No. 1 (ADM1) is one of the most widely used. It includes a range of reaction kinetic equations and is highly regarded for modeling the AD process.

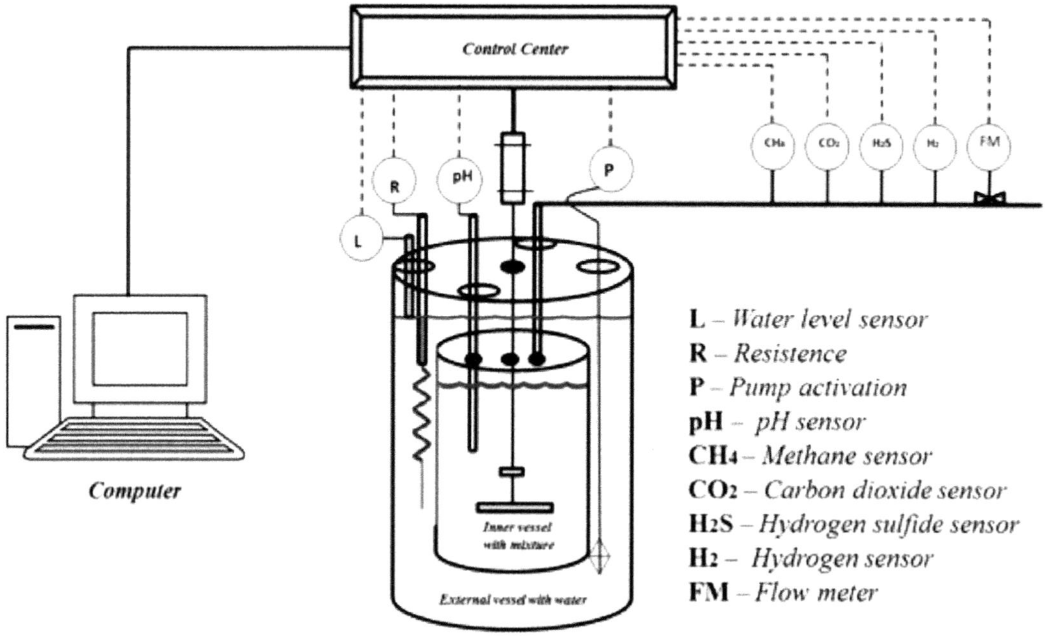

FIGURE 11.8 Complete representation of automated anaerobic biodigester. *Source: From C.D.S. Guimarães, D.R.D.S. Maia, E.G. Serra, Construction of biodigesters to optimize the production of biogas from anaerobic codigestion of food waste and sewage. Energies 11 (4) (2018) 870. https://doi.org/10.3390/en11040870.*

However, calibrating ADM1 presents significant challenges due to the complexity of microbial species and their metabolic pathways [108]. Due to variations in microbial species and complex metabolic pathways, traditional mathematical models often struggle to accurately estimate performance. However, data-driven models, such as those based on AI, offer a viable alternative to process-based modeling, which relies on intricate physico-chemical processes. AI-based models are developed using historical data of system variables and can be employed for real-time operation of AD plants by leveraging online data [109].

Case studies of automated biogas plants

AD primarily involves the biological conversion of organic matter into biogas. The main product of this process is biogas, which consists mainly of CH_4 and carbon dioxide (CO_2). Additionally, the process produces a nutrient-rich digestate as a byproduct. Biogas can be utilized in various ways, including generating heat through combustion, producing both heat and electricity via combined heat and power (CHP) plants, generating electricity with turbines, converting it into natural gas by separating CO_2, or creating fuels by increasing the CH_4 concentration or using CO_2-assisted catalytic reforming [110]. Unlike other renewable energy sources, electricity generation from biogas is not affected by weather conditions. Biogas plants can generate electricity flexibly based on demand, as they can store both feedstock and biogas for use as needed [111]. In biogas plants, organic materials are converted into biogas through anaerobic conditions and a combination of physical and biochemical processes. A typical biogas plant features a biomass storage area, pretreatment facilities, a digester for biochemical reactions, biogas processing units, and digestate processing units. In Europe, substrates such as energy crops, agricultural residues, bio and MSW, industrial waste, and sewage are commonly used in biogas plants. Despite being a well-established technology, AD faces challenges in conventional commercial biogas plants, including poor system stability and low CH_4 production efficiency. Issues like process instability can lead to decreased biogas output, acidification, foaming, or even complete system failure. For instance, Danish centralized biogas plants frequently experience process instability that results in a 20%–30% loss in biogas production. Additionally, foam formation, observed in 80% of plants studied in Germany, Denmark, and America, can reduce biogas yield by 30%–50% and occasionally cause total process failure. Traditional methods for monitoring and correlating operational parameters are often inadequate. As AI continues to advance and integrate into various sectors, its incorporation into biogas plant operations is expected to become crucial for improving performance and stability [112–116].

Biofuel production from organic waste

Biofuel production derived from OW materializes as a sustainable procedure that transforms diverse organic substances into fuels abundant in energy. This strategy effectively tackles issues in waste management while presenting a substitute for conventional fuels, thereby bolstering environmental sustainability and fortifying energy stability. The standard procedure encompasses pivotal stages such as gathering, treatment, transformation, and application.

Biofuel production from OW is an innovative approach to managing waste while generating renewable energy. This process involves converting organic materials, such as kitchen waste, agricultural residues, and livestock manure, into biofuels like biogas, biodiesel, and bioethanol. This not only helps in waste management but also reduces reliance on fossil fuels and mitigates environmental pollution.

Types of biofuels from organic waste

Biogas

Biogas is produced through the AD of OW by microorganisms in the absence of oxygen. This process generates a mixture of CH_4 and carbon dioxide, which can be used as a renewable energy source for heating, electricity, and even as a vehicle fuel.

Biodiesel

Biodiesel is derived from OW such as used cooking oil, animal fats, and agricultural residues. The process involves transesterification, where fats or oils react with an alcohol to produce biodiesel and glycerol. Biodiesel can be used in diesel engines and is a cleaner alternative to conventional diesel fuel.

Bioethanol

Bioethanol is produced by fermenting the sugars and starches found in OW materials like crop residues, kitchen waste, and certain types of industrial waste. This biofuel can be blended with gasoline to reduce emissions and improve fuel efficiency in vehicles.

Biohydrogen

Biohydrogen is generated through the fermentation of OW by specific bacteria. This biofuel is considered a clean energy source as it produces only water when combusted. Biohydrogen can be used in fuel cells to generate electricity or as a direct fuel for vehicles.

Syngas (synthesis gas)

Syngas is produced through the gasification of OW, which involves heating the waste material at high temperatures in the presence of a controlled amount of oxygen. The resulting gas mixture, primarily composed of hydrogen, carbon monoxide, and carbon dioxide, can be used to produce electricity, heat, or as a feedstock for producing other biofuels.

By converting various types of OW into these biofuels, we can effectively manage waste while generating renewable energy, contributing to environmental sustainability and reducing reliance on fossil fuels.

Automation in biofuel production

Automation in biofuel production entails the utilization of cutting-edge technologies like IoT, sensors, and control systems to elevate the efficiency and effectiveness of the biofuel production process. By incorporating these technologies, it aids in the real-time monitoring and control of multiple parameters, thereby guaranteeing the maintenance of optimal conditions for the generation of biofuels.

Internet of Things integration

The IoT plays a crucial role in automating biofuel production. IoT devices, such as sensors and actuators, are used to monitor and control environmental conditions like temperature, moisture, and pH levels in real-time. This ensures that the biofuel production process is efficient and consistent.

Real-time monitoring

Automation allows for real-time monitoring of the biofuel production process. Sensors collect data on various parameters, which is then analyzed to ensure that the conditions are optimal for the production of biofuels like biogas, biodiesel, and bioethanol. This helps in maintaining the quality and yield of the biofuels.

Control systems

Advanced control systems, such as Programmable Logic Controllers (PLCs) and software like MCGS, are used to automate the biofuel production process. These systems can regulate the input and output of materials, control the fermentation or digestion processes, and ensure that the production is carried out efficiently.

Automation in biofuel production streamlines tasks such as waste segregation, monitoring, and control, reducing the reliance on human labor. This not only cuts labor costs and minimizes errors but also boosts system efficiency, productivity, and processing speed, resulting in increased biofuel output and better resource utilization. Automation systems gather substantial data during production, which can be analyzed to detect patterns, optimize processes, and facilitate informed decision-making. Additionally, the application of ML algorithms and deep learning techniques can further improve the efficiency and yield of biofuel production.

Automated biofuel production systems help in reducing the environmental impact of waste management. By efficiently converting OW into biofuels, these systems reduce the amount of waste sent to landfills and lower greenhouse gas emissions. This contributes to a more sustainable and eco-friendly approach to energy production.

By leveraging automation and advanced technologies, biofuel production from OW can be made more efficient, cost-effective, and environmentally friendly.

Technologies for biomass conversion

Biomass conversion methodologies encompass a series of procedures aimed at converting organic matter, including agricultural residues, forest residues, and MSW, into beneficial commodities like biofuels, chemicals, and energy. These methodologies can be classified into distinct categories, each characterized by unique techniques and utilities in the conversion of biomass into practicable forms.

Biomass conversion technologies play a crucial role in tackling the worldwide energy consumption and promoting environmental sustainability through the provision of substitute options for fossil fuels. Biomass conversion into an enriched product involves different biomass sources and varieties, and the determination of the best conversion process for each situation depends on many factors, such as the type and quantity of biomass, the desired product, environmental standards, economic conditions, and available infrastructure. The most common processes used today are of thermochemical and/or biological nature (Fig. 11.9). Thermochemical techniques, such as pyrolysis, have proven to be highly efficient in facilitating the generation of biofuels, as well as heat and electricity, using a wide range of biomass sources [117].

Additionally, advancements in AI and IoT integration have revolutionized the energy sector by optimizing biomass conversion processes, enhancing efficiency, and minimizing

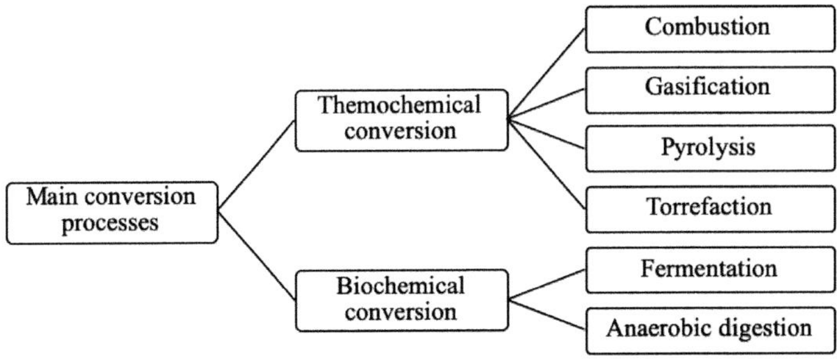

FIGURE 11.9 Main biomass conversion technologies. *Source: From L.C.R. Sá, L.M.E.F. Loureiro, L.J.R. Nunes, A.M.M. Mendes, Torrefaction as a pretreatment technology for chlorine elimination from biomass: a case study using Eucalyptus globulus Labill. Resources 9 (5), 54. https://doi.org/10.3390/resources9050054.*

downtime in power plants [69]. Biomass utilization presents numerous benefits to consider, including its cost-effectiveness due to the abundance of biomass sources available for energy production. Additionally, biomass boasts a lower sulfur content compared to traditional fossil fuels, which contributes to reducing harmful emissions. Furthermore, utilizing biomass for energy generation can significantly decrease greenhouse gas emissions, making it a viable and sustainable option to explore [118].

Here is a summary of the main technologies used in converting biomass.

AD: This process involves the breakdown of OW by microorganisms in the absence of oxygen, producing biogas (a mixture of CH_4 and carbon dioxide) which can be used for heating, electricity, and as a vehicle fuel.

Transesterification: Used primarily for producing biodiesel, this chemical process involves reacting fats or oils with an alcohol to produce biodiesel and glycerol. It is commonly applied to OW such as used cooking oil and animal fats.

Fermentation: This biological process converts sugars and starches from OW into bioethanol. The process is widely used for converting crop residues, kitchen waste, and certain industrial wastes into a renewable fuel that can be blended with gasoline.

Gasification: This thermal process involves heating OW at high temperatures in the presence of a controlled amount of oxygen to produce syngas (a mixture of hydrogen, carbon monoxide, and carbon dioxide). Syngas can be used for electricity generation, heating, or as a feedstock for other biofuels.

Composting: While primarily used for producing compost rather than biofuel, composting OW can be integrated with other processes like AD to enhance biogas production. Automated systems using IoT and sensors can optimize the composting process by monitoring parameters such as temperature, moisture, and pH levels.

Deep learning and IoT: Advanced technologies like deep learning and IoT are used to automate and optimize the biomass conversion process. For instance, IoT devices can monitor and control environmental conditions in real-time, while deep learning algorithms can analyze data to improve process efficiency.

Hyperspectral imaging (HSI): This technology is used for the automatic identification and classification of compostable and biodegradable plastics, which can then be processed into biofuels. HSI helps in sorting materials accurately, thereby improving the efficiency of the biomass conversion process.

By employing these technologies, the conversion of biomass into biofuels can be made more efficient, sustainable, and environmentally friendly.

Challenges and solutions

High initial investment costs: The integration of advanced technologies such as AI and IoT in biomass-to-biofuel conversion requires significant initial investment, which can be a barrier for widespread adoption.

Skill requirements and training: The deployment of sophisticated technologies necessitates skilled personnel, which can be a challenge in regions lacking technical expertise. Training programs and educational initiatives are essential to bridge this gap.

Environmental impact and sustainability: Traditional biomass-to-energy technologies can lead to high pollution levels, deforestation, and degraded forests. Adopting advanced conversion technologies like nanophotocatalysis and pyrolysis can mitigate these environmental impacts by enhancing efficiency and reducing emissions.

Technological advancements: The need for more powerful and efficient nanomaterials for photocatalytic applications is crucial to support the conversion of biomass into energy. Continuous research and development in this area can lead to significant improvements in biofuel production processes.

Feedstock variability: Biomass feedstocks vary widely in their composition and properties, making it challenging to standardize conversion processes. Advanced technologies like pyrolysis and coprocessing can handle a wide variety of feedstocks, producing value-added products and enhancing biofuel yield.

Economic and environmental performance: Choosing the optimal use for biomass involves balancing economic viability and environmental sustainability. Policymakers and industry stakeholders must consider these factors to promote the adoption of biomass energy solutions.

Data privacy and ethical considerations: The use of AI and IoT in biofuel production raises concerns about data privacy and ethical issues. Ensuring equitable access and addressing these concerns is vital for the successful implementation of these technologies.

Waste utilization and reduction: Efficient conversion technologies such as thermal, chemical, and biological processes can transform OW into biofuels, reducing waste and contributing to environmental sustainability.

Integrated approaches, such as the implementation of biorefineries which combine various bio-based processes and the adoption of circular economy concepts that promote the reuse and recycling of resources, play a crucial role in tackling the complex array of challenges faced in the bioenergy sector and in guaranteeing the long-term viability and environmental friendliness of bioenergy production [119].

Moreover, the utilization of waste feedstocks such as waste cooking oils (WCOs) and MSW in the production of biofuels like biodiesel and biojet fuel holds the potential to notably diminish greenhouse gas emissions and play a role in the reduction of waste [2]. Sustained efforts in research and development, coupled with the implementation of supportive policies, are deemed essential in surmounting these obstacles and unlocking the complete capabilities of biofuels as a sustainable energy source [120].

By addressing these challenges through technological innovation, policy support, and capacity building, the production of biofuels from OW can become a viable and sustainable energy solution.

Case studies

Successful implementations of automated organic waste management

San Francisco, United States

San Francisco is renowned for its progressive waste management policies and has implemented various automated systems to enhance efficiency and sustainability. The city aims to achieve zero waste by 2025, making it a leader in OW management.

San Francisco has invested in state-of-the-art automated sorting facilities that use advanced technologies like AI and robotics to efficiently separate OW from other waste streams. These facilities significantly reduce the need for manual sorting, thereby increasing processing speed and accuracy.

The city employs advanced thermochemical and biochemical processes to convert OW into biofuels and other valuable products. These technologies enhance the efficiency of waste-to-energy conversion and reduce greenhouse gas emissions.

San Francisco has achieved remarkable success in the realm of automated OW management, which serves as a clear testament to the seamless integration of cutting-edge technology, well-crafted policy frameworks, and robust community involvement. The waste management infrastructure in this city capitalizes on sophisticated sorting mechanisms, innovative composting solutions, and forward-thinking waste-to-energy projects, all aimed at boosting operational effectiveness and overall efficiency, even in the face of significant financial constraints and infrastructural hurdles [121]. A notable implementation is the use of AI tools to tackle MSWM concerns, which helps in solving complex problems, learning from mistakes, and managing uncertainties [122]. Additionally, the development of fully automated composting machines, which utilize IoT and node microcontrollers, has significantly reduced FW by converting it into high-nutrient flakes suitable for soil enrichment and pest control [123].

San Francisco has made significant strides in reducing landfill waste through the implementation of various automated systems and technologies. One of the key strategies involves the automation of waste auditing processes, which helps in the continuous acquisition of waste generation data at the workstation level. This data-driven approach supports waste management decision-making and has the potential to reduce waste in industrialized building construction facilities by using contour-based image processing algorithms and optimized deep learning classification models to quantify and identify waste materials [124]. Additionally, the city has adopted automated workstations for landfill operators, which include wheel-load scales, video locking systems, recognition modules, and database management systems. These systems automate the accounting of incoming waste, generate detailed reports, and monitor landfill processes to prevent environmental emissions, thereby improving the quality and productivity of landfill management [125].

San Francisco's successful implementation of automated OW management systems serves as a model for other cities. By leveraging advanced technologies, fostering partnerships, and focusing on sustainability, the city has made significant strides toward achieving its zero-waste goals.

Milan, Italy

Milan set out to improve its sustainability metrics by tackling urban OW management. They implemented automated systems for collecting OW directly from households and distributed smart bins with RFID technology to ensure proper waste separation. The collected waste was taken to centralized facilities for sorting and composting, resulting in a significant decrease in landfill waste, effective waste separation, and the production of quality compost for urban agriculture and landscaping. Overcoming initial pushback from residents, the city carried out extensive public education campaigns and smoothly integrated the new systems with existing waste management processes through a phased approach.

Successful implementations of automated OW management in Milan, Italy, can be seen through various innovative approaches and technologies. The Milan Pilot within the Reflow-EU project has been instrumental in collecting data and driving initiatives aimed at enhancing waste management efficiency [126]. One notable example is the development of Arduino-based automatic composting machines designed to maintain optimal temperatures during the composting process, thereby improving the performance of microorganisms and reducing the time required for waste decomposition. These machines operate either on a timed schedule or based on temperature sensors, ensuring consistent compost quality with favorable pH, nitrogen, and carbon levels [127]. Additionally, a fully automated composting machine has been developed to reduce FW by converting it into high-nutrient flakes using heating, cooling, and grinding mechanisms. This machine, integrated with IoT and a mobile application, not only simplifies the composting process but also minimizes odor and notifies users when the compost is ready, making it suitable for both household and industrial use [123].

Milan's automated OW management system has significantly enhanced sustainability by integrating advanced technologies and innovative approaches to waste sorting and disposal. The system employs ML-based object detection techniques to automate waste sorting, which reduces contamination and improves recycling rates, thereby minimizing environmental impact [128].

Milan's waste management has significantly enhanced sustainability through several innovative practices and strategic implementations. One of the key initiatives is the ambitious model adopted by Milan-Bicocca University, which involved removing isolated waste bins from offices and rooms and replacing them with centralized waste collection islands in public spaces. This model required a highly coordinated procedure and communication plan, which not only streamlined waste collection but also promoted recycling and proper waste segregation among the university community [129]. As a result of this centralized waste collection system, Milan-Bicocca University saw a notable decrease in overall waste generation and a marked increase in the amount of recyclable materials diverted from landfills.

In conclusion, Milan's innovative approach to automated OW management has not only significantly enhanced sustainability metrics but also revolutionized waste sorting and disposal practices. By incorporating advanced technologies such as RFID smart bins, Arduino-based composting machines, and ML-based object detection, Milan has successfully reduced landfill waste, improved recycling rates, and minimized environmental impact. The city's successful integration of these systems, coupled with extensive public education campaigns, highlights the importance of strategic implementation and community involvement in achieving sustainable waste management practices. Milan's commitment to sustainability through automated OW management serves as a model for other cities looking to enhance their environmental initiatives and create a more sustainable future.

Seoul, South Korea

Seoul was dealing with more waste management issues as the city grew rapidly and the population increased. They decided to bring in automatic FW collection systems, which included intelligent bins that charged fees based on volume, and sensors that used IoT technology to track waste levels in real-time, making collection routes more efficient. The collected FW was then treated in biogas plants to produce renewable energy. As a result, there was a significant decrease in FW production, improved efficiency in waste collection and processing, and an increase in biogas production, all helping the city reach its renewable energy targets. To cover the costs of the advanced technology, the city used government subsidies and also encouraged the public to get involved by offering financial rewards for reducing waste.

Seoul, South Korea has successfully implemented automated systems for managing OW, utilizing advanced technologies to tackle waste management challenges. One impressive instance is the anaerobic codigestion facility in Seosan City, which handles 320 m^3/day of various OWs like pig manure, FW, excretion, and thickened sewage sludge through a horizontal anaerobic digester. Operating at a mesophilic digestion temperature of 35°C–40°C, this facility achieves a CH_4 content of 68.8% in the produced biogas, showcasing its effectiveness in removing organic matter and producing CH_4 gas [130]. AI is being used more and more in Seoul's MSWM to improve waste processes through better prediction, analysis, and optimization. The use of AI models has proven effective in handling uncertain and incomplete data, leading to more efficient waste management systems [122].

Seoul's OW management system employs a variety of advanced technologies to enhance efficiency and sustainability. One significant approach is the use of large-scale anaerobic codigestion facilities, which process OW such as pig manure, FW, excretion, and thickened sewage sludge to generate biogas. These facilities utilize horizontal anaerobic digesters with a 35-day retention time, maintaining mesophilic conditions at 35°C–40°C and a pH of 7–8, achieving an average CH_4 content of 68.8% in the biogas produced [130]. The system incorporates patented technology for evaporating the liquid phase of fermented waste to produce dry concentrated fertilizers, burning part of the homogeneous mass, and purifying biogas through water to obtain biomethane. This process also involves the collection and combustion of air from production facilities to generate electricity, thereby increasing the yield of biomethane and enhancing thermal and electrical energy production [131].

By incorporating these cutting-edge technologies and innovative strategies into its operations, the city of Seoul has the potential to realize significant improvements in the effectiveness and longevity of its OW management infrastructure. Through the seamless integration of these advanced tools and tactics, Seoul stands to enhance the overall efficiency and environmental sustainability of its waste disposal processes, paving the way for a more streamlined and eco-friendly approach to handling OW materials.

Copenhagen, Denmark

Copenhagen aimed to be a zero-waste city by using advanced automated waste management technologies. They installed underground waste containers with automated suction systems and utilized robots and AI in high-tech sorting facilities to separate OW. This OW was then transformed into biogas and biofertilizers at bioenergy plants. The outcomes included a more effective waste collection system, leading to less traffic congestion and noise, as well as the production of high-quality compost and biofertilizers for local use, making significant progress toward the city's zero-waste target. Despite the initial high costs, Copenhagen managed to maintain system reliability and upkeep by partnering with the public and private sectors and ensuring regular monitoring and upgrades.

Copenhagen, Denmark, has seen several successful implementations of automated OW management systems, contributing significantly to ecological sustainability. One notable initiative is the intelligent kitchen waste composting system that utilizes deep learning and IoT to automate the segregation, composting, and transportation of kitchen waste. This system employs CNNs for waste detection and segregation, and IoT-enabled compost bins to monitor and control the composting process, reducing human labor and operational costs [132]. Another innovative approach is the fully automated composting machine designed to reduce FW efficiently. This machine features heating, cooling, and grinding capabilities, and uses IoT to measure waste and notify users when the composting process is complete, producing high-nutrient flakes suitable for soil enrichment and pest control [123].

Copenhagen has made significant strides in optimizing OW management by integrating innovative technologies and sustainable practices. The city has shifted from traditional waste disposal methods to a circular bioeconomy approach, focusing on converting OW into valuable products. This transition is exemplified by the implementation of multiproduct integrated biorefineries, which convert OWs into chemicals, fuels, and bioenergy, thereby closing the materials and energy loops and reducing reliance on fossil-based resources [133]. Additionally, Copenhagen has leveraged intelligent algorithms and the IMPACT IoT platform to enhance waste logistics management. This system simulates the filling of waste containers and monitors their status in real-time, allowing for timely and efficient waste collection, thus preventing overfilling and reducing unnecessary trips [134].

The various measures and strategies put in place work together as a whole to make a significant impact on the development and maintenance of a more sustainable and efficient OW management system within the city of Copenhagen, ultimately aiding in the achievement of Denmark's overarching environmental objectives and targets. Furthermore, the implementation of these innovative technologies has not only optimized waste collection processes but has also led to a noticeable reduction in greenhouse gas emissions associated with transportation, contributing to Copenhagen's overall sustainability goals.

Comparative analysis of manual versus automated systems of waste management

Automated waste management systems, which make use of IoT technology, sensors, and ML algorithms, have brought about a significant transformation in the waste management industry. This transformation is primarily characterized by enhancements in efficiency, accuracy, and scalability, leading to improved overall performance. One of the key features of these systems is their ability to optimize waste management processes through the incorporation of advanced technologies such as smart dustbins. These smart dustbins are equipped with sensors that can detect fill levels and promptly alert waste collectors when they need to be emptied. In order to gain a comprehensive understanding of the impact of automated waste management systems, a thorough comparative analysis is often carried out. This analysis aims to assess the effectiveness of manual waste management systems in comparison to their automated counterparts across various factors including efficiency, costs, environmental implications, and scalability. Such evaluations are crucial for decision-makers such as policymakers, urban planners, and waste management professionals, as they provide valuable insights into the advantages and limitations associated with each approach. By delving into the nuances of both manual and automated waste management systems, stakeholders can make informed decisions and devise tailored strategies to address the unique challenges faced in their respective urban environments. This is particularly pertinent in densely populated urban areas where effective waste management strategies are essential to accommodate the growing population and ensure sustainability in the long run. Therefore, a thorough understanding of the technological advancements and operational considerations associated with automated waste management systems is imperative for driving positive change and fostering a more sustainable future.

Automated systems significantly improve the accuracy and efficiency of waste management processes. For instance, ML algorithms can enhance the quality and recall of waste classification models from 86.8% and 47.9% to 99.1% and 98.2%, respectively. Manual systems are prone to human error and inefficiencies, which can lead to suboptimal waste management practices [135].

Automated systems eliminate the need for manual scavenging, thereby reducing the risks of occupational hazards such as exposure to toxic gases and suffocation. This promotes a healthier and safer working environment. Manual waste management, especially in the context of sewage cleaning, poses significant health risks and has resulted in fatalities [136].

Automated systems can better manage waste disposal, reducing environmental impact by promoting proper waste management practices. Manual systems may not be as effective in ensuring environmentally sustainable waste disposal, leading to potential environmental degradation.

Automation in waste management, such as the use of microcontrollers for controlling technological and thermochemical processes, ensures better resource and energy efficiency. Manual systems often lack the precision and control needed to optimize resource and energy use, leading to wastage [137].

Automated systems leverage advanced technologies like ML, microcontrollers, and time series algorithms to enhance waste management processes Manual systems rely heavily on human intervention and traditional methods, which may not integrate well with modern technological advancements.

Automated systems are more scalable and adaptable to varying waste management needs, as they can handle large volumes and complex waste types efficiently. Manual systems may struggle with scalability and adaptability, especially as the volume and complexity of waste increase globally [138].

While the initial setup cost for automated systems may be high, the long-term benefits in terms of efficiency, safety, and environmental sustainability can outweigh these costs. Manual systems may have lower initial costs but can incur higher long-term costs due to inefficiencies, health risks, and environmental damage.

Value-added products from organic waste

Numerous research and development efforts are underway globally to convert diverse FWs into valuable products. These initiatives aim to create a practical and economically viable strategy for the bioconversion of FWs into multiple high-value outputs [139].

Organic fertilizers

Organic fertilizers have long been recognized as a viable alternative to chemical fertilizers for enhancing crop yields and reducing CH_4 emissions. FW is utilized both as a source of animal feed and as an organic fertilizer, which is produced through composting and vermicomposting processes [140]. Organic fertilizers are primarily produced through methods such as AD, aerobic composting by microorganisms, chemical hydrolysis (either alkaline or acidic at temperatures of 600°C–1000°C), or the in-situ degradation of natural organic matter. These processes yield various forms of fertilizers, including soil conditioners, composts, soluble bio-waste composts, degraded crops, and mineral and liquid organic fertilizers. Liquid organic fertilizers, in particular, offer advantages for plant growth due to their direct application to the root zone via irrigation systems. They are rapidly absorbed by plants, require smaller quantities per plot, and have a relatively straightforward degradation process. The effectiveness of organic fertilizers can be further improved by incorporating CH_4-oxidizing bacteria. Additionally, organic fertilizers support the growth of microalgae, which can be used as a nutrient medium to cultivate Chlorella for biodiesel production [141]. Biofertilizers can be derived primarily from OW sources such as plant biomass and animal residues, which contain live microbial agents. These agents perform various functions, including aiding nutrient replenishment, improving soil quality, and enhancing soil conditioning. This contributes to more efficient crop production and increased yields [142]. Using FWs to produce bio-fertilizers can boost the population of thermotolerant and thermophilic microorganisms. This, in turn, enhances soil fertility and promotes optimal plant growth [143]. Additionally, developing bio-fertilizers offers socioeconomic and ecological benefits, positively impacting the environment, as well as animal and human health. Bio-fertilizers, depending on the raw materials used, can be produced cost-effectively in various types [144]. High-quality organic-based biofertilizers can be produced from OW through bioconversion technologies to enhance soil fertility and boost crop yields. In recent years, there has been a growing shift from chemical fertilizers to bio-based organic fertilizers due to their benefits, including being free of chemicals and rich in nutrients. These biofertilizers provide essential minerals like nitrogen, potassium, phosphorus, magnesium, and calcium, improving soil organic matter and serving as a safer alternative to synthetic chemical fertilizers [145,146].

Bioplastics and biochemicals

Bioconversion of FW into bioplastics is a valuable waste treatment approach. To produce bioplastics from FW, pretreatment is essential to enhance its biological and physicochemical properties. Common pretreatment strategies include physical hydrolysis, chemical hydrolysis, bio-hydrolysis, and enzymatic hydrolysis. Physical treatments such as heating, grinding, ultrasound, and microwaving are employed to transform FW into fermentable organic compounds [63]. Biological treatments involve using FW as a fermentation substrate for microorganisms. Key biodegradable polymers with notable market potential include polyhydroxyalkanoates (PHA), polyhydroxybutyrate (PHB), polybutylene succinate (PBS), starch blends, polyvinyl alcohol (PVA), and polylactic acid

(PLA) [63,147]. Bioplastics are considered eco-friendly alternatives to traditional plastics, offering substantial benefits in reducing environmental pollution and mitigating greenhouse gas emissions. Unlike conventional plastics, which are nonbiodegradable and contribute to land pollution and the depletion of landfill space, bioplastics are derived from renewable biomass resources. They are produced through microbial systems using OW materials like vegetable oil, fats, and starches from corn or peas. Composed primarily of cellulose, starch, and other biopolymers, bioplastics are biodegradable, recyclable, and free from toxic or hazardous substances. They require less energy for recycling and reduce reliance on fossil fuels, thereby decreasing greenhouse gas emissions [148]. Despite their advantages, scaling up bioplastic production from waste resources presents challenges such as difficulties in handling, pretreatment needs, transportation issues, and the lack of cost-effective conversion technologies compared to petroleum-based plastics [149]. FWs can be utilized to synthesize a diverse range of organic acids, including citric acid, 2,3-butanediol, succinic acid, 3-hydroxypropionic acid, 1,3-propanediol, lactic acid, and VFAs [150].

Animal feed and other by-products

Every day, humanity produces thousands of tons of OW, much of which ends up in landfills. Utilizing this waste as a raw material for animal feed presents an intriguing and underexplored opportunity [151]. This approach not only mitigates the environmental impact of feed production but also enhances resource efficiency. By incorporating vegetable by-products into feed formulations, feed producers can gain a competitive edge through the use of more sustainable materials, thereby reducing dependence on conventional raw materials [152].

Feeding FW to monogastric animals like poultry and swine can improve their incorporation of dietary fatty acids into products like meat and eggs [153]. In a study by Cho et al., broilers fed with varying levels of dried leftover food (DLF) showed numerically higher DHA and significantly higher EPA levels in their meat [154]. Another study by Hossein indicated that high levels of dehydrated restaurant waste in chicken diets reduced PUFA content and increased SFA content in the meat [155]. Kojima's experiment with laying hens showed increased PUFA and decreased SFA in egg yolk with higher kitchen waste inclusion [156]. The fatty acid composition of poultry meat depends on the fatty acid content of the FW, which can vary based on its source. Careful selection of FW is necessary to ensure high-quality feed and avoid off-flavors from residues like fish. More research is needed to assess the nutritional value of FW for animal diets.

Additionally, feeding swine with FW has a long history, stemming from early human settlements. This practice continues worldwide, including in 27 US states as of 2020 [157]. Research shows pigs can incorporate dietary fatty acids from FW into meat, enhancing PUFA content. Choe et al. found that pigs fed restaurant waste had higher PUFA and lower SFA in their meat compared to conventional diets [158]. Environmental benefits include reduced land use and lower carbon emissions, as converting FW to swine feed is more sustainable than AD or composting. Studies suggest it can significantly reduce pork production's environmental impact in the EU [159].

Economic and environmental benefits

The shift from traditional waste management techniques to automated systems for handling OW offers significant economic and environmental benefits. Automated waste management streamlines operations, cuts costs, and helps reduce environmental impacts, especially by lowering greenhouse gas emissions. Additionally, these advanced systems aid in resource recovery, supporting sustainability by turning waste into valuable resources. This section delves into the various advantages of automation in OW management, with a focus on cost-effectiveness, environmental effects, resource recovery, and real-life examples of successful implementations. By thoroughly examining these factors, we seek to highlight the transformative power of automation in moving toward a circular economy and a sustainable future. One key aspect to consider when implementing automated systems for OW management is the potential for increased efficiency in waste sorting and processing, leading to higher recycling rates and reduced landfill usage.

Cost efficiency of automated systems

Automated waste management systems save a lot of money by cutting labor expenses, boosting efficiency, and lowering waste disposal costs. They improve collection routes, decrease fuel usage, and enhance sorting accuracy, all of which result in reduced operational expenses. Furthermore, automation helps prolong the life of waste management infrastructure by enhancing maintenance and reducing wear and tear.

Automated systems minimize the need for human labor by utilizing technologies such as IoT and sensors, which perform tasks like waste detection and segregation automatically. The integration of ML algorithms and smart technologies significantly enhances the efficiency of waste management processes, leading to higher classification accuracy and recall rates. By improving waste segregation and recycling rates, automated systems reduce the volume of waste sent to landfills, thereby lowering disposal costs. Automated systems use real-time data to optimize waste collection routes, reducing the distance traveled and fuel consumption for waste collection vehicles.

Advanced technologies such as image processing and ML improve the accuracy of waste sorting, ensuring that recyclable materials are correctly identified and processed. The overall operational expenses are reduced due to the combined effects of lower labor costs, optimized routes, and improved sorting accuracy. Automation enhances the maintenance of waste management infrastructure by providing real-time data on equipment status, which helps in timely maintenance and reduces wear and tear.

Automated waste management systems leverage advanced technologies such as IoT, sensors, and ML to significantly enhance the efficiency and accuracy of waste management processes. These systems minimize the need for human labor by automating tasks like waste detection and segregation, utilizing IoT devices embedded with sensors to monitor and manage e-waste collection, sorting, and disposal in real-time [160].

The integration of ML algorithms, such as the Improved Particle Swarm Optimization with Deep Learning-based Municipal Solid Waste Management (IPSODL-MSWM) approach, further refines the classification of waste materials, achieving an optimal accuracy of 99.45% [135]. This high level of accuracy ensures that recyclable materials are correctly identified and processed, thereby improving recycling rates and reducing the volume of waste sent to landfills.

Additionally, the use of multi-agent deep reinforcement learning algorithms helps in predicting waste fill levels and optimizing waste collection routes, which reduces the distance traveled and fuel consumption for waste collection vehicles [161]. Image processing and deep learning techniques, such as those employed in the NoWa smartphone app, enhance the accuracy of waste sorting, ensuring that items are correctly categorized as recyclable or compostable [162].

The overall operational expenses are reduced due to the combined effects of lower labor costs, optimized routes, and improved sorting accuracy. Furthermore, the use of real-time data from IoT devices aids in the maintenance of waste management infrastructure by providing timely information on equipment status, which helps in reducing wear and tear through timely maintenance [160,163]. By employing these advanced technologies, automated systems not only improve the efficiency and sustainability of waste management but also contribute to significant cost savings and environmental benefits.

By focusing on these essential areas, automated waste management systems have the ability to not just reduce costs, but also play a significant role in fostering more environmentally friendly and effective waste management procedures for the benefit of both the present and future generations.

Reduction in greenhouse gas emissions

One major plus of using automated OW management is the big decrease in greenhouse gas emissions. These automated systems make waste processing more efficient by redirecting OW away from landfills. In landfills, OW would break down and release CH_4, a strong greenhouse gas. These systems encourage composting and AD, turning OW into biogas, a renewable energy source. This helps decrease the need for fossil fuels and lowers emissions even more.

Automated systems efficiently redirect OW away from landfills to reduce greenhouse gas emissions. These systems promote composting and AD processes, converting waste into biogas for renewable energy. By accurately categorizing waste, they enhance processing efficiency and contribute to emission reductions. The biogas produced helps decrease carbon footprint, support sustainable energy initiatives, and replace fossil fuels, further reducing emissions. These systems align with sustainable waste management practices, benefiting the environment and communities.

Automated systems play a crucial role in redirecting OW from landfills, significantly reducing greenhouse gas emissions by promoting composting and AD processes. These systems utilize advanced technologies such as deep learning, IoT, and AI to efficiently categorize and process waste, thereby enhancing the overall efficiency of waste management operations [164,165]. For instance, an intelligent kitchen waste composting system can automatically segregate compostable and noncompostable waste, reducing the need for human labor and optimizing the composting process [164]. Composting FW not only mitigates CH_4 and N_2O emissions but also offers a substantial reduction in overall greenhouse gas emissions compared to traditional landfilling methods [166]. Additionally, integrating biogas production with photovoltaic systems for self-consumption can further enhance the sustainability and efficiency of renewable energy generation, reducing electricity costs and reliance on fossil fuels [167]. The biogas produced from AD of agricultural and FW can be used for decentralized electrification, providing a renewable energy source that displaces fossil fuel-based electricity and reduces emissions. This approach aligns with sustainable waste management practices, promoting circularity and resource recovery while maintaining environmental health [165]. By leveraging these automated systems, communities can achieve significant environmental benefits, including reduced carbon footprints and support for sustainable energy initiatives, ultimately contributing to a healthier and more sustainable future [168].

The integration of automated systems in OW management is seen as a crucial step toward creating a more sustainable and environmentally conscious future. By incorporating technologies like deep learning, IoT, and AI, these systems improve the efficiency of waste processing, leading to reduced reliance on landfills and decreased greenhouse gas emissions. Intelligent composting systems streamline the segregation and composting of kitchen waste, while technologies such as AD and photovoltaic systems enable the production of biogas and renewable energy, respectively. These innovations not only support decentralized electrification but also promote a circular economy by recovering valuable resources from waste. Communities embracing these advancements can achieve significant environmental benefits, reduce their carbon footprints, and contribute to sustainable energy and environmental stewardship on a global scale.

Resource recovery and sustainability

Automation helps in managing OW better by enhancing the precision and effectiveness of waste sorting, leading to resource recovery. Cutting-edge sorting tools guarantee the successful separation of OW from other waste streams, allowing it to be transformed into useful products like compost, biogas, and biofertilizers. This not only promotes sustainable farming methods but also plays a role in a circular economy by converting waste into valuable resources.

Automation in waste sorting improves precision and effectiveness, leading to better resource recovery and more efficient waste management operations. OW can be transformed into valuable products like compost, biogas, and biofertilizers, reducing landfill volume and creating useful by-products. These products support sustainable farming, enriching soil, reducing the need for chemical fertilizers, and enhancing soil health. Automated systems promote a circular economy by repurposing waste into valuable resources, closing the loop on waste disposal. Composting and AD of OW reduce greenhouse gas emissions, contributing to climate change mitigation and environmental sustainability. The synergy between biogas production and photovoltaic systems enhances renewable energy generation efficiency and sustainability, reducing electricity costs and reliance on fossil fuels. Automated systems in OW management offer economic benefits by lowering waste management costs and creating new economic opportunities through renewable energy and soil amendments production.

Automation in waste sorting significantly enhances precision and effectiveness, leading to improved resource recovery and more efficient waste management operations. The integration of IoT devices and sensors in smart e-waste management systems allows for real-time monitoring and optimization of collection and disposal processes, which can be further enhanced by ML to distinguish between different types of waste, such as metallic and plastic components, thereby maximizing resource recovery and minimizing environmental impact [160]. AI plays a crucial role in waste logistics, reducing transportation distances, costs, and time, while also improving the accuracy of waste identification and sorting, which ranges from 72.8% to 99.95% [137]. In the context of OW, AI technologies can optimize food production and supply chains, redistribute excess food, and support circular economy initiatives, thereby maximizing resource efficiency and minimizing environmental impact. The transformation of OW into valuable products like compost, biogas, and biofertilizers not only reduces landfill volume but also supports sustainable farming by enriching soil and reducing the need for chemical fertilizers, thus enhancing soil health and contributing to climate change mitigation through reduced greenhouse gas emissions [169]. The synergy between biogas production and photovoltaic systems further enhances renewable energy generation efficiency and sustainability, reducing electricity costs and reliance on fossil fuels [170]. Additionally, the construction sector can benefit from digitalization and automation in waste management, as demonstrated by the RECON-MATIC project, which aims to develop novel tools and methodologies for sustainable construction and demolition waste management [171].

Overall, automated systems in OW management offer significant economic benefits by lowering waste management costs and creating new economic opportunities through the production of renewable energy and soil amendments, thereby promoting a circular economy and closing the loop on waste disposal.

Case studies and success stories

Real-life instances where automated systems for handling OW have been effective offer helpful lessons on their financial and ecological advantages. Examples from places like San Francisco, Milan, Seoul, and Copenhagen show how automation has revolutionized waste management methods. These achievements underscore the notable savings in expenses, decrease in emissions, and successful retrieval of resources, providing viable blueprints for other urban areas and districts to adopt.

San Francisco: pioneering waste management automation

San Francisco has implemented advanced AI and IoT technologies to optimize waste collection and sorting processes. This has led to significant cost savings and improved resource recovery rates. The city's automated systems have also contributed to substantial reductions in greenhouse gas emissions by enhancing the efficiency of waste logistics and recycling operations

San Francisco's implementation of advanced AI and IoT technologies in waste management has yielded significant benefits, including cost savings, improved resource recovery rates, and reductions in greenhouse gas emissions. The integration of IoT devices and sensors allows for real-time monitoring and management of waste collection, sorting, and disposal, optimizing these processes and minimizing environmental impact [160]. AI applications in waste logistics have been shown to reduce transportation distances by up to 36.8%, cost savings by up to 13.35%, and time savings by up to 28.22%, which directly contribute to enhanced efficiency and reduced emissions [137]. The use of multi-agent deep reinforcement learning algorithms to analyze IoT-collected data helps predict waste fill levels and optimize collection routes, further improving operational efficiency and reducing unnecessary trips [161]. These technologies collectively contribute to a more sustainable waste management system by improving the accuracy of waste sorting, enabling the recycling of materials such as metals and plastics, and converting plastic waste into bio-fuel through pyrolysis, thus reducing the overall carbon footprint [160].

The comprehensive use of AI and IoT in San Francisco's waste management not only optimizes resource recovery but also plays a crucial role in mitigating the environmental impact of waste, aligning with the city's sustainability goals.

Milan: efficient organic waste segregation

Milan's waste management system utilizes AI algorithms to segregate OW efficiently. The city has seen a marked improvement in waste sorting accuracy, leading to higher recycling rates and reduced landfill usage. This has not only minimized environmental impact but also generated economic benefits through resource recovery and reduced waste management costs

Efficient OW segregation in Milan can be significantly enhanced by leveraging advanced technologies such as IoT, deep learning, and ML. The integration of IoT in waste management systems, as proposed by Dr Shabina Modi, can automate the process of waste sorting and monitoring, reducing human intervention and operational costs while promoting environmental conservation and resource optimization [172]. Additionally, the use of deep learning algorithms, particularly CNNs, can improve the accuracy of waste categorization, as demonstrated by Yash Bobde M and colleagues. They highlight the potential of models like EfficientNet-B3 in intelligent waste identification and recycling applications, which can be crucial for distinguishing OW from other types [173]. Furthermore, the study by Sivaranjani S and team emphasizes the role of IoT, Arduino, and AI in creating an advanced waste segregation model that utilizes sensors and imaging devices for real-time data generation and efficient monitoring, achieving high accuracy in material classification [173].

By integrating these advanced technologies, Milan can achieve a more efficient and sustainable OW segregation system, addressing the challenges of traditional methods and contributing to better waste management practices.

Seoul: smart waste bins and data-driven decision making

Seoul has deployed smart waste bins equipped with sensors and AI to monitor waste levels and optimize collection routes. This system has resulted in a 28.22% reduction in collection time and a 13.35% decrease in costs. The data-driven approach has also enhanced the city's ability to manage waste more sustainably and efficiently

Seoul's implementation of smart waste bins and data-driven decision-making represents a significant advancement in urban waste management, leveraging cutting-edge technologies to address the challenges posed by rapid urbanization and population growth. The integration of smart bins equipped with various sensors, such as waste level, humidity, temperature, CO_2, and CH_4 sensors, enables real-time monitoring and efficient waste collection, thereby reducing operational costs and environmental impact [172]. These smart bins utilize advanced data analysis techniques and IoT technologies, such as ThingSpeak and WiFi ESP8266, to provide a technological backbone for informed decision-making [174]. Additionally, the use of deep convolutional neural networks in smart waste segregation systems enhances the sorting and categorization of waste, addressing issues related to improper disposal and contamination. The deployment of these intelligent systems not only improves the efficiency of waste management but also promotes sustainability and environmental health by minimizing the spread of diseases and reducing pollutants [175]. Furthermore, the application of educational big data in instructional decision-making can be paralleled in waste management, where data-driven strategies can optimize operations and enhance outcomes [176].

This holistic approach not only addresses immediate waste management needs but also aligns with broader goals of smart city development and environmental sustainability.

Copenhagen: integrating artificial intelligence for sustainable waste management

Copenhagen's waste management strategy incorporates AI to improve waste sorting and recycling processes. The city has achieved significant emissions reductions and cost savings by optimizing waste logistics and enhancing resource recovery. The use of AI has also facilitated better monitoring and decision-making, contributing to the city's overall sustainability goals.

Copenhagen's integration of AI for sustainable waste management can draw on several innovative approaches highlighted in recent research. The use of CNNs combined with IoT for classifying decomposable and nondecomposable waste has shown high accuracy, enabling real-time monitoring and analysis, which can significantly enhance waste management efficiency and environmental sustainability [177]. Additionally, the rapid increase in e-waste due to technological advancements necessitates advanced management strategies. AI and IoT can play a crucial role in addressing these challenges by improving the efficiency of e-waste management systems, although implementation hurdles such as resource requirements and labor intensity must be considered [178]. AI's potential extends to sustainable supply chain management, where it can enhance transparency, optimize resource use, and promote ethical practices, thereby reducing waste and supporting sustainability goals [179].

By integrating these AI-driven methodologies, Copenhagen can develop a comprehensive and adaptive waste management system that not only addresses current challenges but also anticipates future needs, ensuring a sustainable urban environment.

These case studies provide a detailed examination and analysis of the significant and far-reaching effects that automation has had on the methods and strategies employed in waste management. They serve as concrete examples showcasing how this technological advancement has brought about a profound and positive change in the way waste is handled and processed. Moreover, these studies present actionable and replicable frameworks and approaches that can be adopted by various urban areas and territories seeking to enhance their waste management systems. By offering practical and tangible models, these case studies demonstrate a clear path for other cities and regions to emulate and implement within their own waste management practices, ultimately leading to more efficient and sustainable solutions.

Future trends and innovations

OW management is undergoing significant advancements with numerous trends and innovations focused on improving sustainability and efficiency. These advancements are driven by the need to address the environmental impact of waste and the opportunities to convert waste into valuable resources.

Advances in automation technologies

The automation of OW management encompasses various innovative technologies that streamline the collection, sorting, and processing of waste. Smart sensors, automated sorting systems, and robotics are at the forefront of these advancements. Smart sensors are employed to monitor waste composition, moisture levels, and temperature, enabling real-time data collection and analysis [180]. Automated sorting systems, equipped with advanced imaging and ML algorithms, enhance the precision and speed of waste segregation, ensuring that OW is efficiently separated from recyclables and contaminants [74].

Robotic applications have also made significant strides in OW management. Robots are used for tasks such as waste collection, sorting, and composting, reducing the reliance on manual labor and improving operational efficiency [181]. For instance, robotic arms equipped with AI-driven vision systems can identify and sort different types of organic and inorganic waste with high accuracy, significantly reducing contamination rates [182].

Integration of artificial intelligence and machine learning

The integration of AI and ML into waste management systems offers predictive analytics and operational efficiency. ML algorithms analyze vast datasets to predict waste generation patterns, optimize collection routes, and enhance recycling rates [183]. AI-driven systems can also identify contaminants in OW streams, improving the purity of recycled materials [74]. The use of AI in real-time monitoring and decision-making processes ensures that waste management operations are adaptive and responsive to changing conditions [184].

Policy and regulatory support

Effective policy and regulatory frameworks are crucial for the successful implementation of automation in OW management. Governments worldwide are increasingly recognizing the importance of sustainable waste management practices and are enacting supportive policies [185]. Public-private partnerships and funding initiatives further bolster the deployment of advanced waste management systems [186]. Policies that support research and development, provide financial incentives, and establish standards for waste management practices are essential for driving innovation and adoption. For instance, the European Union's circular economy action plan emphasizes the importance of using advanced technologies to improve waste management and resource efficiency [187]. Such policies encourage the development and implementation of automation technologies, thus improving the overall efficiency and sustainability of waste management systems.

Future directions in organic waste management

The future of OW management lies in the continued integration of cutting-edge technologies and innovative practices. Emerging trends such as the use of block chain for waste traceability and the development of biorefineries for converting OW into high-value products are set to redefine the industry [188]. The emphasis on circular economy principles will drive the development of closed-loop systems where waste is continually repurposed, reducing the reliance on landfills and incineration [189]. Additionally, community engagement and education will play a pivotal role in promoting sustainable waste management practices [190]. The development of more efficient and effective waste-to-energy technologies, such as AD and composting, will also play a significant role in the future of OW management [181]. The integration of automation technologies, AI, and supportive policies will pave the way for more efficient, sustainable, and economically viable waste management systems.

Conclusion

The future of waste to value added

AI models, such as ANN, can predict the gravimetric composition and specific weight of MSW using socioeconomic data, filling gaps caused by restrictions on field sampling. AI algorithms like CNN can accurately classify waste into biodegradable and nonbiodegradable categories, enhancing waste segregation and recycling efforts.

Advanced technologies like AI and the IoT can optimize the placement of dustbins and the routing of garbage collectors, reducing littering and improving collection efficiency. Genetic algorithms can be employed to find the most efficient routes for garbage collection, addressing challenges such as overcrowded roads and narrow streets.

IoT-enabled sensors can monitor CH_4 production in waste bins, providing real-time data for better waste management and potential biogas production. Cloud-based infrastructures can facilitate real-time monitoring and analysis, enabling responsive waste management frameworks.

The integration of AI and IoT in e-waste management can address the growing volume of electronic waste, offering more efficient and sustainable solutions. Emerging trends in smart e-waste management include the use of advanced methods to mitigate environmental and health risks associated with e-waste.

AI-driven waste management systems contribute to poverty reduction, improved health and well-being, environmental conservation, economic growth, and social justice. These systems align with sustainable development principles, paving the way for a cleaner and more sustainable future.

Despite the benefits, challenges such as high costs, data privacy concerns, and skill gaps need to be addressed for broader AI adoption in waste management. Future research should focus on the socioeconomic impacts of AI in waste management, exploring employment patterns, income distribution, and equitable resource distribution.

By harnessing the power of AI and IoT technologies, the potential for revolutionizing the waste management industry is immense. These cutting-edge technologies have the capability to convert waste materials into valuable resources, thereby playing a crucial role in advancing sustainability and enhancing operational efficiency within a wide range of sectors.

References

[1] Dharmendra (2022). Organic waste: generation, composition and valorisation. Available from: https://doi.org/10.1016/b978-0-323-85792-5.00024-1.

[2] J.A. Alim, S. Asian Trisnawati, V.F. Wistamuger, V.M. Situmorang, Peningkatan pemahaman mayarakat pemberdayaan limbah organik menjadi pupuk organik, Unri Conference Series (2022). Available from https://doi.org/10.31258/unricsce.3.276-279.

[3] G.-Y. Chen, H.-N. Cao, J.-Z. Dong, M. Nan, Y. Xiangfen, H. Congxin, et al. (2020). Organic waste treatment system.

[4] G.-X. Ni, Overview of government tools designed to increase a volume of organic municipal solid waste processed into organic fertilisers, International Journal of Green Economics (2022). Available from https://doi.org/10.1504/ijge.2022.125560.

[5] L.R.B. Silva, L. Kardos, Recycling of organic waste: an overview of pálinka distillery mash composting, Journal of Central European Green Innovation (2022). Available from https://doi.org/10.33038/jcegi.3509.

[6] A. Maalouf, P. Agamuthu, Waste management evolution in the last five decades in developing countries – a review, Waste Management & Research 41 (9) (2023) 1420–1434.

[7] L. Viva, F. Ciulli, A. Kolk, G. Rothenberg, Designing circular waste management strategies: the case of organic waste in Amsterdam, Advanced Sustainable Systems 4 (9) (2020) 2000023.

[8] E.A.B. Cárcamo, R. Peñabaena-Niebles, Opportunities and challenges for the waste management in emerging and frontier countries through industrial symbiosis, Journal of Cleaner Production 363 (2022) 132607.

[9] S.V. Akram, R. Singh, A. Gehlot, M. Rashid, A.S. AlGhamdi, S.S. Alshamrani, et al., Role of wireless aided technologies in the solid waste management: a comprehensive review, Sustainability 13 (23) (2021) 13104.

[10] M.D. Meena, M.L. Dotaniya, B.L. Meena, P.K. Rai, R.S. Antil, H.S. Meena, et al., Municipal solid waste: opportunities, challenges and management policies in India: a review, Waste Management Bulletin 1 (1) (2023) 4–18.

[11] I.R. Abubakar, K.M. Maniruzzaman, U.L. Dano, F.S. AlShihri, M.S. AlShammari, S.M.S. Ahmed, et al., Environmental sustainability impacts of solid waste management practices in the global South, International Journal of Environmental Research and Public Health 19 (19) (2022) 12717.

[12] A.E. Adeniran, A.T. Nubi, A.O. Adelopo, Solid waste generation and characterization in the University of Lagos for a sustainable waste management, Waste Management 67 (2017) 3–10.

[13] Z. Abdullah, M.S. Salleh, K.N.I.K. Ismail, Survey of household solid waste management and waste minimization in Malaysia: awareness, issues and practices, International Journal of Environmental & Agriculture Research (IJOEAR) 3 (12) (2017) 38–48.

[14] S.A. Sarkodie, P.A. Owusu, Impact of COVID-19 pandemic on waste management, Environment, Development and Sustainability 23 (5) (2021) 7951–7960.

[15] H.B. Sharma, K.R. Vanapalli, V.S. Cheela, V.P. Ranjan, A.K. Jaglan, B. Dubey, et al., Challenges, opportunities, and innovations for effective solid waste management during and post COVID-19 pandemic, Resources, Conservation and Recycling 162 (2020) 105052.

[16] A. Tripathi, V.K. Tyagi, V. Vivekanand, P. Bose, S. Suthar, Challenges, opportunities and progress in solid waste management during COVID-19 pandemic, Case Studies in Chemical and Environmental Engineering 2 (2020) 100060.

[17] M.S. Ayilara, O.S. Olanrewaju, O.O. Babalola, O. Odeyemi, Waste management through composting: challenges and potentials, Sustainability 12 (11) (2020) 4456.

[18] W. Fadhullah, N.I.N. Imran, S.N.S. Ismail, M.H. Jaafar, H. Abdullah, Household solid waste management practices and perceptions among residents in the East Coast of Malaysia, BMC Public Health 22 (2022) 1–20.

[19] Z.T. Jagun, D. Daud, O.M. Ajayi, S. Samsudin, A.J. Jubril, M.S.A. Rahman, Waste management practices in developing countries: a socio-economic perspective, Environmental Science and Pollution Research 30 (55) (2023) 116644–116655.

[20] J.K. Debrah, D.G. Vidal, M.A.P. Dinis, Raising awareness on solid waste management through formal education for sustainability: a developing countries evidence review, Recycling 6 (1) (2021) 6.

[21] M. Baawain, A. Al-Mamun, H. Omidvarborna, W. Al-Amri, Ultimate composition analysis of municipal solid waste in Muscat, Journal of Cleaner Production 148 (2017) 355–362.

[22] A. Siddiqi, M. Haraguchi, V. Narayanamurti, Urban waste to energy recovery assessment simulations for developing countries, World Development 131 (2020) 104949.

[23] R. De Bercegol, J. Cavé, A. Nguyen Thai Huyen, Waste municipal service and informal recycling sector in fast-growing Asian cities: co-existence, opposition or integration? Resources 6 (4) (2017) 70.

[24] U.A. Ofon, U.U. Ndubuisi-Nnaji, I.M. Udo, E.S. Udofia, O. Fatunla, S.E. Shaibu, Biogas production potential from anaerobic co-digestion of food waste and animal manure, Journal of Chemical Society of Nigeria (2024). Available from https://doi.org/10.46602/jcsn.v49i1.949.

[25] G. Lim, S.T. Thed, Exploitation of plant-derived food waste for valuable products using bioprocesses: a narrative review, Food Research (2022). Available from https://doi.org/10.26656/fr.2017.6(s1).009.

[26] K. Kalkanis, D. Alexakis, E. Kyriakis, K. Kiskira, J. Lorenzo-Llanes, N.J. Themelis, et al., Transforming waste to wealth, achieving circular economy, Circular Economy and Sustainability (2022). Available from https://doi.org/10.1007/s43615-022-00225-2.

[27] M. Alfè, V. Gargiulo, M. Porto, R. Migliaccio, A.L. Pera, M. Sellaro, et al., Pyrolysis and gasification of a real refuse-derived fuel (RDF): the potential use of the products under a circular economy vision, Molecules (Basel, Switzerland) (2022). Available from https://doi.org/10.3390/molecules27238114.

[28] S.M. Herbstritt, S.L. Fathel, T.L. Richard, Waste to worth: a case study of the biogas circular economy in Pennsylvania, Journal of the ASABE (2023). Available from https://doi.org/10.13031/ja.14889.

[29] K. Kalkanis, D. Alexakis, E. Kyriakis, K. Kiskira, J. Lorenzo-Llanes, N.J. Themelis, et al., Transforming waste to wealth, achieving circular economy, circular economy and sustainability (2022). Available from https://doi.org/10.1007/s43615-022-00225-2.

[30] M. Abdallah, M.A. Talib, S. Feroz, Q. Nasir, H. Abdalla, B. Mahfood, Artificial intelligence applications in solid waste management: a systematic research review, Waste Management 109 (2020) 231–246.

[31] B. Fang, J. Yu, Z. Chen, A.I. Osman, M. Farghali, I. Ihara, et al., Artificial intelligence for waste management in smart cities: a review, Environmental Chemistry Letters 21 (4) (2023) 1959–1989.

[32] U. Soni, A. Roy, A. Verma, V. Jain, Forecasting municipal solid waste generation using artificial intelligence models—a case study in India, SN Applied Sciences 1 (2) (2019) 162.

[33] N. Abou Baker, P. Szabo-Müller, U. Handmann, Transfer learning-based method for automated e-waste recycling in smart cities, EAI endorsed Transactions on Smart Cities 5 (16) (2021) e1.

[34] M. Koskinopoulou, F. Raptopoulos, G. Papadopoulos, N. Mavrakis, M. Maniadakis, Robotic waste sorting technology: toward a vision-based categorization system for the industrial robotic separation of recyclable waste, IEEE Robotics & Automation Magazine 28 (2) (2021) 50–60.

[35] W.L. Mao, W.C. Chen, H.I.K. Fathurrahman, Y.H. Lin, Deep learning networks for real-time regional domestic waste detection, Journal of Cleaner Production 344 (2022) 131096.

[36] G.V. Gurram, N.C. Shariff, R.L. Biradar, A secure energy aware meta-heuristic routing protocol (SEAMHR) for sustainable IoT-wireless sensor network (WSN), Theoretical Computer Science 930 (2022) 63–76.

[37] L.M. Joshi, R.K. Bharti, R. Singh, P.K. Malik, Real time monitoring of solid waste with customized hardware and Internet of Things, Computers and Electrical Engineering 102 (2022) 108262.

[38] R. Sembiring, U. Hasanah, Transition to a circular economy-inspired waste management system, World Journal Of Advanced Research and Reviews (2023). Available from https://doi.org/10.30574/wjarr.2023.19.1.1304.

[39] A. Verardi, P. Sangiorgio, A. Blasi, C.G. Lopresto, V. Calabrò, Bioconversion of crop residues using alternative fermentation-based approaches, Frontiers in Bioscience (2023). Available from https://doi.org/10.31083/j.fbe1503017.

[40] N. Yanqoritha, Influence of physico-chemical and bioactivators for composting of traditional market vegetable waste, Jurnal Penelitian Pendidikan IPA (JPPIPA) (2023). Available from https://doi.org/10.29303/jppipa.v9i4.3238.

[41] G. Badagliacca, G. Testa, S.L. Malfa, V. Cafaro, E.L. Presti, M. Monti, Organic fertilizers and bio-waste for sustainable soil management to support crops and control greenhouse gas emissions in mediterranean agroecosystems: a review, Horticulturae (2024). Available from https://doi.org/10.3390/horticulturae10050427.
[42] L. Peng, R. Tang, G. Wang, R. Ma, Y. Li, G. Li, et al., Effect of aeration rate, aeration pattern, and turning frequency on maturity and gaseous emissions during kitchen waste composting, Environmental Technology and Innovation (2023). Available from https://doi.org/10.1016/j.eti.2022.102997.
[43] (2022). Optimization of composting methods for efficient use of cassava waste, using microbial degradation. Available from: https://doi.org/10.21203/rs.3.rs-2037029/v1.
[44] J.L.S. Pereira, T. Costa, V. Figueiredo, F. Marques, A. Perdigão, I.P. Brás, et al., Influence of different agro-food waste on ammonia and greenhouse gas emissions during composting, Agronomy (2024). Available from https://doi.org/10.3390/agronomy14010220.
[45] B.A. Parra-Orobio, J. Soto-Paz, Evaluation of biochar as an additive in the co-composting of green waste and food waste, Sustainability (2023). Available from https://doi.org/10.3390/su15097437.
[46] S. Kaza, L. Yao, and P. Bhada-Tata, 2018. Trends in solid waste management.
[47] V. Yadav, S. Karmakar, Sustainable collection and transportation of municipal solid waste in urban centers, Sustainable Cities and Society 53 (2020) 101937.
[48] N. Laurieri, A. Lucchese, A. Marino, S. Digiesi, A door-to-door waste collection system case study: a survey on its sustainability and effectiveness, Sustainability 12 (14) (2020) 5520.
[49] V. Ibáñez-Forés, M.D. Bovea, C. Coutinho-Nóbrega, H.R. de Medeiros-García, R. Barreto-Lins, Temporal evolution of the environmental performance of implementing selective collection in municipal waste management systems in developing countries: A Brazilian case study, Waste Management 72 (2018) 65–77.
[50] T.R.P. Ramos, C.S. de Morais, A.P. Barbosa-Povoa, The smart waste collection routing problem: alternative operational management approaches, Expert Systems with Applications 103 (2018) 146–158.
[51] M.M. Hoque, M.T.U. Rahman, Landfill area estimation based on solid waste collection prediction using ANN model and final waste disposal options, Journal of Cleaner Production 256 (2020) 120387.
[52] S. Das, S.H. Lee, P. Kumar, K.H. Kim, S.S. Lee, S.S. Bhattacharya, Solid waste management: scope and the challenge of sustainability, Journal of Cleaner Production 228 (2019) 658–678.
[53] M.A. Hannan, M.H. Lipu, M. Akhtar, R.A. Begum, M.A. Al Mamun, A. Hussain, et al., Solid waste collection optimization objectives, constraints, modeling approaches, and their challenges toward achieving sustainable development goals, Journal of Cleaner Production 277 (2020) 123557.
[54] M. Akhtar, M.A. Hannan, R.A. Begum, H. Basri, E. Scavino, Backtracking search algorithm in CVRP models for efficient solid waste collection and route optimization, Waste Management 61 (2017) 117–128.
[55] M. Abdallah, M. Adghim, M. Maraqa, E. Aldahab, Simulation and optimization of dynamic waste collection routes, Waste Management & Research 37 (8) (2019) 793–802.
[56] M.A. Hannan, M.A. Al Mamun, A. Hussain, H. Basri, R.A. Begum, A review on technologies and their usage in solid waste monitoring and management systems: issues and challenges, Waste Management 43 (2015) 509–523.
[57] F. Habibi, E. Asadi, S.J. Sadjadi, F. Barzinpour, A multi-objective robust optimization model for site-selection and capacity allocation of municipal solid waste facilities: a case study in Tehran, Journal of Cleaner Production 166 (2017) 816–834.
[58] A. Singh, Remote sensing and GIS applications for municipal waste management, Journal of Environmental Management 243 (2019) 22–29.
[59] Y. Pujara, P. Pathak, A. Sharma, J. Govani, Review on Indian Municipal Solid Waste Management practices for reduction of environmental impacts to achieve sustainable development goals, Journal of Environmental Management 248 (2019) 109238.
[60] T.R. Zolnikov, R.C. da Silva, A.A. Tuesta, C.P. Marques, V.R.N. Cruvinel, Ineffective waste site closures in Brazil: a systematic review on continuing health conditions and occupational hazards of waste collectors, Waste Management 80 (2018) 26–39.
[61] M. Pini, F. Lolli, E. Balugani, R. Gamberini, P. Neri, B. Rimini, et al., Preparation for reuse activity of waste electrical and electronic equipment: environmental performance, cost externality and job creation, Journal of Cleaner Production 222 (2019) 77–89.
[62] S. Elagroudy, M.A. Warith, M. El Zayat, Municipal Solid Waste Management and Green Economy, Global Young Academy, Berlin, Germany, 2016.

[63] Y.F. Tsang, V. Kumar, P. Samadar, Y. Yang, J. Lee, Y.S. Ok, et al., Production of bioplastic through food waste valorization, Environment International 127 (2019) 625–644.
[64] M. Rotthong, M. Takaoka, K. Oshita, P. Rachdawong, S.H. Gheewala, T. Prapaspongsa, Life cycle assessment of integrated municipal organic waste management systems in Thailand, Sustainability 15 (1) (2022) 90.
[65] K.H. Kim, E. Kabir, S.A. Jahan, Airborne bioaerosols and their impact on human health, Journal of Environmental sciences 67 (2018) 23–35.
[66] E. Franchitti, E. Pascale, E. Fea, E. Anedda, D. Traversi, Methods for bioaerosol characterization: limits and perspectives for human health risk assessment in organic waste treatment, Atmosphere 11 (5) (2020) 452.
[67] D. Misra, G. Das, T. Chakrabortty, D. Das, An IoT-based waste management system monitored by cloud, Journal of Material Cycles and Waste Management 20 (3) (2018) 1574–1582.
[68] F. Ferrari, R. Striani, S. Minosi, R. De Fazio, P. Visconti, L. Patrono, et al., An innovative IoT-oriented prototype platform for the management and valorisation of the organic fraction of municipal solid waste, Journal of Cleaner Production 247 (2020) 119618.
[69] G. Gaustad, E. Olivetti, R. Kirchain, Improving aluminum recycling: a survey of sorting and impurity removal technologies, Resources, Conservation and Recycling 58 (2012) 79–87.
[70] A. Jank, W. Müller, I. Schneider, F. Gerke, A. Bockreis, Waste separation press (WSP): a mechanical pretreatment option for organic waste from source separation, Waste Management 39 (2015) 71–77.
[71] T.L. Hansen, J. La Cour Jansen, Å. Davidsson, T.H. Christensen, Effects of pre-treatment technologies on quantity and quality of source-sorted municipal organic waste for biogas recovery, Waste Management 27 (3) (2007) 398–405.
[72] G. Bonifazi, S. Serranti, Recycling technologies, Encyclopedia of Sustainability Science and Technology, Springer Science+ Business Media, LLC, 2012, pp. 1–60.
[73] V. Luciani, G. Bonifazi, P. Rem, S. Serranti, Upgrading of PVC rich wastes by magnetic density separation and hyperspectral imaging quality control, Waste Management 45 (2015) 118–125.
[74] S.P. Gundupalli, S. Hait, A. Thakur, A review on automated sorting of source-separated municipal solid waste for recycling, Waste Management 60 (2017) 56–74.
[75] S. Koyanaka, K. Kobayashi, Automatic sorting of lightweight metal scrap by sensing apparent density and three-dimensional shape, Resources, Conservation and Recycling 54 (9) (2010) 571–578.
[76] I. Vegas, K. Broos, P. Nielsen, O. Lambertz, A. Lisbona, Upgrading the quality of mixed recycled aggregates from construction and demolition waste by using near-infrared sorting technology, Construction and Building Materials 75 (2015) 121–128.
[77] A. Kumar, S.K. Verma, Design and development of e-smart robotics-based underground solid waste storage and transportation system, Journal of Cleaner Production 343 (2022) 130987.
[78] R. Sarc, A. Curtis, L. Kandlbauer, K. Khodier, K.E. Lorber, R. Pomberger, Digitalisation and intelligent robotics in value chain of circular economy oriented waste management – a review, Waste Management 95 (2019) 476–492.
[79] A.G. Satav, S. Kubade, C. Amrutkar, G. Arya, A. Pawar, A state-of-the-art review on robotics in waste sorting: scope and challenges, International Journal on Interactive Design and Manufacturing (IJIDeM) 17 (6) (2023) 2789–2806.
[80] S.Z. Diya, R.A. Proma, M.N. Islam, T.T. Anannya, A. Al Mamun, R. Arefeen, et al., 2018, Developing an intelligent waste sorting system with robotic arm: a step towards green environment. In 2018 International Conference on Innovation in Engineering and Technology (ICIET) (pp. 1–6). IEEE.
[81] T. Kiyokawa, J. Takamatsu, S. Koyanaka, Challenges for future robotic sorters of mixed industrial waste: a survey, IEEE Transactions on Automation Science and Engineering 21 (1) (2022) 1023–1040.
[82] A. Namoun, A. Tufail, M.Y. Khan, A. Alrehaili, T.A. Syed, O. BenRhouma, Solid waste generation and disposal using machine learning approaches: a survey of solutions and challenges, Sustainability 14 (20) (2022) 13578.
[83] H.L. Vu, K.T.W. Ng, A. Richter, J. Li, S.A. Hosseinipooya, Impacts of nested forward validation techniques on machine learning and regression waste disposal time series models, Ecological Informatics 72 (2022) 101897.
[84] H. Ma, J. Liu, J. Zhang, J. Huang, Estimating the compressive strength of cement-based materials with mining waste using support vector machine, decision tree, and random forest models, Advances in Civil Engineering 2021 (1) (2021) 6629466.

[85] T. Vafeiadis, A. Nizamis, N. Alexopoulos, C. Ntinas, T. Mastos, A. Papadopoulos, et al., 2018, Data analytics and data modeling for predictive maintenance and automated waste management: an integrated (IoT-Industry 4.0) sensor based solution to improve factory procedures. In *International Conference on Internet of Things and Smart Manufacturing*.

[86] T.P. da Costa, J. Gillespie, X. Cama-Moncunill, S. Ward, J. Condell, R. Ramanathan, et al., A systematic review of real-time monitoring technologies and its potential application to reduce food loss and waste: key elements of food supply chains and IoT technologies, Sustainability 15 (1) (2022) 614.

[87] E. Likotiko, D. Petrov, J. Mwangoka, U. Hilleringmann, Real time solid waste monitoring using cloud and sensors technologies, The Online Journal of Science and Technology-January 8 (1) (2018).

[88] Nair, R.R., Prabhu, P., Panchal, C., Sarnot, L., Bhingarkar, S., Savyanavar, A., et al., 2024, An automated system for waste segregation using deep learning algorithms. In 2024 3rd International Conference for Innovation in Technology (INOCON) (pp. 1–6). IEEE.

[89] R. Gupta, Z.H. Ouderji, Uzma, Z. Yu, W.T. Sloan, S. You, Machine learning for sustainable organic waste treatment: a critical review, NPJ Materials Sustainability 2 (1) (2024) 5.

[90] O. Staub, Revolutionizing the waste supply chain: blockchain for social good, Online, Accessed 1 (22) (2019) 2021.

[91] K. Bułkowska, M. Zielińska, M. Bułkowski, Implementation of blockchain technology in waste management, Energies 16 (23) (2023) 7742.

[92] G. Mohanty, Evaluation of automated portable vermicomposting bin, International Journal For Science Technology And Engineering (2023). Available from https://doi.org/10.22214/ijraset.2023.53670.

[93] D. Mignogna, P. Ceci, C. Cafaro, G. Corazzi, P. Avino, Production of biogas and biomethane as renewable energy sources: a review, Applied Sciences 13 (18) (2023) 10219.

[94] J. Pavičić, K. Novak Mavar, V. Brkić, K. Simon, Biogas and biomethane production and usage: technology development, advantages and challenges in Europe, Energies 15 (8) (2022) 2940.

[95] W. Dessie, X. Luo, M. Wang, L. Feng, Y. Liao, Z. Wang, et al., Current advances on waste biomass transformation into value-added products, Applied Microbiology and Biotechnology 104 (11) (2020) 4757–4770.

[96] R. Mrosso, A.C. Mecha, J. Kiplagat, Characterization of kitchen and municipal organic waste for biogas production: effect of parameters, Heliyon 9 (5) (2023) e16360.

[97] P.K. Dikshit, S.K. Padhi, L. Pattanaik, A. Khan, A. Ranjan, S. Sadhu, A critical review on nanotechnological advancement in biogas production from organic waste, Biomass Conversion and Biorefinery (2023) 1–23.

[98] M. Rasapoor, B. Young, R. Brar, A. Sarmah, W.Q. Zhuang, S. Baroutian, Recognizing the challenges of anaerobic digestion: critical steps toward improving biogas generation, Fuel 261 (2020) 116497.

[99] B.V. Ayodele, M.A. Alsaffar, S.I. Mustapa, An overview of integration opportunities for sustainable bioethanol production from first-and second-generation sugar-based feedstocks, Journal of Cleaner Production 245 (2020) 118857.

[100] T.G. Induchoodan, I. Haq, A.S. Kalamdhad, Factors affecting anaerobic digestion for biogas production: a review. Advanced Organic, Waste Management (2022) 223–233.

[101] A. Cheon, J. Sung, H. Jun, H. Jang, M. Kim, J. Park, Application of various machine learning models for process stability of bio-electrochemical anaerobic digestion, Processes 10 (1) (2022) 158.

[102] I.A. Cruz, W. Chuenchart, F. Long, K.C. Surendra, L.R.S. Andrade, M. Bilal, et al., Application of machine learning in anaerobic digestion: perspectives and challenges, Bioresource Technology 345 (2022) 126433.

[103] D.P. Van, T. Fujiwara, B.L. Tho, P.P.S. Toan, G.H. Minh, A review of anaerobic digestion systems for biodegradable waste: configurations, operating parameters, and current trends, Environmental Engineering Research 25 (1) (2020) 1–17.

[104] G.S. Fanourgakis, K. Gkagkas, E. Tylianakis, G. Froudakis, A generic machine learning algorithm for the prediction of gas adsorption in nanoporous materials, The Journal of Physical Chemistry C 124 (13) (2020) 7117–7126.

[105] L. Alejo, J. Atkinson, V. Guzmán-Fierro, M. Roeckel, Effluent composition prediction of a two-stage anaerobic digestion process: machine learning and stoichiometry techniques, Environmental Science and Pollution Research 25 (2018) 21149–21163.

[106] P. Kazemi, C. Bengoa, J.P. Steyer, J. Giralt, Data-driven techniques for fault detection in anaerobic digestion process, Process Safety and Environmental Protection 146 (2021) 905–915.

[107] M. Carlini, S. Castellucci, A. Mennuni, S. Selli, Simulation of anaerobic digestion processes: validation of a novel software tool ADM1-based with AQUASIM, Energy Reports 6 (2020) 102–115.

[108] D.J. Batstone, D. Puyol, X. Flores-Alsina, J. Rodríguez, Mathematical modelling of anaerobic digestion processes: applications and future needs, Reviews in Environmental Science and Bio/Technology 14 (2015) 595–613.

[109] F. Piadeh, K. Behzadian, A.M. Alani, A critical review of real-time modelling of flood forecasting in urban drainage systems, Journal of Hydrology 607 (2022) 127476.

[110] R. Kapoor, P. Ghosh, M. Kumar, V.K. Vijay, Evaluation of biogas upgrading technologies and future perspectives: a review, Environmental Science and Pollution Research 26 (2019) 11631–11661.

[111] M. Heiker, M. Kraume, A. Mertins, T. Wawer, S. Rosenberger, Biogas plants in renewable energy systems—a systematic review of modeling approaches of biogas production, Applied Sciences 11 (8) (2021) 3361.

[112] S. Cinar, S.O. Cinar, N. Wieczorek, I. Sohoo, K. Kuchta, Integration of artificial intelligence into biogas plant operation, Processes 9 (1) (2021) 85.

[113] N. Scarlat, J.F. Dallemand, F. Fahl, Biogas: developments and perspectives in Europe, Renewable Energy 129 (2018) 457–472.

[114] D. Wu, L. Li, Y. Peng, P. Yang, X. Peng, Y. Sun, et al., State indicators of anaerobic digestion: a critical review on process monitoring and diagnosis, Renewable and Sustainable Energy Reviews 148 (2021) 111260.

[115] H. Cheng, Y. Li, G. Guo, T. Zhang, Y. Qin, T. Hao, et al., Advanced methanogenic performance and fouling mechanism investigation of a high-solid anaerobic membrane bioreactor (AnMBR) for the co-digestion of food waste and sewage sludge, Water Research 187 (2020) 116436.

[116] D. Wu, X. Peng, L. Li, P. Yang, Y. Peng, H. Liu, et al., Commercial biogas plants: review on operational parameters and guide for performance optimization, Fuel 303 (2021) 121282.

[117] B.R. Patra, F. Pattnaik, J. Podder, S. Nanda, A.K. Dalai. Advances in pyrolysis and copyrolysis technologies for biomass conversion to solid and liquid biofuels (2024). Available from: https://doi.org/10.1016/b978-0-443-15377-8.00005-9.

[118] P. Juneja, Biomass waste conversion technologies for its utilization and energy generation in India: a perspective, IOP Conference Series (2024). Available from https://doi.org/10.1088/1755-1315/1285/1/012010.

[119] A. Albahnasawi, M. Eyvaz, M.Y.D. Alazaiza, N. Arslanoğlu, E. Gürbulak, S. Alhout, et al. (2024). Biomass waste and bioenergy production: challenges and alternatives. Available from: https://doi.org/10.1007/978-3-031-52485-1_3.

[120] A.A.A. Samad, R.D. Shree, Alternate fuel biodiesel, International Journal of Innovative Research in Information Security (2023). Available from https://doi.org/10.26562/ijiris.2023.v0903.23.

[121] (2024). Revolutionizing urban waste management in San Francisco: the role of technology-driven solutions in advancing circular economy practices. International Journal of Business and Technology Management. doi:https://doi.org/10.55057/ijbtm.2024.6.1.41.

[122] N. Kumari, S. Pandey, A.K. Pandey, Role of artificial intelligence in municipal solid waste management, British Journal of Multidisciplinary and Advanced Studies (2023). Available from https://doi.org/10.37745/bjmas.2022.0180.

[123] N.A. Fakharulrazi, F. Yakub, M.N. Baba, L.F. Zhao, A. Rafiq, H.H. Yudu, et al. Designing an automated composter for food waste management with the implementation of Internet of Things (2020). Available from: https://doi.org/10.30880/JSUNR.2020.01.02.002.

[124] P. Martinez, O. Mohsen, M. Al-Hussein, R. Ahmad, Vision-based automated waste audits: a use case from the window manufacturing industry, The International Journal of Advanced Manufacturing Technology (2022). Available from https://doi.org/10.1007/s00170-022-08730-2.

[125] T.G. Sereda, S.N. Kostarev. Development of the automated workstation for the operator of the solid municipal wastes landfill (2021). Available from: https://doi.org/10.1088/1755-1315/677/4/042107.

[126] Dotdotdot. (2022). REFLOW – Milan Foody Zero Waste Pilot Data. Available from: https://doi.org/10.5281/zenodo.6772787.

[127] M. Oktasa, S.K. Kartika, Organic waste compost processing using arduino-based automated machines, Jurnal Ilmiah Poli Rekayasa (2022). Available from https://doi.org/10.30630/jipr.17.2.231.

[128] A. Sharma, P. Vaishnav, P.A. Raj, N. Yadav, B. Pandey. Automated waste: sorting and rewards system a technological solution for sustainable waste management (2024). Available from: https://doi.org/10.48001/joegwr.2024.1120-25.

[129] L. Xiao, L. Xiao, Solid waste management approach at the university through living labs and communication strategies: case studies in italy and portugal, Sustainability (2022). Available from https://doi.org/10.3390/su14095240.

[130] Y. Park, J. Khim, J.D. Kim, Application of a full-scale horizontal anaerobic digester for the co-digestion of pig manure, food waste, excretion, and thickened sewage sludge, Processes (2023). Available from https://doi.org/10.3390/pr11041294.

[131] V.M. Nikolaeva, A.I. Borisov. The technology of processing and recycling organic waste (2020). Available from: https://doi.org/10.1088/1755-1315/548/5/052035.

[132] Intelligent kitchen waste composting system via deep learning and IoT (2023). Available from: https://doi.org/10.21203/rs.3.rs-2978777/v1.

[133] L. Carina, G. Julien, R. Ana, S. Carvalho, K.V. Gernaey, Optimal conversion of organic wastes to value-added products: toward a sustainable integrated biorefinery in Denmark, Frontiers in chemical engineering (2022). Available from https://doi.org/10.3389/fceng.2022.837105.

[134] A. Burduk, D. Lapczynska, J. Kochańska, K. Musiał, W.I.K. Dorota, Waste management with the use of heuristic algorithms and Internet of Things technology, Sensors (2022). Available from https://doi.org/10.3390/s22228786.

[135] R. Udayakumar, R. Elankavi, V.R. Vimal, R. Sugumar, Improved particle swarm optimization with deep learning-based municipal solid waste management in smart cities, RGSA: Revista de Gestão Social e Ambiental (2023). Available from https://doi.org/10.24857/rgsa.v17n4-022.

[136] Alerting and detection of toxic gases in sewage using IOT. International Journal for Science Technology and Engineering (2023). Available from https://doi.org/10.22214/ijraset.2023.51784.

[137] B. Fang, Z.-H. Chen, A.I. Osman, M. Farghali, I. Ihara, E.H. Hamza, et al., Artificial intelligence for waste management in smart cities: a review, Environmental Chemistry Letters (2023). Available from https://doi.org/10.1007/s10311-023-01604-3.

[138] S. Rajabov, Utilizing digital technologies for waste management, E3S web of conferences (2023). Available from https://doi.org/10.1051/e3sconf/202338101096.

[139] S. Dahiya, A.N. Kumar, J.S. Sravan, S. Chatterjee, O. Sarkar, S.V. Mohan, Food waste biorefinery: Sustainable strategy for circular bioeconomy, Bioresource Technology 248 (2018) 2–12.

[140] M.K.I. Khan, M. Asif, Z.U. Razzaq, A. Nazir, A.A. Maan, Sustainable food industrial waste management through single cell protein production and characterization of protein enriched bread, Food Bioscience 46 (2022) 101406.

[141] Z. Liu, T.S. de Souza, B. Holland, F. Dunshea, C. Barrow, H.A. Suleria, Valorization of food waste to produce value-added products based on its bioactive compounds, Processes 11 (3) (2023) 840.

[142] Y. Shang, M.K. Hasan, G.J. Ahammed, M. Li, H. Yin, J. Zhou, Applications of nanotechnology in plant growth and crop protection: a review, Molecules (Basel, Switzerland) 24 (14) (2019) 2558.

[143] Q. Lu, Y. Zhao, X. Gao, J. Wu, H. Zhou, P. Tang, et al., Effect of tricarboxylic acid cycle regulator on carbon retention and organic component transformation during food waste composting, Bioresource Technology 256 (2018) 128–136.

[144] T. Mahami, W. Togby-Tetteh, D.I. Kottoh, L. Amoakoah-Twum, E. Gasu, S.N.Y. Annan, et al., Microbial food safety risk to humans associated with poultry feed: the role of irradiation, International Journal of Food Science 2019 (1) (2019) 6915736.

[145] Q. Wang, X. Ren, Y. Sun, J. Zhao, M.K. Awasthi, T. Liu, et al., Improvement of the composition and humification of different animal manures by black soldier fly bioconversion, Journal of Cleaner Production 278 (2021) 123397.

[146] V.K. Badhwar, S. Singh, B. Singh, Biotransformation of paper mill sludge and tea waste with cow dung using vermicomposting, Bioresource Technology 318 (2020) 124097.

[147] A. Tripathi, S. Narayanan, Potassium doped graphitic carbon nitride with extended optical absorbance for solar light driven photocatalysis, Applied Surface Science 479 (2019) 1–11.

[148] L. Alaerts, M. Augustinus, K. Van Acker, Impact of bio-based plastics on current recycling of plastics, Sustainability 10 (5) (2018) 1487.

[149] S.K. Bhatia, S.V. Otari, J.M. Jeon, R. Gurav, Y.K. Choi, R.K. Bhatia, et al., Biowaste-to-bioplastic (polyhydroxyalkanoates): Conversion technologies, strategies, challenges, and perspective, Bioresource Technology 326 (2021) 124733.

[150] V. Ashokkumar, G. Flora, R. Venkatkarthick, K. SenthilKannan, C. Kuppam, G.M. Stephy, et al., Advanced technologies on the sustainable approaches for conversion of organic waste to valuable bioproducts: Emerging circular bioeconomy perspective, Fuel 324 (2022) 124313.

[151] D. San Martin, S. Ramos, J. Zufia, Valorisation of food waste to produce new raw materials for animal feed, Food Chemistry 198 (2016) 68–74.

[152] R.O. Myer, J.H. Brendemuhl, D.D. Johnson, Evaluation of dehydrated restaurant food waste products as feedstuffs for finishing pigs, Journal of Animal Science 77 (3) (1999) 685–692.

[153] A. Georganas, E. Giamouri, A.C. Pappas, G. Papadomichelakis, F. Galliou, T. Manios, et al., Bioactive compounds in food waste: A review on the transformation of food waste to animal feed, Foods 9 (3) (2020) 291.

[154] Y. Cho, G. Lee, J. Jang, I. Shin, K. Myung, K. Choi, et al., Effects of feeding dried leftover food on growth and body composition of broiler chicks, Asian Australas. J. Anim. Sci. 17 (2004) 386–393.

[155] S. Hossein, Growth Performances, Carcass Yield and Meat Quality of Free Range Village Chickens Fed on Diet Containing Dehydrated Processed Food Waste. Master's Thesis, Universiti Putra Malaysia, Seri Kembangan, Malaysia, 2015.

[156] S. Kojima, Dehydrated kitchen waste as a feedstuff for laying hens, International Journal of Poultary Science 4 (2005) 689–694.

[157] M.L. Westendorf, Z.C. Dong, P.A. Schoknecht, Recycled cafeteria food waste as a feed for swine: nutrient content digestibility, growth, and meat quality, Journal of Animal Science 76 (1998) 2976–2983.

[158] J. Choe, K.M. Moyo, K. Park, J. Jeong, H. Kim, Y. Ryu, et al., Meat quality traits of pigs finished on food waste, Korean Journal of Food Science Animal Resources (2017).

[159] R. Salemdeeb, E.K. Zu Ermgassen, M.H. Kim, A. Balmford, A. Al-Tabbaa, Environmental and health impacts of using food waste as animal feed: a comparative analysis of food waste management options, Journal of Cleaner Production 140 (2017) 871–880.

[160] M. Farjana, A. Fahad, M.M. Islam, An IoT- and cloud-based e-waste management system for resource reclamation with a data-driven decision-making process, Iot (2023). Available from https://doi.org/10.3390/iot4030011.

[161] A. Jenifer Selvarani, Waste management using IoT, International Journal of Innovative Research in Advanced Engineering (2023). Available from https://doi.org/10.26562/ijirae.2023.v1006.29.

[162] A. Shrivastava, Real time waste classification using deep learning and AV: deep learning and implementation in the frontend, Journal of Student Research (2023). Available from https://doi.org/10.47611/jsrhs.v12i2.4208.

[163] M.C. Meenakshi, L. Prabha, P. Soni, V. Arya, B. Gupta, A. Albeshri, et al., Machine learning-based automatic litter detection and classification using neural networks in smart cities, International Journal on Semantic Web and Information Systems (2023). Available from https://doi.org/10.4018/ijswis.324105.

[164] Intelligent kitchen waste composting system via deep learning and IoT (2023). Available from: https://doi.org/10.21203/rs.3.rs-2978777/v1.

[165] D. Richards, H. Yabar, Promoting energy and resource recovery from livestock waste: case study yuge farm, Japan, Case Studies in Chemical and Environmental Engineering (2023). Available from https://doi.org/10.1016/j.cscee.2023.100299.

[166] T. Pérez, S.E. Vergara, W.L. Silver, Assessing the climate change mitigation potential from food waste composting, Dental Science Reports (2023). Available from https://doi.org/10.1038/s41598-023-34174-z.

[167] C. Roldán-Blay, D. Dasí-Crespo, C. Roldán-Porta, G. Escrivá-Escrivá, E.Q. Cucarella, Dynamic algorithm to boost the synergy between biogas plants and photovoltaic plants, Renewable Energy & Power Quality Journal (2023). Available from https://doi.org/10.24084/repqj21.245.

[168] K. Karunambiga, M. Sathiya, Technological view on smart waste management, International Journal of Computer Applications Technology and Research (2023). Available from https://doi.org/10.7753/ijcatr1203.1008.

[169] H. Onyeaka, P. Tamasiga, U.M. Nwauzoma, T. Miri, U.C. Julliet, O. Nwaiwu, et al., Using artificial intelligence to tackle food waste and enhance the circular economy: maximising resource efficiency and minimising environmental impact: a review, Sustainability (2023). Available from https://doi.org/10.3390/su151310482.

[170] A. Kumar, Artificial intelligence in sustainable development of municipal solid waste management, International Journal For Science Technology And Engineering (2023). Available from https://doi.org/10.22214/ijraset.2023.53247.

[171] E. Koroxenidis, A.K. Katerina, T.T. Theodosiou, Towards the digitalization and automation of circular and sustainable construction and demolition waste management – project RECONMATIC, IOP Conference Series (2023). Available from https://doi.org/10.1088/1755-1315/1196/1/012044.
[172] S. Modi, Automationg waste segregation using iot, Indian Scientific Journal Of Research In Engineering And Management (2024). Available from https://doi.org/10.55041/ijsrem35121.
[173] Y. Bobde, M.B. Jothi, A. Sharma, Iweews: an intelligent waste extractor for efficient waste segregation by using deep learning technique, International Journal of Engineering Applied Science and Technology (2023). Available from https://doi.org/10.33564/ijeast.2023.v08i02.008.
[174] E. Ahmed, R. Idrissi, J. Benabbou, M. Benhra, H. El. (2024). Enhancing sustainability: leveraging sensor technology in smart bins for real-time data analysis. Available from: https://doi.org/10.1007/978-3-031-53824-7_23.
[175] A. Ande, G. Naveen Sundar, R. Thomas, V. Nair, Narmadha (2024). Smart waste bins using DCNN and Internet of Things. doi:https://doi.org/10.1109/ic-etite58242.2024.10493252.
[176] Y. Jiang, Y.-Y. Huang, Q. Lei, Data-driven instructional decision-making: framework, instrument, and process, World Journal of Educational Research (2024). Available from https://doi.org/10.22158/wjer.v11n3p72.
[177] P. Biplov, B. Krishna, B. Shah, N. Paneru, V. Bhattrai, H. Alexander, et al. (2024). Sustainable waste management with AI: waste classification using deep learning and IoT-based analysis of CH4 production. Available from https://doi.org/10.1109/ic-etite58242.2024.10493623.
[178] D. Ishaan, N. Surindra. (2024). Sustainable E-waste management. Practice, progress, and proficiency in sustainability. doi:https://doi.org/10.4018/979-8-3693-1018-2.ch017.
[179] Integrating AI in Sustainable supply chain management: a new paradigm for enhanced transparency and sustainability. 2023 International Journal for Science Technology and Engineering, Available from https://doi.org/10.22214/ijraset.2023.54139.
[180] S. Gundupalli Paulraj, S. Hait, A. Thakur. Automated municipal solid waste sorting for recycling using a mobile manipulator.
[181] D. Nguyen, V. Gadhamshetty, S. Nitayavardhana, S.K. Khanal, Automatic process control in anaerobic digestion technology: a critical review, Bioresource Technology 193 (2015) 513–522.
[182] C. Lubongo, M.A.A. Bin Daej, P. Alexandridis, Recent developments in technology for sorting plastic for recycling: the emergence of artificial intelligence and the rise of the robots, Recycling 9 (2024) 59.
[183] T. Anh Khoa, C.H. Phuc, P.D. Lam, L.M.B. Nhu, N.M. Trong, N.T.H. Phuong, et al., Waste management system using IoT-based machine learning in university, Wireless Communications and Mobile Computing 2020 (2020) 6138637.
[184] M. Reza, M. Hassan, AI-driven solutions for enhanced waste management and recycling in urban areas, International Journal of Sustainable Infrastructure for Cities and Societies 8 (2023) 1–13.
[185] S. Kaza, L. Yao, P. Bhada-Tata, F. Van Woerden, What a Waste 2.0: A Global Snapshot of Solid Waste Management to 2050, World Bank Publications, 2018.
[186] S.A. Ahmed, S.M. Ali, People as partners: facilitating people's participation in public–private partnerships for solid waste management, Habitat International 30 (2006) 781–796.
[187] E. Mazur-Wierzbicka, Towards circular economy—a comparative analysis of the countries of the European Union, Resources 10 (2021) 49.
[188] P.K. Gopalakrishnan, J. Hall, S. Behdad, Cost analysis and optimization of Blockchain-based solid waste management traceability system, Waste Management 120 (2021) 594–607.
[189] M. Geissdoerfer, P. Savaget, N.M.P. Bocken, E.J. Hultink, The Circular Economy – A new sustainability paradigm? Journal of Cleaner Production 143 (2017) 757–768.
[190] D.C. Wilson, C. Velis, C. Cheeseman, Role of informal sector recycling in waste management in developing countries, Habitat International 30 (2006) 797–808.

Further reading

M. Ferdeş, B.Ş. Zăbavă, G. Paraschiv, M. Ionescu, M.N. Dincă, G. Moiceanu, Food waste management for biogas production in the context of sustainable development, Energies 15 (17) (2022) 6268.

D.K. Gude, H. Bandari, A.K.R. Challa, S. Tasneem, Z. Tasneem, S.B. Bhattacharjee, et al., Transforming urban sanitation: enhancing sustainability through machine learning-driven waste processing, Sustainability 16 (17) (2024) 7626.

C.D.S. Guimaraes, D.R.D.S. Maia, E.G. Serra, Construction of biodigesters to optimize the production of biogas from anaerobic co-digestion of food waste and sewage, Energies 11 (4) (2018) 870.

A. Hussain, U. Draz, T. Ali, S. Tariq, M. Irfan, A. Glowacz, et al., Waste management and prediction of air pollutants using IoT and machine learning approach, Energies 13 (15) (2020) 3930.

S. Muhammad, H.P.S. Abdul Khalil, S. Abd Hamid, Y.M. Albadn, A.B. Suriani, S. Kamaruzzaman, et al., Insights into agricultural-waste-based nano-activated carbon fabrication and modifications for wastewater treatment application, Agriculture 12 (10) (2022) 1737.

M. Râpă, R.N. Darie-Niță, G. Coman, Valorization of fruit and vegetable waste into sustainable and value-added materials, Waste 2 (3) (2024) 258–278.

L.C. Sá, L.M. Loureiro, L.J. Nunes, A.M. Mendes, Torrefaction as a pretreatment technology for chlorine elimination from biomass: a case study using Eucalyptus globulus Labill, Resources 9 (5) (2020) 54.

A. Sarker, M.K. Ghosh, T. Islam, M. Bilal, R. Nandi, M.L. Raihan, et al., Sustainable food waste recycling for the circular economy in developing countries, with special reference to Bangladesh, Sustainability 14 (19) (2022) 12035.

S. Sivaranjani, Priyanka, S. Sherllyn, G.R. Deepaharshini, J. Eunice. Green symphony: a brief review of waste segregation techniques, 2024. Available from: https://doi.org/10.47392/irjaeh.2024.0056.

CHAPTER

12

Future of agriculture: smart vertical farming

Mohammed Alsafran, Malik Adil Nawaz, and Deepak Kasote

Agricultural Research Station, Qatar University, Doha, Qatar

Introduction

In the scenario of a growing population and climate change, more efficient production of food crops is required. Hence, besides traditional agricultural practices, adopting modern tools in agriculture is crucial. Over the last few decades, the agriculture sector has undergone a drastic transformation, from open-field farming to indoor farming. This evolution can be classified into four stages based on agricultural practices, from Agriculture 1.0 to Agriculture 4.0. Modern indoor agriculture constantly evolves with technological advancements, digital tools, and data-driven farm operations. Present-day farming, known as smart farming (represented in Agriculture 4.0), utilizes information and communication technology (ICT) such as cloud computing, the Internet of Things (IoT), artificial intelligence (AI), and robotics to enhance farm management. Smart farming aims to improve production quantity and quality by optimizing resource utilization and reducing environmental impact. This approach offers a sustainable path for agriculture by integrating these technologies into farming [1].

Indoor agriculture, also called controlled environment agriculture (CEA), consists of various technologies and growing environments. CEA includes greenhouse and vertical farming (VF) using smart farming tools. VF is the practice of growing crop on multiple levels in a controlled indoor environment. These days, VF has gained attention for its ability to reduce the environmental impact of agriculture by using less water, soil, and fertilizer. It also requires limited space and resources. In VF, plants are grown in a soil-free environment and watered using hydroponic or aeroponic technology. Because of this, VF offers benefits such as consistent yields, protection from pathogens, product consistency, and the ability to grow crops profitably, even not suited to local conditions. Recently production efficiency, including the quality of produce of VF, has significantly increased with the use of the smart vertical farm system (SVFS), where the IoT, AI, and machines are efficiently utilized.

Harnessing Automation and Machine Learning for Resource Recovery and Value Creation
DOI: https://doi.org/10.1016/B978-0-443-27374-2.00012-1

Recently VF has almost transformed into smart vertical farming (SVF). However, this sector is highly dynamic and rapidly changing with recent advances in automation, including the use of IoT and AI tools in this technology. This chapter provides an overview of VF and SVF, including their components and types, and the crops cultivated in VF. It also summarizes recent advancements in automation, including the integration of IoT and AI in VF/SVF.

Vertical framing: components and types

The VF system includes key components such as a multitier system with light-emitting diode (LED) lighting elements, hydroponic system, nutrient solution unit, HVAC (heating, ventilation, and air conditioning) unit, CO_2 fertilization, a soilless medium, farming structure to support the growing units and irrigation, and controlling unit [2]. Another crucial element of VF is the automation system and sensor technologies. Fig. 12.1 illustrates the overall structure of VF and its various components.

VF consists of various growth systems of different scales, users, technologies, locations, and purposes [3]. These systems can be divided into two categories: those with multiple levels of traditional horizontal growing platforms and those where the crop is grown on a vertical surface [3]. VF includes various types, such as plant factories with artificial lighting (PFALs), container farms, in-store farms, and appliance farms. PFALs are purpose-built structures with innovative designs, constructed for large-scale industrial-level vertical

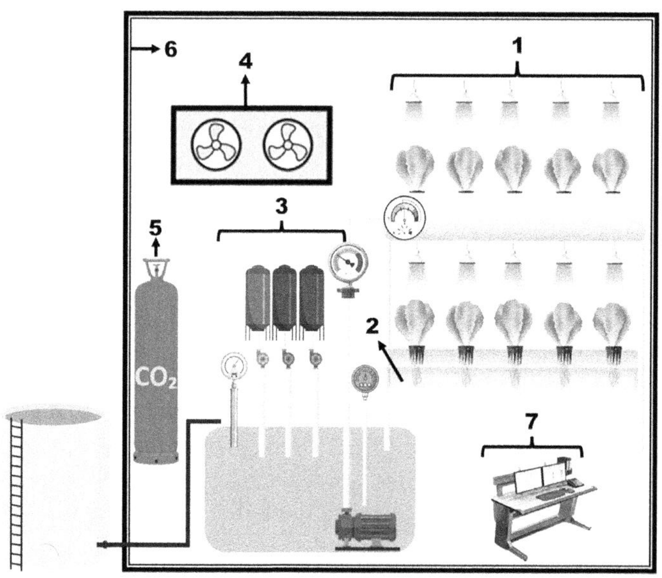

FIGURE 12.1 Schematic illustration of the overall structure of vertical farming and its various components. (1) Multitier system with light-emitting diode lighting elements, (2) hydroponic system, (3) nutrient solution unit, (4) heating, ventilation, and air conditioning unit, (5) CO_2 fertilization, (6) farming structure, and (7) control unit.

farming. In container farms, modified shipping containers with vertically stacked racks, LED lighting, and digitally supervised control systems are used for VF. This VF allows convenient relocation. In-store farms are compact cabinet systems placed in locations such as restaurants and supermarkets, bringing fresh produce closer to consumers. Soil is mainly avoided in VF systems. Based on growing systems, VF systems can also be classified into three types: hydroponic, aeroponic, and aquaponic. Aquaponics utilizes fish to supply nutrients, while hydroponics relies on formulated solutions. Aeroponics involves spraying a nutrient mist directly onto the roots of plants suspended in the air [4].

Suitable crops for vertical farming

VF is considered to be suitable for the cultivation of horticultural crops such as leafy vegetables, baby greens (e.g., spinach, kale, and lettuce), microgreens and herbs (e.g., arugula, radicchio, and basil) [4]. Tomatoes, cucumbers, strawberries, and sprouts (e.g., broccoli, soya bean, and wheat) are also produced using VF [4]. Preferred crops for VF are those that grow quickly, are short in height, and have specific light requirements. Production of crops, such as grapes, peppers, and tree fruit, is also possible with VF. However, certain crops' larger size and extended growth cycle present challenges. Nonedible crops, such as herbs and ornamental flowers, can also be viable options for VF-based production. However, integrating advanced technologies is essential to grow various crops on vertical farms effectively.

Smart vertical farming

VF is a climate-controlled system and thus faces various challenges related to optimization and efficiency. Hence, to achieve optimal performance in VF systems, it is essential to consider several key performance indicators (KPIs) that evaluate productivity, quality, and sustainability aspects [5]. SVF is a system that utilizes automation, hydroponics, machine learning, IoT, and AI, particularly to address the key challenges of VF, including ensuring crop quality, enhancing sustainability, and optimizing resource utilization. In SVF, information and communication technologies are intgrated with machinery, equipment, and sensors to monitor ambient parameters [6]. The pictorial representation of SVF is shown in Fig. 12.2.

Automation

VF generally has high operating expenses and energy costs, including high labor costs. Automating processes in VF is recommended to speed up operations, optimize energy usage, and reduce labor costs [7]. Automation and the use of robots streamline operations, ensuring precision in environmental monitoring and control, and increasing the efficiency of

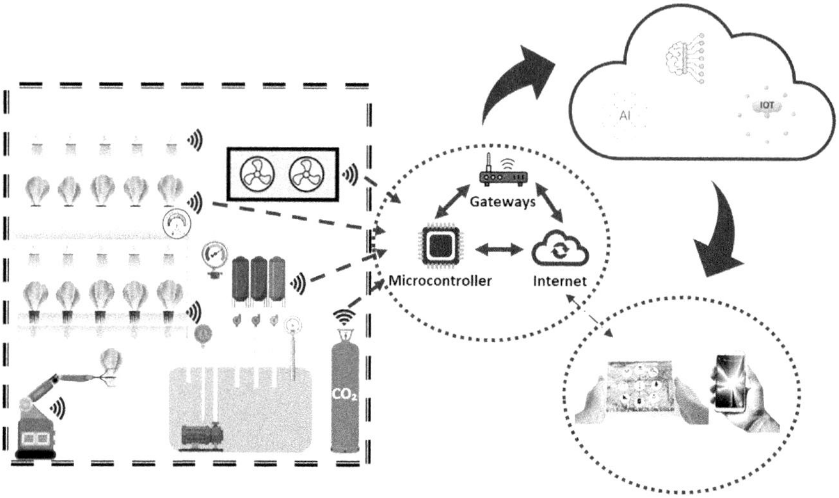

FIGURE 12.2 The pictorial representation of smart vertical farming. Source: *http://www.freepik.com.*

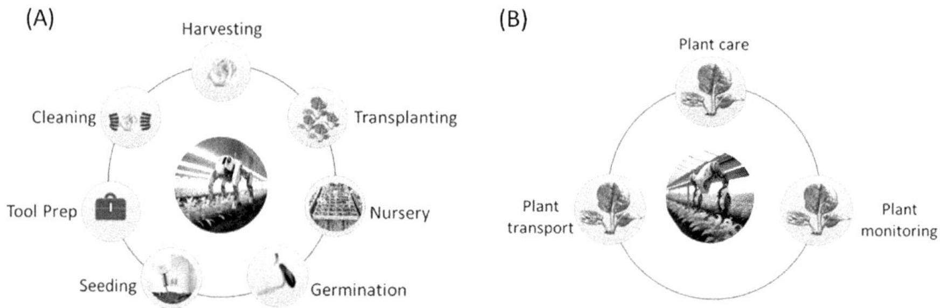

FIGURE 12.3 Automation in vertical farming operations. (A) SANANBIO UPLIFT vertical farming system automates seeding, transplanting, cleaning, and internal transportation. (B) Modular automated crop array online system enables plant care, monitoring, and transportation. Source: *http://www.freepik.com.*

VFs. SANANBIO, a prominent global indoor farming technology provider, has launched UPLIFT, an unmanned AI-powered vertical farming system that automates seeding, germination, transplanting, plant transportation, system cleaning, and harvesting [8] (Fig. 12.3). Wichitwechkarn and coworkers designed validated open-source automation system for vertical farming, named Modular Automated Crop Array Online System (MACARONS). This scalable, customizable, and open-source platform was created for plant care, monitoring, and transportation (Fig. 12.3). It provides precise dosing for each plant, automated data logging of temperature, humidity, and images, and allows for custom behaviors programmed in Python. The system can be monitored and controlled through a user-friendly web interface [7].

Applications of Internet of Things (IoT) and artificial intelligence (AI) technologies

The productivity and quality of agricultural produce of VF can be increased by implementing cutting-edge technologies such as IoT, AI, and machine learning. The use of cutting-edge technologies in VF makes it SVF. In SVF, temperature, humidity, light, and CO_2 levels in the atmosphere are continuously measured, monitored, and controlled around the clock using sensor technology. Similarly, nutrients are periodically injected and adjusted based on the growth stages through the integration of IoT and AI to achieve optimal output [4]. The advancing IoT technologies, which encompass smart sensors, devices, network topologies, big data analytics, and intelligent decision-making, are seen as the key to automating contemporary farming practices like greenhouse farming and VF [1]. In IoT-based VF, the system is designed for monitoring and automating factors such as temperature, humidity, and pH. Various factors, including crop yield, plant health, and accuracy, are considered. The data will be uploaded to the cloud and analyzed by Big Data Analytics using different algorithms for others to utilize [9]. The overview of IoT-based monitoring and control systems for VF is shown in Fig. 12.4.

The IoT-based VF monitoring system can help reduce user burden by providing accurate statistics and analysis. The online system offers immediate access to users and can locate equipment in use and track sensors detecting important input changes. Users can control the system or take automatic action when problems arise. For instance, users can turn off the watering system through a web-based application or water plants when humidity levels are too low. This convenience is expected to boost productivity and decrease water usage in agriculture.

AI involves the development of theories and computer systems that can perform tasks that require human intelligence, such as sensory perception and decision-making. This is usually combined with Cloud Computing, IoT, and big data [10]. Integration of AI with SVF

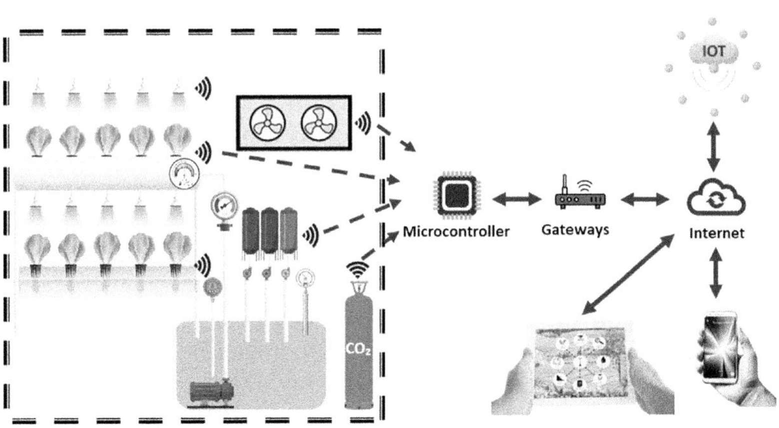

FIGURE 12.4 An overview of IoT-based monitoring and control systems in vertical farming. Source: *http://www.freepik.com.*

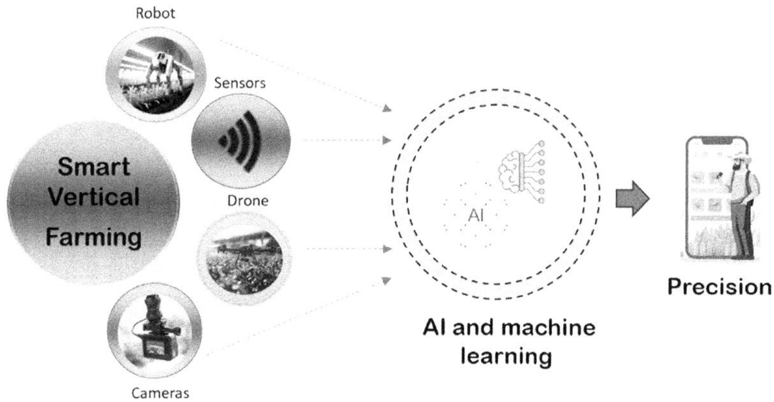

FIGURE 12.5 Integration of AI with smart vertical farming enables precision farming. Source: *http://www.freepik.com.*

enables precision farming and offers distinct advantages, such as monitoring plant growth, automating tasks, and precision (Fig. 12.5) [11].

By incorporating AI-powered technologies into VF farming practices, growers can get data-driven insights about planting schedules, resource allocation, and crop health. AI, powered by machine-learning algorithms, analyses data and makes decisions. AI and machine learning (ML) techniques are still rarely explored in indoor vertical farming systems. AI analyzes physical and imaging sensor data to predict plant health and growth, enabling growers to make informed decisions and increase crop yields. ML, deep learning (DL), and time series analysis are crucial in smart farming. ML algorithms help in crop selection and management, while DL techniques aid in crop selection and forecasting crop production. Time series analysis is utilized for demand forecasting of crop yield [12]. Mokhtar and coworkers predicted the yield of hydroponically grown lettuce (fresh weight) using four ML models, namely, support vector regressor (SVR), extreme gradient boosting (XGB), random forest (RF), and deep neural network (DNN), and demonstrated that among these, the DNN model was promising in predicting fresh lettuce yield. An AI-based lettuce disease detection system was developed in the VF system to accurately identify various lettuce diseases using a Convolutional Neural Network (CNN) image processing model. The model was built using ResNet50 and ImageNet on TensorFlow [13].

Summary

Considering present and future agriculture, VF can be an efficient way to significantly increase plant productivity, especially in areas where limited space and resources are available and environmental conditions are also not favorable. In general, VF is often seen as a complex system to implement due to the sensitivity of growing plants to their environment. Even small changes in the surroundings can greatly affect the efficiency and quality of the farming process. Recent developments in VF, particularly in advanced precision monitoring and automated control systems, have enabled growers to produce fresh and nutritious crops year-round. The integration of VF with IoT, robotics, and AI technologies transforms it into an SVF, which helps optimize crop yields, improve resource efficiency, automate labor-intensive tasks, and precision farming. VF is considered ideal for growing horticultural crops like leafy vegetables, tomatoes, cucumbers, strawberries, and sprouts. However, with smart VF production techniques, it is also possible to grow other crops as well. Though, VF is gaining popularity worldwide for its reliable and sustainable food production. However, VF is still considered a technique that is hard to afford due to high initial costs, a lack of expertise, the need for a controlled growing environment, and high energy consumption for lighting and climate control.

References

[1] K.M. Hosny, W.M. El-Hady, F.M. Samy, Technologies, protocols, and applications of internet of things in greenhouse farming: a survey of recent advances, Information Processing in Agriculture (2024).

[2] T. Van Gerrewey, N. Boon, D. Geelen, Vertical farming: the only way is up? Agronomy 12 (1) (2021) 2.

[3] A.M. Beacham, L.H. Vickers, J.M. Monaghan, Vertical farming: a summary of approaches to growing skywards, The Journal of Horticultural Science and Biotechnology 94 (3) (2019) 277–283.

[4] A. Krishnan, S. Swarna, Robotics, IoT, and AI in the automation of agricultural industry: a review, in: 2020 IEEE Bangalore Humanitarian Technology Conference (B-HTC), IEEE, 2020, pp. 1–6.

[5] N. Grasso, B. Fasciolo, G. Bruno, F. Lombardi, A smart vertical farming system to evaluate productivity, quality, and sustainability of agricultural production, International Conference on Flexible Automation and Intelligent Manufacturing, Springer, 2023, pp. 938–945.

[6] Hegedűs, A. Frankó, P. Varga, S. Gindl, M. Tauber, Enabling scalable smart vertical farming with IoT and machine learning technologies. NOMS 2023-2023 IEEE/IFIP Network Operations and Management Symposium, IEEE, 2023, pp. 1–4.

[7] V. Wichitwechkarn, W. Rohde, R. Choudhary, Design and validation of an open-sourced automation system for vertical farming, HardwareX 16 (2023) e00497.

[8] SANANBIO. Available online: <https://www.sananbio.com/>.[accessed 19.05.24].

[9] J.G. Chand, K. Susmitha, A. Gowthami, K.M. Chowdary, S.K. Ahmed, IOT-enabled vertical farming monitoring system using big data analytics, Second International Conference on Advances in Electrical, Computing, Communication and Sustainable Technologies (ICAECT) 2022 IEEE, 2022, pp. 1–6.

[10] R.R.A. Siregar, K.B. Seminar, S. Wahjuni, E. Santosa, Vertical farming perspectives in support of precision agriculture using artificial intelligence: a review, Computers 11 (9) (2022) 135.

[11] Chowdhury, D.B. P. Argha, M.A. Ahmed, Artificial intelligence in sustainable vertical farming. arXiv preprint arXiv:2312.00030 (2023).

[12] Y. Akkem, S.K. Biswas, A. Varanasi, Smart farming using artificial intelligence: a review, Engineering Applications of Artificial Intelligence 120 (2023) 105899.

[13] M. Siropyan, O. Celikel, O. Pinarer, Artificial intelligence driven vertical farming management system, in: Proceedings of the World Congress on Engineering 2022, 2022.

Further Reading

K. Benke, B. Tomkins, Future food-production systems: vertical farming and controlled-environment agriculture, Sustainability: Science, Practice and Policy 13 (1) (2017) 13–26.

Y.S. Chin, L. Audah, Vertical farming monitoring system using the internet of things (IoT), AIP Conference Proceedings 1883 AIP Publishing, 2017.

A. Csordás, I. Füzesi, The impact of technophobia on vertical farms, Sustainability 15 (9) (2023) 7476.

B. Fasciolo, A.M.M. Awouda, N. Grasso, G. Bruno, P. Chiabert, F. Lombardi, An evaluation of research interests in vertical farming through the analysis of KPIs adopted in the literature, Sustainability 16 (4) (2024) 1371.

S.R. Jaeger, Vertical farming (plant factory with artificial lighting) and its produce: consumer insights, Current Opinion in Food Science (2024) 101145.

M.S.N. Kabir, M.N. Reza, M. Chowdhury, M. Ali, Samsuzzaman, M.R. Ali, et al., Technological trends and engineering issues on vertical farms: a review, Horticulturae 9 (11) (2023) 1229.

A. Mokhtar, W. El-Ssawy, H. He, N. Al-Anasari, S.S. Sammen, Y. Gyasi-Agyei, et al., Using machine learning models to predict hydroponically grown lettuce yield, Frontiers in Plant Science 13 (2022) 706042.

P. Sengodan, An overview of vertical farming: highlighting the potential in Malaysian high-rise buildings, Pertanika Journal of Science & Technology 30 (2) (2022).

M. Swain, Vertical farming trends and challenges: a new age of agriculture using IoT and machine learning, Internet of Things for Agriculture 4.0 (2022) 1–16.

CHAPTER

13

Impact of artificial intelligence for the recycling of organic waste

Malik Adil Nawaz, and Deepak Kasote

Agricultural Research Station, Qatar University, Doha, Qatar

Introduction

Organic solid waste (OSW) comprises solid waste materials with organic components generated through diverse human activities. These sources include agricultural residue (e.g., crop straw and livestock manure), industrial waste such as industrial sludge, and municipal OSW (consisting of green waste, kitchen waste, and plastics). The growing global population, along with economic development and urbanization, has led to a worldwide crisis in the increased generation of OSW [1]. Inadequate and inefficient waste management can result in adverse environmental impacts viz., increased environmental pollution, destruction of ecosystems, negative effects on human health, and depletion of natural resources [2]. This situation has forced governments and autonomous global organizations such as the UN to allocate increased resources toward environmental remediation efforts [3]. Nevertheless, the mitigation of these issues could be significantly improved if a substantial portion of OSW could be efficiently recycled as resources or energy. Enhancing the treatment and recycling efficiency of OSW holds considerable positive implications for environmental protection, resource conservation, economic development, and safeguarding human health.

OSW treatment and recycling methods include classified recycling, incineration, landfill, composting, anaerobic digestion (AD), pyrolysis, and gasification [4]. However, these methods exhibit inherent drawbacks, such as poor efficiency, low accuracy, high costs, and potential environmental and health risks. For example, the prevailing primary separation method for municipal domestic waste involves manual hand-picking, which is not only inefficient but also poses health risks to workers due to harmful substances in the waste [5]. Moreover, waste collection and transportation processes often demand considerable human, material, and financial resources, contributing to air pollution, greenhouse gas emissions, noise, and various other adverse environmental effects [6]. Poorly planned waste management systems can further exacerbate these problems. During incineration, the heterogeneity of OSW creates challenges such as fluctuations in heating characteristics and uneven combustion in the furnace, resulting in reduced incineration efficiency. Moreover, incineration-related parameters, including temperature, ventilation, and flue gas residence time, significantly influence the generation of harmful

Harnessing Automation and Machine Learning for Resource Recovery and Value Creation
DOI: https://doi.org/10.1016/B978-0-443-27374-2.00013-3

substances such as SO_2, HCl, NOx, and dioxins, posing potential environmental and health risks [7]. In the composting process, determining maturity is crucial. However, classical approaches based on physicochemical analyses are time-consuming and expensive. Furthermore, the nonuniformity of solid waste materials leads to significant discrepancies in the results of multiple experiments, resulting in low accuracy. These challenges and defects have significantly impeded the advancement of treatment technology and simultaneously restricted the treatment and resource utilization of OSW.

In recent years, artificial intelligence (AI) has emerged as a popular tool for predicting OSW generation and its efficient management or recycling. The fundamental principle of AI is to use inductive inference to generalize the link between input and output variables, and then to use the associations discovered from empirical data to influence judgments in new situations. Three primary processes are often involved in AI modeling procedures: training, cross-validation, and testing. The AI model is trained by varying the model parameters (not hyperparameters) based on the training dataset. Cross-validation step: by adjusting the model's hyperparameters, the best model is chosen using a validation dataset. Testing step: a test dataset is used to further assess the chosen model's performance. The model is then prepared for use in making predictions. Early in the new millennium, AI was utilized in OWM to forecast the heating value of municipal solid waste (MSW). Nonetheless, AI has drawn a lot of interest and widely studied since 2010. AI has been gradually applied to many OWM treatments and recycling domains, especially since last the decade, and several studies on OWM using various AI models have been published [8,9].

Various studies have reported the numerous benefits of AI such as time-saving, achieving high accuracy in predicting complex nonlinear issues, and notably decreasing the need for manpower and resources in avoidable repetitive trials [10–13]. The advantages of AI mentioned above can address and potentially resolve the limitations of traditional OWM. Moreover, considering that OWM datasets are typically not too huge, the computational expenses of AI training are substantially lower compared to the potential benefits it offers. Therefore, AI shows huge potential for OWM applications and requires further studies.

This chapter offers a current, methodical overview of the different kinds of OSWs, their recycling for use in biomass production, biofuel, and agriculture, as well as the applications of AI in various OWM and recycling, including forecasting waste characteristics, projecting process parameters, and organizing operations. Furthermore, a thorough examination of several AI models and methods that are frequently applied in OSW recycling is carried out, including application domains and stated performance metrics.

Types of organic waste

Organic waste originates from various sources and can be categorized into several main types, as illustrated in Fig. 13.1.

Agricultural residues

Agricultural activities generate substantial amounts of organic residues, including crop residues, animal manure, and agricultural by-products. Crop residues such as stalks, leaves,

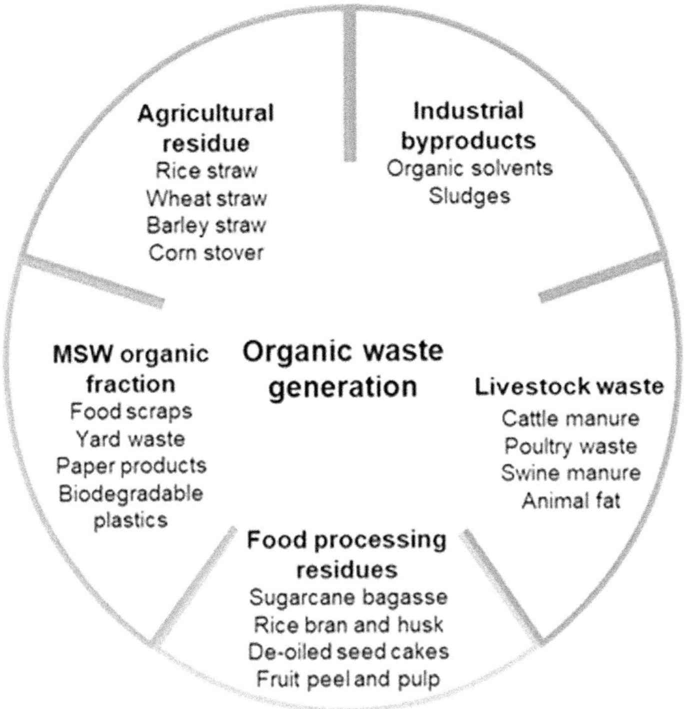

FIGURE 13.1 Different types of organic wastes.

and husks are commonly produced during harvesting and processing operations. Agricultural by-products, such as straw and bran, result from processing agricultural commodities. These organic residues hold potential for various applications, including soil amendment, composting, and bioenergy production [14].

Municipal solid waste organic fraction

The organic fraction of MSW comprises biodegradable materials discarded by households, businesses, and institutions. This includes food scraps, yard waste, paper products, and biodegradable plastics. Despite constituting a significant portion of MSW, organic waste often ends up in landfills, contributing to environmental pollution and greenhouse gas emissions [15]. However, through proper segregation and treatment, organic waste can be converted into valuable products such as compost, biogas, and bio-based chemicals [16].

Food processing residues

Food processing industries generate organic residues during the manufacturing and processing of food products. These residues may include fruit and vegetable peels, trimmings, pulp, and wastewater from cleaning operations [17]. While often considered waste, these by-products possess inherent value and can be utilized for various purposes.

For instance, fruit and vegetable peels can be processed into dietary fibers, antioxidants, and natural colorants. Wastewater streams can undergo treatment to recover valuable nutrients and energy [18].

Livestock waste

Livestock operations produce significant quantities of organic waste, primarily in the form of manure and bedding materials [19]. Manure contains valuable nutrients such as nitrogen, phosphorus, and potassium, but improper management can lead to environmental pollution and odor issues. Proper handling and treatment of livestock waste can mitigate these concerns while harnessing its potential as a fertilizer, soil conditioner, or feedstock for biogas production [20].

Industrial byproducts

Industrial processes generate various organic by-products and waste streams, ranging from wastewater to biomass residues. These include organic solvents, sludges, lignocellulosic materials, and organic-rich effluents [21]. While often challenging to manage due to their diverse compositions and complex characteristics, industrial organic wastes offer opportunities for resource recovery and circular economy initiatives. Innovative approaches such as anaerobic digestion (AD), pyrolysis, and biochemical conversion can transform these wastes into biofuels, biochemicals, and biodegradable materials [22].

Recycling of organic wastes

Organic wastes can be recycled using various processes listed in Fig. 13.2.

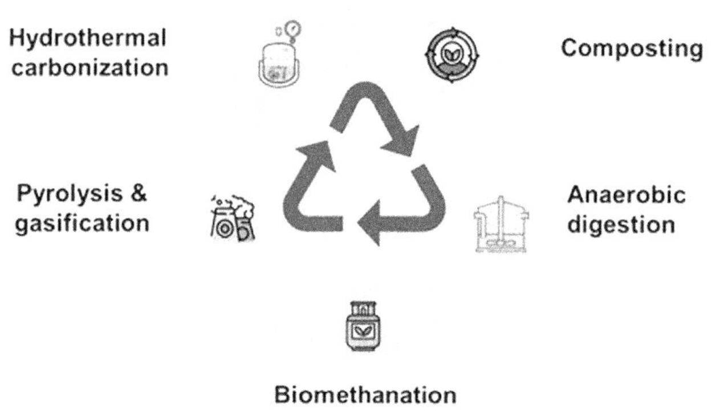

FIGURE 13.2 Various recycling processes of organic wastes.

Composting

Composting is a natural biological process that transforms organic wastes into a nutrient-rich soil amendment known as compost [23]. During composting, microorganisms break down organic matter under controlled conditions of moisture, aeration, and temperature. This process results in the production of stable humus-like material, which can improve soil structure, fertility, and water retention [24]. Compost can be used in agriculture, landscaping, and horticulture to enhance plant growth and soil health while diverting organic waste from landfills.

Anaerobic digestion

AD is a biochemical process that converts organic waste into biogas and digestate in the absence of oxygen [25]. Microorganisms break down organic matter in a sealed, oxygen-free environment, producing biogas primarily composed of methane and carbon dioxide [26]. Biogas can be used as a renewable energy source for heat and power generation, while digestate serves as a nutrient-rich fertilizer. AD is particularly suitable for treating high-moisture organic wastes such as food scraps, manure, and wastewater sludge, offering simultaneous waste treatment and energy recovery [27].

Biomethanation

Biomethanation is a specialized form of AD that focuses on the production of methane-rich biogas from organic wastes. This process involves the use of specific microbial consortia or enzymes to enhance methane production rates and yields [28]. Biomethanation technologies include wet and dry fermentation systems, continuous stirred-tank reactors, and plug-flow digesters. Biomethane produced from organic wastes can be upgraded to natural gas quality and injected into the gas grid or used as a transportation fuel, offering a renewable alternative to fossil fuels [29].

Pyrolysis and gasification

Pyrolysis and gasification are thermochemical processes that convert organic wastes into biochar, syngas, and bio-oil through high-temperature heating in the absence of oxygen (pyrolysis) or with limited oxygen (gasification) [30]. Pyrolysis produces biochar, a carbon-rich solid material with potential applications in soil amendment and carbon sequestration. Gasification generates syngas, a mixture of hydrogen, carbon monoxide, and methane, which can be used for heat and power generation or converted into biofuels and chemicals [31]. These technologies offer an alternative to traditional waste disposal methods, reducing greenhouse gas emissions and promoting resource recovery [32].

Hydrothermal carbonization

Hydrothermal carbonization (HTC) is a thermal conversion process that utilizes high temperature and pressure in a water-rich environment to convert organic wastes into hydrochar, a carbonaceous solid material resembling coal. HTC can process a wide range of organic feedstocks, including sewage sludge, biomass, and organic-rich residues [33]. Autogenous pressure and temperature ranges of 150°C–350°C are commonly used for HTC. As a result, during the HTC reaction, the water content remains in the aqueous phase but its density and dielectric constant drop. Similarly, the strong dehydration and decarboxylation processes cause the feedstock's O and H concentrations to drop. Hydrochar produced through HTC can be used as a soil conditioner, adsorbent, or solid fuel, offering potential applications in agriculture, wastewater treatment, and energy production.

Use of artificial intelligence in organic waste management and recycling

Waste sorting and classification

AI-powered sorting systems utilize computer vision and machine learning algorithms to accurately identify and classify different types of organic wastes. By analyzing visual characteristics such as shape, color, and texture, these systems can segregate organic materials from mixed waste streams with high precision [34]. Automated sorting technologies enable faster processing speeds and higher recovery rates, improving the quality of recyclable materials and reducing contamination levels in organic waste streams [35]. A pictorial representation of the AI-assisted waste collection, sorting, and classification process is presented in Fig. 13.3.

Predictive analytics for waste generation

Several AI algorithms are being utilized in automated organic waste management (OVM). A list of common AI algorithms with their advantages and limitations is presented in Table 13.1. Typically, AI algorithms can analyze historical data on waste generation patterns, environmental factors, and demographic trends to forecast future waste generation rates with greater accuracy. By incorporating real-time data from sensors, IoT devices, and waste management systems, predictive analytics models can optimize waste collection schedules, route planning, and resource allocation [36]. This proactive approach helps municipalities and waste management companies anticipate demand fluctuations and allocate resources more efficiently, minimizing operational costs and environmental impacts.

Optimization of anaerobic digestion and biogas production

AI-based control systems can optimize the operation of AD plants and biogas production facilities by continuously monitoring process variables and adjusting operational parameters in real-time [46]. Machine learning algorithms analyze sensor data, process

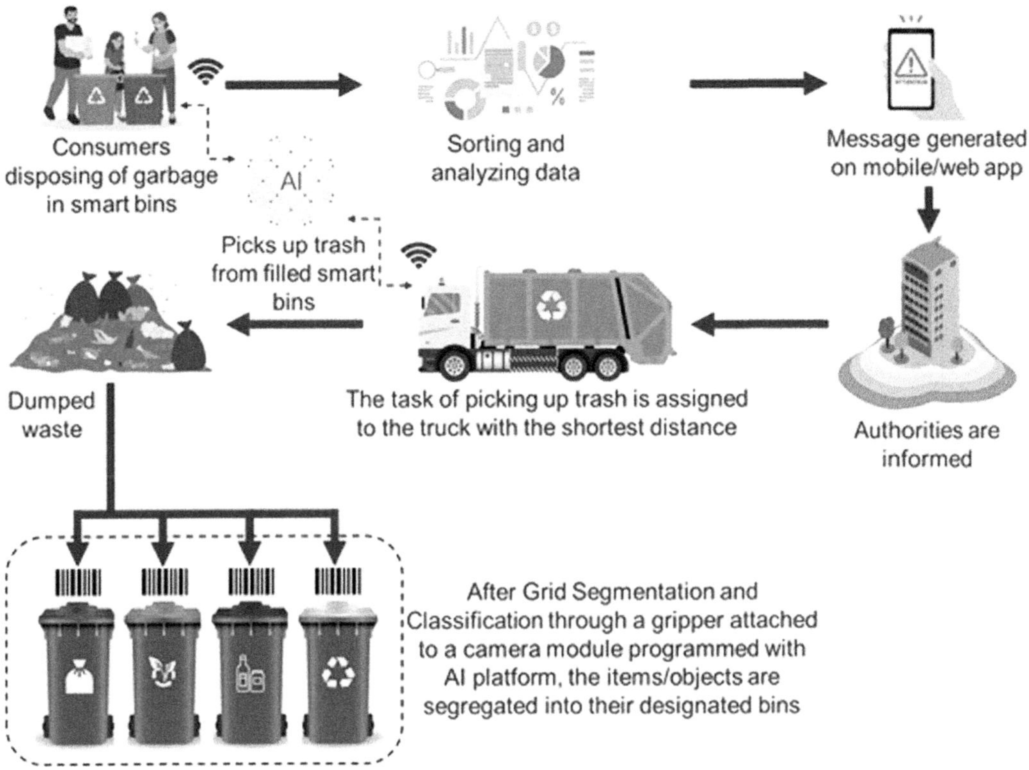

FIGURE 13.3 Artificial intelligence-assisted waste collection, sorting, and classification process.

performance metrics, and external factors to optimize feedstock composition, digestion conditions, and biogas yield. AI-driven process optimization enhances the overall efficiency, stability, and profitability of AD systems, maximizing biogas production while minimizing energy consumption and process downtime [47].

Intelligent waste valorization and resource recovery

AI algorithms facilitate the identification of valuable resources and market opportunities within organic waste streams, enabling the development of innovative valorization pathways and value-added products [48]. By analyzing the chemical composition, nutritional content, and material properties of organic wastes, AI can identify potential applications in bio-based materials, biochemicals, and renewable energy [49]. Intelligent waste valorization strategies leverage AI-driven decision support systems to optimize resource recovery, product design, and supply chain management, creating new revenue streams and business opportunities in the circular economy [50].

TABLE 13.1 Overview of artificial intelligence algorithms utilized in organic waste management.

AI algorithms	Current usage	Advantages	Disadvantages	References
Support vector machines (SVMs) or support vector networks (SVNs)	Garbage segregation system. Discovering organic waste along with satellite measurements. Estimating organic waste recycling.	Suitable for unstructured and semi-structured data. Beneficial for complex and nonlinear datasets. High accuracy for high-dimensional spaces. Efficient for maximum marginal splitting amongst classes.	High computational complexity and requires rigorous training. Unsuitable for noisy datasets. Requires additional scaling features. Unsuitable for large datasets. Requires careful parameter tuning to achieve optimal performance.	Zhang et al. [37]
Decision trees (DT)	Characterization of organic waste. Data mining for waste management.	User-friendly features for data processing and interpretation. Efficient in handling nonlinear datasets. Automatically select the most relevant features, aiding in identifying key factors influencing OVM. Does not require data normalization.	Prone to overlifting especially during complex and large dataset processing resulting in inaccurate outcomes. Unsuitable for regression and prediction of data containing continuous values. Results impacted by noisy dataset. Long-term training requirements. Expensive computational platform. Small changes in the initial dataset may result in significant changes in tree structure, impacting the reliability of data prediction of OVM.	Wang et al. [38]
Genetic algorithms (GA)	Optimization of various aspects of waste collection and supply chain networks. Identification of areas with similar waste composition. Preparation of waste collection routines.	Simple to comprehend. Facilitates multipurpose optimization. Efficient performance for both continuous and discrete datasets. Effectively finds optimal or near-optimal solutions. Efficiently handles local minima and maxima. Easily adaptable to diverse problem domains.	Require population tuning. Do not always guarantee the global optimum due to their reliance on heuristics and multiple searches. Requires significant computational resources and demands additional effort and time. Unable to utilize gradients.	Movahed et al. [39]
Linear regression (LR)	Characterization of organic waste.	Provides a simple and interpretable model for analyzing the impact of various factors on waste generation. Allows for the identification of significant independent variables and the development of models to estimate household waste production effectively.	May not capture the complexity of waste generation accurately, especially in scenarios where waste production is influenced by multiple social, economic, and demographic factors. May not always provide reliable predictions for practical applications across different cities due to the diverse nature of waste generation drivers.	Popli et al. [40]

Radio frequency identification (RFID)	Waste tracking. Work order management. Verification.	Enhances accuracy and mitigates overfitting in DT-based problems. Well-suited for handling both continuous and categorical variables. Does not require additional features scaling. Capable of accommodating nonlinear parameters. Resilient to the influence of data noise. Does not require data normalization.	Requires significant computational power and resources. Rigorous training requirement. Slower computational compared to DT.	
Multilayer perceptron (MLP)	Prediction of high heating value. Waste classification. Prediction of waste generation and cost estimation.	Efficient in solving nonlinear problems. Effective in handling large amounts of input data well. Quick predictions after training. The same accuracy ratio can be achieved with smaller samples.	High complexity and computational cost.	Zhang et al. [41]
Artificial neural network (ANN)	Prediction of high heating value. Waste classification. Prediction of waste generation and cost estimation. Forecasting waste generation platforms. Optimization of waste collection routes.	Effective classification of organic waste into different categories with high accuracy (~91.7%). Well suited for handling complex nonlinear datasets. Versatile and can be applied to various aspects of OWM.	Requires a large amount of high-quality training data. Insufficient or incomplete data can lead to overfitting and reduced model performance. Complex internal structure and decision-making process. Computationally expensive compared to traditional methods.	
Convoluted neural network (CNN)	Waste classification. Waste identification, and sorting. Waste recycling. Real-time waste classification.	Efficient and real-time waste classification. High accuracy.	Complexity of models. High error rates.	Yang et al. [42]
Recurrent neural network (RNN)	Modeling waste disposal rates. Forecasting waste generation. Optimization of waste collection routes.	Efficient handling of sequential data. Efficient capturing of long-term dependencies in data.	Vanishing gradient problem. The complex model with expensive computation.	
K-nearest neighbors (KNN)	Forecasting waste generation. Waste classification.	Efficient waste classification. Simplicity and wide applicability. Real-time classification.	Limited accuracy. Dependence on data quality.	Rajani and Sriramya [43]

(*Continued*)

TABLE 13.1 (CONTINUED)

AI algorithms	Current usage	Advantages	Disadvantages	References
	Optimization of waste disposal and recycling processes.			
Deep neural network (DNN)	Waste classification. Waste localization. Optimization of waste treatment.	Efficient and real-time waste classification. Efficient handling of complex relationships.	Model complexity. High error rates. Dependence on data quality.	Malik et al. [44]
Particle swarm optimization (PSO)	Parameter optimization in the waste management model. Optimization of thermochemical waste treatment technologies.	Requires fewer parameters to tune. Efficient global search.	Limitations for real-time applications. Weak local search ability.	
Back propagation (BP)	Waste classification. Optimization of organic Rankine cycles. Hybrid optimization with particle swarm optimization (PSO). Modeling of leaching behavior of solidified wastes.	Efficient waste classification. Automated waste recognition.	Model complexity. Data dependency.	Feng et al. [45]

Decision support systems for policy and planning

AI-powered decision support systems provide policymakers, urban planners, and waste management professionals with valuable insights and recommendations for optimizing waste management strategies and policy interventions [51]. By integrating data from multiple sources, including socioeconomic indicators, environmental impact assessments, and stakeholder preferences, AI algorithms can model different scenarios, evaluate trade-offs, and identify optimal solutions for OVM and recycling. Decision support systems enable evidence-based decision-making, fostering collaboration, transparency, and innovation in waste management governance.

Potential and challenges of artificial intelligence in the recycling of organic waste

The application of AI in organic waste recycling holds immense promise for revolutionizing waste management practices and advancing toward a circular economy. However, along with its potential benefits, the integration of AI in recycling processes also presents various challenges that need to be addressed. In this section, we explore the potential opportunities offered by AI in organic waste recycling, along with the key challenges that must be overcome for successful implementation.

Potential opportunities

Enhanced sorting and segregation

AI-powered sorting systems can accurately identify and segregate different types of organic waste materials, improving the efficiency and quality of recycling processes [52]. By analyzing visual characteristics and employing machine learning algorithms, these systems can achieve higher accuracy rates and minimize contamination levels in recycled materials.

Optimization of resource recovery

AI algorithms play a crucial role in optimizing the functionality of recycling facilities by maximizing resource recovery from organic waste streams [53]. Through real-time monitoring of process variables and dynamic adjustment of operational parameters, AI-driven control systems significantly enhance material recovery rates, energy efficiency, and overall process performance [54].

Predictive analytics for demand forecasting

AI-based predictive analytics models can indeed forecast future demand for recycled organic materials, enabling better resource planning and allocation in recycling facilities [55]. By analyzing historical data, market trends, and environmental factors, these advanced models can assist in anticipating fluctuations in demand and adjusting production processes accordingly. This proactive approach not only minimizes waste but also optimizes resource utilization, contributing to more efficient and sustainable recycling practices [1].

Innovation in product design and development

Based on the provided sources, AI-driven data analysis indeed plays a crucial role in identifying opportunities for innovation in product design and development using recycled organic materials. By analyzing material properties, chemical composition, and performance characteristics, AI algorithms can facilitate the creation of new value-added products and applications, thereby fostering market growth and diversification.

The integration of AI technologies, such as machine learning algorithms, in materials informatics enables companies to leverage data effectively for innovation [56]. AI can streamline manufacturing processes of bio-based materials by optimizing procedures, detecting redundancies, and enhancing efficiency, leading to cost-effective production and waste reduction [57]. Additionally, AI-powered robotic sorting machines, like YUTA, offer high-speed sorting capabilities for recyclables, with live monitoring and data analysis features, contributing to efficient recycling processes and resource utilization.

Moreover, AI demand forecasting in various sectors, including manufacturing, finance, and healthcare, demonstrates the transformative impact of AI in predicting market trends, optimizing operations, and enhancing resource management. By accurately predicting demand and aligning production levels with actual market needs, AI-driven forecasting minimizes waste, optimizes resource utilization, and supports sustainable growth.

In the context of recycled organic materials, AI algorithms can analyze data on material properties, chemical composition, and performance characteristics to identify innovative opportunities for product design and development. By leveraging AI-driven data analysis, companies can create new value-added products and applications, driving market growth and diversification while promoting sustainability through the use of recycled materials.

Key challenges

Data quality and availability

One of the primary challenges in implementing AI in organic waste recycling is the availability and quality of data. AI algorithms require large volumes of high-quality data for training and validation, which may be lacking or fragmented in the recycling industry. Improving data collection, standardization, and sharing mechanisms is essential to overcome this challenge [58].

Technological complexity and integration

Implementing AI-driven recycling technologies requires integration with existing waste management systems and infrastructure, which can be complex and costly. Overcoming technological barriers and ensuring compatibility with existing processes and equipment is crucial for successful deployment of AI in recycling facilities [59].

Regulatory and ethical considerations

The use of AI in organic waste recycling raises various regulatory and ethical considerations related to data privacy, safety, and transparency. Ensuring compliance with data protection regulations, ethical guidelines, and industry standards is essential to build trust and confidence in AI-driven recycling solutions [60].

Skills and capacity building

Developing and maintaining AI expertise and capabilities within the recycling industry is another challenge that needs to be addressed. Training personnel in AI technologies, data analytics, and machine learning techniques is essential to harness the full potential of AI in organic waste recycling and overcome skills shortages [61].

Summary

This chapter provides a comprehensive review of OSW recycling for agriculture, biofuel, and biomass production, along with the applications of AI in OWM. Traditional OSW treatment methods have limitations such as low efficiency, high costs, and environmental risks, which AI can address by offering computational solutions to manage uncertainty and incomplete data. AI-driven data analysis is essential for identifying opportunities for innovation in product design and development using recycled organic materials. AI-powered sorting systems utilize computer vision and machine learning algorithms to accurately identify and classify various types of organic waste, leading to faster processing speeds and higher recovery rates. The integration of AI with OWM can facilitate the creation of new value-added products OSW.

References

[1] L.A.B. Paes, B.S. Bezerra, R.M. Deus, D. Jugend, R.A.G. Battistelle, Organic solid waste management in a circular economy perspective—a systematic review and SWOT analysis, Journal of Cleaner Production 239 (2019) 118086.

[2] S.B. Wassie, Natural resource degradation tendencies in Ethiopia: a review, Environmental Systems Research 9 (2020) 1–29.

[3] S. Kharola, M. Ram, N. Goyal, S.K. Mangla, O. Nautiyal, A. Rawat, et al., Barriers to organic waste management in a circular economy, Journal of Cleaner Production 362 (2022) 132282.

[4] H. Durak, Comprehensive assessment of thermochemical processes for sustainable waste management and resource recovery, Processes 11 (2023) 2092.

[5] C. Ludwig, S. Hellweg, S. Stucki, Municipal Solid Waste Management: Strategies and Technologies for Sustainable Solutions, Springer Science & Business Media, 2012.

[6] F.O. Ajibade, B. Adelodun, K.H. Lasisi, O.O. Fadare, T.F. Ajibade, N.A. Nwogwu, et al., Environmental pollution and their socioeconomic impacts, Microbe Mediated Remediation of Environmental Contaminants, Elsevier, 2021.
[7] L.M. Ulloa-Murillo, L.M. Villegas, A.R. Rodríguez-Ortiz, M. Duque-Acevedo, F.J. Cortés-García, Management of the organic fraction of municipal solid waste in the context of a sustainable and circular model: analysis of trends in Latin America and the Caribbean, International Journal of Environmental Research and Public Health 19 (2022) 6041.
[8] A.B. Jude, D. Singh, S. Islam, M. Jameel, S. Srivastava, B. Prabha, et al., An artificial intelligence based predictive approach for smart waste management, Wireless Personal Communications 127 (2021) 123456789.
[9] T.-W. Wu, H. Zhang, W. Peng, F. Lü, P.-J. He, Applications of convolutional neural networks for intelligent waste identification and recycling: a review, Resources, Conservation and Recycling 190 (2023) 106813.
[10] R. Roohi, M. Jafari, E. Jahantab, M.S. Aman, M. Moameri, S. Zare, Application of artificial neural network model for the identification the effect of municipal waste compost and biochar on phytoremediation of contaminated soils, Journal of Geochemical Exploration 208 (2020) 106399.
[11] P. Tahmasebi, S. Kamrava, T. Bai, M. Sahimi, Machine learning in geo-and environmental sciences: from small to large scale, Advances in Water Resources 142 (2020) 103619.
[12] X. Zhu, Y. Li, X. Wang, Machine learning prediction of biochar yield and carbon contents in biochar based on biomass characteristics and pyrolysis conditions, Bioresource Technology 288 (2019) 121527.
[13] X. Zhu, X. Wang, Y.S. Ok, The application of machine learning methods for prediction of metal sorption onto biochars, Journal of Hazardous Materials 378 (2019) 120727.
[14] U. De Corato, I. De Bari, E. Viola, M. Pugliese, Assessing the main opportunities of integrated biorefining from agro-bioenergy co/by-products and agroindustrial residues into high-value added products associated to some emerging markets: a review, Renewable and Sustainable Energy Reviews 88 (2018) 326–346.
[15] K.D. Sharma, S. Jain, Overview of municipal solid waste generation, composition, and management in India, Journal of Environmental Engineering 145 (2019) 04018143.
[16] T.R. Johnson, Municipal solid waste management, Routledge Handbook of Environmental Policy in China, Routledge, 2017.
[17] N. Mirabella, V. Castellani, S. Sala, Current options for the valorization of food manufacturing waste: a review, Journal of Cleaner Production 65 (2014) 28–41.
[18] V. Joshi, Fruit and vegetable processing waste management—an overview, International Journal of Food and Fermentation Technology 10 (2020) 67–94.
[19] S. Parihar, K. Saini, G. Lakhani, A. Jain, B. Roy, S. Ghosh, et al., Livestock waste management: a review, Journal of Entomology and Zoology Studies 7 (2019) 384–393.
[20] N. Bhatnagar, D. Ryan, R. Murphy, A. Enright, A comprehensive review of green policy, anaerobic digestion of animal manure and chicken litter feedstock potential—global and Irish perspective, Renewable and Sustainable Energy Reviews 154 (2022) 111884.
[21] S. Doifode, A.G. Matani, A. Matani, Effective industrial waste utilization technologies towards cleaner environment, International Journal of Chemical and Physical Sciences 4 (2015), p. 536.
[22] M. Shahbaz, N. Rashid, J. Saleem, H. Mackey, G. McKay, T. Al-Ansari, A review of waste management approaches to maximise sustainable value of waste from the oil and gas industry and potential for the State of Qatar, Fuel 332 (2023) 126220.
[23] E. Goldan, V. Nedeff, N. Barsan, M. Culea, M. Panainte-Lehadus, E. Mosnegutu, et al., Assessment of manure compost used as soil amendment—a review, Processes 11 (2023) 1167.
[24] T.T.K. Ho, T.H. Le, C.-S. Tran, P.-T. Nguyen, V.-N. Thai, X.-T. Bui, Compost to improve sustainable soil cultivation and crop productivity, Case Studies in Chemical and Environmental Engineering 6 (2022) 100211.
[25] K.F. Adekunle, J.A. Okolie, A review of biochemical process of anaerobic digestion, Advances in Bioscience and Biotechnology 6 (2015) 205.
[26] P.M. Christy, L. Gopinath, D. Divya, A review on anaerobic decomposition and enhancement of biogas production through enzymes and microorganisms, Renewable and Sustainable Energy Reviews 34 (2014) 167–173.
[27] F. Momayez, K. Karimi, M.J. Taherzadeh, Energy recovery from industrial crop wastes by dry anaerobic digestion: a review, Industrial Crops and Products 129 (2019) 673–687.

[28] G. Cwicklinski, R. Miras, J. Pérard, C. Rinaldi, E. Darrouzet, C. Cavazza, Development of a CO_2-biomethanation reactor for producing methane from green H_2, Sustainable Energy & Fuels 8 (2024) 1068–1076.

[29] K. Rajendran, D. Mahapatra, A.V. Venkatraman, S. Muthuswamy, A. Pugazhendhi, Advancing anaerobic digestion through two-stage processes: current developments and future trends, Renewable and Sustainable Energy Reviews 123 (2020) 109746.

[30] R. Ruan, K. Ding, S. Liu, P. Peng, N. Zhou, A. He, et al., Gasification and pyrolysis of waste, Current Developments in Biotechnology and Bioengineering, Elsevier, 2020.

[31] A. Molino, V. Larocca, S. Chianese, D. Musmarra, Biofuels production by biomass gasification: a review, Energies 11 (2018) 811.

[32] D. Puyol, D.J. Batstone, T. Hülsen, S. Astals, M. Peces, J.O. Krömer, Resource recovery from wastewater by biological technologies: opportunities, challenges, and prospects, Frontiers in Microbiology 7 (2017) 2106.

[33] R. Sivaranjanee, P.S. Kumar, G. Rangasamy, A recent advancement on hydrothermal carbonization of biomass to produce hydrochar for pollution control, Carbon Letters 33 (2023) 1909–1933. Available from https://doi.org/10.1007/s42823-023-00576-2.

[34] S.P. Gundupalli, S. Hait, A. Thakur, A review on automated sorting of source-separated municipal solid waste for recycling, Waste Management 60 (2017) 56–74.

[35] C. Lubongo, M.A.B. Daej, P. Alexandridis, Automated sorting technology for plastic waste, Reuse of Plastic Waste in Eco-Efficient Concrete, Elsevier, 2024.

[36] S. Mishra, L. Jena, H.K. Tripathy, T. Gaber, Prioritized and predictive intelligence of things enabled waste management model in smart and sustainable environment, PLoS One 17 (2022) e0272383.

[37] S. Zhang, A.H. Omar, A.S. Hashim, T. Alam, H.A.E.-W. Khalifa, M.A. Elkotb, Enhancing waste management and prediction of water quality in the sustainable urban environment using optimized algorithm of least square support vector machine and deep learning techniques, Urban Climate 49 (2023) 101487.

[38] Y. Wang, T. Huntington, C.D. Scown, Tree-based automated machine learning to predict biogas production for anaerobic co-digestion of organic waste, ACS Sustainable Chemistry & Engineering 9 (2021) 12990–13000.

[39] Z.P. Movahed, M. Kabiri, S. Ranjbar, F. Joda, Multi-objective optimization of life cycle assessment of integrated waste management based on genetic algorithms: a case study of Tehran, Journal of Cleaner Production 247 (2020) 119153.

[40] K. Popli, C. Park, S.-M. Han, S. Kim, Prediction of solid waste generation rates in urban region of Laos using socio-demographic and economic parameters with a multi linear regression approach, Sustainability 13 (2021) 3038.

[41] J. Zhang, C. Li, Y. Yin, J. Zhang, M. Grzegorzek, Applications of artificial neural networks in microorganism image analysis: a comprehensive review from conventional multilayer perceptron to popular convolutional neural network and potential visual transformer, Artificial Intelligence Review 56 (2023) 1013–1070.

[42] L. Yang, H. Nguyen, X.-N. Bui, T. Nguyen-Thoi, J. Zhou, J. Huang, Prediction of gas yield generated by energy recovery from municipal solid waste using deep neural network and moth-flame optimization algorithm, Journal of Cleaner Production 311 (2021) 127672.

[43] J. Rajani, P. Sriramya, Artificial neural networks are compared to K-nearest neighbors in a novel swarm optimization for smart waste management systems employing automated machine learning, Journal of Survey in Fisheries Sciences 10 (2023) 1998–2007.

[44] M. Malik, S. Sharma, M. Uddin, C.-L. Chen, C.-M. Wu, P. Soni, et al., Waste classification for sustainable development using image recognition with deep learning neural network models, Sustainability 14 (2022) 7222.

[45] Y.-Q. Feng, Y.-Z. Liu, X. Wang, Z.-X. He, T.-C. Hung, Q. Wang, et al., Performance prediction and optimization of an organic Rankine cycle (ORC) for waste heat recovery using back propagation neural network, Energy Conversion and Management 226 (2020) 113552.

[46] S. Swami, S. Suthar, R. Singh, A.K. Thakur, L.R. Gupta, V.S. Sikarwar, Integration of anaerobic digestion with artificial intelligence to optimise biogas plant operation, Environment, Development and Sustainability (2023) 1–31.

[47] M. Khan, W. Chuenchart, K. Surendra, S.K. Khanal, Applications of artificial intelligence in anaerobic co-digestion: recent advances and prospects, Bioresource Technology 370 (2023) 128501.

[48] A. Taneja, R. Sharma, S. Khetrapal, A. Sharma, R. Nagraik, B. Venkidasamy, et al., Value addition employing waste bio-materials in environmental remedies and food sector, Metabolites 13 (2023) 624.

[49] A.D.A. Bin Abu Sofian, X. Sun, V.K. Gupta, A. Berenjian, A. Xia, Z. Ma, et al., Advances, synergy, and perspectives of machine learning and biobased polymers for energy, fuels, and biochemicals for a sustainable future, Energy & Fuels 38 (2024) 1593–1617.

[50] H. Onyeaka, P. Tamasiga, U.M. Nwauzoma, T. Miri, U.C. Juliet, O. Nwaiwu, et al., Using artificial intelligence to tackle food waste and enhance the circular economy: maximising resource efficiency and minimising environmental impact: a review, Sustainability 15 (2023) 10482.

[51] L. Andeobu, S. Wibowo, S. Grandhi, Artificial intelligence applications for sustainable solid waste management practices in Australia: a systematic review, Science of The Total Environment 834 (2022) 155389.

[52] M. Reza, AI-driven solutions for enhanced waste management and recycling in urban areas, International Journal of Sustainable Infrastructure for Cities and Societies 8 (2023) 1–13.

[53] M.T. Munir, B. Li, M. Naqvi, Revolutionizing municipal solid waste management (MSWM) with machine learning as a clean resource: opportunities, challenges and solutions, Fuel 348 (2023) 128548.

[54] M. Elahi, S.O. Afolaranmi, J.L. Martinez Lastra, J.A. Perez Garcia, A comprehensive literature review of the applications of AI techniques through the lifecycle of industrial equipment, Discover Artificial Intelligence 3 (2023) 43.

[55] P. Fraga-Lamas, S.I. Lopes, T.M. Fernández-Caramés, Green IoT and edge AI as key technological enablers for a sustainable digital transition towards a smart circular economy: an industry 5.0 use case, Sensors 21 (2021) 5745.

[56] S. Ramakrishna, T.-Y. Zhang, W.-C. Lu, Q. Qian, J.S.C. Low, J.H.R. Yune, et al., Materials informatics, Journal of Intelligent Manufacturing 30 (2019) 2307–2326.

[57] R. Kumar, S.K. Gupta, H.-C. Wang, C.S. Kumari, S.S.V.P. Korlam, From efficiency to sustainability: exploring the potential of 6G for a greener future, Sustainability 15 (2023) 16387.

[58] Z. Ye, J. Yang, N. Zhong, X. Tu, J. Jia, J. Wang, Tackling environmental challenges in pollution controls using artificial intelligence: a review, Science of the Total Environment 699 (2020) 134279.

[59] Z.Q.S. Nwokediegwu, E.D. Ugwuanyi, Implementing ai-driven waste management systems in underserved communities in the USA, Engineering Science & Technology Journal 5 (2024) 794–802.

[60] H. Roberts, J. Zhang, B. Bariach, J. Cowls, B. Gilburt, P. Juneja, et al., Artificial intelligence in support of the circular economy: ethical considerations and a path forward, AI & Society (2022) 1–14.

[61] R. Nishant, M. Kennedy, J. Corbett, Artificial intelligence for sustainability: challenges, opportunities, and a research agenda, International Journal of Information Management 53 (2020) 102104.

CHAPTER

14

Agriculture: the next machine-learning frontier

Al-Maadeed Somaya[1], and Akbari Younes[2]

[1]Qatar University, Doha, Qatar [2]Computer Science and Engineering Department, Qatar University, Doha, Qatar

Introduction and definition of the problem

Many industries are being revolutionized by artificial intelligence (AI), and agriculture is no exception. AI integration in agriculture is changing conventional farming practices, which contributes to the development of a more effective and sustainable food system [1]. Presently, high-performance computing technologies and efficient, reasonably priced sensors are attempting to convert conventional agriculture into a more effective intelligent agriculture. The low cost, easy implementation, and high communication efficiency of internet of things (IoT) technologies present a wonderful chance to gather huge a data in agricultural production, breeding, and cultivation studies. In the meantime, unsupervised machine learning (ML) techniques have been utilized in intelligent farming. To identify the type of crop disease, for example, ML techniques have been applied to the analysis of photographs of diseased leaves. Thus, it is possible to view the application of AI and ML techniques to this significant social need as a revolutionary expansion of the agriculture sector [2].

Even though the field of scientific study on ML techniques for smart agriculture is growing quickly, there are still a number of unanswered concerns that need to be considered and investigated. For example, applying and advancing ML techniques for the identification of plant species, detection of pests, prediction of crop yield, accurate fertilization, smart agricultural IoT, traceability of the security of the food supply chain, food security, and some other critical problems in agriculture [3,4].

AI and ML surveillance systems in agriculture offer insights to monitor crops, detect pests, and assess soil flaws, enabling farmers to plant seeds at the best moment for optimum yield. Many agricultural activities are threatened by weeds. They suffocate pastures, invade crops, reduce agricultural productivity, and, in rare instances, put animals in risk. AI sensors are able to identify locations where weeds are present and identify the most effective pesticide to apply there. AI can predict weather patterns, evaluate crop health, and find pests, illnesses, or undernourished plants. Drones with AI capabilities can help farmers keep

Harnessing Automation and Machine Learning for Resource Recovery and Value Creation
DOI: https://doi.org/10.1016/B978-0-443-27374-2.00014-5

an eye on the condition of their crops. Specialists review the drone's photos and create an assessment of the farm's condition. This aids in pest management for farms. Currently, some farmers are using agricultural robots to reduce labor costs and the workload of workers by using these robots for physical labor [5].

This chapter will provide the most recent developments in software, hardware, pipeline models, experimental technology, ML, and AI algorithms, as well as applications for intelligent agriculture.

Artificial intelligence

AI is the term used to describe the imitation of human intelligence in computers that have been designed to think and behave like people. It includes a broad range of technologies and methodologies that allow machines to sense their surroundings, reason, gain experience, and make choices in order to accomplish particular objectives. AI systems are capable of carrying out activities that usually require human intelligence, such as comprehending natural language, identifying patterns in data, resolving complicated issues, and forming predictions [3].

Many of the applications used now, according to purists, are still quite basic and are not actually AI. AI is commonly defined as a technology that allows machines to mimic a variety of intricate human talents. But this does not really give us anything to work with. Actually, it just changes the terminology used to refer to "AI." It is still unclear exactly what AI is as long as such "complex human skills" are not described. It is hardly unexpected that defining AI is so challenging. After all, it is really a simulation or imitation of human intelligence, which we still do not fully comprehend.

The high-level expert group on AI of the European Commission (EC) has proposed a definition that is comparable [6], stating that: systems that display intelligent behavior by analyzing their environment and taking actions—with some degree of autonomy—to achieve specific goal.

It is important to note that all of the existing applications of AI that are classified as such under this definition are categorized as "weak" or "narrow" [6]. The AI that we are familiar with today is primarily focused on specialized tasks, like speech or object recognition, and is not very related to the entire range of human cognitive abilities that AI covers [7].

AI has multiple branches, such as: ML, deep learning, natural language processing (NLP), computer vision, robotics as shown in Fig. 14.1.

ML is the major core of AI focuses on giving machines the ability to learn from data and get better overtime without needing to be explicitly programmed. Based on prior experiences, ML algorithms are able to recognize patterns and forecast future events. Important concepts in ML are as [7]:

1. Supervised learning: this type of learning uses labeled data to train an algorithm with an already-known desired output.
2. Unsupervised learning: the algorithm is able to find patterns and structures in the dataset by using unlabeled data.

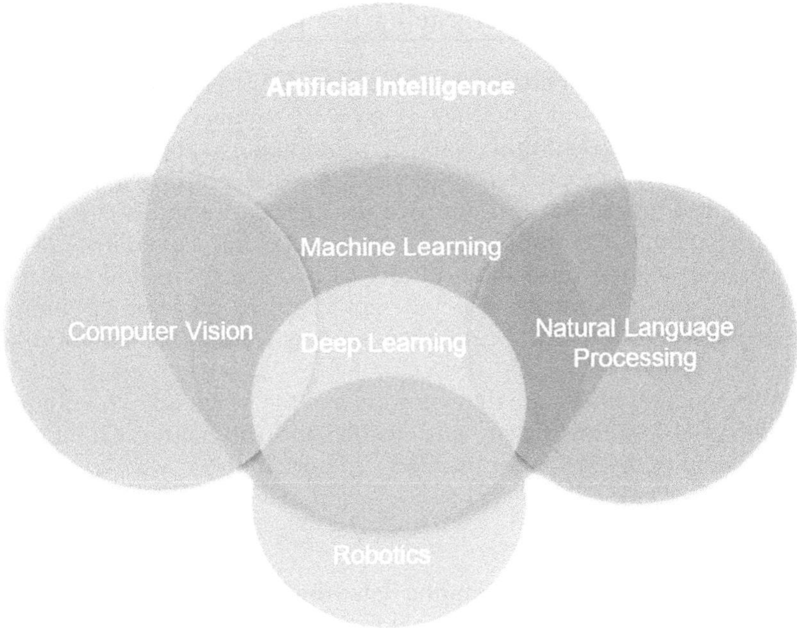

FIGURE 14.1 Types and applications of artificial intelligence.

3. Reinforcement learning: models pick up new skills by interacting with an environment, receiving feedback in the form of rewards or penalties based on its actions, and aiming to maximize cumulative rewards overtime.

On the other hand, deep learning is a subset of ML that processes and interprets complicated data by using multiple-layered artificial neural networks (hence the term "deep"). Deep learning algorithms have proven remarkably effective in applications like NLP, audio recognition, and image recognition [8,9].

NLP is a branch of AI. It is concerned with making machines able to comprehend, translate, and produce human language. Chatbots, sentiment analysis software, language translation tools, and virtual assistants are all powered by NLP algorithms [9]. Another area of AI is computer vision that gives computers the ability to decipher and evaluate visual data from pictures or movies. Applications like picture categorization, object detection, and autonomous driving are made possible by computer vision algorithms, which are able to identify things, recognize faces, and comprehend spatial relationship.

Robotics AI is also used in robotics to give machines the ability to sense their surroundings, organize and carry out actions, and communicate with people and the real world. Numerous industries, including manufacturing, healthcare, logistics, and agriculture, use AI-powered robots. All things considered, AI is a rapidly evolving science that has the power to transform industries, increase efficiency, improve decision-making, and solve difficult social issues. AI technologies are predicted to become more significant in influencing the direction of technology and society as they develop.

Why artificial intelligence in agriculture

AI in agriculture has the potential to completely transform farming practices and increase productivity, sustainability, and efficiency. A growing number of people are realizing that AI can revolutionize the agricultural sector by providing a wide range of advantages that effectively tackle the intricate problems the sector faces. According to [10], AI applications are being widely used these days to improve the agriculture industry's performance and operational automation. For example, prediction models for the overall value of agricultural produce are the most often used AI applications in agriculture, followed by harvesting applications. Even though agriculture is inherently limited by resource limits, AI applications in the management of natural resources (such as water, soil, and land) are currently lacking.

AI applications in agriculture are diverse and have the potential to revolutionize farming practices, making them more efficient, sustainable, and productive. The following subsections describes some of the key AI applications in agriculture.

Precision farming

Farmers can use AI to gather and examine data from different sources, including sensors, satellite photography, and drones, to precisely monitor and manage crops. Based on current data, soil conditions, and crop health, this involves optimizing the use of fertilizers, pesticides, and irrigation. Precision agriculture (PA) cannot exist without AI, which helps farmers allocate resources as efficiently as possible while maximizing output. Farmer data on soil composition, weather patterns, crop health, and insect infestations may be gathered and analyzed in large quantities by utilizing AI technologies like ML, remote sensing (RS), and data analytics. Precise decision-making in a range of agricultural tasks, such as irrigation, fertilization, pesticide application, and harvesting, is made possible by this data-driven methodology. Precision farming avoids waste of inputs, lessens its impact on the environment, and improves crop yields and quality by customizing treatments to particular crop needs and field conditions.

Over the past few decades, agriculture has become increasingly significant to the global economy. The environment is being negatively impacted by the demand on agriculture and animal production to meet the needs of a fast expanding population. Conversely, smart farming technologies are becoming more prevalent in contemporary agriculture to help maximize the output of crops and livestock while reducing expenses and waste. PA is a data-driven, technology-enabled farming management technique that monitors, quantifies, and assesses the requirements of specific fields and crops.

Precision livestock farming (PLF), which relies on the automated monitoring of individual animals, is used, among other things, to monitor animal behavior and their physical environment, as well as for animal development, milk production, and illness detection. Jianbo et al. (2014) [11] provides a quick overview of current scientific and technological trends in Pennsylvania and their application in crop and livestock farming. They conclude that to ensure the accuracy and practicality of the produced systems in commercial settings, additional research is necessary into a number of stages and techniques involved in the creation and operation of PA applications.

In recent years, precision seeding has emerged as the key component and direction of advancement for contemporary seeding technologies. In addition to saving seeds, precision seeding allows you to efficiently adjust the sowing depth, density, and distance. Statistics show that when using precision seeding instead of a traditional drill, output rises by 10%–30% [11].

A variety of sensors, yield and soil monitors, and RS tools—such as imagery from drones, personnel, aircraft, or satellites—are used in PA to observe, measure, and sense data on specific fields and crops [12,13].

Accurate real-time information for decision-making, such as underperforming areas requiring site-specific crop fertilization strategies, can be produced by properly calibrated yield monitors [14]. With the use of sensors and monitoring technologies, PLF gathers information on the health and welfare of the animals. This information helps farmers make more productive and efficient decisions regarding feed, waste, and other inputs. Dynamic Colter depth control systems use a combination of ultrasonic soil surface sensors and Colter position sensors [15].

Crop monitoring and management

AI-driven satellites, drones, and sensors are able to track crops in real-time and identify early indicators of disease, nutrient shortages, and other problems. Farmers may take proactive steps to reduce hazards and optimize crop health and output thanks to this early detection. AI-driven systems can precisely detect and target weeds and pests, which lessens the need for chemical pesticides and herbicides. This preserves crop health while reducing input costs for farmers and minimizing the impact on the environment. AI systems are able to monitor crop health and identify early indicators of disease, nutrient deficits, or insect infestations by analyzing photos and sensor data. Farmers are able to reduce hazards and safeguard agricultural production by acting promptly thanks to this early detection.

Commercial App has been developed using computer vision techniques, to monitor crop health and stress [16]. It is necessary to regularly monitor agricultural crops by first gathering and then analyzing data from different RS devices [17]. Combining several features with classification using ML is a standard method for processing Red, Green, Blue (RGB) images of the crops efficiently [18].

Analyzing biophysical characteristics of crops such as height, Leaf Area Index (LAI), biomass, and chlorophyll content is made easier with the aid of image processing. On the other hand, structural details, crop moisture content, and water availability are provided by radar pictures [19]. For crop monitoring, these two sources are therefore complementary to one another. Because of this, there's been an increasing trend of integrating radar and optical pictures together for crop-type classification jobs [20–22].

Predictive analytics

In order to forecast future trends and results, such as crop yields, market demand, and weather conditions, AI algorithms can examine historical data, weather patterns, and environmental elements. Farmers can use this predictive power to help them make well-informed decisions on crop planting, harvesting, and marketing. The use of chemical herbicides and pesticides is decreased when weeds and pests are precisely identified and targeted by AI-powered systems. This preserves crop health while reducing input costs for farmers and minimizing the impact on the environment. Pre-harvest estimates of important crops are possible, and appropriate steps can be made to handle any challenges associated with potential food scarcity forecasting [23]. In order to predict the fate, transit, and estimation of pollution from the point of discharge into the river until it is absorbed by agricultural crops, a new framework for AI modeling is introduced in [24]. AI models are created to estimate the Chromium concentration in tomato leaves by utilizing data gathered from the case study and investigating various scenarios. Smart farming can also be aided by a variety of smartphone apps, including those that identify diseases, determine fertilizer needs, conduct soil studies, and forecast water requirements [25].

Labor optimization

AI-powered automation devices, such robotic harvesters and autonomous tractors, are able to complete repetitive jobs more quickly and effectively than human laborers. This solves the manpower deficit, lessens the demand for manual labor, this frees up agricultural laborers to concentrate on more expert and calculated tasks.

A low-cost agricultural robot for crop monitoring, pest identification, and fertilizer and pesticide spraying was presented by the authors of Ref. [26]. While labor expenses were reduced by the prototype system's autonomous operation, productivity was marginally lower than that of human workers. Based on IoT technology, an autonomous organic fertilizer mixer was created in Ref. [27] with the goal of improving efficiency and lowering labor costs. The upgraded mixer attempted to significantly simplify the process of mixing organic fertilizer by enabling remote monitoring, updates, and alarms. Some studies built an automated irrigation approach using Arduino equipment to reduce the amount of labor and time needed for irrigation [28].

Resource efficiency

AI helps farms reduce waste and increase resource efficiency by optimizing inputs like water, fertilizers, and energy. In addition to cutting expenses, this minimizes the ecological imprint of agriculture and conserves resources, both of which support environmental sustainability. Because AI makes it possible to implement more productive and ecologically friendly farming methods, it is essential for advancing sustainability in the agricultural sector. AI assists farmers in anticipating and mitigating the effects of natural catastrophes, harsh weather, and climate change on crop productivity using predictive analytics and modeling. Furthermore, early detection of insect outbreaks, nutritional deficits, and crop illnesses is made possible by AI-powered monitoring systems, which reduces the need for chemical inputs and allows for prompt intervention. AI enhances the long-term viability and sustainability of agricultural systems through maximizing resource consumption, decreasing waste, and boosting resilience to environmental shocks.

Around the world, 85% of the freshwater resources are used by the agriculture industry. And as the population grows and the demand for food rises, this percentage is rising quickly [29]. This makes it necessary for us to develop more effective methods to guarantee the wise use of water resources for irrigation. Savitha and Uma Maheshwari (2018) [30] also developed the concept of an automatic and efficient irrigation system by creating remote sensors with Arduino technology, which can increase output by 40% [30].

Another study provided automatic irrigation system that uses fusion of sensors, each with a specific function. For example, the soil moisture sensor measures the amount of moisture in the soil, the temperature sensor measures the temperature, the pressure regulator sensor keeps pressure levels constant, and the molecular sensor improves crop growth [31].

Crop breeding and genetics

AI systems are capable of analyzing genetic data to pinpoint desired crop features and expedite the breeding process. This makes it possible to create new crop varieties that are more resistant to diseases, pests, and climate change, guaranteeing food security in the face of changing difficulties. The modifications to DNA caused by genome editing include insertions, deletions, large fragment substitutions, and single nucleotide substitutions. The following site-specific nucleases cause nucleotide excision: zinc-finger nucleases, transcription activator like effector nucleases, CRISPR-associated protein, and engineered homing endonucleases, also known as mega-nucleases [32,33].

When applied in lab settings, advanced computational methodologies like ML and deep learning, as well as tools and techniques like transcriptome, metabolomics, overexpression, knock-out, RNA interference, and gene editing, could facilitate the successful and efficient validation of gene functions [34,35]. Future farmers will be able to use genomic data to discern genetic variations among individual plants within a crop and use that information to inform breeding decisions.

Supply chain optimization

AI may enhance agriculture's supply chain management and logistics, boosting the effectiveness of the operations involved in distribution, storage, and transportation (Fig. 14.2). This lowers food waste and boosts overall market competitiveness by guaranteeing that crops reach markets and customers on time. In conclusion, AI has several uses in agriculture, ranging from supply chain optimization and predictive analytics to precision farming and crop monitoring. Farmers can meet the challenges of feeding a growing population worldwide while increasing production, profitability, and sustainability by utilizing AI. AI holds immense potential in agriculture by enabling precision farming practices and promoting sustainability [36]. By harnessing the power of AI technologies, farmers can optimize productivity, enhance resource efficiency, and address the challenges of feeding a growing global population while minimizing environmental impact. In general, AI applications in agriculture present chances to boost profitability, sustainability, and production while tackling the problems of feeding a growing world population and adjusting to shifting environmental circumstances [37].

Recent study uses blockchain technology with AI to identify the major obstacles, namely the absence of postharvest facilities and the pressing requirement for improved market

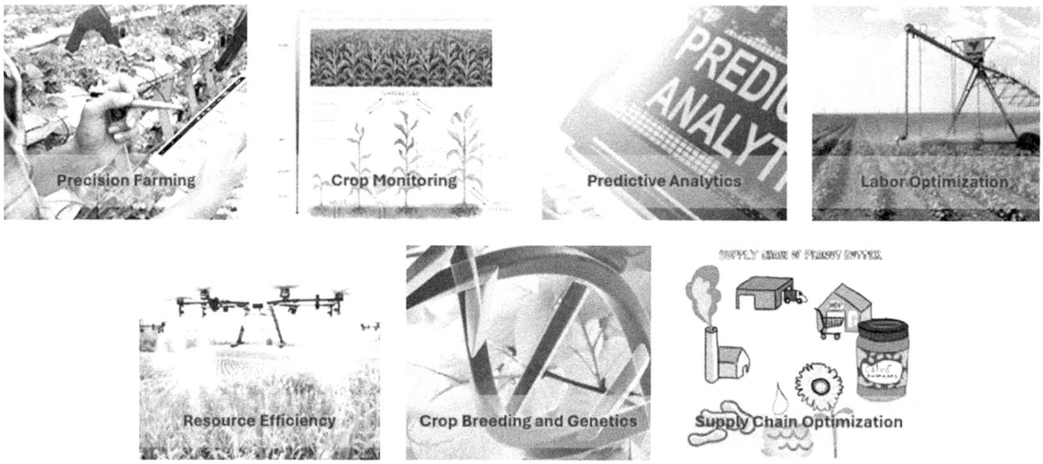

FIGURE 14.2 Artificial intelligence benefits for agriculture and farming industry.

information systems for farmers. This will provide farmers more control over postharvest tasks in order to address these deficiencies [37].

Through the use of algorithms, advanced data analytics, and AI, the entire supply chain can be alerted about how a weather event or other supply chain shock may influence, as well as whether or when product emergencies are likely to arise [38].

Artificial intelligence integration in agriculture machinery

The integration of AI into agriculture machinery, combined with the utilization of IoT and sensor technologies, marks a pivotal advancement in modern farming practices. This section delves into the synergistic relationship between AI, IoT, and sensors in agriculture machinery, exploring how these technologies collaborate to revolutionize various aspects of farming operations. From autonomous vehicles equipped with AI algorithms to IoT-enabled sensors monitoring soil health and crop conditions, the convergence of these technologies holds immense potential to enhance productivity, optimize resource management, and promote sustainability in agriculture. Through a comprehensive examination of AI integration in agriculture machinery and its integration with IoT and sensors, we uncover the transformative impact of these innovations on the future of farming.

Farm management system can be seen as the integration and design and implementation of systems are now being developed [39].

Internet of things and sensors in agriculture

The significance of sensing technology has increased in the context of IoT and industry 4.0. By combining them with AI technology, biosensor, bioinspired, and bioapplicable sensor development can be accelerated. The term "IoT" for agriculture describes how farmers can monitor, control, and optimize many parts of their farming operations by using connected devices and sensors to gather data from farming operations. The most recent models for sensors used in agriculture industry are listed in this section. The combination of sensing technologies and ML shows great promise for using AI-guided sensors in complex models to address uncertainty. Agriculture relies on a variety of sensors to monitor and collect data related to soil, weather, crop health, and environmental conditions. Many agricultural sensors are available that can identify many characteristics, including water level, livestock, plant disease, smoke, flame, electrical conductivity, CO_2, moisture, nitrate, pH, temperature, humidity, light, and weather station. In Table 14.1, some common types of sensors such as soil moisture, temperature and humidity, light, pH, crop health, nutrient, Global Positioning System (GPS), and water quality sensors are explored.

Some of the sensors used in real-time soil moisture monitoring are time domain reflectometry [49–52], frequency domain reflectometry [53,54], resistivity, and capacitive

TABLE 14.1 Sensors used for internet of things in agriculture.

Sensor category	Explanation	Used in studies (references)
Soil moisture	Detecting the soil's moisture content and provide details on the level of soil hydration. Assist farmers in making the most of their irrigation schedules and avoiding over or underwatering crops.	[40,41]
Temperature and humidity	An essential component of plant growth and development, soil and air temperatures. Evaluating risks associated with temperature, including heat stress, disease outbreaks, and frost. Low humidity can cause plants to experience water stress, while high humidity might encourage fungal infections.	[42]
Light	How much and how long crops receive sunshine. Determining the amount of light that is available for photosynthesis so that planting configurations and shading structures can be optimized properly.	[43]
pH	Measuring the soil's acidity or alkalinity. Adjusting the pH levels of the soil.	[44]
Crop health	Evaluate crop health and identify indicators of disease, stress, or insect infestation using a variety of technologies, including thermal imaging and spectroscopy. They give farmers early warnings, which enables prompt disease control and intervention.	[45,46]
Nutrient	Assess the levels of essential nutrients in soil, including potassium, phosphate, and nitrogen. Optimizing nutrient management techniques, such as fertilization and soil additives, and in monitoring soil fertility.	[47]
GPS and navigation	Precision agricultural applications use GPS and navigation sensors to monitor field boundaries, track the location of farm equipment, and build precise maps for planting, spraying, and harvesting processes.	[48]
Water quality	Detecting variables in irrigation water sources like pH, conductivity, turbidity, and dissolved oxygen. Maintaining irrigation water quality and avoiding problems with water, such as pollution or salinity buildup.	[49]

sensors [55]. Recent study report presents the creation of a soil moisture sensor that can monitor soil moisture down to a depth of 50 cm and adapts to field environments. This study introduces the Arduino-based soil moisture sensors, which work best with the standard method of detecting soil moisture. These sensors include the copper based probe and the stainless steel 316L grade (SS316L) probe [56]. Another study introduced modeling and performance analysis of several smart soil moisture sensors using neural network classifier models [57].

Humidity sensors help to control these factors so that crop health is maximized. Most of the studies combine temperature and Humidity sensors with other sensors. Artificial neural network-based system for coal mine safety monitoring was proposed in Ref. [58]. The suggested method computes the miner's temperature, humidity, and heart rate in order to produce a warning and, if required, determine the miner's location. Recent studies use deep learning. Sensors that detect photosynthetically active radiation, determine the amount of light needed for plant growth [59]. A single sensor board, like the MTS400, has sensors for barometric pressure, temperature, relative humidity, and light intensity [60,61].

Although the measured value might not be as accurate as a lab's, the accuracy of soil nutrients and pH is improved by increasing the number of samples [62] there are two kinds of electrochemical sensors in use: ion selective field effect transistor (ISFET) and ion-selective electrode (ISE) [30]. The Nernst equation defines an ISE as a change in potential relative to a reference electrode with a linear response with respect to the target ion. ISEs are able to identify potassium, ammonium, and nitrate that are there. Its primary flaw is the response time lag, which can last for many minutes. ISE and FET are combined to create the ISFET. To allow for chemical control of the ISFET threshold voltage, an ion membrane is layered on top of the FET insulation layer; the measured voltage indicates the specific ion. Due to a few advantages, such as small size, low output impedance, high S/N ratio, faster reaction, and the ability to integrate several ISFETs on a chip, ISFETs are more common than ISEs.

These are but a handful of the numerous kinds of sensors that are employed in agriculture to gather information and facilitate well-informed decision-making for the best possible crop yield and environmental sustainability. Weather stations: these devices use a variety of sensors to measure several aspects of the weather, including humidity, temperature, wind speed, precipitation, and sun radiation. For the purpose of crop management decisions and weather forecasts, they offer extensive data.

Agriculture can become more resilient, sustainable, and efficient by utilizing the combined power of IoT and AI technology. This will enhance farmer profitability and productivity while reducing its negative effects on the environment.

Autonomous farming equipment

Drones, robotic harvesters, and AI-driven autonomous tractors can carry out planting, spraying, and harvesting operations with little assistance from humans. In addition to addressing the manpower deficit, this lowers labor expenses and boosts farm operations' efficiency. The term "agricultural machinery" refers to a broad category of devices made to make different farming and agricultural chores easier. In contemporary agricultural techniques, these machines are essential for boosting productivity, efficiency, and sustainability. Here are a few typical categories of traditional farm equipment: tractors, harvesters, seeders, sprayers, and tillage machinery. Tractors can be fitted with various add-ons and tools using AI to effectively carry out particular duties. Harvesters can be equipped with AI system to automatically gather crops like fruits, vegetables, and grains. To harvest and process crops in a single operation, they can be fitted with systems for cutting, threshing, and cleaning. Planters and seeders devices are used to maximize germination and crop establishment, these devices make sure that seeds are precisely spaced, planted, and positioned. In Table 14.2, some common equipment used for IoT in agriculture such as sprayers and irrigation systems are explored.

These are but a handful of the numerous varieties of agricultural machinery and equipment that farmers and growers might use to satisfy the various demands of contemporary agriculture. In the agricultural industry, every kind of machinery is essential to raising profitability, sustainability, and production.

TABLE 14.2 Autonomous farming equipment for internet of things in agriculture.

Farming equipment	Explanation	Used in studies (references)
Sprayers	Agrochemicals such as fertilizers, herbicides, and pesticides are applied to crops using sprayers. They are crucial for crop protection and fertilizer management and come in a variety of forms, such as aerial, self-propelled, and tractor-mounted sprayers.	[63,64]
Tillage machinery	Before planting, soil is prepared with tillage machinery, which is used for harrowing, cultivating, discing, and plowing. Improving seedbed conditions for the best possible crop growth, breaking up the soil, and controlling weeds.	[64–66]
Irrigation systems	Irrigation systems provide crops with water to augment rainfall and provide sufficient moisture for plant development. These include center pivot systems, drip irrigation, sprinkler systems, and flood irrigation techniques, each designed to meet the unique requirements of various crops and soil types.	[65,67,68]
Livestock equipment	To make it easier to maintain and care for livestock animals like cattle, poultry, and pigs, livestock equipment comprises feeders, waterers, handling facilities, and housing structures.	[69,70]
Grain handling and storage	Harvested grains like wheat, corn, and soybeans are transported, stored, and managed using grain handling equipment, which includes conveyors, elevators, and grain bins. During storage, these devices aid in maintaining grain quality and preventing spoiling.	[71]
Hay and forage machinery	Forage crops including hay, silage, and straw are harvested, processed, and stored using hay and forage machinery. This covers devices like fodder harvesters, tedders, balers, and mowers.	[72]
Precision agriculture technology	With precise monitoring and control of agricultural operations, precision agriculture technology maximizes yields, improves resource efficiency, and optimizes farm management practices. It comprises GPS guidance systems, drones, sensors, and data analytics software.	[73]

TABLE 14.3 Artificial intelligence in agricultural equipment.

AI application	Explanation	Used in studies (references)
Autonomous vehicles	Carry out manual labor-intensive operations including planting, seeding, spraying, and harvesting. Traveling fields, keep an eye on crops by sensors, cameras, and GPS technology.	[74]
Precision farming systems	Examine data from a variety of sources, including satellite photography, weather forecasts, and soil sensors. Monitoring environmental conditions in real-time, as well as applying inputs like water, fertilizer, and pesticides at variable rates.	[75]
Object classification	Detecting items like weeds, pests, and illnesses by examining pictures and videos taken by mounted or drone cameras.	[76]
Predictive maintenance	Tracking the state of farm equipment in real-time, utilizing sensor data analysis to anticipate possible problems before they arise.	[77]
Smart irrigation systems	Determining the best times to water crops by utilizing information from weather forecasts, crop water requirements, and soil moisture content.	[30]
Crop harvesting and sorting	AI-enabled harvesting machinery can recognize and pick ripe crops, arrange produce in accordance with size and quality criteria, and package fruits and vegetables in accordance with market demands.	[78]
Robotic weeding systems	Identify and eradicate weeds from fields. Targeted mechanical or thermal weed management is made possible by these robots' ability to discern between crops and weeds using computer vision and machine learning	[79]
Data analytics and decision support	Examining vast amounts of data gathered from farm equipment, sensors, and outside sources to offer farmers meaningful insights and decision support. This contains suggestions for the best times to plant, rotate crops, and control pests.	[75]

AI, Artificial intelligence.

Artificial intelligence in agricultural equipment

The term "AI" in agriculture machinery describes how AI technologies are incorporated into farming technology to improve automation, accuracy, and efficiency. Table 14.3 shows a few examples of how AI is being used in agricultural equipment.

Farmers may contribute to more sustainable and effective farming methods by increasing productivity, cutting expenses, and minimizing their environmental impact by incorporating AI technologies into agricultural machinery.

Limitations

One major obstacle to the widespread use of AI in agriculture is the lack of straightforward solutions that smoothly integrate and embed AI in agriculture. Most farmers do not have the time or digital know-how to independently research AI solutions. These new AI solutions will need to be incorporated into the legacy systems and infrastructure that farmers now utilize in order to smoothly integrate and embed AI into agriculture [3]. AI is limited to agriculture since it is unable to function outside of its programming. Farmers, especially those in rural regions, are not aware of these technologies and lack technical skills.

AI may pave the road for agriculture to become semiautonomous as awareness grows and technology become more affordable for the typical farmer. A substantial amount of data is needed for AI systems to train their computers and generate reliable predictions. When it comes to land that is mostly used for agriculture, spatial data is easy to collect, but temporal data is more difficult. While crops are being grown, the majority of crop-specific data can only be obtained once a year. Building a solid ML model requires a significant amount of time due to the maturation of data infrastructure. When integrating agricultural AI, there are a few things to consider, like how it will grow, what kinds of data it will need, privacy of the data and whether or not the farmer would eventually have to pay for this service. For its ethical application to be implemented successfully, these challenges must be addressed and made clear to the end user.

Another challenge is the big data available in the digital universe as it is expected to hold more than 44 zettabytes of data by 2020, according to a 2014 study conducted by EMC digital universe with research and analysis by International Data Corporation (IDC). This worry has been allayed by AI, which opens up new possibilities for developing deep learning, ML, and pattern recognition-based new techniques for processing this massive amount of data.

Future directions

AI integration in agriculture is a fast-moving field with a wide range of possible applications that could revolutionize farming methods and solve issues plaguing the sector. The following are some important future directions for AI integration in agriculture:

1. Advanced robotics: agricultural robots of the future will be more intelligent, adaptable, and self-sufficient. More operations, such targeted spraying, selected harvesting, autonomous crop management, and precision planting, will be possible for these robots to accomplish. In farming operations, advanced robotics will enable higher efficiency, lower labor costs, and decrease environmental effect.
2. Edge computing and IoT: as IoT sensors and edge computing devices proliferate on farms, real-time data on crop health, weather patterns, soil conditions, and equipment performance will be gathered. AI systems will examine this data at the edge to offer quick insights and facilitate decision-making in real time, improving resource management and farming techniques.
3. Genomic prediction and breeding: in order to generate crops with better features like yield, climate change resilience, and pest and disease resistance, AI algorithms will be crucial to genomic prediction and breeding programs. AI has the potential to speed up the breeding process and create more productive and sustainable agricultural varieties by evaluating large volumes of genetic data.

4. Predictive analytics and risk management: farmers will be able to anticipate crop yields, market pricing, and weather patterns with greater accuracy as AI-powered predictive analytics develops. With the use of this information, farmers will be able to minimize risks and increase profitability by making well-informed decisions on the planting, harvesting, and marketing of their crops.
5. Autonomous farming systems: these increasingly common fully autonomous farming systems combine robotics, AI-powered machines, and drones to run entire farms with little to no human involvement. With the help of these systems, farming operations will be optimized in all respects, from planting and irrigation to harvesting and logistics, leading to increased production, sustainability, and efficiency.
6. Personalized farming solutions: AI technology will make it possible to provide customized farming solutions that are made to meet the unique requirements of different crops and farms. AI may make tailored recommendations for increasing yields, lowering input costs, and maximizing profits for individual farms by evaluating data on soil composition, climate, crop health, and market demand.
7. Collaborative AI platforms: farmers, academics, agronomists, and industry stakeholders will be able to share data, insights, and best practices through the emergence of collaborative AI platforms. These platforms will promote the sharing of knowledge, teamwork on research projects, and the creation of creative answers to prevalent problems in agriculture.
8. Ethical and responsible AI usage: the importance of ethical and responsible AI usage will increase as AI becomes increasingly integrated into agriculture. This includes tackling issues with data privacy, prejudice, and algorithmic governance in addition to guaranteeing accountability, justice, and openness in AI algorithms and decision-making processes. All things considered, the application of AI to agriculture has enormous potential to advance farming techniques' resilience, sustainability, and productivity—thereby promoting environmental stewardship and global food security. Farmers may use AI to solve the intricate problems facing the agricultural sector and build a more lucrative and sustainable future for the sector by embracing new technology and innovations.

Conclusions

In summary, the application of AI to agriculture is a revolutionary development in agricultural methods that gives hitherto unseen chances to boost food production efficiency, sustainability, and productivity. Farmers can address the complex issues facing the agricultural industry and build a more prosperous and sustainable future for agriculture by utilizing AI technologies like advanced robotics, edge computing, genomic prediction, predictive analytics, autonomous farming systems, personalized farming solutions, collaborative platforms, and ethical AI use.

PA, which maximizes agricultural techniques based on real-time data and predictive insights to improve resource management, lessen environmental impact, and increase crop yields, is made possible by AI-driven advancements. AI-powered autonomous agricultural systems provide increased automation and efficiency in farm operations, lowering labor costs and decreasing human error. AI-enhanced genomic prediction and breeding programs speed up the creation of crop varieties with better qualities, such resilience, yield, and disease resistance, promoting food security and sustainable agriculture.

Moreover, collaborative AI platforms foster creativity and group action to address shared agricultural concerns by facilitating knowledge exchange and collaboration among farmers, academics, agronomists, and industry stakeholders. But as AI is used more and more in agriculture, it is critical to give ethical and responsible AI use top priority in order to protect data security and privacy and to ensure algorithmic decision-making that is transparent, equitable, and accountable. All things considered, AI integration in agriculture has great potential to transform farming methods and help the sector overcome its many complicated problems. Farmers can use AI to build a more resilient, sustainable, and egalitarian food system and secure a better future for agriculture and the earth by embracing new technologies and ideas.

References

[1] L. Zhang, et al., IoT and blockchain in precision agriculture: enhancing food supply chain transparency, Computers and Electronics in Agriculture (2023) 193.

[2] V. Dharmaraj, C. Vijayanand, Artificial intelligence (AI) in agriculture, International Journal of Current Microbiology and Applied Sciences 7 (2018) 2122–2128.

[3] M. Javaid, et al., Understanding the potential applications of artificial intelligence in agriculture sector, Advanced Agrochem 2 (2023) 15–30.

[4] S. Mor, S. Madan, R.S. Mor, Carbon footprint and agricultural supply chain performance: evidence from apple production, Journal of Cleaner Production (2024) 427.

[5] R. Kumar, et al., Artificial intelligence: new technology to improve Indian agriculture, International Journal of Chemical Studies 8 (2020) 2999–3005.

[6] H. Sheikh, C. Prins, E. Schrijvers, Mission AI: The new system technology, Springer Nature, 2023, p. 410.

[7] Y. Himeur, et al., Video surveillance using deep transfer learning and deep domain adaptation: towards better generalization, Engineering Applications of Artificial Intelligence 119 (2023) 105698.

[8] J. Chai, H. Zeng, A. Li, E.W.T. Ngai, Deep learning in computer vision: a critical review of emerging techniques and application scenarios, Machine Learning with Applications 6 (2021) 100134.

[9] P.M. Nadkarni, L. Ohno-Machado, W.W. Chapman, Natural language processing: an introduction, Journal of the American Medical Informatics Association 18 (5) (2011) 544–551.

[10] V. Sachithra, L.D.C.S. Subhashini, How artificial intelligence uses to achieve the agriculture sustainability: Systematic review, Artificial Intelligence in Agriculture 8 (2023) 46–59.

[11] Z. Jianbo, X. Junfang, Z. Yong, Z. Shun, Design and experimental study of the control system for precision seed-metering device, International Journal of Agricultural and Biological Engineering 7 (3) (2014) 13–18.

[12] R. Alfred, J.H. Obit, C.P.Y. Chin, H. Haviluddin, Y. Lim, Towards paddy rice smart farming: a review on big data, machine learning, and rice production tasks, IEEE Access 9 (2021) 50358–50380.

[13] McFadden, J., Njuki, E. and Griffin, T., 2023. Precision Agriculture in the Digital Era: Recent Adoption on US Farms. U.S. Department of Agriculture Economic Research Service EIB-248.

[14] A. Monteiro, S. Santos, P. Gonçalves, Precision agriculture for crop and livestock farming—brief review, Animals 11 (2021) 2345.

[15] G.S. Hundal, et al., Exploring barriers to the adoption of internet of things-based precision agriculture practices, Agriculture 13 (2023) 163.

[16] J.Q. Qiao, H.J. Wu, R. Huo, X.W. Gao, R. Borriss, Stimulation of plant growth and biocontrol by Bacillus amyloliquefaciens subsp. plantarum FZB42 engineered for improved action, Chemical and Biological Technologies in Agriculture 1 (2014) 1–14.

[17] A. Das, et al., Machine learning model ensemble for predicting sugarcane yield through synergy of optical and SAR remote sensing, Remote Sensing Applications 30 (2023) 100962.

[18] Y. Yuan, B.L. Ton, W.J. Thomas, J. Batley, D. Edwards, Supporting crop plant resilience during climate change, Crop Science 63 (4) (2023) 1816–1828.

[19] A. Veloso, S. Mermoz, A. Bouvet, T. Le Toan, M. Planells, J.F. Dejoux, et al., Understanding the temporal behavior of crops using Sentinel-1 and Sentinel-2-like data for agricultural applications, Remote sensing of environment 199 (2017) 415–426.

[20] S. Chung, L.E. Breshears, J.Y. Yoon, Smartphone near infrared monitoring of plant stress, Computers and Electronics in Agriculture 154 (2018) 93–98.

[21] P. Karmakar, et al., A novel fusion approach in the extraction of kernel descriptor with improved effectiveness and efficiency, Multimedia Tools and Applications 80 (2021) 14545–14564.

[22] P. Karmakar, et al., Crop monitoring by multimodal remote sensing: a review, Remote Sensing Applications 33 (2024) 101093.

[23] M.W. Liu, M. Ozdogan, X. Zhu, Crop type classification by simultaneous use of satellite images of different resolutions, IEEE Transactions on Geoscience and Remote Sensing 52 (2014) 3637–3649.

[24] A. Montazeri, et al., A novel AI-based approach for modelling the fate, transportation and prediction of chromium in rivers and agricultural crops: a case study in Iran, Ecotoxicology and Environmental Safety 263 (2023) 115269.

[25] S. Mandal, et al., Adaption of smart applications in agriculture to enhance production, Smart Agricultural Technology (2024).

[26] A.S.A. Ghafar, et al., Design and development of a robot for spraying fertilizers and pesticides for agriculture, Materials Today: Proceedings (2021).

[27] A. Hadi Ishak, et al., Autonomous fertilizer mixer through the internet of things (IoT), Materials Today: Proceedings (2021).

[28] K. Jha, et al., A comprehensive review on automation in agriculture using artificial intelligence, Artificial Intelligence in Agriculture (2019).

[29] T. Talaviya, D. Shah, N. Patel, H. Yagnik, M. Shah, Implementation of artificial intelligence in agriculture for optimisation of irrigation and application of pesticides and herbicides, Artificial Intelligence in Agriculture 4 (2020) 58–73.

[30] M. Savitha, O.P.U. Maheswari, Smart Crop Field Irrigation in IOT Architecture Using Sensors, International Journal of Advanced Research in Computer Science (2018).

[31] K. Varatharajalu, J. Ramprabu, Wireless irrigation system via phone call & SMS, International Journal of Engineering and Advanced Technology 8 (2) (2018) 397–401.

[32] G. Nerkar, S. Devarumath, M. Purankar, A. Kumar, R. Valarmathi, R. Devarumath, et al., Advances in Crop Breeding Through Precision Genome Editing, Front Genet (2022).

[33] J. Smith, S. Grizot, S. Arnould, A. Duclert, J.C. Epinat, P. Chames, et al., A combinatorial approach to create artificial homing endonucleases cleaving chosen sequences, Nucleic Acids Research 34 (2006).

[34] R.R. Mir, et al., High-throughput phenotyping for crop improvement in the genomics era, Plant Science (2019).

[35] E. Salsman, A. Kumar, W. AbuHammad, A.O. Abbasabadi, M. Dobrydina, S. Chao, et al., Development and validation of molecular markers for grain cadmium in durum wheat, Molecular Breeding (2018).

[36] A. Mardenli, D. Sackmann, S. Rhein, Current challenges in agricultural supply chains: an empirical assessment based on expert perspectives, Transportation Research Procedia, Elsevier B.V, 2023, pp. 66–76.

[37] T. Kalimuthu, et al., Intelligent decision-making framework for agriculture supply chain in emerging economies: research opportunities and challenges, Computers and Electronics in Agriculture, (2024).

[38] E.M. El Bhilat, A. El Jaouhari, L.S. Hamidi, Assessing the influence of artificial intelligence on agri-food supply chain performance: the mediating effect of distribution network efficiency, Technological Forecasting and Social Change 200 (2024).

[39] P. Rajak, A. Ganguly, S. Adhikary, S. Bhattacharya, Internet of Things and smart sensors in agriculture: Scopes and challenges, J Agric Food Res 14 (2023).

[40] J.R. Millan-Almaraz, et al., FPGA-based fused smart sensor for real-time plant-transpiration dynamic estimation, Sensors 10 (2010).

[41] I. Molina, et al., Characterizing olive grove canopies by means of ground-based hemispherical photography and spaceborne RADAR data, Sensors 11 (2011).

[42] A. Montagnoli, et al., Estimating forest aboveground biomass by low density lidar data in mixed broadleaved forests in the Italian pre-alps, Forest Ecosystems 2 (2015).
[43] J.N. Schuster, M.J. Darr, R.P. McNaull, Performance benchmark of yield monitors for mechanical and environmental influences. In: 2017 ASABE Annual International Meeting (2017).
[44] D.J. Cocovi-Solberg, M. Rosende, M. Miró, Automatic kinetic bioaccessibility assay of lead in soil environments using flow-through microdialysis as a front end to electrothermal atomic absorption spectrometry, Environmental Science & Technology 48 (2014).
[45] Z. Li, et al., Real-time monitoring of plant stresses via chemiresistive profiling of leaf volatiles by a wearable sensor, Matter 4 (2021) 2553–2570.
[46] S.D. Calisgan, et al., Micromechanical switch-based zero-power chemical detectors for plant health monitoring, Journal of Microelectromechanical Systems 29 (2020).
[47] M.Y. Kim, et al., Highly stable potentiometric sensor with reduced graphene oxide aerogel as a solid contact for detection of nitrate and calcium ions, Journal of Electroanalytical Chemistry 897 (2021).
[48] M. Reger, J. Stumpenhausen, H. Bernhardt, Evaluation of LiDAR for the Free Navigation in Agriculture, AgriEngineering 4 (2022).
[49] A.K., Eddy covariance flux tower: a promising technique for greenhouse gases measurement, Advances in Plants & Agriculture Research 7 (2017).
[50] H. He, et al., A review of time domain reflectometry (TDR) applications in porous media, Advances in Agronomy (2021).
[51] S. Lee, H.K. Yoon, Hydraulic conductivity of saturated soil medium through time-domain reflectometry, Sensors 20 (2020).
[52] W. Skierucha, A. Wilczek, A. Szypłowska, C. Sławiński, K. Lamorski, A TDR-based soil moisture monitoring system with simultaneous measurement of soil temperature and electrical conductivity, Sensors (Switzerland) 12 (2012).
[53] A. Dwevedi, et al., Soil sensors: detailed insight into research updates, significance, and future prospects, New Pesticides and Soil Sensors (2017).
[54] Y. Linmao, et al., FDR soil moisture sensor for environmental testing and evaluation, Physics Procedia 25 (2012).
[55] C. Escriba, et al., Toward smart soil sensing in v4.0 agriculture: a new single-shape sensor for capacitive moisture and salinity measurements, Sensors 20 (2020).
[56] J.C. Songara, J.N. Patel, Calibration and comparison of various sensors for soil moisture measurement, Measurement (Lond) 197 (2022).
[57] S. Verma, R. Pahuja, Recalibration and performance comparison of soil moisture sensors using regression and neural network characteristic models, Materials Today: Proceedings (2021).
[58] D. Gao, J. Xin, F. Zhang, A decision tree algorithm for forest fire prediction based on wireless sensor networks, International Journal of Embedded Systems 13 (2020).
[59] K. Paul, S.S. Chatterjee, P. Pai, A. Varshney, S. Juikar, V. Prasad, et al., Viable smart sensors and their application in data driven agriculture, Comput Electron Agric 198 (2022).
[60] S.M. Abd El-Kader, B.M. Mohammad El-Basioni, Precision farming solution in Egypt using the wireless sensor network technology, Egyptian Informatics Journal 14 (2013).
[61] H. Moreno, et al., Evaluation of vineyard cropping systems using on-board rgb-depth perception, Sensors 20 (2020).
[62] M.A. Ali, L. Dong, J. Dhau, A. Khosla, A. Kaushik, Perspective—electrochemical sensors for soil quality assessment, Journal of the Electrochemical Society 167 (2020).
[63] F. Javier García-Ramos, et al., Applicability of a 3D laser scanner for characterizing the spray distribution pattern of an air-assisted sprayer, Journal of Sensors 2018 (2018).
[64] F. Javier García-Ramos, et al., Analysis of the air flow generated by an air-assisted sprayer equipped with two axial fans using a 3D sonic anemometer, Sensors 12 (2012).
[65] L. Jia, X. Zhou, Q. Wang, Effects of agricultural machinery operations on PM2.5, PM10 and TSP in farmland under different tillage patterns, Agriculture 13 (2023).
[66] A.Y. Nesmiyan, V.A. Chernovolov, A.M. Semenihin, V.P. Zabrodin, S.L. Nikitchenko, A review of assessment of the machinery tillage tools' performance for higher crop production efficiencies, Research on Crops (2018).
[67] S. Abba, J.W. Namkusong, J.A. Lee, M.L. Crespo, Design and performance evaluation of a low-cost autonomous sensor interface for a smart IoT-based irrigation monitoring and control system, Sensors 19 (2019).

[68] M.N. Rajkumar, S. Abinaya, V.V. Kumar, Intelligent irrigation system - An IOT based approach. In: IEEE International Conference on Innovations in Green Energy and Healthcare Technologies - 2017, IGEHT 2017.
[69] S. Mishra, S.K. Sharma, Advanced contribution of IoT in agricultural production for the development of smart livestock environments, Internet of Things 22 (2023).
[70] E. Anastasiou, A.T. Balafoutis, S. Fountas, Applications of extended reality (XR) in agriculture, livestock farming, and aquaculture: a review, Smart Agricultural Technology (2023).
[71] S.R. Tushar, M.F.Bin Alam, S.M. Zaman, J.A. Garza-Reyes, A.B.M.M. Bari, C.L. Karmaker, Analysis of the factors influencing the stability of stored grains: Implications for agricultural sustainability and food security, Sustainable Operations and Computers 4 (2023).
[72] M. Zawada, M. Nijak, J. Mac, J. Szczepaniak, S. Legutko, J. Gościańska-Łowińska, et al., Control and Measurement Systems Supporting the Production of Haylage in Baler-Wrapper Machines, Sensors 23 (2023).
[73] S. Atalla, et al., IoT-enabled precision agriculture: developing an ecosystem for optimized crop management, Information 14 (2023).
[74] L. Emmi, et al., Exploiting the internet resources for autonomous robots in agriculture, Agriculture 13 (2023).
[75] C. Liang, T. Shah, IoT in agriculture: the future of precision monitoring and data-driven farming, Eigenpub Review of Science and Technology 7 (2023).
[76] R. Singh, A. Jain, S. Sharma, Enhancing agricultural productivity through blockchain and IoT, Computers and Electronics in Agriculture 192 (2023).
[77] M.A. Sami, T.A. Khan, Forecasting failure rate of IoT devices: A deep learning way to predictive maintenance, Computers and Electrical Engineering 110 (2023).
[78] N.N. Thilakarathne, M.S.A. Bakar, P.E. Abas, H. Yassin, Towards making the fields talks: A real-time cloud enabled IoT crop management platform for smart agriculture, Front Plant Sci 13 (2023).
[79] U. Weiss, P. Biber, Plant detection and mapping for agricultural robots using a 3D LIDAR sensor, Robotics and Autonomous Systems 59 (2011).

CHAPTER

15

Recycling robots to tackle electrical waste

Samia Adil, Duc Hoa Pham, Yangziaozhe Jiang, Shuaishuai Zhu, and Jaehwan Kim

Creative Research Center for Nanocellulose Future Composites, Department of Mechanical Engineering, Inha University, Incheon, Republic of Korea

Introduction

Electrical and electronic waste, or e-waste, refers to a broad category of discarded electrical or electronic goods without the intention of reuse. E-waste generally includes old computers, smartphones, televisions, refrigerators, washing machines, and electronic toys [1]. The increasing global demand for electrical and electronic equipment, driven by technological advancements and rising disposable incomes, has significantly increased e-waste. This e-waste mainly ends up in landfills, contributing to resource depletion and environmental degradation. In 2019, 53.6 million tons of e-waste was generated globally, which is projected to reach 74.7 million tons by 2030. E-waste contains up to 60 different metals, including copper, gold, silver, palladium, platinum, and rare earth elements [2]. Recovering and recycling these metals from e-waste reduces reliance on mining, minimizes environmental pollution, helps conserve diminishing natural resources, and unlocks significant economic opportunities [3].

E-waste often ends up in landfills due to inadequate recycling infrastructure and improper disposal practices. E-waste contains hazardous materials, such as lead, mercury, cadmium, and brominated flame retardants, posing significant environmental and health risks if not appropriately managed [3,4]. To mitigate the impacts of e-waste, it is essential to promote responsible consumption, improve recycling infrastructure, implement effective e-waste management policies and regulations, and raise awareness among consumers about properly recycling it [3,5]. Additionally, manufacturers can design products with longevity, repairability, and recyclability to minimize e-waste generation. Recycling e-waste involves several stages; the first is collection and sorting, followed by disassembly, where devices are dismantled, and then preprocessing, which includes separating different materials. In the final and end-processing stage, advanced technologies refine these materials to recover valuable metals and components while safely managing hazardous substances, ensuring

Harnessing Automation and Machine Learning for Resource Recovery and Value Creation
DOI: https://doi.org/10.1016/B978-0-443-27374-2.00015-7

effective recycling and minimal environmental impact [6]. However, current recycling technologies often struggle with inefficiencies due to the labor-intensive nature of manual sorting and the inability to handle the complexity of materials in e-waste effectively. These limitations lead to lower recovery rates of valuable materials and increased environmental contamination [7]. To address these challenges, Integrating robotic technologies into e-waste recycling facilities can enhance the efficiency, accuracy, and sustainability of recycling processes while minimizing the environmental and health risks associated with improper e-waste disposal. Current robotic technologies for e-waste management utilize advanced sensors, artificial intelligence, and machine learning to automate the sorting and processing of e-waste with high precision and speed. These robots can identify and separate various materials, including metals, plastics, and hazardous components, which increases recycling efficiency and reduces manual labor. Their adaptability and continuous learning capabilities allow them to improve sorting accuracy over time, addressing some of the inefficiencies of traditional recycling methods [8].

Furthermore, continuous research and development efforts are needed to address the unique challenges of recycling e-waste effectively. This chapter introduces e-waste, covering its classification, composition, stages of recycling, recent trends in e-waste management, and how robotic technologies can be utilized for e-waste recycling. The last part of the chapter includes the possibilities for implementing robots in recycling e-waste in terms of sorting and separation, disassembling and dismantling, component recovery, quality control and inspection, material handling, transportation, data management, and reporting.

Electrical waste

As the introduction mentions, e-waste refers to a broad category of electronic equipment ranging from small gadgets, such as electronic toys and mobile phones, to large equipment used in industrial manufacturing setups [2]. Everything that requires energy or has a circuit board can eventually become electronic garbage at the end of life. Generally, e-waste is classified into six categories, as shown in Fig. 15.1A and Table 15.1 [2,9]. This classification is based on the size, weight, life span, and material composition of the electronic devices.

Consequently, the collection, logistical procedures, and recycling technologies differ for each category for effective recycling or safe disposal. In today's world, technological developments have shifted industries towards automation, which has increased reliance on electrical and electronic equipment. Similarly, the digital revolution and advancement in computing and processing technologies with affordability have prompted consumers to opt for the latest and superior products while discarding the older products more frequently [2,9]. All these technological advancements have steered to an exponential increase in e-waste.

Worldwide, over two billion metric tonnes of waste are generated, with e-waste having a massive portion of 63.3 million tonnes, which is anticipated to increase to 74.7 million tonnes by 2030. The e-waste generation experienced a significant rise of 18% from 2019 to 2024 [10]. In 2019, only 17% of total e-waste was appropriately collected and recycled. Consequently, 83% of e-waste produced in 2019 lacked documentation and is therefore prone to illegal

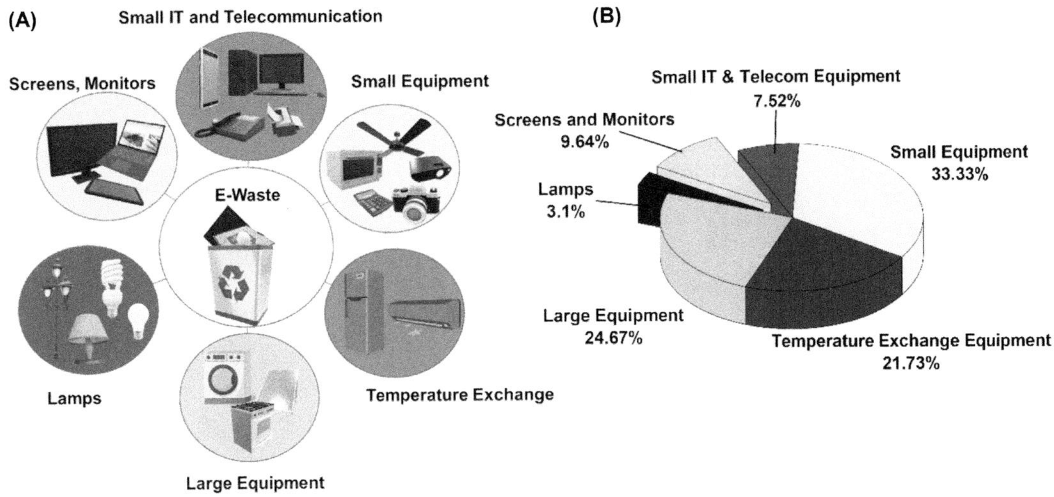

FIGURE 15.1 (A) Categories of e-waste and (B) global e-waste generation in 2022 by types [2,10].

TABLE 15.1 The six categories of e-waste [2,10].

	Categories	Items
1.	Temperature exchange equipment	Freezers, heat pumps, air conditioners, refrigerators, etc.
2.	Screens and monitors	Tablets, laptops, monitors, notebooks, televisions, etc.
3.	Lamps	LED lamps, fluorescent lamps, high-intensity discharge lamps, etc.
4.	Large equipment	Photovoltaic panels, washing machines, copying machines, electric stoves, clothes dryers, large printing machines, etc.
5.	Small equipment	Microwaves, vacuum cleaners, kettles, radios, cameras, toys, calculators, scales, ventilation equipment, electronic tools, medical devices, electric shavers, etc.
6.	Small IT and telecommunication equipment	Personal computers, routers, printers, mobile phones, telephones, GPS, calculators, etc.

dumping or open burning, posing severe threats to human health and the ecosystem [3]. Among various categories of e-waste, small equipment, and large equipment are ranked as the top two e-waste contributors, as shown in Fig. 15.1B [10]. In the continental comparison, Asia has emerged as the largest e-waste generator, accounting for 48.63% of the total e-waste in 2022, followed by America (23.26%), Europe (21.16%), Africa (5.82%), and Oceania (1.13%) as shown in Fig. 15.2A [10]. Although Asia had the highest e-waste generation among continents, its per capita waste generation (6.4 kg/inh) was considerably lower than that of Europe, Oceania, and the Americas, owing to its vast population (4.40 billion), as shown in Fig. 15.2B.

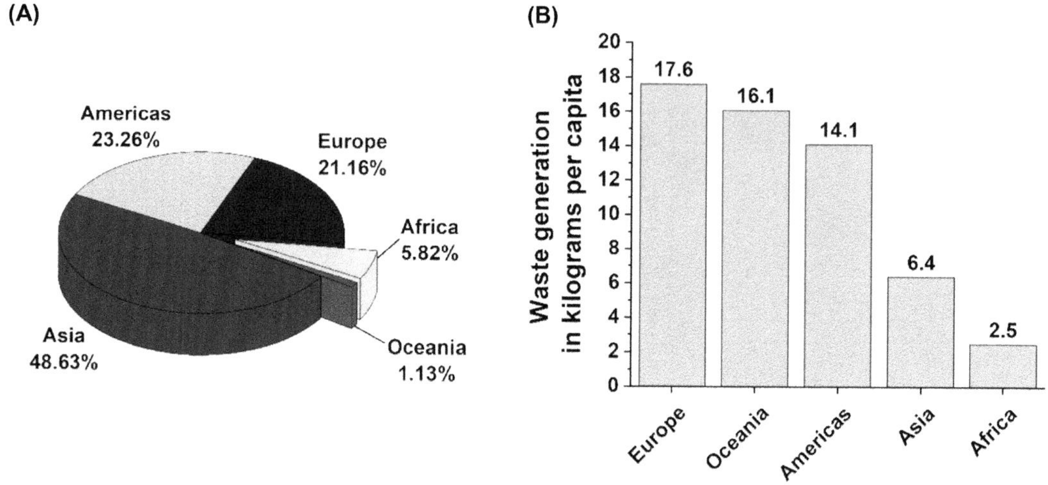

FIGURE 15.2 (A) Generation of electronic waste worldwide in 2022 by region and (B) global per capita e-waste generation in 2022 by region [10].

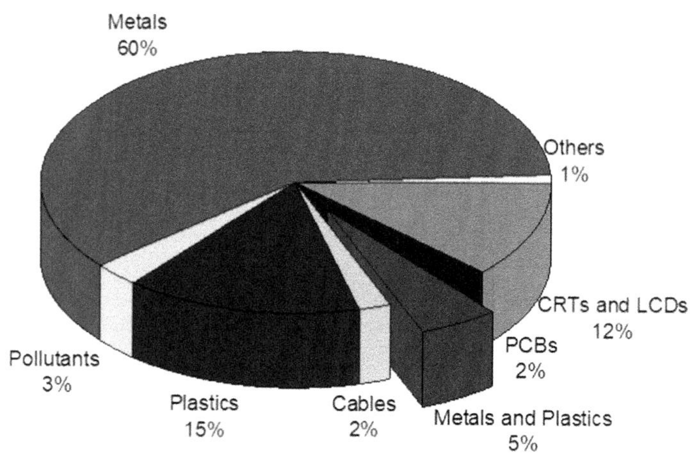

FIGURE 15.3 Composition of e-waste [11].

On the contrary, Africa has exhibited the lowest e-waste generation rate at 2.5 kg/inh due to limited access to electronics compared to other continents [10].

E-waste is composed of both precious and hazardous materials. The composition of e-waste depends on many factors, such as types of electronic devices, their model, manufacturer, manufacturing year, and age. Generally, e-waste includes metals (~60%), plastics (~15%), cathode ray tube (CRT) and liquid crystal display (LCD) monitors (~12%), pollutants (~3%), printed circuit boards (PCBs, ~2%), cables (~2%), a mix of metal and plastic materials (~5%), and other miscellaneous materials as illustrated in Fig. 15.3 [11].

Most e-waste comprises metals, including precious metals, rare earth elements (REE), toxic metals, base metals, and metalloids. Precious metals, such as silver, gold, palladium, and platinum, and rare earth elements, such as europium, terbium, neodymium, yttrium, gadolinium, dysprosium, lanthanum, cerium, praseodymium, and samarium, are found in e-waste. Other metals found in e-waste include aluminum, beryllium, carbon, copper, iron, nickel, lead, tin, zinc, cadmium, thallium, mercury, americium, arsenic, bismuth, boron, antimony, cobalt, ruthenium, gallium, indium, germanium, manganese, barium, niobium, lithium, rhodium, tantalum, selenium, thorium, titanium, and vanadium [4,11–13]. Most PCBs used in electronics contain metals (40%), plastics (30%), and ceramics (30%) [3,5,7]. Precious metals and REE are essential in electronics due to their thermal stability and conducting properties. The demand for REE and precious metals in high-tech products is increasing despite their limited availability in the earth's crust. This accelerating demand for REEs has resulted in economic pressures and price fluctuations across the globe. For example, producing a single mobile phone requires approximately 60 to 64 elements, with metals comprising 50 to 60% of the total weight of cell phones and laptops [13]. Similarly, a single computer production utilizes a diverse array of over 30 elements [5,13]. Likewise, 15% of e-waste includes plastics, primarily used in casings for electronic items and cables for electrical wire insulation. In contrast, glass is used in the old CRT monitors and modern flat panel displays [11].

Apart from valuable metals, rechargeable and disposable batteries contain substances, such as lead, cadmium, mercury, and brominated flame retardants, which pose serious environmental risks that require careful recycling management of e-waste. Lead from lead-acid batteries and CRT monitors is a significant source of toxicity in e-waste. Cadmium, primarily used in batteries, light-sensitive resistors, and corrosion-resistant alloys, further adds to the toxicity of e-waste. Mercury is used mainly in fluorescent tubes and devices like tilt switches [3,4]. Most electronic products utilize toxic brominated flame retardants, such as octabrominated diphenyl ethers and penta-brominated diphenyl ethers, for fire resistance. Among the other hazardous components of electronic waste, Americium (Am) is recognized as a carcinogenic radioactive source. Furthermore, toxic beryllium oxide, found in central processing units (CPUs) and various electronic components, enhances the toxicity and complexity of recycling and managing electronic waste [3,4].

Despite the potential benefits of reducing global metal demand and landfills, only 17 percent of e-waste was fully recycled in 2019 [14]. Hazardous elements found in e-waste endanger human health and the ecosystem if inappropriately handled during recycling or disposal. These compounds incinerate into soils and water supplies, thus contaminating ecosystems and perhaps endangering humans and wildlife [12]. Similarly, poor e-waste management causes the release of harmful gases and other pollutants during burning, which results in air pollution [4]. In addition to the gigantic amount of e-waste produced worldwide, the complexity of electronic equipment creates difficulties for efficient resource recovery and recycling. Electronics production utilizes an enormous number of rare and valuable materials, which raises demand and may result in unsustainable practices and the depletion of resources [11]. Furthermore, in some areas, the unregulated recycling industry exposes workers to hazardous practices and conditions, increasing the hazards to their health and safety [14].

Recycling technologies of electrical wastes

Recycling technologies for e-waste have different stages designed to recover valuable materials and reduce the environmental burden of discarded electronics. There are four main sequential steps of e-waste recycling: (i) Collection and sorting, (ii) Disassembly/dismantling, (iii) Preprocessing, and (iv) End-processing (including smelting, leaching, and refining) [6].

As mentioned, all four steps should operate and interact holistically to achieve the e-waste recycling objectives, stated as follows [15]:

- Handle hazardous or toxic components present in e-waste in an environmentally friendly way while avoiding secondary and tertiary emissions.
- Optimize the recovery efficiency of valuable materials.
- Establish businesses that are economically, environmentally, and socially sustainable (Table 15.2 demonstrates the sustainability attributes).

Sustainability attributes are employed as economic, environmental, and social criteria to ensure comprehensive assessment and improvement. Economically, these attributes focus on cost-efficiency and profitability. Environmentally, they emphasize reducing waste, lowering emissions, and conserving resources. Socially, they address community impacts, equitable practices, and overall quality of life. The four main stages of e-waste recycling are described below, detailing how each step contributes to the efficient recovery of materials and the reduction of environmental impact, in line with the sustainability attributes of economic, environmental, and social criteria.

Collection and sorting

The collection and sorting step of e-waste is the first step that will determine the efficiency of a recycling chain. At this stage, discarded electronic devices are gathered and categorized

TABLE 15.2 Sustainability attributes as economic, environmental, and social criteria [15].

Fields	Attributes
Economic	Low net costs
	Low capital costs
	Increased potential for local economic
Environmental	Low electricity usage
	Low fuel usage for transportation
	Low freshwater usage
	Low emissions
	High metal recovery rates
Social	Job creation
	Low health and safety impacts

based on their type, material composition, and potential for reuse or recycling. Effective sorting is crucial for maximizing material recovery and ensuring that hazardous components are handled appropriately, setting the foundation for efficient downstream recycling processes. Collection methods can include curbside pickup, drop-off centers, or special e-waste recycling programs. Sorting can be manual or automated, using technologies, such as optical sensors, magnetic separation, or robotics to classify materials, such as metals, plastics, and hazardous components. This ensures efficient processing in later recycling stages [16,17].

One of the main factors that play a critical role in the holistic recycling process, besides collection policies and methods, is the cooperation of citizens. In countries with a tradition of waste sorting at the source, like Japan, the e-waste recycling chain tends to yield a higher quantity and quality of recycled products than countries lacking such practices [18].

Disassembly/dismantling

Disassembly or dismantling is the recycling chain's most costly and time-consuming operation. During this stage, e-waste undergoes manual disassembly and classification based on its properties and toxicity. Depending on the condition of the valuable components, they may either be reused or directed to efficient recycling processes, while the hazardous components will be removed, stored, and treated safely [16].

Metallic and nonmetallic materials in e-waste that are not separated in the earlier stages are subjected to mechanical processes, such as shredding, crushing, pulverizing, grinding, and ball-milling. These methods reduce the material's size and facilitate the detachment of components for more efficient separation and recovery. The e-waste components in a complex and nonhomogeneous form present a significant challenge for recycling efforts. Therefore this step is essential to enhance the efficiency of the subsequent processes [1].

Preprocessing

Various equipment and techniques are employed to classify and separate powders or bulk materials based on particle size, density, magnetic properties, or electrical characteristics. These include screeners, classifiers, shakers, separators, and other specialized devices. Screeners, such as drums and rotary sifters, sieve particles based on size, while air classifiers or cyclones separate particles using airflow. Concentrating tables and density separators, such as jigging equipment and spiral concentrators, segregate materials based on density. Electrostatic separators, such as charge particles, separate conductors from dielectrics, while floatation systems use air bubbles to separate hydrophobic and hydrophilic particles. Magnetic separators utilize magnetic fields to separate ferromagnetic materials from nonmagnetic ones. Mechanical classifiers, such as rake, spiral, and bowl classifiers, mechanically separate materials based on size or density. Trommels, large rotary drums with openings, separate coarse materials from finer ones. Water classifiers, such as elutriators and hydrocyclones, use settling or flow in water to classify materials based on size or shape. Fig. 15.4 illustrates the preprocessing processes for printed circuit boards (PCB).

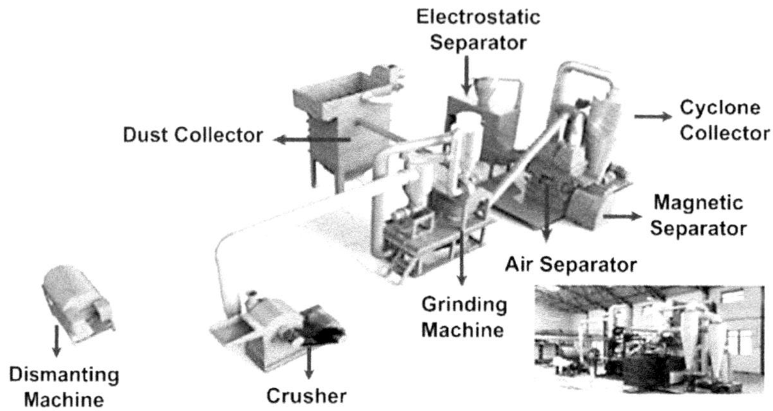

FIGURE 15.4 A schematic representation of preprocessing for PCB [19].

Researchers are exploring combinations of separation techniques and enhancing existing methods, such as screw threads in floatation tanks or improved electrode designs for electrostatic separators. Vacuum methods, such as vacuum pyrolysis, recover metals from e-waste without polluting wastewater. These techniques demonstrate the potential for extracting value from residual materials, even in industries like automotive recycling [20].

End-processing

Several techniques, such as smelting, leaching, pyrolysis, and refining, are utilized at this stage. Pyrolysis is one of the most well-researched chemical recycling methods, involving the breakdown of long-chain polymers into smaller molecules in an oxygen-free environment. This process produces valuable products, such as oil, char, and syngas, which can be used as energy sources or chemical feedstock, as shown in Fig. 15.5 [16].

The most widely applied process for end processing is metallurgy, which includes hydrometallurgy, pyrometallurgy, and biometallurgy [6]. Pyrometallurgy is a traditional attractive method that involves various processes, such as smelting, drossing, sintering, melting, and gas-phase reactions [21]. This technique treats the materials under very high temperatures to obtain a desired metal mixture. Despite its effectiveness in recovering metals, pyrometallurgy often generates toxic by-products and greenhouse gases, necessitating stringent environmental controls and innovative approaches to mitigate its impact [17].

Hydrometallurgy is utilized for its attractive advantages, such as reduced harmful emissions, less energy usage, and reduced operating expenses. Hydrometallurgical processes involve three sequential steps: leaching, solution concentration and purification, and metal recovery. Leaching uses specific types of reagents (usually acidic or basic solution) depending on the desired metal, which will convert the valuable metals to their ionic solution. Fig. 15.6 illustrates the disassembly, pyrometallurgy, and hydrometallurgy of Li-ion batteries. After leaching, the leach liquor is concentrated and purified using precipitation or cementation, electrowinning, solvent extraction, and ion exchange. Finally, the metal will be fully recovered and ready for sale as raw materials in the metal recovery step.

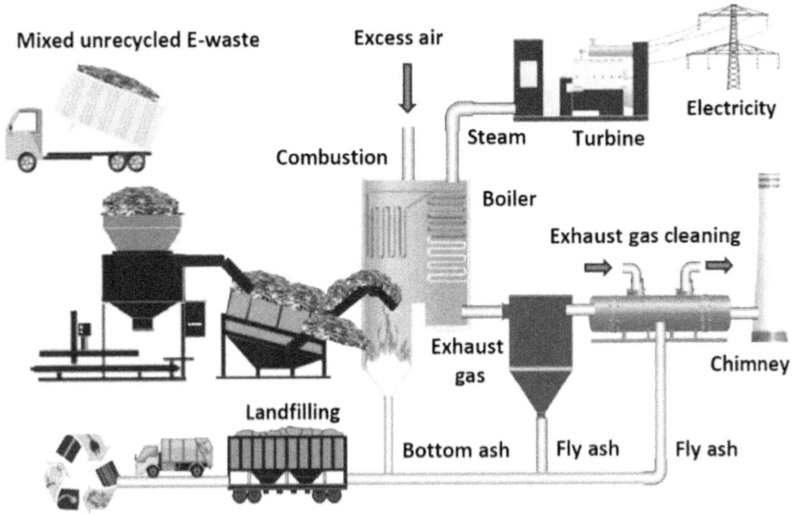

FIGURE 15.5 Schematic representation of the pyrolysis process used in e-waste recycling [16].

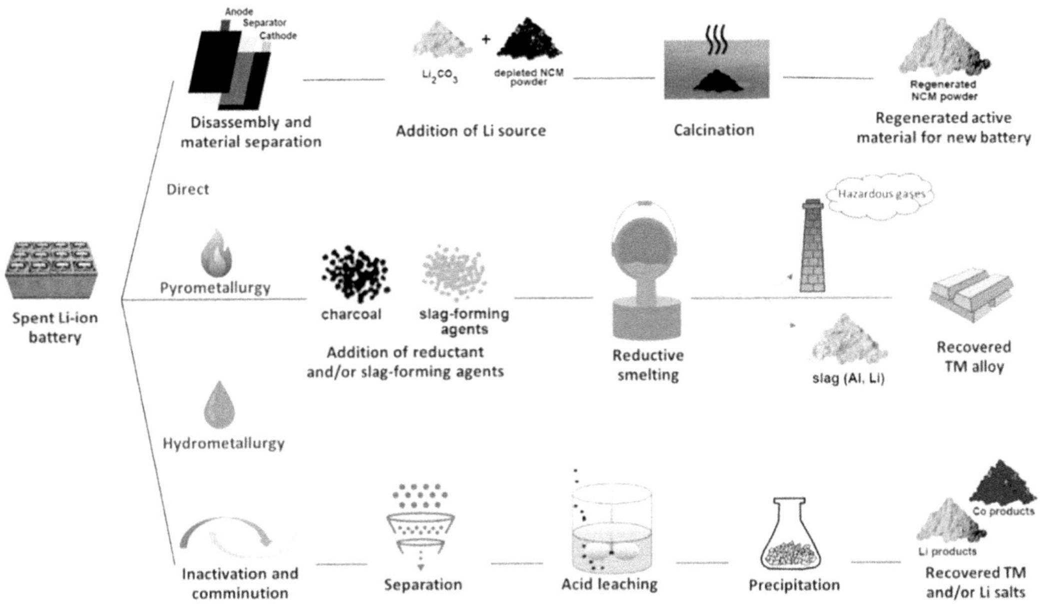

FIGURE 15.6 Common recycling methods for recovering active materials from Li-ion batteries include direct, pyrometallurgical, and hydrometallurgical approaches [22].

Biometallurgy is a novel technology that has gained recent attention from researchers due to its sustainability. This process uses microorganisms, such as algae, bacteria, and fungi, to capture and convert solid metal into soluble and extractable liquor. Like hydrometallurgy,

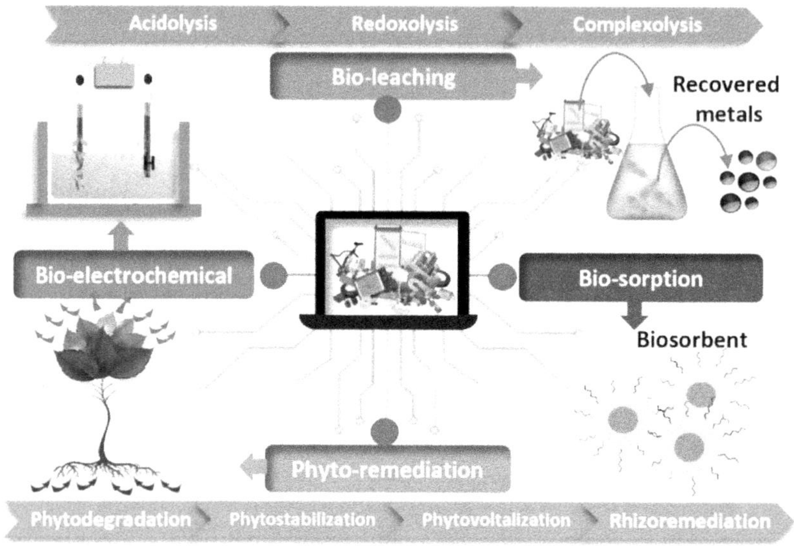

FIGURE 15.7 Biological techniques utilized in e-waste recycling [16].

the resulting mixture containing precious metals undergoes concentration and refinement to obtain recycled metal, ready for use in a new production cycle, as shown in Fig. 15.7.

Robot technologies

As described in the introduction, the inevitable consequence of population growth, urbanization, and economic development has increased waste generation. Hence, a substantial proportion of e-waste is not recycled and is shipped to developing countries for disposal. These countries often lack the infrastructure and resources to safely recycle material and protect their workers' health and safety. Therefore there is an urgent need for technologies that can automatically and safely dismantle e-waste and recover its high-value components. Robot technologies, in this regard, are crucial for e-waste recycling because they offer precision in handling intricate and hazardous components, ensuring the safe and efficient disassembly of devices. They also automate labor-intensive processes, increasing processing speed, reducing human exposure to toxic materials, and improving overall recycling efficiency and resource recovery in a scalable, cost-effective manner. Robotic technologies utilize sophisticated sensors and control systems to improve the precision and efficiency of sorting, disassembly, and material recovery, overcoming obstacles and enhancing overall recycling processes. Fig. 15.8 illustrates an industrial robot with different sensors and control algorithms, showcasing its potential utilization for e-waste recycling [23] Robot technologies employed at various stages in e-waste recycling are described below, outlining how these autonomous systems enhance efficiency at each step, from automated

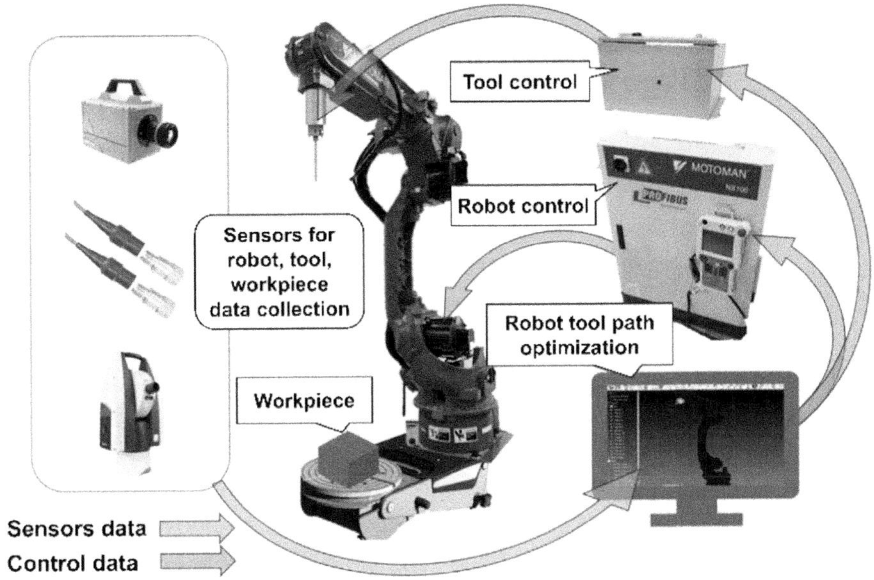

FIGURE 15.8 An industrial robot with different sensors [23].

sorting and separation to disassembly, component recovery, and final processing, ensuring safety and precision in handling complex materials.

Disassembling and dismantling

One of the most common types of e-waste nowadays is the cell phone. The lifespan of cell phones is among the shortest of any consumer electronics product, with more than a billion cell phones sold each year globally. Some of these phones are resold, but many are discarded due to damage from human or nonhuman causes. These discarded cell phones enter landfills or incinerators, releasing toxic materials into the air, water, and soil. Notably, discarded cell phones contain high-value metals, such as gold, platinum, tantalum, titanium, and tungsten, as well as the rare earth metals neodymium, praseodymium, and dysprosium, all of which are worth recycling. The first cell phone recycling robot to be tested and implemented in actual recycling operations was Apple's Liam robotic system, launched in 2016, followed by an improved version, Daisy, in 2018. Motivated by environmental and economic considerations, the robot is now widely used in Apple after-sales service centers in developed and a few developing countries. Electric vehicles, which have become a staple of the transportation industry in recent years, contain many electronic devices and generate a lot of e-waste. Unlike cellular phones, the production of electric vehicles utilizes large-sized 6-degree-of-freedom industrial robots with many actuators and hydraulics for drilling, milling, cutting, and gripping [24]. The most challenging e-waste in electric vehicles, typically lithium-ion batteries, necessitates more rigorous processes to extract valuable materials and separate environmentally damaging materials for subsequent safe disposal.

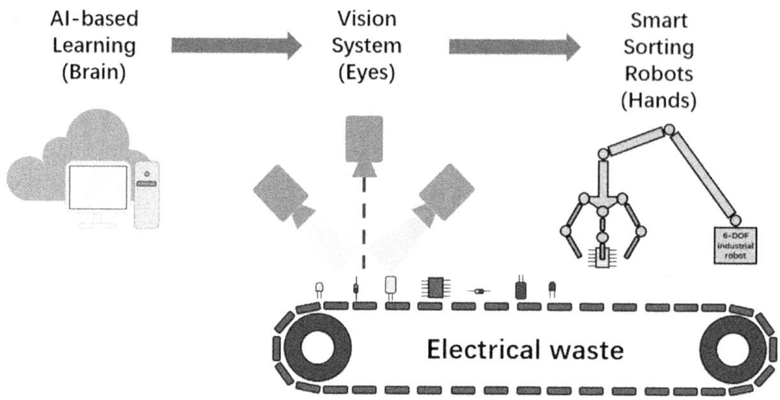

FIGURE 15.9 Robotic e-waste recycling system with artificial intelligence and machine vision technology [8].

In recent years, systems that employ artificial intelligence and machine vision technologies have made their path toward recycling e-waste [8], as shown in Fig. 15.9. These systems have a significant impact on recycling and dismantling practices. However, the complete automation of machines for electronic waste recycling still faces significant technological challenges. Consequently, strategies in which machines and humans collaborate and share dismantling tasks offer the most considerable promise in the short term. It would reduce the heavy reliance on human labor in many monotonous and often hazardous recycling practices and increase the amount of material recovered. Furthermore, multi-criteria assessment work on robotics would stimulate the wider use of robots in most areas of the recycling industry.

Sorting and separation

After disassembling and dismantling, individual e-waste materials need to be sorted and separated. A typical robotic waste-sorting system comprises a pick-and-place mechanism integrated with a conveyor belt. Guided by a vision system equipped with multiple cameras and specialized algorithms, the robot identifies objects within its field of view, determines their positions, and executes the pickup task. The efficiency of the waste-sorting process relies heavily on the accuracy of the computer vision algorithms in distinguishing various waste types, as well as the adaptability of the robotic gripper in handling diverse objects. The system typically analyzes detected items based on color, shape, and volume to determine the most appropriate gripper for each task, as shown in Fig. 15.10.

Many automated robotic systems have been developed for sorting e-waste based on different detection mechanisms. These robots separate individual materials using various methods, depending on the nature of the waste. As shown in Table 15.3, eddy current, magnetic, laser-induced breakdown spectroscopy (LIBS), dual-energy X-ray transmission (DE-XRT), hyperspectral, optical recognition, etc., are reported separation and detection methods [26,27]. However, for some electronic components containing multiple materials, such as PCBs, CRTs, and LCDs, the above techniques make it difficult to separate them from e-waste. With Industry 4.0 and artificial intelligence (AI) technologies development,

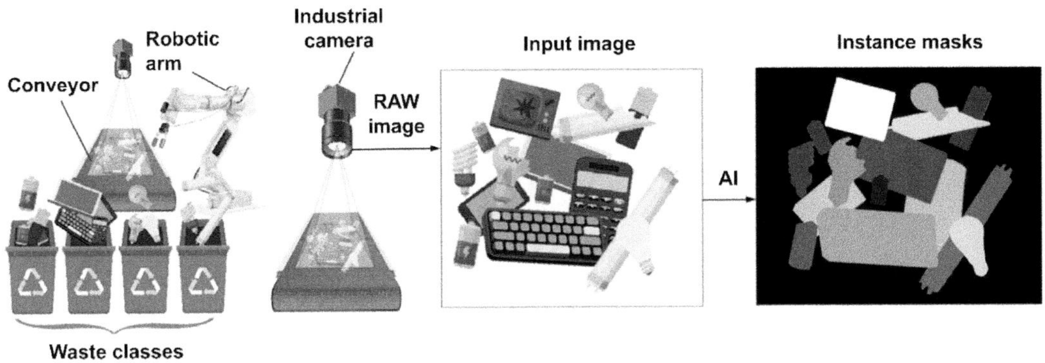

FIGURE 15.10 A standard robotic waste separation setup [25].

TABLE 15.3 Separation and detection methods of e-waste.

Category	Subcategory	Detection method	References
Recyclable electronic component	PCBs, CRTs, LCDs, etc.	AI vision-based cognitive robotic (Algorithms: CNN, R-CNN [30], etc)	[31]
Hazardous material	Pb, mercury, arsenic, barium, beryllium	LIBS	[32]
Ferrous metal	Iron, steel, cast iron, wrought iron, and their alloys, etc.	Magnetic	[33]
Nonferrous metal	Cu, Mg, aluminum alloy, etc.	DE-XRT	[34]
	Cu, Al, Pb, Zn, brass, etc.	Hyperspectral	[35]
	Cu, Al, Mg, Zn, Ni, etc.	Optical	[32,36]
Nonmetallic	HDPE, LDPE, PP, PVA, etc.	LIPS/LIBS	[32,37]
	PVA, etc.	X-ray	[38]
	Lead, ceramic, etc.	UV	[39]
	Colored glass, etc.	Optical	[36]

AI-based and vision-based cognitive robotic systems can potentially address the challenge of complexity. As the current mainstream, AI high-resolution image recognition system refers to continuous machine learning through cases to improve its recognition accuracy [28]. The MARBLE prototype is a waste collection robot that autonomously navigates roadways and dynamically plans paths using a suite of sensors, including LIDAR, ultrasonic sensors, depth cameras, and GPS, to detect and avoid static and moving obstacles. A robotic arm, integrated with a depth camera, enables precise object detection and controlled gripping for

FIGURE 15.11 A prototype robot "MARBLE" for waste collection [29].

waste collection. The robot features a built-in compactor that transfers collected waste from the basket to a designated storage container. The basket itself is engineered and programmed for the temporary storage of waste and efficient extraction once emptied, as shown in Fig. 15.11 [29].

Robotics, automation, and AI also transform e-waste recycling by enhancing workplace safety. These technologies enable controlled, enclosed processing environments that minimize worker exposure to hazardous materials. By automating and streamlining operations, they also support the transition from a linear model of acquisition and disposal to a circular economy focused on resource recovery and reuse. This approach boosts operational efficiency and significantly reduces the risks associated with handling hazardous components in e-waste. For example, When a flat panel display lamp is damaged, e-waste handlers may be exposed to mercury vapor and dust, posing significant health risks due to the toxic properties of these substances. In this regard, a robot developed by ecocycle can safely handle flat panel displays (FPDs) in Australia ensuring efficient recycling while minimizing human contact with hazardous substances, such as mercury and lead [40].

Beyond its autonomous, efficient, and safe performance, AI in robotic e-waste recycling also incorporates a dynamic learning capability. As shown in Fig. 15.12, the knowledge base (KB) and the detection system module can be adapted to different objects and requirements, ensuring flexibility and precision in processing diverse e-waste materials. It constantly modifies strategies and learning based on needs and feedback. Sensors composed of cameras serve as the eyes, and the robot execution unit as the body. Cognitive robotics generally consists of a workbench, disassembly unit/grabbing unit, robot arm module, and detection sensor [41].

Fig. 15.13 depicts the commercial AI vision robotic system developed by the company ZenRobotics for sorting waste. They created a belt-mounted robotic arm sortation system with a sorting speed of more than 65 picks/min, 1.5 times faster than a human sorter.

Applying a similar procedure to e-waste, the AI vision-based cognitive robot can detect and separate recyclable electronic components. Meanwhile, hazardous materials are also separated from e-waste as soon as possible through the corresponding sensor. The remaining waste can be separated using the magnetic method to separate ferrous metals and an eddy current to separate nonferrous metals and nonmetallic materials. The modular

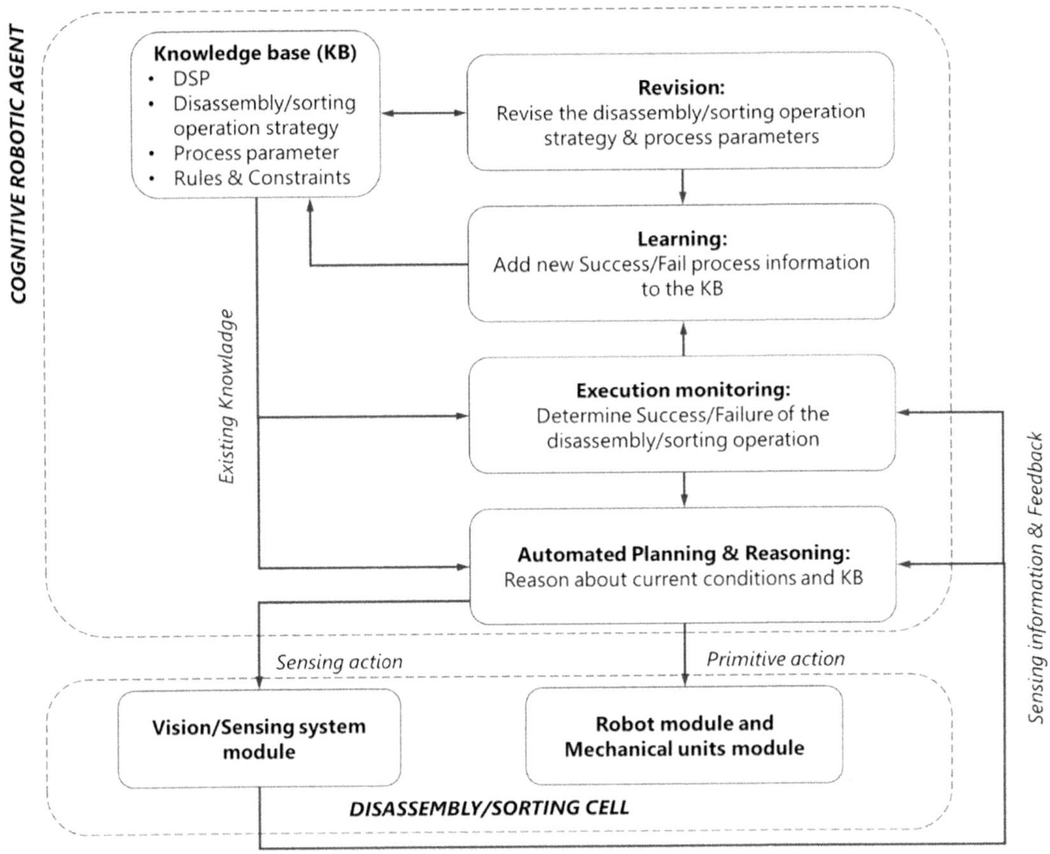

FIGURE 15.12 AI-based cognitive robotic system frame [41].

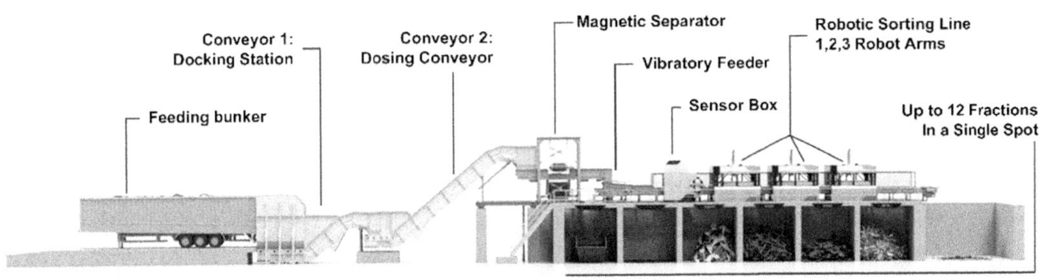

FIGURE 15.13 AI-based robotic automotive systems workflow (ZenRobotics).

sorting robot equipped with different sensors can complete the detailed sorting. However, for various irregular e-waste, the grabbing procedures are different and complicated. For light materials, vacuum trapping is a simple and cost-effective option.

TABLE 15.4 Main robot AI grabbing methods and algorithms.

Category	Feature	Algorithm	References
2D plane grabbing	Single grabbing angle	RGB based	[42]
		Depth based	[43]
		RGB and depth based	[44]
6-DoF space grabbing	Easily affected by environmental noise, limited data acquisition from a single angle	GPD in point clouds	[45]
		PointNet GPD	[46]
Shape completion	The effect of shape completion is excellent but cannot be used universally	3D CNN-based shape completion	[47]
		RGB- and depth-based shape completion	[48]
Humanoid gripper	Limited environmental adaptability	Neural Networks Grasplt based	[49]

On the contrary, more complex materials need multirobot systems and appropriate robot modules according to the situation. The combination of several algorithms may achieve maximum efficiency and benefits. The recent main grabbing methods are shown in Table 15.4.

Component recovery

Some e-waste components, such as palladium, rhodium, ruthenium, iridium, gold, silver, and platinum, are rare and expensive. Therefore component recovery has become one of the most critical procedures for recycling e-waste. For example, Waste PCBs contain up to 28% of precious metals [50]. Thus recycling and reusing these metals in environmentally friendly and low-cost methods is very important. The PCBs contain thermosetting resins that dissolve below 200°C through ester exchange reactions. Subsequently, electronic components with retained electronic properties can be easily separated from the circuit board. Next, cognitive robots can easily remove components, such as chips, capacitors, etc., from PCBs. After that, the circuit boards are shredded and, like other materials, further recycled through chemical processing and physical separation [51].

Quality control and inspection

Machine vision is a technology that enables the extraction of information from images with minimal or no human intervention. The information extracted from images can be as simple as acceptance/rejection signals or as complex as colors, textures, and shape features. This technology controls machines and guides industrial processes to solve real-world problems in industrial automation and other application areas.

TABLE 15.5 Electronic component feature extraction method.

Information extraction methods		References
SIFT descriptor and bag-of-words feature extractor	Used to extract features of electronic components	[52]
SVM and artificial neural network classifiers	Used to classify electronic components or electronic assemblies	[53]
Quantitative methods, object roundness, and correlation matching	Identify electronic components (length, width, and number of terminals of recycled electronic components and terminals are considered features)	[54]
Compare the input image to a standard image stored in the database	Evaluate the relevance of input images	
Compare configuration files	Classify the electronic components used according to the number of terminals.	
Blob measurement method	Classification of color-coded resistors based on power rating	[55]
Optical character recognition method	Identify components from the printed information on the component surface	[56]

Most machine vision systems implemented in the electronics industry provide acceptance or rejection signals, which are used to guide pick-and-place robots. Components are positioned and visually inspected to ensure they are intact and picked correctly. Visual inspection involves feature extraction. The machine extracts features, such as length, width, and contrast of electronic components for selection. The extracted features are usually used to categorize the components to ensure that the components are intact and picked up correctly. Commonly used feature extraction methods are shown in Table 15.5.

Visual recognition based on artificial intelligence is essential for using robots for e-waste recycling. Without fast and reliable product recognition, automated recycling will not be able to cope with the increasing volume and variety of e-waste.

Material handling and transportation

Primarily, hazardous materials, such as mercury, lead, radioactive materials, and batteries, should be transported in special containers and disposed of carefully. Hazardous material delivery robots are specifically designed to transport toxic materials, and their routes are allocated away from busy areas. The remaining parts, such as metals, are refined and purified through quality inspection, plastics are recycled into plastic pellets, and circuit boards are reused or chemically processed to extract precious metals. Next is the material transportation step. The massive amount of recycled products makes manual operations challenging to complete. Thus various transportation robotic systems are used, such as autonomous mobile robots, which can independently navigate the environment without human intervention [57], automated storage and retrieval systems that store and retrieve goods in a warehouse controlled by a computer [58], automated guided vehicles that human operators guide the robots [59], system vehicle loops designed to handle/sort large quantities of materials within warehouses, reducing load unit transfer time, shuttle cars are used to connect different areas of an automated storage system. The Internet of Things (IoT), based on Industry 4.0, integrates the sorting, transportation, and entry and exit management of materials into one station.

Data management and reporting

The trend toward recycling electronic waste is becoming more pronounced as the rate of discarded electronic devices has increased. The increased recycling rate is good news, but at the same time, the increased volume of recycled waste is a huge concern. Autonomous sorting, quality control, inventory management, and maintenance task robots have improved operational efficiency, resource optimization, and environmental sustainability. They typically include the following components:

- Automated sorting and data collection: Robotic systems equipped with sensors and cameras can effectively identify and separate different types of materials. This automated sorting of electrical waste for various types and quantities of materials contributes to comprehensive data management.
- Quality control and inventory management: Robots perform quality control checks throughout recycling to ensure that recycled materials meet specific standards. By detecting defects, contamination, or damage levels, the overall quality of recycled products can be significantly improved. In addition, the robotic system tracks inventory levels of raw and recycled materials in real time, optimizing supply chain management and supporting efficient resource allocation.
- Reporting and analytics: The robotic system generates detailed reports and analytics based on data collected during recycling. These reports provide insight into key performance indicators, such as recycling rates, waste diversion metrics, and environmental impact assessments.
- Data security and integration with IoT and AI: Robotic systems ensure data security and compliance throughout the recycling process, and integration with IoT devices and AI systems enhances data management capabilities. This integration enables predictive maintenance, intelligent decision-making, and continuous process optimization, further improving the efficiency and sustainability of electrical waste recycling operations.

Summary

E-waste includes a wide range of electrical or electronic devices, ranging from small gadgets to large industrial equipment, which become electronic garbage at their end of life. E-waste is categorized based on size, weight, lifespan, and material composition. Technological advances have accelerated e-waste generation, with over 63 million tonnes produced annually and projections reaching 74.7 million tonnes by 2030. Despite this, only a tiny fraction is adequately managed and recycled, while most e-waste is illegally dumped or incinerated, posing severe health and environmental risks. E-waste contains valuable metals, including precious and rare earth elements, such as gold, silver, palladium, and platinum, and rare earth elements, such as neodymium and europium. It also contains hazardous materials, such as lead, mercury, cadmium, arsenic, and brominated flame retardants. The improper handling of these toxic materials can lead to significant ecological damage and health hazards.

The e-waste recycling process starts with collection and sorting, where electronic waste is gathered from several sources and sorted into different categories based on type, size, and material composition. This step is decisive for organizing and preparing the waste for further treatment. Next is the disassembly/dismantling stage, in which e-waste devices are manually or mechanically disassembled to separate their components, such as circuit boards, batteries, and casings. At this stage, valuable components and hazardous materials are separated. Following this, the preprocessing stage includes shredding and grinding to reduce the size of the components, making them ready for further processing. Additionally, toxic and hazardous materials are often removed during this stage. Finally, in the end-processing stage, the materials are treated to recover valuable or precious metals and other resources using pyrolysis, chemical leaching, smelting, etc. All the recycling stages of e-waste aim to maximize resource recovery while minimizing environmental impact.

Robotic technologies improve e-waste recycling by automating the sorting and disassembly of electronic devices, which enhances efficiency and safety. Advanced robots with AI and sensors can precisely identify and separate components, such as metals, plastics, and circuit boards. These technologies reduce manual labor and workers' exposure to hazardous materials. Additionally, robotics streamline the recycling process, enabling faster processing and higher recovery rates of valuable resources. Robotic systems contribute to a more effective and environmentally friendly e-waste management process.

Robotic technologies also significantly enhance quality control and data management in e-waste recycling. Robots can accurately identify and sort different materials by utilizing advanced sensors and AI, ensuring each component is handled according to its specific requirements. This precision helps maintain high-quality standards and reduces contamination in the recovered materials. Additionally, robotics systems can collect and analyze data in real time, providing valuable insights into recycling, such as the quantity and type of materials processed. The data management capability allows for continuous monitoring and optimization of operations, leading to more efficient resource recovery and improved overall performance of e-waste recycling. Thus integrating robotic technology into e-waste recycling facilities can achieve better quality control and data-driven decision-making for efficiently handling complex sorting and processing tasks, ensuring higher recovery rates and reduced environmental impact.

References

[1] J. Van Yken, N.J. Boxall, K.Y. Cheng, A.N. Nikoloski, N.R. Moheimani, A.H. Kaksonen, E-waste recycling and resource recovery: a review on technologies, barriers and enablers with a focus on oceania, Metals (Basel) 11 (2021) 1313. Available from https://doi.org/10.3390/met11081313.

[2] A. Kumar, M. Holuszko, D.C.R. Espinosa, E-waste: an overview on generation, collection, legislation and recycling practices, Resources, Conservation and Recycling 122 (2017) 32–42. Available from https://doi.org/10.1016/j.resconrec.2017.01.018.

[3] R.S. Dave, B.M. Shah, R.D. Tipre, E-waste: metal pollution threat or metal resource? Journal of Advanced Research in Biotechnology 1 (2016) 1–14. Available from https://doi.org/10.15226/2475-4714/1/2/00103.

[4] S.R. Lim, J.M. Schoenung, Toxicity potentials from waste cellular phones, and a waste management policy integrating consumer, corporate, and government responsibilities, Waste Management 30 (2010) 1653–1660. Available from https://doi.org/10.1016/j.wasman.2010.04.005.

[5] A. Das, A. Vidyadhar, S.P. Mehrotra, A novel flowsheet for the recovery of metal values from waste printed circuit boards, Resources, Conservation and Recycling 53 (2009) 464–469. Available from https://doi.org/10.1016/j.resconrec.2009.03.008.

[6] M. Kaya, A.M. Tita, Electronic waste recycling in maintaining a circular economy, Electronic Waste Management, Wiley, 2023, pp. 301–316. Available from https://doi.org/10.1002/9781119891543.ch22.

[7] K. Palanisamy, R.G. Subburaj, Integration of electronic waste management: a review of current global generation, health impact, and technologies for value recovery and its pertinent management technique, Environmental Science and Pollution Research 30 (2023) 63347–63367. Available from https://doi.org/10.1007/s11356-023-26719-6.

[8] E. Simaei, S. Rahimifard, AI-based decision support system for enhancing end-of-life value recovery from e-wastes, International Journal of Sustainable Engineering 17 (2024) 1–17. Available from https://doi.org/10.1080/19397038.2024.2306293.

[9] K. Liu, Q. Tan, J. Yu, M. Wang, A global perspective on e-waste recycling, Circular Economy 2 (2023) 10028. Available from https://doi.org/10.1016/j.cec.2023.100028.

[10] Statista, Global e-waste generation by type 2015–2022. Retrieved January 29, 2025, from https://www.statista.com/statistics/499912/ewaste-generation-worldwide-by-type/.

[11] D.R. Tipre, B.R. Khatri, S.C. Thacker, S.R. Dave, The brighter side of e-waste-a rich secondary source of metal, Environmental Science and Pollution Research 28 (2021) 10503–10518. Available from https://doi.org/10.1007/s11356-020-12022-1/Published.

[12] O. Tsydenova, M. Bengtsson, Chemical hazards associated with treatment of waste electrical and electronic equipment, Waste Management 31 (2011) 45–58. Available from https://doi.org/10.1016/j.wasman.2010.08.014.

[13] B. Tansel, From electronic consumer products to e-wastes: global outlook, waste quantities, recycling challenges, Environment International 98 (2017) 35–45. Available from https://doi.org/10.1016/j.envint.2016.10.002.

[14] M. Heacock, C.B. Kelly, K.A. Asante, L.S. Birnbaum, Å.L. Bergman, M.N. Bruné, et al., E-waste and harm to vulnerable populations: a growing global problem, Environmental Health Perspectives 124 (5) (2016) 550. Available from https://doi.org/10.1289/ehp.1509699.

[15] Mathias, H. Christian, E.M.M. Christina, M. Federico, W. Feng, M. Esther, et al., Market potential of innovative e-waste recycling technologies in developing countries, in: R'09 Twin World Congress and World Resources Forum, Davos, Switzerland: Empa Materials Science and Technology, 2009.

[16] A.M. Elgarahy, M.G. Eloffy, A.K. Priya, A. Hammad, M. Zahran, A. Maged, et al., Revitalizing the circular economy: an exploration of e-waste recycling approaches in a technological epoch, Sustainable Chemistry for the Environment 7 (2024) 100124. Available from https://doi.org/10.1016/j.scenv.2024.100124.

[17] S. Gulliani, M. Volpe, A. Messineo, R. Volpe, Recovery of metals and valuable chemicals from waste electric and electronic materials: a critical review of existing technologies, RSC Sustainability 1 (2023) 1085–1108. Available from https://doi.org/10.1039/d3su00034f.

[18] S.K. Ghosh, Electronic waste management in the Asia Pacific Region, Electronic Waste Management, The Royal Society of Chemistry, 2019, pp. 166–187. Available from https://doi.org/10.1039/9781788018784-00166.

[19] M. Kaya, Recovery of metals and nonmetals from electronic waste by physical and chemical recycling processes, Waste Management 57 (2016) 64–90. Available from https://doi.org/10.1016/j.wasman.2016.08.004.

[20] R. Kellner, Integrated approach to e-waste recycling, Electronic Waste Management, The Royal Society of Chemistry, 2008, pp. 111–160. Available from https://doi.org/10.1039/9781847559197-00111.

[21] K.S. Maiti, I. Khatun, S.R. Hansda, D. Ghosh, Current advances in recycling of electronic wastes, Sustainable Management of Electronic Waste, Wiley, 2024, pp. 341–373. Available from https://doi.org/10.1002/9781394166923.ch17.

[22] Z.J. Baum, R.E. Bird, X. Yu, J. Ma, Lithium-ion battery recycling—overview of techniques and trends, ACS Energy Letters 7 (2022) 712–719. Available from https://doi.org/10.1021/acsenergylett.1c02602.

[23] M. Makulavičius, S. Petkevičius, J. Rožėnė, A. Dzedzickis, V. Bučinskas, Industrial robots in mechanical machining: perspectives and limitations, Robotics 12 (2023) 160. Available from https://doi.org/10.3390/robotics12060160.

[24] M. Barwood, J. Li, T. Pringle, S. Rahimifard, Utilisation of reconfigurable recycling systems for improved material recovery from e-waste, Procedia CIRP vol. 29, Elsevier B.V, 2015, pp. 746–751. Available from https://doi.org/10.1016/j.procir.2015.02.071.

[25] D. Nišić, B. Lukić, Z. Gordić, U. Pantelić, A. Vukićević, E-waste management in Serbia, focusing on the possibility of applying automated separation using robots, Applied Sciences (Switzerland) 14 (2024) 5685. Available from https://doi.org/10.3390/app14135685.

[26] R. Bogue, Robots in recycling and disassembly, Industrial Robot 46 (2019) 461–466. Available from https://doi.org/10.1108/IR-03-2019-0053.

[27] S.P. Gundupalli, S. Hait, A. Thakur, A review on automated sorting of source-separated municipal solid waste for recycling, Waste Management 60 (2017) 56–74. Available from https://doi.org/10.1016/j.wasman.2016.09.015.

[28] L.S. Dalenogare, G.B. Benitez, N.F. Ayala, A.G. Frank, The expected contribution of Industry 4.0 technologies for industrial performance, International Journal of Production Economics 204 (2018) 383–394. Available from https://doi.org/10.1016/j.ijpe.2018.08.019.

[29] V.B. Justo, A. Gupta, T.F. Umland, D. Göhlich, Minimum energy utilization strategy for fleet of autonomous robots in urban waste management, Robotics 12 (2023) 159. Available from https://doi.org/10.3390/robotics12060159.

[30] M. Koskinopoulou, F. Raptopoulos, G. Papadopoulos, N. Mavrakis, M. Maniadakis, Robotic waste sorting technology: toward a vision-based categorization system for the industrial robotic separation of recyclable waste, IEEE Robotics & Automation Magazine 28 (2021) 50–60. Available from https://doi.org/10.1109/MRA.2021.3066040.

[31] K. Naito, A. Shirai, S.I. Kaneko, G. Capi, Recycling of printed circuit boards by robot manipulator: a deep learning approach, in: IEEE International Symposium on Robotic and Sensors Environments, ROSE 2021 - Proceedings, Institute of Electrical and Electronics Engineers Inc., 2021. Available from https://doi.org/10.1109/ROSE52750.2021.9611773.

[32] D.W. Hahn, N. Omenetto, Laser-induced breakdown spectroscopy (LIBS), part II: review of instrumental and methodological approaches to material analysis and applications to different fields, Applied Spectroscopy 66 (2012) 347–419. Available from https://doi.org/10.1366/11-06574.

[33] X.N. Zhu, C.C. Nie, S.S. Wang, Y. Xie, H. Zhang, X. Lyu, et al., Cleaner approach to the recycling of metals in waste printed circuit boards by magnetic and gravity separation, Journal of Cleaner Production 248 (2020) 119235. Available from https://doi.org/10.1016/j.jclepro.2019.119235.

[34] K. Neubert, H. Wotruba, Investigations on the detectability of rare-earth minerals using dual-energy X-ray transmission sorting, Journal of Sustainable Metallurgy 3 (2017) 3–12. Available from https://doi.org/10.1007/s40831-016-0069-1.

[35] P. Tatzer, M. Wolf, T. Panner, Industrial application for inline material sorting using hyperspectral imaging in the NIR range, Real-Time Imaging 11 (2005) 99–107. Available from https://doi.org/10.1016/j.rti.2005.04.003.

[36] J. Huang, T. Pretz, Z. Bian, Intelligent solid waste processing using optical sensor based sorting technology, 3rd International Congress on Image and Signal Processing 2010 (2010) 1657–1661. Available from https://doi.org/10.1109/CISP.2010.5647729, IEEE.

[37] E.R.K. Neo, Z. Yeo, J.S.C. Low, V. Goodship, K. Debattista, A review on chemometric techniques with infrared, Raman and laser-induced breakdown spectroscopy for sorting plastic waste in the recycling industry, Resources, Conservation and Recycling 180 (2022). Available from https://doi.org/10.1016/j.resconrec.2022.106217.

[38] F. Bezati, D. Froelich, V. Massardier, E. Maris, Addition of tracers into the polypropylene in view of automatic sorting of plastic wastes using X-ray fluorescence spectrometry, Waste Management 30 (2010) 591–596. Available from https://doi.org/10.1016/j.wasman.2009.11.011.

[39] X. He, B. Xi, Z. Wei, X. Guo, M. Li, D. An, et al., Spectroscopic characterization of water extractable organic matter during composting of municipal solid waste, Chemosphere 82 (2011) 541–548. Available from https://doi.org/10.1016/j.chemosphere.2010.10.057.

[40] L. Korycki, AI and robotics recycle e-waste at Ecocycle. June 6, 2022. https://wastemanagementreview.com.au/ai-and-robotics-recycle-e-waste-at-ecocycle/ (accessed 09.09.24).

[41] S. Vongbunyong, S. Kara, M. Pagnucco, Basic behaviour control of the vision-based cognitive robotic disassembly automation, Assembly Automation 33 (2013) 38–56. Available from https://doi.org/10.1108/01445151311294694.

[42] I. Lenz, H. Lee, A. Saxena, Deep learning for detecting robotic grasps, International Journal of Robotics Research 34 (2015) 705–724. Available from https://doi.org/10.1177/0278364914549607.

[43] J. Mahler, J. Liang, S. Niyaz, M. Laskey, R. Doan, X. Liu, et al., Dex-Net 2.0: Deep Learning to Plan Robust Grasps with Synthetic Point Clouds and Analytic Grasp Metrics (2017).

[44] S. Kumra, C. Kanan, Robotic Grasp Detection using Deep Convolutional Neural Networks (2017). https://doi.org/10.0/Linux-x86_64.

[45] A. ten Pas, M. Gualtieri, K. Saenko, R. Platt, Grasp pose detection in point clouds, The International Journal of Robotics Research 36 (2017) 1455–1473. Available from https://doi.org/10.1177/0278364917735594.

[46] H. Liang, X. Ma, S. Li, M. Gorner, S. Tang, B. Fang, et al., PointNetGPD: detecting grasp configurations from point sets, International Conference on Robotics and Automation (ICRA) 2019 (2019) 3629–3635. Available from https://doi.org/10.1109/ICRA.2019.8794435, IEEE.

[47] J. Varley, C. Dechant, A. Richardson, J. Ruales, P. Allen, Shape Completion Enabled Robotic Grasping (2017). https://doi.org/10.0/Linux-x86_64.

[48] X. Yan, M. Khansari, J. Hsu, Y. Gong, Y. Bai, S. Pirk, et al., Data-Efficient Learning for Sim-to-Real Robotic Grasping using Deep Point Cloud Prediction Networks (2019).

[49] M. Liu, Z. Pan, K. Xu, K. Ganguly, D. Manocha, Generating grasp poses for a high-DOF gripper using neural networks, IEEE/RSJ International Conference on Intelligent Robots and Systems (IROS) 2019 (2019) 1518–1525. Available from https://doi.org/10.1109/IROS40897.2019.8968115, IEEE.

[50] Z. Chen, M. Yang, Q. Shi, X. Kuang, H.J. Qi, T. Wang, Recycling waste circuit board efficiently and environmentally friendly through small-molecule assisted dissolution, Scientific Reports 9 (2019). Available from https://doi.org/10.1038/s41598-019-54045-w.

[51] Workshop Materials on WEEE Management in Taiwan 2024. Workshop Materials on WEEE Management in Taiwan, Printed Circuit Board Recycling Methods (Handout 10). Accessed at: https://www.epa.gov/sites/default/files/2014-05/documents/handout-10-circuitboards.pdf.

[52] Y. Iwahori, Y. Takada, T. Shiina, Y. Adachi, M.K. Bhuyan, B. Kijsirikul, Defect classification of electronic board using dense SIFT and CNN, Procedia Comput Science 126 (2018) 1673–1682. Available from https://doi.org/10.1016/j.procS.2018.08.110.

[53] G. Zhai, J. Chen, S. Wang, K. Li, L. Zhang, Material identification of loose particles in sealed electronic devices using PCA and SVM, Neurocomputing 148 (2015) 222–228. Available from https://doi.org/10.1016/j.neucom.2013.10.043.

[54] N. Hayashi, S. Koyanaka, T. Oki, Verification of algorithm for automatic detection of electronic devices mounted on waste printed circuit boards, Journal of the Air & Waste Management Association 72 (2022) 420–433. Available from https://doi.org/10.1080/10962247.2022.2044408.

[55] W. Zhao, S.R. Gurudu, S. Taheri, S. Ghosh, M.A.M. Sathiaseelan, N. Asadizanjani, PCB component detection using computer vision for hardware assurance, Big Data and Cognitive Computing 6 (2022). Available from https://doi.org/10.3390/bdcc6020039.

[56] D. Katz Bonello, Y. Iano, U.B. Neto, Text recognition in PCBS: an object character recognition (OCR) algorithm, International Journal of Development Research 10 (2020) 1–7. Available from https://doi.org/10.37118/ijdr.19567.07.2020.

[57] A. Löcklin, F. Dettinger, M. Artelt, N. Jazdi, M. Weyrich, Trajectory prediction of workers to improve AGV and AMR operation based on the manufacturing schedule, Procedia CIRP vol. 107, Elsevier B.V., 2022, pp. 283–288. Available from https://doi.org/10.1016/j.procir.2022.04.046.

[58] R. Manzini, M. Gamberi, A. Regattieri, Design and control of an AS/RS, International Journal of Advanced Manufacturing Technology 28 (2006) 766–774. Available from https://doi.org/10.1007/s00170-004-2427-6.

[59] J. Mehami, M. Nawi, R.Y. Zhong, Smart automated guided vehicles for manufacturing in the context of Industry 4.0, Procedia Manufaturing 26 (2018) 1077–1086. Available from https://doi.org/10.1016/j.promfg.2018.07.144.

CHAPTER

16

Machine learning for sustainable development in electronics

Iraklis Varlamis[1], Ilias Panagiotopoulos[1], Christos Chronis[1], George Dimitrakopoulos[1], and Faycal Bensaali[2]

[1]Department of Informatics and Telematics, Harokopio University of Athens, Athens, Greece
[2]Department of Electrical Engineering, College of Engineering, Qatar University, Doha, Qatar

Introduction

Machine learning (ML) is a rapidly advancing method poised to bring significant advantages through its widespread application in diverse fields, including energy, infrastructure, agriculture, and electronics. The cutting-edge ML models play a crucial role in predicting and forecasting, and in revolutionizing various industries, and their integration in the domain of sustainable development holds significant promise [1].

This chapter delves deep into the relationship between ML and sustainable practices in the electronics sector, aiming to explore the synergies that can drive positive environmental and economic impacts. The overarching goal is to provide a comprehensive understanding of how ML applications can be strategically employed to address the challenges posed by the electronics industry and contribute to advancing sustainable development goals. Emphasis is given to the potential of ML to enhance decision-making processes, optimize resource utilization, and contribute to long-term environmental and social well-being [2]. By harnessing the power of data-driven insights, ML becomes an enabler for informed and sustainable practices, offering innovative solutions to issues ranging from e-waste management [3] to greener electronics [4]. This fusion of ML and sustainable development is not merely a technological advancement but a transformative approach that has the potential to reshape the electronics industry and contribute meaningfully to a more resilient and equitable future.

Against the backdrop of an ever-evolving electronics landscape, the significance of adopting sustainable practices becomes increasingly pronounced. It is thus imperative to embrace sustainability in the electronics industry, overcoming all the potential barriers and considering the sector's substantial environmental footprint and the urgent need for responsible and eco-friendly approaches [5]. As electronic devices become ubiquitous in modern life, understanding the environmental repercussions of their manufacturing, usage,

Harnessing Automation and Machine Learning for Resource Recovery and Value Creation
DOI: https://doi.org/10.1016/B978-0-443-27374-2.00016-9

and disposal becomes paramount. Sustainable practices address the ecological impact and hold potential economic benefits, fostering innovation and resilience in the face of resource scarcity. By scrutinizing the relevance of sustainable practices in electronics, we explore how environmental care is connected to economic viability and explain how a holistic approach based on ML can propel sustainability in the electronics industry.

The chapter begins with an exploration of the foundational principles of ML, highlighting key concepts that form the backbone of its applications in the electronics sector. Subsequently, the challenges confronting the electronics industry in the pursuit of sustainability are thoroughly examined, focusing on the environmental impact of manufacturing, e-waste management, and resource depletion challenges. The focus then shifts to the transformative applications of ML in sustainable electronics, showcasing its role in predictive maintenance, manufacturing process optimization, energy efficiency enhancement, and intelligent waste sorting. Case studies and success stories are presented to illustrate real-world implementations, followed by an analysis of barriers, limitations, and ethical considerations. The chapter concludes by projecting future trends and innovations, envisioning the integration of ML into circular economy models, and offering policy implications and recommendations for a more sustainable electronics industry. This comprehensive exploration aims to equip readers with insights into the potential and challenges of ML in shaping a sustainable future for the electronics sector.

Foundations of machine learning in electronics

ML is a subset of artificial intelligence that develops models and computer systems that learn from historical data and previous experiences without the need for explicit programming. ML algorithms enable intelligent machines to recognize patterns, make predictions, and continuously improve their performance. In the context of electronics and sustainable development, ML holds the potential to revolutionize various facets, from design and manufacturing to maintenance and optimization. By harnessing the power of data analytics and computational learning, ML equips electronic devices and systems to evolve intelligently, adapting to changing conditions and user behaviors. At the same time, ML's capacity to control and analyze data throughout the entire electronics production, usage, and recycling life cycle ensures a comprehensive and sustainable approach to managing information and resources across the industry [6].

Various models, techniques, and algorithms are crucial in boosting ML performance. In supervised ML, models are trained on labeled data and can then assist in pattern recognition in tasks like image and speech recognition. In unsupervised ML, data are explored to find patterns without the existence of predefined labels and are usually applied in clustering and anomaly detection tasks. In reinforcement ML, software agents learn through trial and error from their interaction with an environment. Algorithms and models like decision trees, neural networks, and support vector machines become pivotal in processing data and enabling the extraction of valuable insights and predictions for improved functionality.

Supervised ML, with its capacity for pattern recognition based on labeled data, offers compelling applications in enhancing electronics sustainability. One notable application is in predictive maintenance, where ML models trained on historical performance data of electronic components can accurately forecast potential failures [7]. This proactive approach minimizes downtime and extends the operational lifespan of electronic devices. Additionally, in the electronics manufacturing process, supervised ML is instrumental in quality control [8,9]. By analyzing labeled datasets of defect-free and defective products, ML models can identify and rectify flaws during production, reducing waste and resource consumption. Moreover, supervised ML facilitates efficient energy management in electronic systems by predicting optimal usage patterns and regulating power consumption, contributing to energy efficiency and environmental conservation.

Unsupervised ML introduces a new dimension to electronics' sustainability by exploring data patterns without predefined labels. In the context of electronic waste management, unsupervised ML algorithms excel in clustering and categorizing diverse materials from discarded electronic devices [10]. This facilitates the automated sorting of electronic waste for recycling, streamlining the recovery of valuable materials and minimizing environmental impact. Additionally, unsupervised ML plays a crucial role in anomaly detection, by identifying irregularities or potential malfunctions in electronic systems. This application enhances the overall reliability of electronic devices and reduces the likelihood of unexpected failures, contributing to both sustainability and user satisfaction.

In the realm of electronics sustainability, reinforcement ML is employed for dynamic energy management. Electronic devices equipped with reinforcement ML algorithms adapt their power consumption based on real-time environmental conditions and user behavior, contributing to energy efficiency [11]. Moreover, reinforcement ML is instrumental in the development of smart electronic systems that can autonomously adjust their functionalities to meet Sustainable Development Goals [12]. For instance, in intelligent waste sorting and recycling, reinforcement ML enables robotic systems to learn optimal sorting strategies over time, improving the efficiency of the recycling process and reducing the environmental footprint of electronic waste disposal [13].

Challenges in electronics for sustainable development

The electronics industry faces numerous sustainability challenges, including the environmental impact of manufacturing, e-waste management issues, resource depletion, and high energy consumption (see Fig. 16.1). Traditional manufacturing practices often result in significant carbon emissions, chemical pollutants, and resource depletion. The extraction and processing of raw materials, coupled with energy-intensive production methods, contribute to environmental degradation. Addressing these challenges necessitates a paradigm shift toward sustainable manufacturing practices [14].

ML integrating machine learning models into electronics manufacturing enables data-driven optimizations that enhance efficiency, reduce waste, and curtail the overall environmental impact of electronics production. For instance, ML algorithms in production planning can analyze historical production data, demand forecasts, and supply chain information to optimize scheduling and resource allocation. The use of ML in life cycle assessments can provide a more accurate and dynamic understanding of the environmental

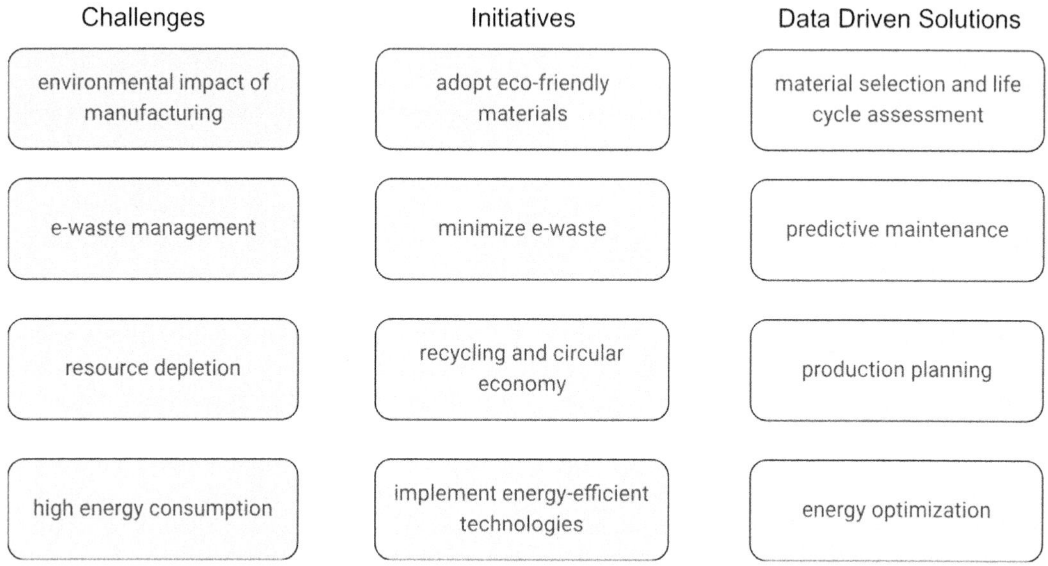

FIGURE 16.1 Challenges, initiatives, and machine learning solutions for sustainable development in electronics.

impact of materials and manufacturing processes and enable better decision-making in selecting materials and optimizing production methods for reduced environmental footprint. This data-driven approach ensures that manufacturing facilities operate at peak efficiency, minimizing idle time and reducing energy consumption. Moreover, ML can enhance quality control by continuously monitoring production lines for defects or irregularities [15]. Automated inspection systems, powered by ML, can swiftly identify faulty components, minimizing defective units' production and reducing electronic waste. Additionally, predictive maintenance models can anticipate equipment failures, allowing for timely repairs and preventing unscheduled downtime [16]. This extends the machinery's lifespan and contributes to the overall efficiency and sustainability of electronics manufacturing.

The proliferation of electronic devices has led to a corresponding surge in electronic waste (e-waste), presenting a significant challenge for sustainable development. Managing the disposal and recycling of obsolete electronics is intricate due to the presence of hazardous materials and complex components. Inadequate e-waste management poses risks to both human health and the environment. Developing effective and scalable e-waste recycling solutions requires innovative approaches. ML applications, particularly in intelligent waste sorting and material identification, can streamline the recycling process. Automated systems can accurately categorize electronic components, facilitating their efficient recovery for reuse and recycling [17]. Moreover, raising awareness about responsible e-waste disposal practices and establishing comprehensive recycling infrastructures are essential to addressing the e-waste management challenges.

The electronics industry faces a dual challenge concerning resource depletion and energy consumption. The relentless demand for electronic devices leads to the depletion of finite resources, exacerbating concerns about sustainability. Additionally, the energy-intensive nature of electronics manufacturing and usage contributes to carbon emissions and environmental strain. To mitigate these challenges, there is a pressing need for innovations in materials sourcing, with an emphasis on sustainable and recyclable materials [18]. Furthermore, optimizing energy consumption throughout the life cycle of electronic devices is crucial [19]. ML algorithms can play a pivotal role in predicting energy usage patterns, optimizing manufacturing processes, and enhancing the overall energy efficiency of electronic systems. By addressing resource depletion and energy consumption challenges collaboratively, the electronics industry can transition toward a more sustainable and environmentally conscious trajectory.

Overall, ML is a powerful ally in addressing challenges related to the electronics industry's sustainable development. The integration of ML minimizes environmental footprint by reducing waste, optimizing resource usage, and enhancing energy efficiency, thereby paving the way for a more sustainable future.

Applications of machine learning for sustainable electronics

ML offers many predictive models and techniques for extracting patterns from data. It also offers reinforcement learning solutions that allow systems to respond to varying conditions efficiently. All these options support the development of a wide range of applications that promote sustainability in the electronics industry.

Predictive maintenance

Predictive maintenance of electronics is a game-changer for sustainability. By leveraging historical performance data and real-time sensor information, predictive maintenance models can forecast potential failures in electronic components before they occur. Data range from vibration data and acoustic signals to electrical signals and time-series sensor temperature measurements and other operating conditions. The most popular ML techniques applied in predictive maintenance for various equipment types include linear regression, Gaussian process regression, support vector machines, random forests, neural networks, k-nearest neighbors, and deep learning methods, which aim to predict the expected lifespan of electronic devices or early diagnose a faulty behavior [20]. Such proactive approaches minimize downtime, extend the lifespan of devices, and reduce the need for frequent replacements. Consequently, reducing electronic waste aligns with sustainability goals, fostering a circular economy where resources are conserved through efficient maintenance practices. This predictive capability is especially crucial in critical systems, such as those used in energy infrastructure or in medical devices, where reliability is paramount [16].

Optimize the manufacturing process

ML also is pivotal in optimizing electronics manufacturing processes, contributing to sustainability across the entire production life cycle. Through data analysis, ML algorithms identify inefficiencies and bottlenecks in manufacturing workflows [21]. This information enables proactive adjustments, streamlining processes, and minimizing resource wastage. Additionally, ML facilitates demand forecasting by analyzing market trends, allowing manufacturers to adjust production levels accordingly [22]. This adaptive manufacturing approach not only reduces excess inventory and associated waste but also ensures a more responsive and sustainable supply chain.

Enhance energy efficiency

Enhancing energy efficiency is a key focus in sustainable electronics, and ML offers innovative solutions to achieve this goal. ML algorithms analyze usage patterns, environmental conditions, and real-time data from electronic devices to optimize energy consumption [23]. This adaptive approach ensures that devices operate at their most energy-efficient levels, reducing overall power consumption and environmental impact. Whether in consumer electronics or industrial applications, the integration of ML for energy management aligns with sustainability initiatives, promoting responsible resource use and minimizing the carbon footprint of electronic systems.

Support recycling

Finally, ML is used for intelligent waste sorting and recycling of electronics. Automated systems equipped with ML algorithms can accurately identify and sort various electronic components from discarded devices [24]. This not only streamlines the recycling process but also enables the recovery of valuable materials, contributing to a more circular and sustainable approach. ML applications, such as image recognition and material categorization, enhance the efficiency of e-waste recycling facilities, ensuring that electronic components are properly recovered and repurposed, thus minimizing environmental impact and fostering responsible waste management practices [17].

Case studies and success stories

According to the International Roadmap for Devices and Systems (IRDS[1]), the research in sustainable electronics focuses on the development of ultra-high energy efficiency in the main technological building blocks of modern systems: data centers on the one side and sensors on the other. Almost zero power electronics will reduce the energy consumption of data centers, which today correspond to about 0.8% and 1% of the global electricity demand. Low-power computing paradigms such as memory-centric architectures, quantum computing, and spiking neural networks may target low energy consumption while guaranteeing high information processing efficiency [25].

Managing huge amounts of data continuously collected by sensors, transmitted to the cloud, or processed in the edge pushes the limits of current electronics, and requests for a holistic approach encompassing advancements in both hardware and algorithms.

[1] https://irds.ieee.org/editions/2023.

The opportunities to decrease the power consumption of computing units build in the simultaneous improvement across all levels of the technological ecosystem, spanning process steps, circuits, architecture, software, and algorithms. Bringing data processing operations closer to the edge is expected to minimize energy-intensive data transport. Collaboration within the electronics community, involving technology architects, circuit designers, software developers, and electronic systems providers, is crucial for pioneering breakthrough solutions. Innovative techniques such as ultra-thin film wafer bonding and 2D material growth, along with 3D technologies and advanced packaging, contribute to resource efficiency. The proposal to explore a silicon-based approach for quantum computing, incorporating low-temperature CMOS technology and error-correction codes, exemplifies a multifaceted strategy for ushering in a new era of responsible and sustainable electronics [4].

Several companies and initiatives are at the forefront of integrating ML into sustainable electronics, exemplifying successful implementations that prioritize environmental responsibility. One notable company is Google, which has leveraged ML algorithms (Graph Networks for Materials Exploration—GNoME) and DeepMind to discover new materials, including layered materials, which are promising systems for electronics and energy storage [26]. IBM and Lenovo are two other prominent players actively incorporating ML for sustainable practices. Their AI-driven solutions have enhanced the efficiency of data centers and addressed energy consumption challenges [27].

Electric vehicle companies, such as Tesla, have also embraced ML to enhance energy efficiency and sustainability. They utilize advanced algorithms for battery management and autonomous driving features, contributing to electric transportation's overall ecological footprint reduction [28].

These examples showcase a growing trend where ML is not just a technological advancement but a key enabler for sustainable practices across various sectors, emphasizing the role of innovative companies and initiatives in driving positive environmental impact.

Integrating ML in sustainable electronics yields positive environmental and economic impacts [29]. Environmentally, ML algorithms contribute significantly to energy efficiency, particularly in data centers where dynamic control algorithms optimize cooling systems. This reduces operational costs and aligns with global efforts to curb carbon emissions. Sustainable electronics, guided by ML, also emphasize resource conservation, promoting longer device life cycles and the use of eco-friendly materials. These initiatives mitigate electronic waste by optimizing manufacturing processes, contributing to a more circular economy. Additionally, ML facilitates the seamless integration of renewable energy sources into electronic systems, ensuring more efficient management and utilization of clean energy.

From an economic perspective, adopting sustainable electronic practices driven by ML leads to tangible cost savings for businesses [30]. Companies benefit from reduced operational expenses related to energy consumption and waste management, enhancing their bottom line. Moreover, embracing ML for sustainability positions businesses as innovators and market leaders. Consumers increasingly value environmentally conscious products, driving demand and creating new market opportunities for forward-thinking companies. Beyond financial gains, adopting ML for sustainability helps businesses navigate evolving environmental regulations effectively, reducing the risk of penalties and enhancing their reputation with stakeholders. This dual impact underscores the transformative potential of ML in reshaping industries toward a more sustainable and economically viable future.

Barriers and limitations

Using ML in sustainable electronics can bring many advantages but never comes without challenges and in several cases has limitations to overcome.

Data availability

The first challenge concerns the availability and quality of data, which is crucial for training any ML model. In the context of sustainable electronics, obtaining comprehensive datasets can be problematic due to several factors. Firstly, the lack of standardization in data collection processes makes compiling a unified and consistent dataset for all tasks difficult. Information on electronic devices, materials, and manufacturing processes may not be readily available or easily accessible. Additionally, the evolving nature of sustainable practices and technologies in the electronics industry means that historical datasets might not accurately represent current conditions. As sustainability initiatives progress and technologies advance, there is a constant need for updated and relevant data to train ML models effectively [31].

Regulatory and social barriers

The cost of electronics reuse is usually high and in certain instances, regulations may hinder rather than promote the objective of encouraging the reuse of discarded electrical and electronic equipment. These obstacles, impede the reuse sector businesses and require the appropriate legislative amendments to be overcome [32,33].

As ML becomes increasingly integral to sustainable electronics, ethical considerations and social impacts demand careful examination. Using algorithms in decision-making processes may inadvertently perpetuate biases or lead to unintended consequences. Striking a balance between technological advancement and ethical responsibility is crucial. Over the past decade, international trade in discarded electronics (e-waste) has raised concerns due to the hazards associated with their materials. Ethical electronics repair, reuse, repurposing, and recycling attempt to bring principles of good and right action into the electronics life cycle. The respective ethics for using ML and AI must also be considered when developing ML solutions that concern the sustainability of electronics development.

Technological and infrastructural constraints further impede the integration of ML into sustainable electronics. Limited access to advanced technologies and inadequate infrastructure in certain regions may hinder the widespread adoption of ML applications, impacting their effectiveness in promoting sustainability. Overcoming these barriers requires concerted efforts in data standardization, ethical algorithm development, and addressing technological and infrastructural disparities [34].

Future trends and innovations

The future of sustainable development in electronics is intricately tied to the continuous evolution of ML technologies. This section delves into exploring emerging technologies that hold promise for reshaping the landscape of sustainable electronics. From advancements in reinforcement learning to the integration of edge computing, understanding the potential of these emerging technologies is crucial for envisioning a future where ML plays a pivotal role in driving sustainability initiatives within the electronics industry.

Reinforcement learning (RL) stands out as a game-changer, which enables machines to make sequential decisions based on trial and error, thus opening avenues for autonomous optimization of resource usage in electronic devices. By allowing systems to learn from interactions with their environment, RL can revolutionize energy management in electronics, ensuring devices adapt dynamically to usage patterns, enhancing efficiency, and minimizing environmental impact. Concurrently, the rise of edge computing emerges as a transformative force. Edge computing involves processing data closer to the source rather than relying solely on centralized cloud systems. This reduces latency and contributes to energy efficiency, a critical factor in sustainable electronics. As devices become more capable of processing information locally, the demand for cloud resources diminishes, leading to a more eco-friendly and responsive electronic ecosystem. Additionally, the intersection of ML with quantum computing and neuromorphic engineering unveils further possibilities for advanced computing architectures to tackle complex problems in electronics sustainability. These technologies collectively mark a paradigm shift, positioning ML as a catalyst for a sustainable, technologically advanced future in the electronics industry.

As the intersection of ML and sustainable electronics matures, predicting future developments in sustainable practices becomes imperative. This section assesses the trajectory of environmentally conscious methodologies within electronics manufacturing, usage, and disposal. By analyzing ongoing research, industry trends, and global initiatives, we aim to forecast the adoption of innovative practices that align with ecological goals, creating a roadmap for integrating ML in shaping a sustainable future for the electronics sector.

Anticipation of technological advances is essential for staying ahead in the ever-evolving landscape of sustainable electronics. This section explores potential breakthroughs in materials science, energy efficiency, and manufacturing processes that could revolutionize the industry. Integrating ML algorithms with these technological advancements can enhance resource efficiency, reduce environmental impact, and propel the electronics sector toward a more sustainable and technologically advanced future. By examining current research trajectories and industry collaborations, we aim to provide insights into the transformative technologies that may define the future of sustainable development in electronics.

Integration of machine learning into circular economy models

ML is critical in analyzing complicated problems, through operationally viable algorithms, resulting in accurate solutions and decisions. In this context, ML technologies can assist in developing a circular economy model at a quicker rate than other technologies. The ability of ML to analyze vast datasets and extract meaningful insights makes it a powerful enabler for the development of circular economy models in the electronics industry. This integration can accelerate the transition toward sustainable practices by addressing key aspects of resource management and consumption [35].

ML technologies are pivotal in advancing circular economy principles by optimizing resource utilization, minimizing waste, and fostering sustainable consumption patterns. Through real-time monitoring and data analysis, ML algorithms can assess the entire life cycle of electronic products, from production to disposal. This comprehensive understanding enables the identification of inefficiencies, areas for improvement, and opportunities for circularity, contributing to the overall sustainability of the electronics supply chain.

One of the significant contributions of ML to circular economy models lies in its ability to identify opportunities for resource recovery and value creation. ML algorithms can analyze data related to product end-of-life, facilitating the identification of components suitable for recycling or repurposing. This reduces electronic waste and creates new avenues for extracting value from the reuse of discarded materials. Integrating ML into recycling processes allows the electronics industry to move toward a more sustainable and resource-efficient circular economy.

The synergy between ML and circular economy principles is particularly evident in electronics. ML algorithms can optimize the disassembly of electronic devices, and sorting components for reuse or recycling with high precision. Additionally, ML-powered predictive maintenance ensures extended product lifecycles, reducing the need for frequent replacements and minimizing the environmental impact of electronic waste. The incorporation of ML into circular economy practices thus emerges as a strategic approach to enhance the sustainability of the electronics industry, aligning with global efforts toward environmental conservation. By leveraging the capabilities of ML, organizations can move beyond traditional linear economic models, embracing circularity, resource efficiency, and sustainability. As technology continues to evolve, the collaboration between ML and circular economy principles promises to create a more resilient and environmentally responsible electronics ecosystem.

Conclusion

This chapter explored the intersection between ML and sustainable development in electronics. In this direction, it verified the foundational role of ML in optimizing manufacturing processes, enhancing energy efficiency, and supporting recycling practices, which showcases versatility in addressing challenges within the electronics industry. The successful application of ML in predictive maintenance and its pivotal role in fostering a circular economy model further underscores its transformative impact on sustainable development.

Looking ahead, the future of the integration of ML into sustainable development in electronics is promising. As technology continues to advance, there is a clear trajectory toward increasingly sophisticated ML applications in addressing the industry's evolving challenges. The identified barriers and limitations, such as data availability and regulatory challenges, serve as focal points for future research and innovation. By overcoming these hurdles, the electronics sector can harness the full potential of ML to not only meet current sustainability goals but also to set new standards for environmental management.

In conclusion, the convergence of ML and sustainable development in electronics marks a paradigm shift in industry practices. The way toward a more sustainable future depends on the continued collaboration between technological innovation and environmental responsibility. As organizations strive to integrate ML into their operations, they pave the way for a resilient, resource-efficient, and environmentally conscious electronic ecosystem that aligns with the broader global agenda for sustainable development.

AI disclosure

During the preparation of this work, the author(s) used ChatGPT in order to improve the use of English and the readability of the text. After using this tool/service, the author(s) reviewed and edited the content as needed and take(s) full responsibility for the content of the publication.

References

[1] K.K. Hiran, D. Khazanchi, A.K. Vyas, S. Padmanaban, Machine Learning for Sustainable Development, De Gruyter, India, 2021, pp. 1–201. Available from https://www.degruyter.com/document/doi/10.1515/9783110702514/html?langen, https://doi.org/10.1515/9783110702514.

[2] V. Bhatnagar, S. Sharma, A. Bhatnagar, L. Kumar, Role of machine learning in sustainable engineering: a review, IOP Conference Series: Materials Science and Engineering 1099 (1) (2021) 012036. Available from https://doi.org/10.1088/1757-899x/1099/1/012036.

[3] P. Ramya, R. V, B.R. M, E-waste management using hybrid optimization-enabled deep learning in IoT-cloud platform, Advances in Engineering Software 176 (2023). Available from https://doi.org/10.1016/j.advengsoft.2022.103353.

[4] J.R. Lequepeys, M. Duranton, S. Bonnetier, S. Catrou, R. Fournel, T. Ernst, et al., Overcoming the data deluge challenges with greener electronics, in: ESSCIRC 2021 - IEEE 47th European Solid State Circuits Conference, Proceedings, Institute of Electrical and Electronics Engineers Inc., France, 2021, pp. 7–14. Available from http://ieeexplore.ieee.org/xpl/mostRecentIssue.jsp?punumber=9567713, https://doi.org/10.1109/ESSCIRC53450.2021.9567836.

[5] R.R. Menon, V. Ravi, An analysis of barriers affecting implementation of sustainable supply chain management in electronics industry: a Grey-DEMATEL approach, Journal of Modelling in Management 17 (4) (2022) 1319–1350. Available from https://doi.org/10.1108/JM2-02-2021-0042, http://www.emeraldgrouppublishing.com/jm2.htm.

[6] R. Seidel, A. Mayr, F. Schafer, D. Kisskalt, J. Franke, Towards a smart electronics production using machine learning techniques, in: Proceedings of the International Spring Seminar on Electronics Technology 2019-, IEEE Computer Society, Germany, 2019. Available from http://ieeexplore.ieee.org/xpl/conferences.jsp, https://doi.org/10.1109/ISSE.2019.8810176.

[7] C. Duan, J.P. Janet, F. Liu, A. Nandy, H.J. Kulik, Learning from failure: predicting electronic structure calculation outcomes with machine learning models, Journal of Chemical Theory and Computation 15 (4) (2019) 2331–2345. Available from https://doi.org/10.1021/acs.jctc.9b00057, http://pubs.acs.org/journal/jctcce.

[8] P. Chaudhary, A. Sinha, S. Jain, Advanced computational techniques for sustainable computing, Advanced Computational Techniques for Sustainable Computing (2022) 1–322. Available from https://doi.org/10.1201/9781003046431.

[9] N. Asadizanjani, M. Tehranipoor, D. Forte, Counterfeit electronics detection using image processing and machine learning, Journal of Physics: Conference Series 787 (1) (2017) 012023. Available from https://doi.org/10.1088/1742-6596/787/1/012023.

[10] F.M. Assef, M.T.A. Steiner, E.Pd Lima, A review of clustering techniques for waste management, Heliyon 8 (1) (2022). Available from https://doi.org/10.1016/j.heliyon.2022.e08784, http://www.journals.elsevier.com/heliyon/.

[11] T. Yang, L. Zhao, W. Li, A.Y. Zomaya, Reinforcement learning in sustainable energy and electric systems: a survey, Annual Reviews in Control 49 (2020) 145–163. Available from https://doi.org/10.1016/j.arcontrol.2020.03.001, https://www.journals.elsevier.com/annual-reviews-in-control.

[12] B.G. Kim, Y. Zhang, M. Van Der Schaar, J.W. Lee, Dynamic pricing and energy consumption scheduling with reinforcement learning, IEEE Transactions on Smart Grid 7 (5) (2016) 2187–2198. Available from https://doi.org/10.1109/TSG.2015.2495145.

[13] C.B. Kristensen, F.A. Sørensen, H.B. Nielsen, M.S. Andersen, S.P. Bendtsen, S. Bøgh, Towards a robot simulation framework for E-waste disassembly using reinforcement learning, Procedia Manufacturing 38 (2019) 225–232. Available from https://doi.org/10.1016/j.promfg.2020.01.030, http://www.journals.elsevier.com/procedia-manufacturing.
[14] K. Mathiyazhagan, A. Gnanavelbabu, N.K. N, V. Agarwal, A framework for implementing sustainable lean manufacturing in the electrical and electronics component manufacturing industry: an emerging economies country perspective, Journal of Cleaner Production 334 (2022) 130169. Available from https://doi.org/10.1016/j.jclepro.2021.130169.
[15] A.A.R.M.A. Ebayyeh, A. Mousavi, A review and analysis of automatic optical inspection and quality monitoring methods in electronics industry, IEEE Access 8 (2020). Available from https://doi.org/10.1109/access.2020.3029127.
[16] T. Zonta, C.A. da Costa, R. da Rosa Righi, M.J. de Lima, E.S. da Trindade, G.P. Li, Predictive maintenance in the Industry 4.0: a systematic literature review, Computers and Industrial Engineering 150 (2020). Available from https://doi.org/10.1016/j.cie.2020.106889, https://www.journals.elsevier.com/computers-and-industrial-engineering.
[17] H. Li, Z. Jin, S. Krishnamoorthy, E-waste management using machine learning, in: ACM International Conference Proceeding Series, Association for Computing Machinery, China, 2021, pp. 30–35. Available from http://portal.acm.org/, https://doi.org/10.1145/3469968.3469973.
[18] O. Osibanjo, I.C. Nnorom, G.U. Adie, M.B. Ogundiran, A.A. Adeyi, Global Management of Electronic Wastes: Challenges Facing Developing and Economy-in-Transition Countries Metal Sustainability: Global Challenges, Consequences, and Prospects, Wiley Blackwell, Nigeria, 2016, pp. 52–84. Available from http://onlinelibrary.wiley.com/book/10.1002/9781119009115, https://doi.org/10.1002/9781119009115.ch3.
[19] G. Fettweis, E. Zimmermann, ICT energy consumption-trends and challenges, in: Proceedings of the 11th International Symposium on Wireless Personal Multimedia Communications, Finland, 2008, Vol. 2, No. 4, p. 6.
[20] T.P. Carvalho, F.A.A.M.N. Soares, R. Vita, Rd.P. Francisco, J.P. Basto, S.G.S. Alcalá, A systematic literature review of machine learning methods applied to predictive maintenance, Computers and Industrial Engineering 137 (2019). Available from https://doi.org/10.1016/j.cie.2019.106024, https://www.journals.elsevier.com/computers-and-industrial-engineering.
[21] Y. Li, S. Carabelli, E. Fadda, D. Manerba, R. Tadei, O. Terzo, Machine learning and optimization for production rescheduling in Industry 4.0, The International Journal of Advanced Manufacturing Technology 110 (9-10) (2020) 2445–2463. Available from https://doi.org/10.1007/s00170-020-05850-5.
[22] C.F. Chien, Y.S. Lin, S.K. Lin, Deep reinforcement learning for selecting demand forecast models to empower Industry 3.5 and an empirical study for a semiconductor component distributor, International Journal of Production Research 58 (9) (2020) 2784–2804. Available from https://doi.org/10.1080/00207543.2020.1733125, http://www.tandfonline.com/toc/tprs20/current.
[23] A. Alsalemi, Y. Himeur, F. Bensaali, A. Amira, C. Sardianos, C. Chronis, et al., A micro-moment system for domestic energy efficiency analysis, IEEE Systems Journal 15 (1) (2021) 1256–1263. Available from https://doi.org/10.1109/JSYST.2020.2997773, https://ieeexplore.ieee.org/xpl/mostRecentIssue.jsp?punumber4267003.
[24] A. Senthilselvi, V. Sellam, S.A. Alahmari, S. Rajeyyagari, Accuracy enhancement in mobile phone recycling process using machine learning technique and MEPH process, Environmental Technology and Innovation 20 (2020) 101137. Available from https://doi.org/10.1016/j.eti.2020.101137.
[25] Y. Zhang, K. Shan, X. Li, H. Li, S. Wang, Research and technologies for next-generation high-temperature data centers – state-of-the-arts and future perspectives, Renewable and Sustainable Energy Reviews 171 (2023). Available from https://doi.org/10.1016/j.rser.2022.112991.
[26] A. Merchant, S. Batzner, S.S. Schoenholz, M. Aykol, G. Cheon, E.D. Cubuk, Scaling deep learning for materials discovery, Nature 624 (7990) (2023) 80–85. Available from https://doi.org/10.1038/s41586-023-06735-9, https://www.nature.com/nature/.
[27] L. Brochard, V. Kamath, J. Corbalán, S. Holland, W. Mittelbach, M. Ott, Energy-Efficient Computing and Data Centers, John Wiley & Sons, 2019, p. 2019.
[28] M. Ahmed, Y. Zheng, A. Amine, H. Fathiannasab, Z. Chen, The role of artificial intelligence in the mass adoption of electric vehicles, Joule 5 (9) (2021) 2296–2322. Available from https://doi.org/10.1016/j.joule.2021.07.012.
[29] T. Ahmad, D. Zhang, C. Huang, H. Zhang, N. Dai, Y. Song, et al., Artificial intelligence in sustainable energy industry: status Quo, challenges and opportunities, Journal of Cleaner Production 289 (2021) 125834. Available from https://doi.org/10.1016/j.jclepro.2021.125834.

[30] R. Pillai, B. Sivathanu, M. Mariani, N.P. Rana, B. Yang, Y.K. Dwivedi, Adoption of AI-empowered industrial robots in auto component manufacturing companies, Production Planning and Control 33 (16) (2022) 1517–1533. Available from https://doi.org/10.1080/09537287.2021.1882689, http://www.tandf.co.uk/journals/titles/09537287.asp.

[31] M. Matthess, S. Kunkel, B. Xue, G. Beier, G, Supplier sustainability assessment in the age of Industry 4.0–Insights from the electronics industry, Cleaner logistics and supply chain 4 (2022) 100038.

[32] N. Milovantseva, C. Fitzpatrick, Barriers to electronics reuse of transboundary e-waste shipment regulations: an evaluation based on industry experiences, Resources, Conservation and Recycling 102 (2015) 170–177. Available from https://doi.org/10.1016/j.resconrec.2015.07.027.

[33] S. Svensson-Hoglund, J.L. Richter, E. Maitre-Ekern, J.D. Russell, T. Pihlajarinne, C. Dalhammar, Barriers, enablers and market governance: a review of the policy landscape for repair of consumer electronics in the EU and the U.S, Journal of Cleaner Production 288 (2021). Available from https://doi.org/10.1016/j.jclepro.2020.125488, https://www.journals.elsevier.com/journal-of-cleaner-production.

[34] R. Evans, W.J.V. Vermeulen, Governing electronics sustainability: meta-evaluation of explanatory factors influencing modes of governance applied in the electronics value chain, Journal of Cleaner Production 278 (2021) 122952. Available from https://doi.org/10.1016/j.jclepro.2020.122952.

[35] P. Bhattacharjee, I. Howlader, M.A. Rahman, H.M.M. Taqi, M.T. Hasan, S.M. Ali, et al., Critical success factors for circular economy in the waste electrical and electronic equipment sector in an emerging economy: implications for stakeholders, Journal of Cleaner Production 401 (2023). Available from https://doi.org/10.1016/j.jclepro.2023.136767.

CHAPTER

17

Automated sorting of recyclable domestic waste

Jaehwan Kim, Muhammad Latif, Jongmin Song, Giseok Park, and Jieun Lee

Creative Research Center for Nanocellulose Future Composites, Department of Mechanical Engineering, Inha University, Incheon, Republic of Korea

Introduction

Nowadays, when waste is collected, it is separated into recyclables and nonrecyclables. The separation has the effect of increasing the recycling rate of waste and reducing the amount of time spent separating recyclable waste, which can save a lot of money. However, even if the waste is collected separately into recyclable and nonrecyclable waste, recyclable waste is still mixed with various types of waste, and it cannot be recycled unless it is separated into homogeneous recyclable waste, so it must be separated into homogeneous recyclable waste. Common recyclable materials include paper and cardboard, plastics, glass, metals, textiles, electronics, organic waste, tires and batteries.

In the past, such sorting was done by installing a long conveyor, dividing the area according to the type of recyclable waste to be sorted, assigning people to each area, and transferring the collected recyclable waste to the conveyor while manually sorting and separating the selected recyclable waste in each area. As a result, an unnecessary workforce is wasted for manual reclassification of recyclable waste, and the economic efficiency of the workforce is in question. Efficient sorting of these materials is very important for recycling them. This chapter explains how we can sort these wastes automatically. There are several methods and technologies for sorting recyclable waste, including mechanical sorting, optical sorting, magnetic separation, air separation, robotics, RFID technology and smart bins. A combination of these technologies may be used in an integrated sorting facility to achieve optimal results. These technologies can be optimized by using data analysis and artificial intelligence. The effectiveness of the system is also explained, including public awareness and proper labeling of recyclable waste.

Harnessing Automation and Machine Learning for Resource Recovery and Value Creation
DOI: https://doi.org/10.1016/B978-0-443-27374-2.00017-0

Recycle domestic wastes

Nowadays, when waste is collected, it is separated into recyclables and nonrecyclables. The separation has the effect of increasing the recycling rate of waste and reducing the amount of time spent separating recyclable waste, which can save a lot of money. However, even if the waste is collected separately into recyclable and nonrecyclable waste, recyclable waste is still mixed with various types of waste, and it cannot be recycled unless it is separated into homogeneous recyclable waste, so it must be separated into homogeneous recyclable waste. Common recyclable materials include paper and cardboard, plastics, glass, metals, textiles, electronics, organic waste, tires and batteries.

Types of wastes

The rise in population, economic expansion, urbanization, and high living standards have significantly accelerated waste production worldwide, especially in developing nations, and have become universal environmental issues [1,2]. Waste is generally any material, substance, or by-product resulting from human activity that is no longer beneficial or needs to be discarded. It is mainly generated from various sources, including industries, households, agriculture, construction sites, institutional, and healthcare sectors. There are four significant types of waste: municipal solid waste (MSW), industrial waste, agricultural waste, and hazardous waste [3]. MSW is usually generated from residential or commercial areas, including households, schools, hotels, offices, and healthcare facilities. The considerable elements of MSW are food waste, plastics, paper, metals, electronics, and glasses. Industrial waste includes various materials, including packaging, oils, resins, paints, solvents, metals, ceramics, stones, rubber, leather, straws, cloth, and wood. Agricultural wastes include livestock wastes, agro-industrial by-products, and agricultural crop residues. Hazardous waste emerges as a by-product across various industrial, farming, and manufacturing processes and from nuclear facilities, hospitals, and healthcare institutions. Major producers of industrial hazardous waste include chemical, petrochemical, petroleum, metal, wood chemical treatment, pulp and paper, leather, textile, and energy production factories (such as coal-fired, nuclear, and petroleum production facilities). Domestic waste, the most critical component of MSW, refers explicitly to waste generated from households or residential areas and typically includes food waste, packaging materials, paper, plastics, rags, glass, metals, and textiles. Fig. 17.1 presents a breakdown of the types and sources of solid waste.

The MSW generation rate fluctuates from country to country and season. These variations are closely linked to each country's economic development and activity levels. In 2016, 2 billion tons of MSW were generated globally. However, the World Bank's report (What a Waste 2.0) suggests that this number will rise to 2.59 billion tons by 2030 and to 3.40 billion by 2050 [5]. As shown in Fig. 17.2, by 2030, high-income countries are projected to have the slowest growth in waste generation because they have reached a stage of economic development where material consumption is less closely tied to GDP growth. Low-income countries are subjected to substantial economic and population growth, with waste levels expected to triple by 2050 [5].

The waste materials found in MSW are categorized into different categories through a standard waste audit based on their compositions. Fig. 17.3A shows the various categories of waste compositions. The biggest waste category globally is food and green waste, which account for

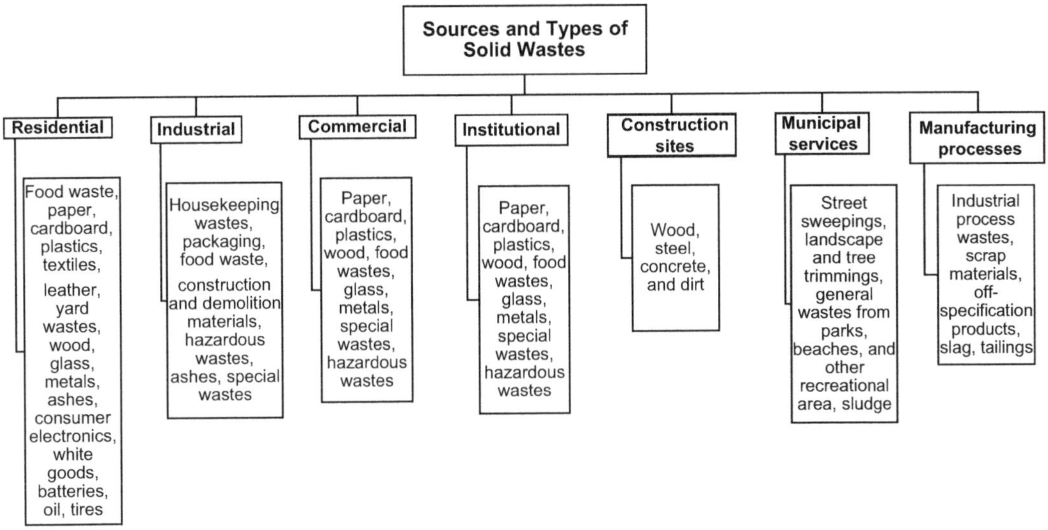

FIGURE 17.1 Sources and types of solid wastes [4]. *Source: From. S. Suthar, P. Singh, Household solid waste generation and composition in different family size and socio-economic groups: a case study, Sustainable Cities and Society 14 (1) (2015) 56–63, https://doi.org/10.1016/j.scs.2014.07.004.*

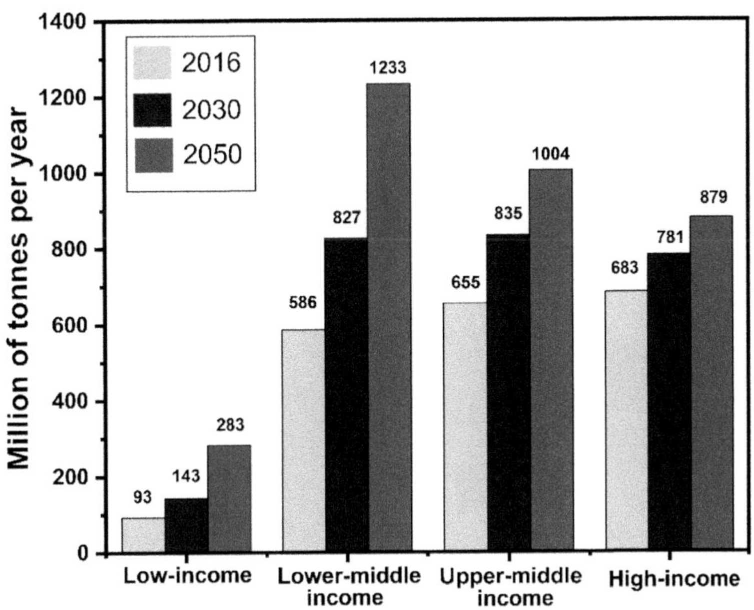

FIGURE 17.2 The projected municipal solid waste generation by country's income group. *Source: From S. Kaza, L.C. Yao, P. Bhada-Tata, F. Van Woerden, What a Waste 2.0: A Global Snapshot of Solid Waste Management to 2050, World Bank, Washington, DC, 2018, https://doi.org/10.1596/978-1-4648-1329-0.*

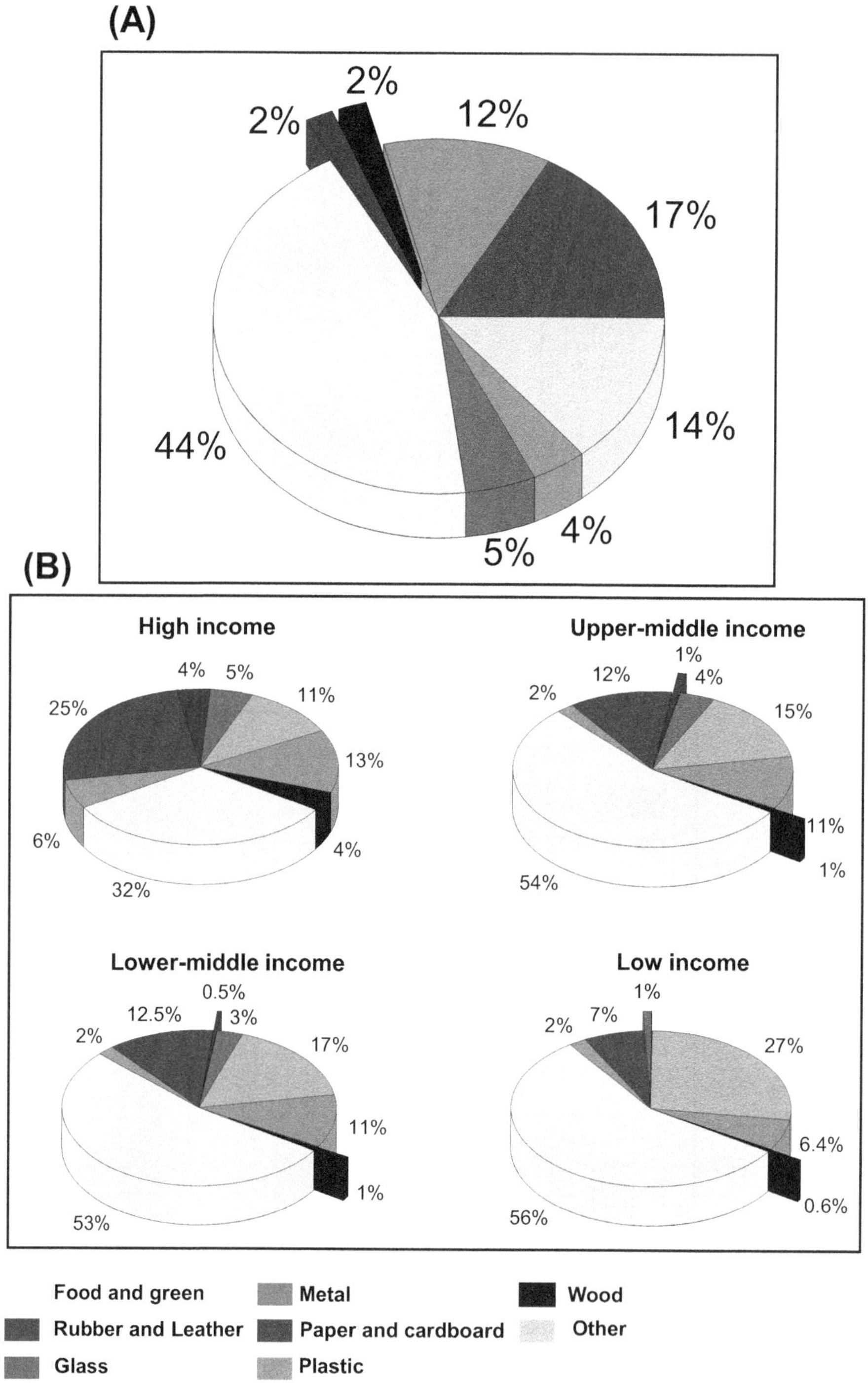

FIGURE 17.3 Waste composition (A) globally and (B) by country's income level. *Source: From S. Kaza, L.C. Yao, P. Bhada-Tata, F. Van Woerden, What a waste 2.0: a global snapshot of solid waste management to 2050, World Bank, Washington, DC, 2018, https://doi.org/10.1596/978-1-4648-1329-0.*

about 44% of the total MSW. Dry recyclables such as plastic, paper and cardboard, metal, and glass constitute an additional 38% of the total MSW. However, waste composition varies significantly depending on the income level of the countries, as shown in Fig. 17.3B. The food and green category decreases as a country's income rises and vice versa. In higher-income countries, consumed goods typically contain more materials like paper and plastic than those in lower-income countries.

Characteristics of wastes

When waste is collected, it is separated into recyclables and nonrecyclables. Separation increases the recycling rate of waste and reduces the time spent separating recyclable waste, which can save much money. However, even if the waste is collected separately into recyclable and nonrecyclable waste, recyclable waste is still mixed with various types of waste, and it cannot be recycled unless it is separated into homogeneous recyclable waste, so it must be separated into homogeneous recyclable waste. The recycling facility for the mixed MSW aims to recover about 15% of the waste stream as usable materials. However, the remaining 85% of waste primarily consists of organic matter, which can produce refuse-derived fuel and compost, which is beneficial for soil health and plant growth [6]. Common recyclable materials include paper, cardboard, plastics, glass, metals, textiles, electronics, organic waste, tires, and batteries [7]. After discounting for contaminated materials, roughly 60% of all waste, including paper, plastics, glass, metals, and organic waste, holds the potential for recycling [8]. The remaining 40% of waste is disposed of in landfills, as shown in Fig. 17.4. About 19% undergo materials recovery through recycling and composting, and 11% are treated through modern incineration. However, 33% of waste is still openly dumped around the world. Waste disposal methods vary considerably based on income and region. In lower-income countries, open dumping is prevalent due to the lack of landfill infrastructure. Approximately 93% of waste is burned or disposed of on roads, open land, or waterways in these nations. In contrast, high-income countries dumped only 2% of MSW.

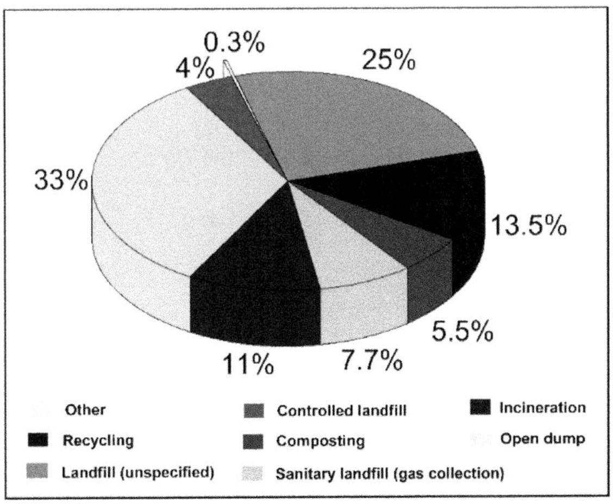

FIGURE 17.4 Global waste treatment and disposal. *Source: From S. Kaza, L.C. Yao, P. Bhada-Tata, F. Van Woerden, What a waste 2.0: a global snapshot of solid waste management to 2050, World Bank, Washington, DC, 2018, https://doi.org/10.1596/978-1-4648-1329-0.*

Automated sorting

This section explains how we can sort these wastes automatically. Several methods and technologies for sorting recyclable waste include mechanical sorting, optical sorting, magnetic separation, air separation, robotics, RFID technology, and smart bins. Combining these technologies may be used in an integrated sorting facility to achieve optimal results. These technologies can be optimized by using data analysis and artificial intelligence.

Automated waste categorization

At a time when the issue of waste is on the rise in environmental concerns, automated waste sorting systems are also in the focus. Fig. 17.5 is shown from data obtained from CEWEP and EEA. Considering the current waste situation, the waste growth rate is 7% per year. Considering the future waste reducing, a waste growth rate of 4.5% per year is depicted. Fig. 17.5 shows the information about the total waste production. The total waste is increasing with the day, and waste classification and efficient utilization are becoming a challenge for both developed and developing countries. There is a lot of interest in automated sorting of waste because if we can sort this waste efficiently and do the sorting based on characteristics automatically, we can shift to resource recycling.

Automated waste sorting methods

Mechanical sorting methods use a variety of techniques and equipment to process waste and can be categorized according to the type and nature of the waste. This waste sorting method significantly improves the efficiency of waste management. The process of mechanical sorting can be divided into three stages. The first is the initial sorting stage, where large waste articles can be sorted manually or semiautomatically. In order to facilitate the overall waste disposal

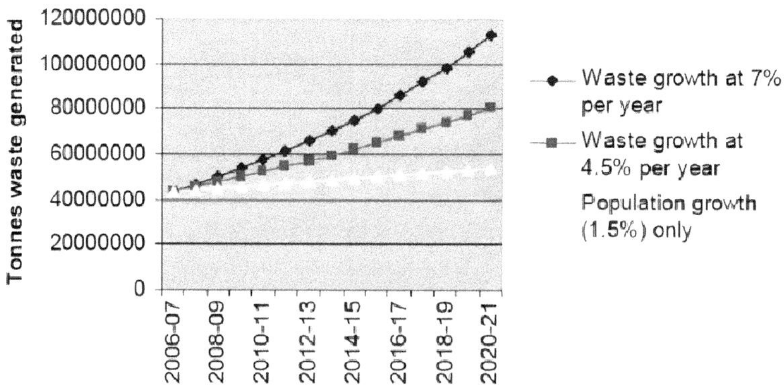

FIGURE 17.5 Comparative waste generation 2006–07 to 2020–21.

process, this step can reduce the burden on the sorting system and increase the efficiency of subsequent sorting operations by presorting large waste or nonrecyclable items in advance. Even though it is a mechanical sorting, sometimes waste needs to be precleaned before the actual process, so the first initial sorting step requires a way to size it into a suitable form using manual or semiautomated methods. Then, granular mechanical sorting uses a variety of mechanical sorting techniques to segment further the waste that has been preprocessed through initial sorting. This step involves sorting based on the physical characteristics of the waste, such as size, shape, density, and weight, using screening techniques to sort based on physical size by varying the size of the sieve's holes. A Trommel screen, also known as trommeling, is a rotating cylindrical separator that separates particles based on size. The separation is based on a rotating perforated drum with holes that gradually increase in size along the length of the cylinder. Depending on the design of the device, smaller materials may fall to the rear of the process, while larger particles may fall later. The advantages of trommel screens are that they are simple in function and require relatively little capital, as well as low operating and maintenance costs. An important parameter in trommel screen separation is the rotation speed. The material must remain in the drum for at least two minutes [9,10]. Finally, waste can be separated based on density and weight, either by vibration or by utilizing an inclined plane to sort it into fine, light, and heavy. When using a gravity sorter, heavier materials will fall to the bottom, while lighter materials will remain on top. Similarly, density separation can be based on density, which utilizes water or a reference liquid to sort waste by utilizing density differences.

In optical sorting the automated process of sorting solid products using cameras or lasers, which can recognize an object's color, size, shape, structural characteristics, and more, depending on the type of sensor used and the software in the image processing system. It consists of four main components: a feeding system, an optical system, image processing software, and a separation system. Optical sorting is used in a variety of industries, including food processing, recycling, mining, pharmaceuticals, agriculture, and more. Fig. 17.6 shows an optical sorting system [11]. 3D scanning using triangulation combines a camera attached to a conveyor belt with a laser beam to make judgments.

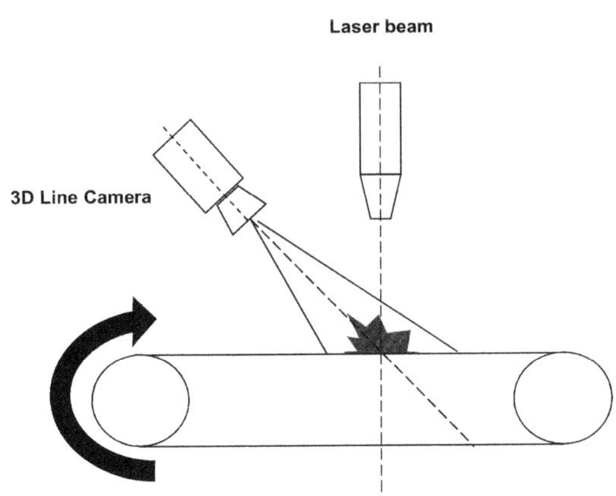

FIGURE 17.6 Installation of 3D visual sensors. *Source: From J. Huang, T. Pretz, Z. Bian, Intelligent solid waste processing using optical sensor based sorting technology, Proceedings - 2010 3rd International Congress on Image and Signal Processing, CISP 2010, vol. 4, 2010, pp. 1657–1661, https://doi.org/10.1109/CISP.2010.5647729.*

In triangulation scanning, an LED successively emits light in red, blue, and green wavelengths to identify the color of a sample by comparing it to a reference chromaticity, which attenuates the light emitted by the LED depending on the color of the sample. Secondly, UV sorting technology is used to distinguish between mixed glass scrap and opaque and other specialty glasses. Depending on the nature of the glass specimen, UV light is absorbed and weakened. The intensity of the attenuated UV light is then compared to a standardized reference value to classify.

Magnetic sorting or magnetic separation is a method of separating materials based on their magnetic properties, which utilizes the magnetic susceptibility of certain materials. Magnetic sorting is widely used to separate ferrous and nonferrous metallic materials, offering the advantages of high efficiency, low maintenance requirements, and the ability to separate materials without the need for complex equipment effectively. In addition, magnetic sorting is environmentally friendly and does not require the use of chemicals or consumables. For example, a magnetic drum measures the magnetic susceptibility of the waste to separate the ferrous portion from the nonferrous and other mixed waste portions. When mixed waste is fed, a strong magnetic flux attracts and holds ferromagnetic materials. Fig. 17.7 shows a magnetic sorting system [12]. A magnetic head pulley transports the mixed waste near the magnetic head pulley via a conveyor belt. The magnetic belt retains the ferromagnetic waste portion, and the other nonferromagnetic portion is discharged. The eddy current method uses a rotating drum-type separator. The external magnetic flux repels the nonmagnetic electrically conductive metal parts from the mixed metal waste.

Air classification is a method of separating materials based on their aerodynamic properties. It relies on the principle of how different materials react to airflow and air currents. Air classification is used to efficiently separate materials with different properties in industries such as recycling, mineral processing, agriculture, and food processing. Lighter particles are blown over a greater distance, while heavier particles are blown over a shorter distance. Air classification offers a number of advantages, including high efficiency, minimal water usage, and the ability to process materials with

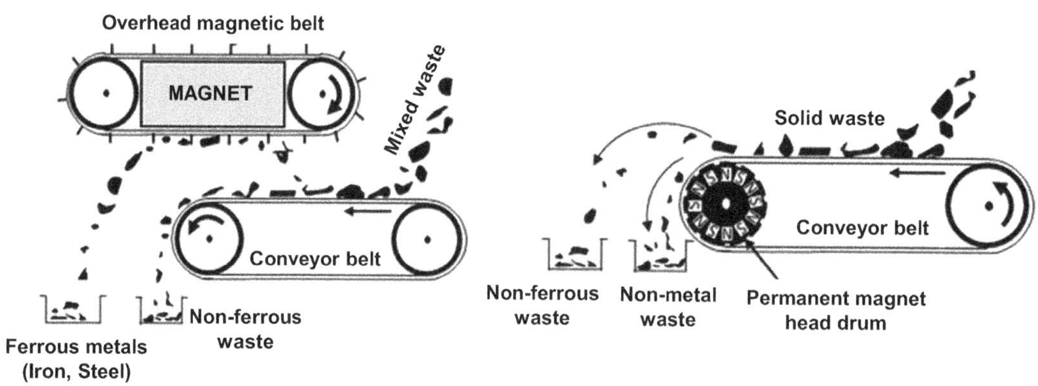

FIGURE 17.7 Magnetic overhead belt sort and eddy current sort. *Source: From S.P. Gundupalli, S. Hait, A. Thakur, A review on automated sorting of source-separated municipal solid waste for recycling, Waste Management 60 (2017) 56–74, https://doi.org/10.1016/j.wasman.2016.09.015.*

varying densities and particle sizes. It can also handle materials that are difficult to separate by other methods, such as plastics, minerals, seeds, and agricultural products. Air classification is also environmentally friendly and does not require the use of chemicals or consumables.

Robotic sorting technology combines a variety of advanced technologies, such as artificial intelligence, machine learning, and computer vision, to effectively sort waste. Robotic sorting technology has a number of advantages over manual human sorting in terms of accuracy, speed, and safety. In mechanical sorting, the products that were previously sorted by the screening trommel sorting method are moved to the robotic sorting operation via a conveyor belt [13]. Robotic sorting technology works as follows: through computer vision, robotic sorting systems identify the type and nature of waste. High-resolution cameras and sensors are used to detect and analyze various features of the waste, such as size, shape, and texture, in detail. The sorted data is processed through artificial intelligence algorithms (AI) to sort waste items by the category to which each item of waste belongs, which is then physically separated by a robotic arm. AI robotic arms can accurately sort waste items of different shapes and sizes in a range that humans cannot, and are faster and more accurate than humans, reducing waste sorting time and increasing recycling rates. In addition, human health and safety are improved by protecting humans from hazardous waste and reducing the amount of time spent working in harsh environments. With machine learning, the system is able to sort waste with the ability to improve over time through continuous learning, allowing the system to adapt to different types of waste. In addition, it can quickly adapt to changes in waste disposal regulations or new types of waste.

Smart bins utilize a variety of technologies to separate waste automatically. In the process, it sorts it based on parameters set in the sensors in advance. There is the reactive sensor technology that can sort the waste into biodegradable and nonbiodegradable waste, etc. For this method, moisture sensors can be utilized to sort wet and dry automatically. Next, there is the RLC metal detection circuit technology that can automatically sort metallic and nonmetallic waste. The sorting in this is based on the fact that the inductance of the coil increases whenever an object is placed near the coil, and if the change in inductance exceeds a threshold, the waste is sorted as metallic, and the waste is sorted towards the metallic waste through a slope in the bin. [14] Sensor-based sorting systems allow smart bins to manage waste more efficiently, increasing the recovery rate of recyclable resources, reducing recycling errors, and contributing to an improved municipal waste management system. A similar system is how smart bins incorporate IoT. Using an RFID board, you can easily build an Arduino Mega with an RFID reader on each trash can, which reads the ID inside the tag to enable remote transmission. These remote servers contain information about the amount of trash in each bin, as well as the location of the bin and the type of trash it contains [15]. In addition, we created a smart bin with an embedded system that we named the automated teller machine trash bin. The system uses image-detecting sorting sensors to sort waste that is considered trash [16]. These smart bins not only utilize a variety of sensors to sort trash automatically but also can detect fire hazards that can sometimes occur inside the bin. In addition, the automated sorting system can help narrow down the materials that need to be sorted manually.

References

[1] B. Seng, H. Kaneko, K. Hirayama, K. Katayama-Hirayama, Municipal solid waste management in Phnom Penh, capital city of Cambodia, Waste Management & Research: The Journal for a Sustainable Circular Economy 29 (2011) 491–500. Available from https://doi.org/10.1177/0734242X10380994.

[2] L.A. Guerrero, G. Maas, W. Hogland, Solid waste management challenges for cities in developing countries, Waste Management 33 (2013) 220–232. Available from https://doi.org/10.1016/j.wasman.2012.09.008.

[3] T. Noor, A. Javid, A. Hussain, S.M. Bukhari, W. Ali, M. Akmal, et al., Types, sources and management of urban wastes, Urban Ecology 2020 (2020) 239–263. Available from https://doi.org/10.1016/B978-0-12-820730-7.00014-8.

[4] Suthar S., Singh P. Household solid waste generation and composition in different family size and socio-economic groups: a case study. Sustainable Cities and Society 2015;14:56–63.

[5] S. Kaza, L.C. Yao, P. Bhada-Tata, F. Van Woerden, What a waste 2.0: a global snapshot of solid waste management to 2050, Urban Development, World Bank, Washington, DC, 2018.

[6] Williams P.T. Waste Treatment and Disposal. 1st Ed. Wiley, 2005.

[7] A. Ahsan, M. Alamgir, M.M. El-Sergany, S. Shams, M.K. Rowshon, N.N.N. Daud, Assessment of municipal solid waste management system in a developing country, Chinese Journal of Engineering 2014 (2014) 561935. Available from https://doi.org/10.1155/2014/561935.

[8] R.M. Harrison, R.E. Hester, R. Harrison, R. Hester (Eds.), Environmental and Health Impact of Solid Waste Management Activities, The Royal Society of Chemistry, 2002, p. 18.

[9] S.T. Lau, W.H. Cheung, C.K. Kwong, C.P. Wan, K.K.H. Choy, C.C. Leung, et al., Removal of Batteries from Solid Waste Using Trommel Separation, Elsevier, 2005.

[10] Y.S. Chen, S.S. Hsiau, H.Y. Lee, Y.P. Chyou, C.J. Hsu, Size Separation of Particulates in a Trommel Screen System, Elsevier, 2010.

[11] Huang J., Pretz T., Bian Z. Intelligent Solid Waste Processing Using Optical Sensor Based Sorting Technology. IEEE 2010.

[12] S.P. Gundupalli, S. Hait, A. Thakur, A review on automated sorting of source-separated municipal solid waste for recycling, Waste Management 60 (2017) 56–74. Available from https://doi.org/10.1016/j.wasman.2016.09.015.

[13] M. Koskinopoulou, F. Raptopoulos, G. Papadopoulos, N. Mavrakis, M. Maniadakis, Robotic waste sorting technology: toward a vision-based categorization system for the industrial robotic separation of recyclable waste, IEEE Robotics & Automation Magazine 28 (2021) 50–60. Available from https://doi.org/10.1109/MRA.2021.3066040.

[14] J. Sakshi, L. Hardika, A. Ishika, G.M. Lodha, Survey paper on 'Smart Dustbin Management System', Journal of Advancement in Software Engineering and Testing 2 (2019) 1–4. Available from https://doi.org/10.5281/zenodo.3381771.

[15] A. Suvarnamma, J.A. Pradeepkiran, Smart Bin system with waste tracking and sorting mechanism using IoT, Cleaner Engineering and Technology 5 (2021) 100348. Available from https://doi.org/10.1016/j.clet.2021.100348.

[16] Sunny M.S.H., Dipta D.R., Hossain S., Faruque H.M.R., Hossain E. Design of a convolutional neural network based smart waste disposal system. 2019 1st International Conference on Advances in Science, Engineering and Robotics Technology (ICASERT), Dhaka, Bangladesh, 2019.

CHAPTER

18

Intelligent waste management: a comprehensive review of machine learning and deep learning applications in advanced recycling

Adiba Tabassum Chowdhury[1], Muhammad E.H. Chowdhury[2], Md. Sakib Bin Islam[2], Mehrin Newaz[1], Abdus Salam[3], Faizul Sayem[1], Mazhar Hasan-Mia[4], Mohammad Nashbat[4], Molla Ehsanul Majid[5], Amith Khandakar[2], Saad Bin Abul Kashem[6], and Azad Ashraf[4]

[1]Department of Electrical and Electronic Engineering, University of Dhaka, Dhaka, Bangladesh [2]Department of Electrical Engineering, Qatar University, Doha, Qatar [3]Department of Electrical and Computer Engineering, Rajshahi University of Engineering and Technology, Rajshahi, Bangladesh [4]Department of Chemical Engineering, University of Doha for Science and Technology, Doha, Qatar [5]Department of Computer Applications, Academic Bridge Program, Qatar Foundation, Doha, Qatar [6]Department of Computing Science, AFG College with the University of Aberdeen, Doha, Qatar

Introduction

A poetic voyage is embarked upon in the symphony of waste handling, where the cutting edge is blended with time-honored traditions. Here, amidst the ever-changing trash scene, an urgent need is felt, pushing us to the forefront of creativity. As the symbiotic compilation of intelligence and technology is observed through the prism of Ref. [1], the fundamentals of waste management are redefined. A well-organized story unfolds as the journey is undertaken, guiding us through the halls of history, development, and the infinite possibilities of intelligent waste management. The discussion progresses below as the domains of understanding and learning are explored.

Harnessing Automation and Machine Learning for Resource Recovery and Value Creation
DOI: https://doi.org/10.1016/B978-0-443-27374-2.00018-2

Background and context of traditional waste management practices

Against the backdrop of mass consumerism and industrialization's explosion, the 19th century revealed a tapestry laced with threads of economic upheaval and material deprivation. During these turbulent times, recycling—which had its roots deepened, especially during the turmoil of World Wars I and II—emerged out of need as a practical solution to resource shortages and economic crises. The necessity of conservation sprang from the furnace of war, when countries had to deal with the harsh reality of shortage and rationing and had to come up with creative ways to manage their resources [2].

The 20th century marked a pivotal shift in attitudes toward waste as environmental awareness rose. Recycling evolved from a necessity to a symbol of ethical stewardship and a reflection of socioeconomic culture [3]. However, the effectiveness of current approaches is still hampered by their reliance on manual procedures and human interaction, even with the progress achieved in mainstreaming recycling activities [4]. The need for sophisticated solutions that can automate a large pile of trash detection according to types, as shown in Fig. 18.1, and recycling operations are becoming more important due to rising garbage and efficiency needs. Advances in technology and sustainability drive research into innovative waste management, with machine learning (ML) and neural networks (NNs) poised to revolutionize efficiency and environmental stewardship [5].

Evolving landscape: the need for advanced solutions

Waste management is evolving and requires cutting-edge solutions. Traditional methods, once efficient, are now struggling to meet industrialization, population growth, and urbanization. The current emphasis on sustainability and environmental stewardship has prompted a reassessment of traditional methods [6]. Given the looming threat of climate change, proactive resource recovery and waste reduction are crucial for mitigating its impacts. Various waste management techniques, including incineration, recycling, and landfilling, are frequently used

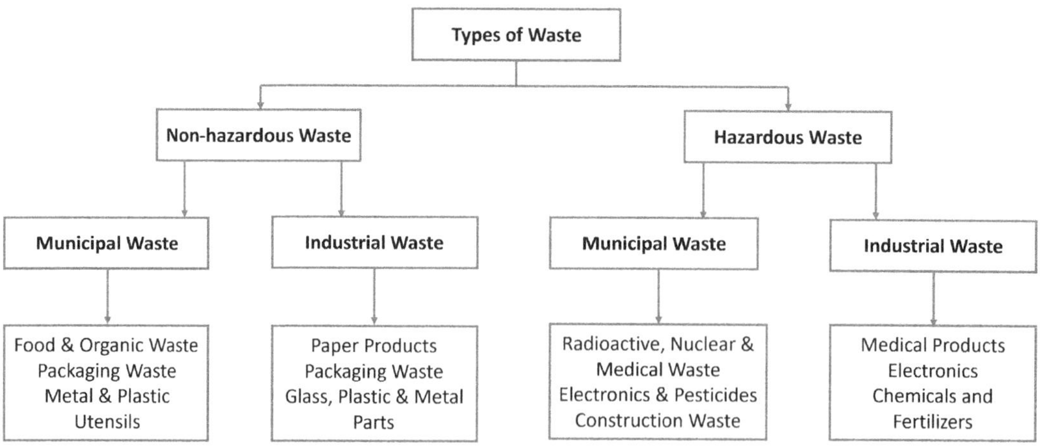

FIGURE 18.1 Types of waste to be detected by waste management system.

in these systems. Collection services aim for proper disposal through scheduled pickups, categorizing waste into organic, recyclables, and general waste for efficient processing.

Despite these efforts, escalating waste complexity underscores the need for advanced management solutions [7]. Considering this, there is a greater need than ever for creative solutions. Cutting-edge technologies like artificial intelligence (AI), NN, and ML offer promising approaches to tackle modern waste management challenges. These technologies improve trash sorting, processing, and recycling efficiency and inform decision-makers on resource allocation and environmental impact [8]. A comprehensive strategy that combines technical innovation with community involvement, legislative changes, and sustainable behaviors is essential while navigating this changing terrain. For advanced solutions, beyond technology, fostering environmental consciousness and cross-sector collaboration is very essential. By achieving this, we can create a more sustainable, fair, and resilient future for future generations [9].

Unveiling machine learning: revolutionizing waste management

ML revolutionizes trash management by transforming our approach and understanding of waste. Its algorithms and predictive capabilities unlock new possibilities, analyzing vast datasets on waste composition, generation rates, and disposal trends to reveal insights and address inefficiencies. Like industrious students, these algorithms are always learning and adapting, which allows them to optimize waste management procedures with previously unheard-of accuracy [10].

The impact of ML can be seen in every aspect of trash management, from recycling to collection. Thanks to predictive analytics, waste collection services can more precisely predict demand, which also helps them plan routes and timetables to reduce operating expenses and their negative effects on the environment [11]. ML algorithms transform sorting procedures in recycling facilities by accurately identifying recyclable items with never-before-seen precision. Furthermore, by constantly modifying parameters in response to real-time data inputs, ML improves energy conversion efficiency in waste-to-energy facilities [12].

ML enhances garbage management by fostering sustainability and efficiency, facilitating a shift toward a circular economy by identifying waste reduction and diversion opportunities, thereby minimizing waste and conserving resources. Additionally, ML reduces the environmental impact of trash disposal by streamlining waste management procedures, opening the door to a future that is both greener and more robust [13]. ML in trash management is a paradigm shift that boosts efficiency and creativity in a changing industry. The possibilities and uses of ML show that it is more than just a tool—it is a lighthouse pointing the way to a sustainable and wealthy future [14].

Navigating neural networks: a deep dive into smart solutions

NNs shine as innovative lights in the waste management technology frontier, helping us navigate the maze of obstacles. These sophisticated systems, which draw inspiration from the intricacy of the human brain, are remarkably capable of learning, adapting, and optimizing solutions on their own [15]. NNs are pivotal in waste management, processing vast data to identify patterns, predict trends, and enhance decision-making. Acting as virtual brains, they analyze sensor, camera, and database inputs, orchestrating efficient sorting, recycling, and resource allocation. By leveraging deep learning and algorithms, NNs discern waste stream complexities, identifying recyclables with unprecedented accuracy and speed [16].

NNs enable waste management systems to adapt in real time to shifting demands and conditions, surpassing basic categorization tasks. By leveraging NNs, professionals can enhance environmental stewardship, optimizing collection routes, minimizing landfill use, and maximizing resource recovery, heralding a new era in waste management [17]. We uncover a tapestry of possibilities as we thread through the complex pathways of NNs in trash management, from predictive models that foretell future patterns in garbage creation to smart bins that automatically sort recyclables.

Defining objectives: the scope and purpose of the review

Amidst extensive literature on waste management technology, this section aims to clarify our study's goals and scope, combining rigorous analysis with creative inquiry to uncover our project's essence. Our primary objective is to elucidate the transformative and practical impacts of applying ML and NNs in waste management, guiding readers through their complex applications [18]. Our goal is to assess how these advanced technologies contribute to reducing waste generation, optimizing recycling processes, and minimizing environmental harm, with a focus on sustainability and efficiency.

Additionally, by providing insights into the possibilities and constraints faced in the real world when adopting intelligent waste management technologies [19], this study seeks to close the knowledge gap between theory and practice. We want to give stakeholders, policymakers, and academics a comprehensive grasp of this emerging subject's present state-of-the-art and future directions by integrating existing literature and critically analyzing empirical findings.

Our project is driven by an urgent need for innovative solutions to the escalating global waste crisis. By detailing the benefits and potential pitfalls of ML and NNs in waste management, we aim to empower decision-makers with the knowledge to enact positive changes toward a more sustainable and resilient future. The following sections delineate the domains of ML in waste management, NNs in waste management, advanced recycling techniques, decision-making in waste management, integration of ML, NNs in waste decision system, and challenges in this track.

Machine learning in waste management

ML has become instrumental in revolutionizing waste management practices worldwide. By leveraging advanced algorithms to analyze extensive datasets, ML offers opportunities to optimize processes across the waste lifecycle. This section provides an overview of ML applications in waste management, highlighting its role in enhancing efficiency, accuracy, and sustainability throughout the waste management chain.

Unpacking machine learning: concepts and algorithms

Within the realm of waste management, the integration of ML necessitates a comprehensive understanding of its underlying concepts and algorithms. This subsection elucidates these foundational aspects, providing a scholarly exploration of the fundamental principles governing ML methodologies. By establishing this groundwork, we aim to facilitate a more nuanced examination of ML applications in waste management, thereby enhancing our comprehension of its efficacy in addressing the multifaceted challenges within the domain.

Various ML techniques, including support vector regression (SVR), random forest (RF), artificial neural networks (ANN), K-nearest neighbors (KNN), extra gradient boosting (XGBoosting), light gradient boosting machine (LightGBM), extra trees (ETS), among others, have been utilized in waste generation prediction [20]. These algorithms have unique strengths and limitations. For instance, LightGBM is fast, memory-efficient, and performs well with large datasets, but may overfit smaller datasets. KNN is simple and interpretable yet struggles with scaling and is outlier-sensitive [21]. SVR excels in high-dimensional spaces and memory efficiency but struggles to scale with large datasets. ETS offers remarkable speed over RF but can face higher variance from its random splits [22].

In Table 18.1, a comprehensive overview of the ML and deep learning (DL) models employed in the reviewed papers is provided. In conclusion, exploring machine learning principles and algorithms in this subsection provides a foundational understanding vital for further discussions. Examining supervised and unsupervised learning and algorithms like decision trees and NNs sets the stage for grasping ML's broader applications.

Time travel through data: historical applications in waste management

This section initiates a retrospective exploration of waste management practices. This investigation traverses' historical contexts to unveil the progression of methodologies, strategies, and technological interventions employed in addressing waste-related challenges. By scrutinizing past applications and approaches, this section offers invaluable insights into the evolutionary trajectory of waste management practices, laying the foundation for examining contemporary and prospective advancements in the field.

In the 19th century, recycling arose due to material scarcity and economic upheavals, notably during the First and Second World Wars [34]. Environmental awareness surged in the 20th century, with recycling becoming integral to global socioeconomic practices [35]. Nonetheless, contemporary recycling relies heavily on human intervention, highlighting the need for intelligent systems to automate processes and enhance waste detection [36].

TABLE 18.1 Proposed machine learning (ML) and deep learning (DL) models for smart waste management.

Article	Model	Approach
Ahmed et al. [3]	Deep learning time series forecasting (DLTSF)	Long short-term memory (LSTM) architecture in conjunction with a DLTSF network was employed for forecasting Al-basaten MSW. This model underwent training, validation, and testing phases using a real-world dataset comprising sorted and weighted waste data from individual zones.
Ali et al. [7]	Linear regression	Employed as a function algorithm to evaluate the correlation coefficient between variables and to predict the performance of Municipal Solid Waste Management (MSWM).
	Multilayer perceptron neural network	Trained using training data obtained from the process network synthesis (PNS) model and evaluated using testing data.
Bircanoglu et al. [2]	RecycleNet	RecycleNet utilizes deep convolutional neural network architectures such as Inception-Resnet and Inception-v4. These architectures are trained without pretrained weights.
Cha et al. [23]	Artificial neural net (ANN)	Discusses the selection of the artificial neural network (multilayer perceptron) (ANN [MLP]) algorithm as one of the predictive models developed.
Cheema et al. [5]	Visual geometry group with 16 layers (VGG16)	The VGG16 algorithm is chosen due to its suitability in terms of computation and accuracy for object detection and classification processes.
Cubillos [24]	LSTM neural network	The paper utilizes a multisite LSTM neural network to predict household waste generation rates using a long-term database. The model is applied to historical weekly waste weight data from households in Herning, Denmark, spanning from 2011 to 2018.
Kang et al. [25]	Resnet-34	The Resnet-34 architecture, which undergoes further optimization through three key aspects. These optimizations encompass the multifeature fusion of input images, the feature reuse within residual units, and the introduction of a novel activation function.
Rutqvist et al. [26]	Artificial neural network (ANN)	The methods under investigation comprise the current manually designed model and its adapted version, alongside conventional machine learning algorithms.
Kumar et al. [27]	YOLOv3	The YOLOv3 algorithm implemented within the Darknet neural network framework trains a dataset aimed at waste segregation.
Sirawattananon et al. [28]	Resnet-50	A robotic automation system is suggested, leveraging deep learning techniques to facilitate accurate waste separation within recycling categories. The ResNet-50 architecture has been employed for waste classification in this context.
Gómez et al. [29]	Convolutional neural networks (CNNs)	Object detection and recognition tasks.
	Generative adversarial networks (GANs)	Data synthesis and augmentation.

(*Continued*)

Table 18.1 (Continued)

Article	Model	Approach
Hossen et al. [30]	GCDN-net	The authors present a multistage machine learning strategy employing the GCDN-Net model, trained on a dataset comprising 25,000 images, to precisely categorize various types of garbage.
Jin et al. [31]	MobileNetV2	Incorporating an enhanced MobileNetV2 model featuring an attention mechanism and transfer learning showcases the utilization of sophisticated ML principles in tackling the complexities associated with garbage detection and classification.
Zeng et al. [32]	PublicGarbageNet	The paper employs CNNs for garbage classification, utilizing multitask learning to classify both major and subcategories simultaneously. Optimization techniques, including backbone selection, data augmentation, learning rate optimization, and label smoothing, are utilized to improve model performance.
Ziouzios et al. [33]	CNN	The paper introduces a novel methodology using CNN to identify common waste materials in real-time on moving belts in waste collection facilities.

Traditional methods prioritize disposal over waste reduction and resource efficiency, leading to challenges like environmental pollution and hazards.

Conventional waste management methodologies frequently depend on manual sorting and disposal techniques, which can result in inefficiencies and environmental implications. Given the escalating rates of urbanization and waste production, there exists a requisite for sophisticated systems that harness technological advancements [37].

The retrospective analysis of historical applications in waste management underscores the evolution of methodologies and technologies utilized to tackle waste-related challenges. By tracing the trajectory of past practices, this section illuminates the advancements made over time and provides valuable lessons for contemporary and future endeavors in waste management. Through this exploration, we gain a deeper understanding of the complexities inherent in waste management processes and pave the way for informed decision-making and innovative solutions in addressing current and emerging waste management issues.

Riding the wave: recent advances and trends in machine learning for waste management

Recent applications of ML algorithms across a spectrum of disciplines have demonstrated promising outcomes. Noteworthy examples include the utilization of ML algorithms for predicting tropospheric ozone levels in semiarid climate regions [38], identifying visuospatial disorders in dementia patients through virtual reality [39], and forecasting the stability of hard rock pillars [40].

These instances underscore the adaptability and efficacy of ML algorithms in addressing diverse predictive challenges.

In practical contexts, ML algorithms have practical utility in estimating municipal waste (MW) quantities and enhancing prediction accuracy. Gopalan et al. [41] used an ensemble voting regression algorithm, combining ML techniques like RF, adaptive boosting, and gradient boosting machines, outperformed individual models, offering precise MW generation predictions. A study by Yogeashwaran [13] utilized DL models, including VGG16, MobileNet, Xception, and Custom CNN, to develop an end-to-end waste classification application. The results showed the potential of ML in improving waste management and implementing automated garbage disposal systems. These findings highlight ML algorithms' potential as valuable tools for practitioners and decision-makers in waste management planning and execution.

Recent ML applications emphasize ML algorithms' integration with Internet of Things (IoT) in waste management [42]. For instance, a cost-effective IoT sensor node architecture was proposed for waste management in smart cities, utilizing ML to optimize waste collection processes [43]. Additionally, ML-IoT fusion in smart transportation and ML prediction in 5G IoT enhance congestion management, showcasing ongoing advancements in waste management and IoT.

Smart bins and beyond: real-world implementations and success stories

In practical settings, the integration of ML into waste management has been exemplified by initiatives conducted at academic institutions like Ton Duc Thang University in Vietnam. Utilizing ML and graph theory, their framework tested within the university's operations, predicted trash bin fill levels and optimized collection routes through low-cost IoT architecture [17]. This deployment demonstrated the efficacy of intelligent waste bins in enhancing collection operations, emphasizing cost and resource efficiency. It also indicated the potential for extending ML-driven waste management solutions to smart cities, showcasing practical implementation and inherent benefits of ML in waste management systems.

Regarding real-world applications, Gopalan et al. [41] proposed a ML-based intelligent waste management system for precise waste collection and prediction. The Android app for real-time data collection explains the system's implementation. The document also discusses creating a waste classification and prediction database using ML algorithms.

As we conclude this section, it becomes evident that smart bins represent just the beginning of a broader paradigm shift toward intelligent and data-driven approaches to waste management, promising a future marked by continued innovation and positive environmental impact.

Neural networks in waste management

NNs possess transformative potential within the intricate realm of waste management, merging innovation with efficiency. To comprehensively understand their role in intelligent

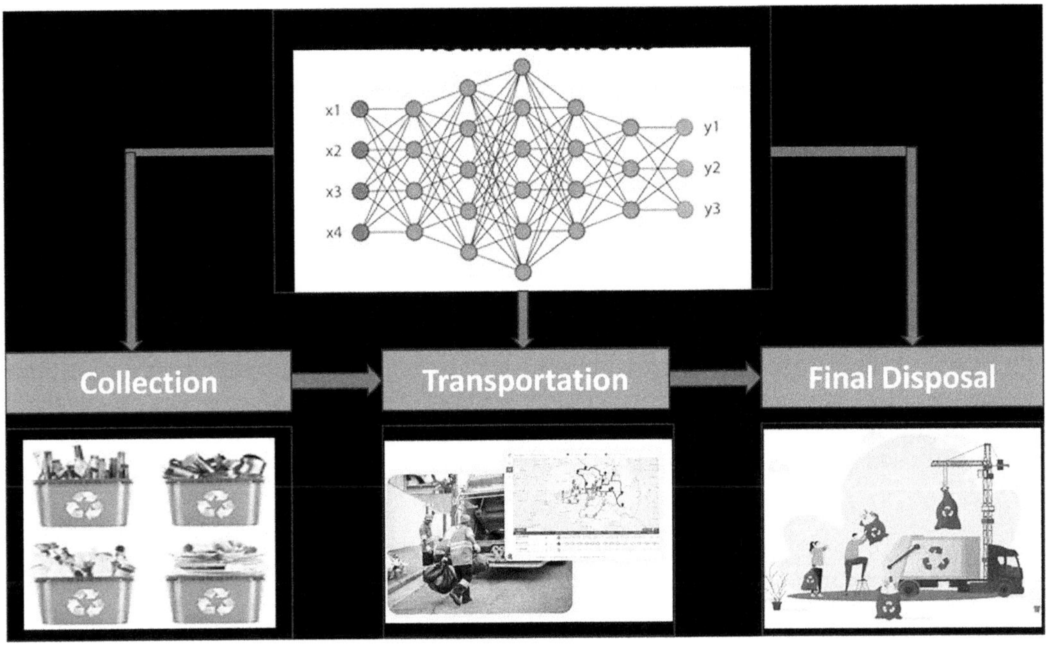

FIGURE 18.2 Neural networks application in trash management.

waste management, it is essential to explore the fundamental components of these marvels of computation. Delving into the intricate workings of neurons and the diverse array of NN types paves the way for grasping their significance in revolutionizing recycling practices [44]. Here, we explore the neural innovations' hallways, revealing state-of-the-art uses of NN, as illustrated in Fig. 18.2, that could revolutionize garbage sorting and recycling. The four subsections will highlight different aspects of the NN-trash management relationship.

Building blocks: understanding the basics of neural networks

The cornerstone of modern AI is NNs, which emulate the human brain's intricate processes. Comprising interconnected layers of nodes or neurons, NNs excel in learning and prediction across diverse fields like waste management, adeptly tackling complex, nonlinear challenges [45]. The three layers of a NN typically include the input, hidden, and output layers. Raw data enter the input layer and undergo processing through interconnected neurons across layers, with each connection assigned a weight representing its strength. Neurons apply activation functions to inputs, introducing nonlinearity and enabling the representation of intricate data correlations within the network [24].

The application of convolutional neural networks (CNNs), which are well known for their effectiveness in image processing tasks, is particularly relevant to waste management. CNNs excel in processing visual data, leveraging hierarchical features extracted by convolutional layers to recognize fine details in images. This capability streamlines trash sorting, enhances recycling efficiency, and reduces contamination, thus contributing to sustainable waste management practices [46]. NNs, with topologies like feedforward, convolutional, and recurrent networks are highly valuable tools in waste management. They enhance recycling facilities' efficiency and accuracy by learning from data, modeling complex relationships, and making informed decisions, thus improving trash processing capabilities [30].

Sorting the future: types of neural networks in waste management

In waste management, NNs prove effective for sorting and classifying waste products. Various types of NNs enhance the efficiency and accuracy of garbage sorting by learning patterns from extensive datasets [32]. We go over a few different kinds of NNs that are frequently used in trash management in Table 18.2.

Neural innovations: cutting-edge applications in waste sorting and recycling

NNs revolutionized waste management recently, particularly in garbage sorting and recycling. This segment accompanied by Table 18.3 explores recent advancements, showcasing how NNs transform waste management procedures. Fig. 18.3 summarizes cutting-edge solutions in garbage sorting and recycling.

Success in layers: case studies highlighting neural network achievements

NNs have shown to be incredibly effective in many waste management applications, providing novel answers to challenging problems. By combining sophisticated algorithms with large datasets, NNs have greatly improved the accuracy and efficiency of waste management procedures [56]. The following discussion explores prominent case studies demonstrating the effective use of NNs in waste management:

- Optimization of waste collection routes: One widely recognized case study uses NNs to optimize waste collection routes. A NN-based system was built to dynamically optimize garbage collection routes by examining historical data on waste creation, population density, and traffic patterns. Through constant learning and adaptation to real-time data, the system successfully lowers fuel usage, shortens collecting times, and lessens its negative effects on the environment [26,57].
- Categorization of recyclable items: Automating the categorization of recyclable items in waste streams has largely been made possible by NNs. CNNs were used in a ground-breaking study to precisely detect and separate various recyclable items, such as metals, glass, and plastics, from mixed trash streams. In addition to increasing recycling

TABLE 18.2 Neural networks utilized in intelligent waste management.

NN type	Description	Application in waste management
CNNs [47]	Specialized in image identification, making it perfect for waste item classification using visual information.	Separating items into recyclable and nonrecyclable categories using pictures that sensors and cameras have captured.
Recurrent neural networks (RNNs) [48]	Ideal for managing consecutive data and temporal correlations, and highly beneficial for analyzing the composition of garbage.	Forecasting changes in waste composition over time using sensor readings and previously collected information.
Generative adversarial networks (GANs) [16]	Create artificial data that mimics real-world samples; this might help enhance small datasets for garbage sorting.	Creating synthetic waste material photos to improve classification model training datasets.
Deep reinforcement learning (DRL) [49]	Used to optimize garbage sorting procedures by contact with the environment, allowing for continual learning and adaptability.	Directing robotic sorting systems to modify their sorting tactics in response to immediate input.
Self-organizing maps (SOMs) [50]	Unsupervised neural networks for waste material visualization and grouping according to attributes.	Arrange waste materials in groups on a map to show their differences and similarities.
Transfer learning models [41]	Assist pretrained models in transferring their expertise to new garbage sorting jobs using sparsely labeled data.	Sorting answers related to trash handling more quickly by leveraging preexisting models that have been trained on massive datasets.
Feedforward neural network (FFNN) [22]	Conventional neural networks include layers of neurons connected exclusively to other levels in the network.	Sorting waste materials according to traits that have been removed is appropriate for easier sorting jobs.
Artificial neural network (ANN) [51]	A general term for neural networks that are modeled after the composition and operations of the human brain.	Used for a number of waste management activities, such as optimization, prediction, and categorization.
AlexNet [25]	Renowned for its performance in the ImageNet Large Scale Visual Recognition Challenge, this deep CNN architecture is a pioneer.	Used especially for large-scale datasets, for waste management picture classification applications.
VGG16 [52]	It has a consistent and simple structure, with 16 layers that include tiny convolutional filters.	Deep CNN architecture is suitable for fine-grained waste material categorization, offering good accuracy with enough processing resources.
ResNet [18]	CNN design that addresses the vanishing gradient issue using residual connections, allowing for the training of extremely deep networks.	Useful for garbage sorting jobs that demand a high degree of precision and network depth.
Inception-ResNet [18]	Hybrid CNN architecture that combines ResNet's residual learning with the multiscale feature extraction of Inception modules.	Gives trash categorization jobs a balance between accuracy and computational efficiency.

(*Continued*)

Table 18.2 (Continued)

NN type	Description	Application in waste management
DenseNet [53]	CNN's densely linked layer design makes it easier to reuse features and solves the vanishing gradient issue.	Appropriate for garbage sorting jobs where accuracy depends on information flowing between layers.
MobileNet [13]	CNN design with depth-wise separable convolutions for efficiency, tailored for mobile devices.	Utilized in waste management applications, including mobile robotic platforms, when computational resources are scarce.
Custom CNN [13]	Personalized CNN is designed for particular applications. It can be trained and developed by particular needs.	Utilized for garbage categorization, recognizing various waste kinds from pictures that waste management systems' sensors or cameras have taken.
EfficientNet [53]	A scalable CNN architecture that uses less processing and fewer parameters to achieve cutting-edge accuracy.	EfficientNet optimizes performance and resource consumption for garbage sorting systems through item recognition and categorization.
Xception [1]	A deep convolutional neural network design that preserves performance by using fewer parameters and computation thanks to its depth-wise separable convolutions.	Xception can be used in waste management to help sort and dispose of garbage by performing object identification and segmentation activities.
R-CNN [53]	CNN based on Region. It first suggests areas, then classes the things within those zones.	To facilitate effective garbage collection and segregation in smart waste systems, R-CNN may be utilized for waste detection and localization in pictures.
YOLO [27]	You only look once (YOLO), a real-time object identification system can identify things in pictures or live video broadcasts.	YOLO may be used in smart trash bins to identify and categorize different kinds of waste in real-time, which makes automated waste sorting possible.
SSD [1]	Another real-time object identification system that predicts several bounding boxes and their associated class probabilities at once is the Single Shot Multibox Detector (SSD).	By precisely identifying and localizing trash objects, SSD can help with waste management by improving the effectiveness of garbage collection and sorting.
YOLACT [1]	Real-time instance segmentation is a well-known feature of the Yet Another Efficient Object Detection System (YOLACT).	YOLACT may be applied to waste instance segmentation, which allows various waste items in smart waste systems to be precisely identified and sorted.
HRNet [1]	More precise object localization results from High-Resolution Network (HRNet), which upholds high-resolution representations over the whole network.	HRNet can enhance waste management systems and facilitate effective trash sorting by offering more accurate object location and segmentation.

(*Continued*)

Table 18.2 (Continued)

NN type	Description	Application in waste management
SegNet [1]	An encoder-decoder architecture with deep convolutions intended for semantic pixel-wise segmentation.	Semantic segmentation of trash objects in photos may be accomplished with SegNet, which helps distinguish and classify various garbage categories.
DeepLab [1]	Atrous convolution is used by DeepLab, a cutting-edge deep learning model for semantic picture segmentation, to capture object context at various sizes.	To improve the accuracy of garbage sorting systems, DeepLab may be used in waste management to precisely categorize waste objects based on their semantic content.

efficiency, this automated sorting procedure also makes resource recovery easier and reduces landfill trash [58].

- Predictive maintenance of waste processing equipment: By facilitating predictive maintenance procedures, NNs have completely changed how waste recycling facilities accomplish maintenance. Recurrent neural networks (RNNs) analyze sensor data from different equipment components to predict probable faults or failures before they happen. In waste management facilities, predictive maintenance algorithms have decreased maintenance costs, decreased downtime, and improved operational dependability [28,47].
- Landfill optimization and gas emission prediction: Predicting gas emissions and optimizing landfills are two more fascinating uses for NNs. NN models can precisely anticipate methane emissions and optimize landfill management techniques by combining data on trash composition, meteorological conditions, and landfill operations. This preemptive strategy reduces greenhouse gas emissions, mitigates harm to the atmosphere, and makes the most use of landfill space [59].
- Assessment of consumer behavior for garbage reduction: NNs are essential for comprehending how consumer behavior affects garbage production. NN-based algorithms can identify patterns and preferences driving waste creation by examining social media interactions, purchase habits, and demographic data. The insights gained from these studies help businesses and policymakers create focused actions and successfully support waste reduction programs [54–56].

The aforementioned case studies serve as prime examples of the diverse ways in which NNs have advanced waste management techniques. These creative solutions, which use AI and ML, have enormous potential to promote resource efficiency, sustainability, and environmental stewardship in waste management systems [60].

Decision-making in waste management

Several real-life successful scenarios in this field have demonstrated the potential of using intelligent methodologies and deep machine learning techniques to optimize waste sorting and

TABLE 18.3 Cutting-edge techniques in waste classifying and processing.

Introduction	Application	Impact
Automated waste sorting systems [54]	Automated trash sorting systems are essential for optimizing waste management procedures by effectively separating recyclables from mixed waste streams. Convolutional neural networks (CNNs), complex algorithms that can perform picture identification and classification tasks, are the brains behind these systems. Using visual cues, optical sorting machines, which use the power of CNNs, can reliably recognize and separate various recyclable materials. This automated method improves recycling operations' overall efficiency and sorting accuracy. Fig. 3.2 shows trash categories for sorting the recyclables with images.	Increases the precision and effectiveness of sorting in recycling processes.
Smart recycling bins [44]	Since smart recycling bins use neural network algorithms to improve and incentivize recycling practices, they represent a paradigm change compared to traditional garbage-collecting techniques. Using behavioral analytic techniques to comprehend customers' recycling patterns, these intelligent bins offer incentives and individualized feedback to promote sustainable practices. Furthermore, as they have real-time data analytics built in, smart bins optimize garbage collection routes, saving money and lessening their negative environmental effects.	Promotes environmentally friendly behaviors and streamlines garbage collection routes, which reduces costs and improves the environment.
Predictive maintenance in recycling facilities [31]	Recyclers may handle their equipment more proactively thanks to neural network-driven predictive maintenance. Neural network algorithms analyze sensor data and historical maintenance logs to predict likely equipment failures. This makes it possible to take quick action to prevent costly downtimes. This data-driven approach improves equipment reliability and maintenance schedules, which leads to significant cost savings and operational advantages.	Lowers maintenance costs and improves equipment dependability in recycling facilities.
Waste composition analysis [55]	Accurate categorization of waste streams is crucial to ensure the efficacy of recovered materials and to design effective recycling processes. Neural networks are essential for waste composition analysis because they may be integrated with chemometric models and spectroscopic approaches. Neural networks can identify and quantify various components found in trash samples by using complex pattern recognition algorithms. This allows recycling facilities to improve materials' recuperation and quality control.	Improves material recovery and quality control in recycling plants.

classification processes. This section provides insights into applying ML and DL techniques to accurately predict waste generation and optimize waste management processes and DL object classifiers in waste collection planning, specifically for e-waste. The section also explores the limitations of conventional trash decision-making and the use of ML and NNs.

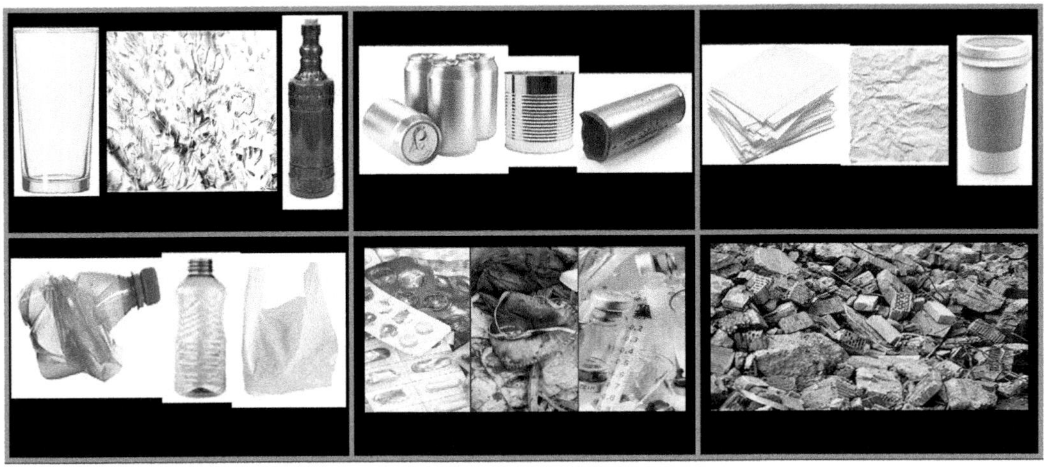

FIGURE 18.3 Waste categorical images for smart waste sorting using neural networks (NNs).

The decision dilemma: challenges in traditional waste management decision-making

Waste management traditionally includes manual collection and disposal without sorting, reduction, reuse, or recycling. Traditional waste decision-making faces several challenges [49]. Firstly, there is often insufficient government regulation and budget for waste management. Secondly, families may not fully understand the importance of waste management. One major challenge is that waste classification becomes complex when domestic waste is mixed with other types of waste, making it challenging to identify and classify with specific sensors alone [51]. Additionally, the speed of garbage classification and waste detection accuracy becomes significantly poor due to manual labor in waste identification.

Another challenge is the lack of reliable and accessible methods for estimating waste generation based on limited data. Forecasting waste generation is crucial for effective waste management and planning of waste management facilities [24,61]. Without accurate estimates, traditional approaches may not effectively capture the complex and nonlinear relationships between socioeconomic and demographic variables and waste generation [62]. Furthermore, this inadequate waste sorting can lead to improper disposal, increased contamination, and reduced recycling efficiency, impacting environmental sustainability [9].

These challenges hinder effective waste classification, which is crucial for later treatment and reuse. Relying solely on manual garbage classification is labor-intensive, inefficient, and detrimental to human health. To address these challenges and improve waste decision-making, the integration of ML and DL techniques has shown great potential [44]. This advanced waste decision-making has shown promising results in real-life scenarios.

Beyond the status quo: role of machine learning in decision support

ML plays a vital role in waste decision-making by automating waste sorting and classification processes. ML techniques, including ANN, are used to develop sophisticated classifiers by combining various features extracted from waste images [60]. These models accurately determine the type of waste based on relevant information extracted from large datasets. ML models have been successfully applied in waste management, contributing to the efficient separation and recycling of materials.

One of the main applications of ML in decision support is waste bin-level detection. ANN is commonly used in this area, as it has fault tolerance and can simulate complex and nonlinear relationships in multivariate systems [63]. SVM is also proficient in maximizing class separations and minimizing classification errors, making them useful for waste bin level detection [45,64,65].

Another important application is in process output prediction. Linear regression analysis (LR) is often employed to predict waste generation and other process parameters in waste management. LR is easy to interpret and has low computational costs, although it may not be suitable for modeling nonlinear data. Decision trees (DT) are another ML technique used in decision support. They are particularly useful for expert systems and have easy interpretation of results and low computational costs. However, DT are prone to data overfitting [57].

Regarding software platforms, various options are available for implementing ML models in waste management decision support systems (DSS). Some popular platforms include MATLAB®, IBM SPSS, R, WEKA, Scikit-learn, Caffe, KNIME, Microsoft Azure, Neural Designer, and Tableau [49]. These platforms offer different advantages and disadvantages, such as ease of use, integration capabilities, and data handling capabilities. ML models and software platforms help improve existing waste management schemes by providing accurate predictions, optimizing routes, and identifying illegal waste disposal sites [6,66]. Utilizing ML in decision support can make waste management processes more efficient, cost-effective, and sustainable. Overall, ML plays a crucial role in decision support for waste management, offering advanced computational approaches and enhancing the efficiency of waste management processes.

Smart choices: integrating neural networks for informed decision-making

NNs have been successfully applied in various domains of waste management, including predicting waste characteristics, waste generation, waste classification, biogas generation, and optimal waste collection routes [11,29]. Their ability to handle incomplete or uncertain datasets and tackle complicated tasks makes them effective in modeling processes with nonlinear behavior. Therefore integrating NNs can contribute to informed decision-making in waste management by providing accurate predictions and optimizing various aspects of the process [52].

CNNs are a type of NN that is particularly well-suited for image classification tasks, such as garbage classification. CNNs have emerged as essential tools for image analysis in intelligent waste identification and recycling (IWIR) applications, including recyclable material identification, trash pollution detection, and solid waste classification [1]. CNNs have been progressively applied to solve problems related to IWIR and have shown promise in addressing waste management challenges [18,30,51]. The study highlights the importance of establishing a sustainable waste management system and the urgent need for efficient and accurate IWIR technologies.

The distribution of CNN models (described in Sections 3.1 and 6.2) used in these tasks shows a growing trend in their application in IWIR. However, challenges and limitations still exist, such as data scarcity and quality. While there are publicly available datasets, they may not meet the requirements of various IWIR applications. Efforts are needed to efficiently collect large quantities of high-quality data for training CNN models in IWIR. It aims to guide further research and deploy CNNs in addressing waste management challenges.

Case in point: improved decision-making scenarios in waste management

Improved decision-making scenarios in waste management can be achieved by applying artificial intelligence (AI) techniques. AI models have the potential to revolutionize waste management processes by providing accurate predictions, optimizing resource allocation, and enhancing overall efficiency.

One key aspect where AI can contribute to improved decision-making is waste forecasting. AI models, such as ANN, support the accurate prediction of waste characteristics and generation rates [55,56]. These models can detect patterns and trends by analyzing historical data, enabling waste management authorities to make informed decisions regarding waste collection and disposal strategies. Furthermore, AI models can optimize waste collection routes. Through techniques like genetic algorithms and expert systems, AI can analyze vast amounts of data, including geographic information systems (GIS), to identify the most efficient routes and schedules for waste collection vehicles [59]. This optimization minimizes fuel consumption, reduces greenhouse gas emissions, and lowers collection costs [67,68].

Additionally, AI can facilitate better waste management planning. By simulating scenarios with different AI models, decision-makers can assess the potential impacts of various waste management strategies [20,67]. This allows for identifying cost-effective solutions, integrating recycling and recovery processes, and optimizing waste treatment facilities. Overall, implementing AI techniques in waste management can lead to improved sustainability, cost-effectiveness, and ecological benefits. By leveraging AI models, waste management authorities can make more informed decisions, optimize resource allocation, and contribute to a greener and more efficient waste management system.

Advanced recycling techniques

This section examines cutting-edge techniques and technologies transforming how trash is managed. Conventional procedures often have difficulties dealing with the complexities of contemporary trash, consequently demanding the use of advanced alternatives. This section explores cutting-edge sorting technologies and revolutionary recycling processes, focusing on the pioneering of innovation. It emphasizes the potential and problems associated with sustainable waste solutions.

Recycling realities: a snapshot of traditional methods

According to various studies, conventional waste management methods face several challenges. Gómez et al. [29] emphasize the significance of effective waste management in response to the growing production of solid domestic waste (SDW) and the long-lasting ecological consequences of nonorganic waste. Although recycling technologies are important, they have limits that require the development of more effective trash disposal alternatives. Gulghane et al. [69] highlight that the building and infrastructure sectors substantially contribute to environmental impact, resource use, and trash production. Conventional waste management systems suffer from enforcement gaps, limited construction agency awareness, and communication barriers, exacerbating the issue. Fig. 18.4 depicts the sequential process of conventional recycling, from collection to recycling.

FIGURE 18.4 Traditional recycling process.

Breaking the mold: introduction to advanced recycling techniques

Advanced recycling encompasses pyrolysis, gasification, solvolysis, and microwave treatment, effectively breaking down old plastics into original constituents, specialized polymers, and feedstocks for new plastic production. Vital for meeting recycled plastics demand, it aligns with circular economy goals and reduces plastic waste's environmental impact. Amidst conventional waste management as shown in Fig. 18.4 and its limitations, advanced recycling methods are increasingly vital for sustainable plastic management.

According to Hasan et al. [4], the rise in population, urbanization, and lifestyle changes make implementing more advanced waste management systems necessary. Their proposal integrates IoT and DL for an intelligent garbage management system, aiming to boost efficiency and recycling. Fig. 5.2 depicts e-waste sorting and recycling steps, which are crucial for environmental sustainability. The flowchart guides plastic, glass, and metal segregation for specialized recycling.

Machine learning's green thumb: enhancing recycling processes

The application of ML and NNs significantly contributes to the improvement of recycling operations and in making an advanced waste recycling process as illustrated in Fig. 18.5. In the study, Hossen et al. [30] suggest a multistage ML methodology that uses deep NNs to

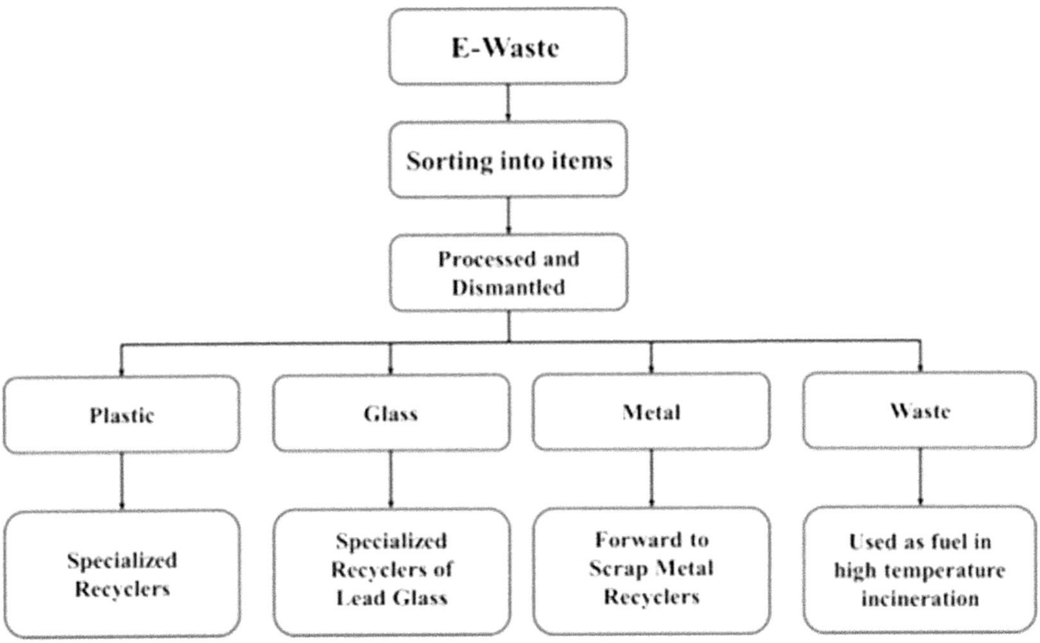

FIGURE 18.5 Advanced waste recycling process.

automate garbage sorting and recycling systems, enhancing accuracy and efficiency. Jassim et al. [63] emphasize the significance of a precise assessment of waste production rates to plan and execute waste management strategies effectively. They propose using advanced ML models such as multilayer perceptron artificial neural networks (MLP-ANN) and SVR for predictive purposes. Huang and Koroteev [70] highlight the significance of accurate data gathering and introduce the ML-driven predictive analytic framework (MLDPAF) as a means to enhance waste management procedures using NNs and ML algorithms. Table 18.4 showcases significant research advancements in using ML methods to tackle issues in waste management. These studies cover predicting trash production, improving recycling, and refining waste categorization. They aim to boost waste management efficiency, accuracy, and sustainability by integrating ML algorithms and modern technology, offering innovative solutions for environmental challenges.

Innovation showcase: examples of cutting-edge recycling technologies

Recycling innovations demonstrate modern systems' potential. Zeng et al. [32] present PublicGarbageNet, an advanced DL framework for accurate trash categorization, showing how sophisticated technologies may change waste management. Jin et al. [31] describe a DL-based machine vision system for trash identification and categorization that improves accuracy and efficiency over standard waste management methods. These state-of-the-art technologies offer inventive methods to address growing waste accumulation and promote environmental sustainability. By leveraging technology, these solutions aim to optimize recycling processes, minimize waste generation, and foster a circular economy. These innovative approaches are designed to mitigate the challenges of refuse collection while advancing ecological sustainability, encapsulating a proactive response to contemporary environmental concerns. Instances of state-of-the-art recycling technologies:

- Plastic-to-fuel conversion systems: Cutting-edge technologies facilitate the transformation of plastic waste into practical fuels, including diesel, gasoline, and petroleum, via depolymerization and pyrolysis. These systems not only mitigate plastic pollution but also offer an alternative energy source, thereby making a positive contribution toward a more sustainable energy framework.
- Advanced classification robots: The integration of robotics and AI is bringing about a major change in the way recycling facilities handle waste classification. By integrating AI algorithms into sophisticated sorting devices, distinct recyclable materials can be effectively identified and segregated from combined waste streams, thereby increasing recycling rates and decreasing contamination.
- Chemical recycling: Involves the utilization of technologies that decompose plastic waste into its molecular components, thereby enabling the manufacturing of new plastic products while ensuring that their quality remains unchanged. These novel procedures provide an environmentally conscious substitute for conventional mechanical recycling techniques while also aiding in the mitigation of plastic refuse accumulation.

TABLE 18.4 Significant studies using machine learning to transform waste management, address difficulties, provide solutions, and use sophisticated technology for efficiency and sustainability.

Author and year	Challenges addressed	Proposed solutions	Models used
Ahmed et al. 2022 [3]	Inefficiencies in trash collection, lack of data-driven planning, limited recycling optimization, and income-based waste compositions	ML/deep learning (DL) integration for accurate forecasting, waste-to-energy technology optimization, web-based decision tools	DLTSF with LSTM
Akanbi et al. 2020 [48]	Manual construction waste calculation, limited old building data, inability to estimate waste for particular projects	Accurate salvage and trash estimates, building information modeling (BIM) integration, and industry practice improvements.	DNN, CNN, RNN, and deep belief networks
Ascher et al. 2022 [10]	Low landfill disposal areas, inefficient operations, low recycling rates and poor hazardous waste management	Tool development for the web utilizing life cycle assessment (LCA) and life cycle costing (LCC) methodology and decision-making approaches	ANNs, regression, SVMs, and CNNs
Bircanoglu et al. 2018 [2]	Low-precision garbage sorting requires plenty of personnel	Pretraining on massive datasets refines RecycleNet design	Inception-Resnet, Inception-v4, DenseNet, and RecycleNet
Carrera et al. 2023 [8]	Slow sorting, manual categorization mistakes, and higher costs	ML plastic categorization using spectroscopic technology	Logistic regression, passive aggressive classifier, perceptron, SGD classifier, linear SVC, SVM algorithm
Cha et al. 2022 [23]	Traditional approaches fail with small datasets	Larger datasets, ML algorithms exploration, and improved data processing	ANN (MLP), SVMR
Chen et al. 2023 [71]	Garbage sorting is inaccurate, laborious, and expensive	Explore sophisticated positioning and recognition methods	YOLOv5s, ShuffleNet v2, depth separable convolution techniques
Chin et al. 2022 [66]	Incineration due to contamination high treatment costs, recyclability difficult to define, lack of connection	Enhanced data-driven tools that complement grading and Pinch principles	KNN, decision trees
Chu et al. 2018 [46]	Insufficient municipal solid waste (MSW) management legislation and funding. Poor recycling technology, and high manual categorization costs	Waste sorting using multilayer hybrid system (MHS) improves household education	MLP, CNN
Cubillos 2020 [24]	Low data reliability, high waste generation rate fluctuation, insufficient historical data, and practical measurement restrictions	Adding data sources long short-term memory (LSTM) parameter optimization, predicting horizon extension	LSTM, MLP

- Intelligent waste management systems: Intelligent waste management systems are created by integrating IoT devices and sensors into waste containers and collection vehicles. These systems optimize collection routes, enable data-driven decision-making to enhance operational efficiency and resource utilization, and monitor refuse levels in real-time.
- Biodegradable packaging materials: Represent environmentally sustainable alternatives to conventional plastic packaging, thanks to advancements in this field. By virtue of their inherent biodegradability, these materials reduce the ecological repercussions of packaging detritus and encourage the use of environmentally sustainable packaging alternatives.

The examples listed above merely symbolize a portion of the extensive array of state-of-the-art recycling technologies that are propelling progress in the domain of waste management. By fostering collaboration among research institutions, industry, and government, and adopting these technologies, we can lay the foundation for a circular and sustainable economy.

Integration of machine learning, neural networks, and decision-making

Combining ML, NNs, and DSS can create sophisticated decision-making systems. ML algorithms can analyze large datasets, find patterns, and predict to aid decision-making. ML subset NNs excel at image and audio recognition. However, DSS organizes and displays data for decision-making. Combining ML and NNs' predictive power with DSS's decision-making assistance can help organizations make more data-driven decisions. These technologies work well together. ML and NNs can spot patterns in massive datasets that humans cannot. A DSS can then clearly present this information, emphasizing key insights and decision-maker recommendations. DSS can also create interactive models and simulations to help decision-makers compare outcomes. This is crucial in complex decision-making situations with unclear consequences.

With ML and NNs' predictive capability and DSS's decision-making assistance, companies may construct powerful decision-making tools to make data-driven judgments.

Power trio: synergies and complementarities between machine learning, neural networks, and decision support

ML methods, including advanced neural networks, significantly improve waste decision-making. These data-driven models analyze vast data on garbage creation rates, composition, collection routes, and truck capacity, enhancing efficiency [22]. CNNs, advanced DL models, have the ability to automatically classify garbage into several categories with an accuracy of over 95%. They do this by analyzing visual patterns present in images [25]. Long short-term memory (LSTM) networks are used to forecast the daily amounts of garbage generated in different areas by analyzing the patterns and connections in past data [63]. In addition to analytics, the actionable insights derived by ML models provide crucial inputs for the development of improved DSS, resulting in a beneficial feedback loop [32,65]. For example, the use of LSTM predictions would allow for the real-time adjustment of collection routes and timetables, considering the limitations of vehicle capacity. CNN waste identification aids targeted recycling strategies. IoT monitors waste logistics, allowing timely adjustments and improvements for enhanced waste management efficiency [49]. The integration of AI and optimization algorithms tackles key challenges in waste management, like routing inefficiencies, lack of segregation, and limited visibility [18,20]. ML and NNs, along with data-driven decision support, have the potential to create a sustainable circular economy in the garbage sector.

Tech Tango: navigating challenges in integrating advanced technologies

Emerging technologies such as AI, internet of things sensors, robots, and data analytics are revolutionizing the waste management business. For example, AI can optimize collection routes, sensors can check fill levels in real time, and robots can sort recyclables automatically. However, incorporating these improvements entails overcoming challenges. Legacy infrastructure, different data sources, aversion to automation, and a lack of digital skills are all significant barriers. Waste management organizations must create strong IT roadmaps to phase in digital solutions, engage in worker reskilling, encourage cross-functional cooperation, and collaborate with technology vendors. Before large-scale rollouts, an agile, phased integration method may pilot-test innovations to demonstrate their effectiveness.

With proper planning and change management, the trash industry may reap new technology's efficiency and sustainability advantages while negotiating complicated integration issues. The results of recent studies indicate that there is an increasing emphasis on boosting conventional recycling methods by using innovative and automated alternatives. Traditional recycling procedures, which include sorting, cleaning, and reprocessing waste materials [65], have been extensively used; nevertheless, they have limits in terms of efficiency, accuracy, and scalability [25]. These issues are intended to be solved by advanced recycling approaches, which use technologies such as ML, DL, and computer vision to automate trash identification, classification, and sorting procedures [28].

Advanced recycling approaches include image-based smart trash sorting systems that employ DL models such as ResNet-50 to categorize waste items [28] accurately. These systems are examples of some of the real-world applications of these techniques. These systems enable automatic waste classification into categories such as Polyethylene Terephthalate, plastic, metal, and rubbish, improving recycling rates and efficiency. According to Dubey et al.'s research from 2020, the incorporation of cutting-edge methods with Internet of Things devices, such as smart bins equipped with computer vision capabilities, serves to further improve garbage collection and management.

However, some obstacles must be overcome before improved recycling processes can be used. One of the most important things to improve the accuracy and resilience of ML models is to collect varied picture data from various object locations. Furthermore, to apply this on a broad scale, it is important to have sorting machines that are fully functional and that include classification models in distributed smart recycling systems. In addition, there are major obstacles in the areas of economic viability, legal and legislative frameworks, technical complexity, and scalability [16,30]. Table 18.5 shows the integration of some advanced technologies to address challenges regarding waste management.

Despite these obstacles, more sophisticated recycling systems have the potential to produce major advantageous outcomes. They can manage complicated waste streams that are difficult to recycle using conventional methods. Additionally, they can foster a more circular economy, which will reduce their influence on the environment. In addition, the use of highly developed methods in conjunction with GIS can potentially optimize trash collection routes according to the number of recyclables and rubbish collected [46].

Table 5.1 describes the conventional and high-tech methodologies for waste management. The research that has been done on the topic emphasizes the increasing need to use sophisticated recycling processes to improve waste management practices. ML, DL, and computer vision are examples of sophisticated approaches that provide creative ways to automate and enhance garbage identification, classification, and sorting operations. Traditional recycling systems, on the other hand, have their limits. It is essential to overcome obstacles relating to technology, cost, infrastructure, and legislation to ensure the broad acceptance and success of these strategies. However, the actual application of these techniques in real-life settings is still in its infancy.

Real-world impact: success stories of integrated waste management systems

Integrated waste management systems have reduced trash and promoted sustainability in many places. Recycling works report that San Francisco recycles and composts 80% of its garbage. City actions have decreased pollution, generated employment, and cut disposal costs. In the 1970s, Curitiba, Brazil, started sustainable garbage management. In Curitiba, waste-to-energy projects and public education have reduced landfill methane emissions and increased recycling to over 70% [68]. Modern garbage management in Malmö, Sweden, includes sorting and a biogas plant that creates power from organic waste. This integrated approach reduces carbon emissions and boosts recycling.

Successful advanced waste decision-making uses ML to forecast trash creation and improve waste management [20]. Ensemble learning-based categorization methods predict weekly residential garbage, allowing quicker and cheaper waste collection fleet routing. IoT has increased trash management efficiency and resource usage, reducing bin overflow fourfold and garbage collection trips 18% [19]. The paper discusses e-waste collection plans using DL object classifiers. Smartphone-based image recognition algorithms can detect and categorize e-waste from user images, simplifying information sharing between consumers and garbage collection firms. This method allows garbage collection firms to automatically categorize e-waste from submitted photos with 90%–97% accuracy. The research by Gómez et al. [29] showed that ML can improve municipal solid waste (MSW) management, composting, anaerobic digestion, landfill, incineration, and pyrolysis. ML models for waste treatment prediction, optimization, and categorization save time, improve accuracy, and minimize resource usage. Ihsanullah et al. [60] has improved waste management decision-making.

TABLE 18.5 From conventional to high-tech: the evolution of recycling methods.

Topic	Traditional recycling	Advanced recycling	Real-life examples	Challenges	Benefits
Description	Conventional methods like sorting, cleaning, and reprocessing waste materials	Innovative techniques using technologies like machine learning, deep learning, computer vision	Image-based smart waste sorting systems, smart bins with computer vision, waste route optimization using GIS	Collecting diverse data, integrating classification models, technological complexity, scalability, cost, regulations	Address complex waste streams, improve efficiency and accuracy, promote circular economy, reduce environmental impact
Techniques	Paper, plastic, glass, metal recycling	Chemical recycling, biological treatments, advanced sorting	Convolutional neural networks like ResNet-50 for waste classification	Lack of standards, varying waste composition, complexity of industrial waste sorting	Automate waste detection, classification, sorting, increase recycling rates
Limitations	Time-consuming, labor-intensive, prone to errors	Implementation in real-life scenarios still in early stages, knowledge gap between waste management and data science	Largest waste image datasets (LSWID) for training models	Insufficient government regulations, public awareness, high cost of manual classification	Overcome limitations of traditional recycling methods

They used ANNs to estimate MSW production rates in various areas of India, revealing regional disparities. Waste management operations like MSW energy recovery gas yield prediction have been optimized using AI. AI has also been used to automate garbage sorting and recycling and recover precious materials. Table 18.6 summarizes some success stories in the field of integrated waste management systems.

Ref. [45] automating garbage sorting and recycling using ML algorithms is a successful example. CNNs and SVMs have been used to categorize garbage with great accuracy. Another effective example is image recognition-based plastic bottle recycling categorization

TABLE 18.6 Success stories of integrated waste management systems.

Author(s)	Successful scenarios	Techniques used	Benefits
Hasan et al. [4]	Smart waste management and categorization system	IoT, convolutional neural networks (CNN)	Real-time data monitoring, automated waste sorting
Huang and Koroteev [70]	IoT-based waste management, mapping disposal routes	IoT, artificial neural networks, machine learning	Efficient waste collection, improved fuel efficiencies
Jiang et al. [72]	Predicting odor gas generation from domestic waste	Random forest, XGBoost, LightGBM	Data-driven method for understanding and predicting odor production
Jin et al. [31]	Garbage detection and classification	Deep learning, transfer learning, MobileNetV2	High garbage classification accuracy (90.7%), real-time performance
Zeng et al. [32]	Garbage classification	PublicGarbageNet framework	Improved waste classification accuracy (96.35%)
Ziouzios et al. [33]	Advanced recycling techniques	Convolutional neural networks (CNN)	Automated detection and classification of recyclable materials
Namoun et al. [20]	Predicting waste generation, optimizing waste management processes	Ensemble learning, IoT	Faster and cost-effective routing plans, reduced trips, minimized traffic congestion, decreased air pollution
Nowakowski and Pamula [19]	Waste collection planning for e-waste	Deep learning object classifiers, image recognition	Automated recognition and classification of e-waste, accuracy ranging from 90% to 97%
Ihsanullah et al. [60]	Predicting MSW generation rates, optimizing waste management processes, waste sorting and recycling	Artificial neural networks (ANN), AI techniques	Insights into regional differences, accurate forecasting, automated sorting and recovery of valuable materials
Ozdemir et al. [45]	Automating waste sorting, improving recycling rates	Convolutional neural networks (CNN), support vector machines (SVM)	High classification accuracy, improved recycling processes

with good color identification accuracy. CNNs have been used for intelligent garbage detection and recycling [1]. CNNs have been utilized to build image analysis systems for recyclable item identification, garbage pollution detection, and solid waste categorization. CNNs in IWIR may help create a sustainable waste management system and improve the environment. ML algorithms have predicted garbage creation using meteorological, demographic, and socioeconomic data [44]. Trash bin recognition and route optimization using ML algorithms enhance trash collection efficiency and save expenses. ML algorithms in trash categorization and recycling promote resource use and sustainability in waste management. Hasan et al. [22] described a smart trash management system incorporating IoT for real-time monitoring and CNN models for automatic garbage categorization, improving urban waste operations. Huang and Koroteev [70] examine IoT-based garbage monitoring, mapping complicated disposal pathways with increased efficiency, and using ML models to estimate waste generation. To estimate home waste odor gas emission, hence facilitating data-driven odor management techniques, Jiang et al. [72] used ML models such as RF and XGBoost. Another study emphasized the PublicGarbageNet framework's potential for accurate trash categorization across key categories, demonstrating the influence of sophisticated NNs on efficient waste management. Ziouzios et al. [33] compared old manual procedures with modern automated recycling based on CNNs, providing real-world CNN systems for recyclable material identification despite implementation costs and skill constraints. Table 6.2 summarizes the success stories of integrated waste management systems. Success in integrated waste management inspires others to adopt them. Sustainable and circular cities may minimize waste, boost the economy, and enhance citizens' lives.

Challenges and future directions

The combination of ML and NNs offers both exciting prospects and difficult obstacles in the dynamic field of waste management decision-making and recycling techniques. This section explores the complex topic of intelligent waste management, outlining the present roadblocks to the smooth integration of ML and NNs and speculating on potential future paths that might revolutionize the industry. As we set out to sort through the intricacies and steer clear of obstacles to a more sustainable and effective approach to trash management and recycling, we also tackle ethical issues.

Obstacles ahead: current challenges in applying machine learning and neural networks to waste management

The application of NNs and ML to waste management techniques can significantly transform resource optimization, sustainability, and efficiency. Nevertheless, several obstacles are in the way of achieving this potential, which must be overcome to be successfully implemented [56]. Table 18.7 lists some of the major challenges that impede the smooth integration of ML and NNs in recycling and garbage management.

TABLE 18.7 Current challenges and solutions in smart waste management techniques.

Challenges	Description	Solutions
Data scarcity and quality issues [73]	Training ML algorithms is hampered by the lack and unpredictability of high-quality data. Models with inconsistent labeling and annotation are biased.	Improving data gathering techniques and investigating alternate data sources to address the lack of available data. Improving the quality of your data by using strict validation and verification procedures. Using strategies like transfer learning and data augmentation to lessen the impact of small datasets.
Complexity of waste classification [16]	Waste classification encompasses a variety of materials, each with its special qualities. Particulate or tainted waste streams make precise identification more difficult.	Creating advanced categorization algorithms that can manage the intricacy of waste streams. To increase classification accuracy, including sophisticated feature engineering and multimodal sensing approaches. Using adaptive learning techniques to adjust to changing waste compositions and categorization specifications.
Logistical constraints and operational limitations [67]	Investments in infrastructure and logistical planning are necessary for the deployment of ML systems. In practical contexts, scalability and dependability present issues.	Investing in logistics and infrastructure to enable the use of sensor-based technologies. Providing scalable machine learning techniques that can function dependably in dynamic waste management settings. Establishing stringent maintenance and monitoring procedures guarantees that ML systems continue functioning.
Interpretability and explainability [10]	Since ML algorithms are black-box, they are difficult to understand and comprehend. Developing trust with stakeholders requires an understanding of algorithmic judgments.	Creating understandable and transparent machine learning models that shed light on the decision-making process. To improve interpretability, using strategies like feature significance analysis and model explainability. Working together with stakeholders to create communication plans that promote comprehension of and confidence in machine learning-based solutions.

The ethical landscape: balancing technological advancements with environmental responsibility

It is crucial to traverse the complex ethical terrain surrounding technology developments, especially in the areas of ML and NNs, to optimize waste management procedures. While these innovations promise efficiency and environmental benefits, they also raise significant ethical concerns. Accountability, transparency, and justice ensure equitable distribution of technology's advantages. Building stakeholder confidence demands openness in algorithmic decision-making, fostering trust among waste management experts, legislators, and the public. It is imperative to navigate these ethical considerations to realize the full potential of intelligent waste management systems [59].

Furthermore, ethical discussions on intelligent waste management tech focus on environmental responsibility. While ML and NNs can streamline recycling and reduce trash, their wider environmental impacts require scrutiny. Every tech lifecycle phase affects the environment, from material extraction to energy usage. To maximize AI's waste management potential, minimizing ecological footprint is crucial [11,53,54].

Visionary paths: potential future developments and research frontiers

Beyond the present difficulties and moral dilemmas discussed in Sections 7.1 and 7.2, the innovative route of intelligent waste management offers a variety of prospective future advancements and research directions. New possibilities to advance waste management toward more sustainable and efficient methods arise as technology advances and our understanding of the issue grows. Integrating ML and NNs with edge computing, blockchain, and IoT offers growth potential [33,64]. Linked systems optimize resource allocation, using blockchain for transparent data management and IoT sensors for real-time waste stream monitoring. Edge computing enables local data processing, reducing latency and enhancing waste management process responsiveness for quicker responses. Garbage sorting and recycling may transform with automation and robots [69]. AI-enabled systems recognize and separate items, boosting sorting precision. Autonomous drones monitor disposal sites, while collaborative robots enhance efficiency in waste processing, promising more efficient procedures [62].

Future research in intelligent waste management should prioritize holistic methods, considering social, economic, and environmental concerns alongside technology breakthroughs [74]. Collaborative frameworks involving stakeholders from academia, industry, government, and communities can cocreate sustainable solutions. Minimizing waste creation and maximizing resource recovery may be achieved through the circular economy concept and lifecycle-based waste management systems [11,53,62].

It is possible to fully realize the potential of ML and NNs to transform waste management methods by adopting innovative approaches and investigating uncharted research territories. We can create a future where trash is seen as a resource to be maximized and used for the good of society and the environment rather than a problem to be solved through multidisciplinary collaboration, innovation, and environmental stewardship.

Conclusion

The convergence of ML and NN technology is poised to bring about a significant revolution in the waste management industry. This research has highlighted the enormous potential of intelligent systems to transform waste management practices' effectiveness, sustainability, and environmental impact through thoroughly investigating their applications in trash sorting, recycling, and decision-making processes. Combining cutting-edge technology with trash management offers a potential route toward a more sustainable and environmentally friendly future. It can improve sorting accuracy, streamline recycling procedures, and support well-informed decision-making. Notwithstanding the optimism, one must acknowledge the enormous obstacles that still need to be overcome, including the lack of data, the complexity of the models, and the requirement for multidisciplinary cooperation and regulatory frameworks. It will need coordinated efforts by academics, decision-makers, and industry stakeholders to address these issues. However, the goal of intelligent waste management as a pillar of sustainable development may be accomplished with persistent innovation, teamwork, and commitment, opening the door for a healthier, cleaner Earth for future generations.

References

[1] T.-W. Wu, H. Zhang, W. Peng, F. Lü, P.-J. He, Applications of convolutional neural networks for intelligent waste identification and recycling: a review, Resources, Conservation and Recycling 190 (2023) 106813. Available from https://doi.org/10.1016/j.resconrec.2022.106813.

[2] C. Bircanoğlu, M. Atay, F. Beşer, Ö. Genç, M.A. Kızrak, RecycleNet: intelligent waste sorting using deep neural networks, presented at the 2018 Innovations in intelligent systems and applications (INISTA), IEEE, 2018, pp. 1–7.

[3] A.K.A. Ahmed, A.M. Ibraheem, M.K. Abd-Ellah, Forecasting of municipal solid waste multi-classification by using time-series deep learning depending on the living standard, Results in Engineering vol. 16, (2022) 100655.

[4] M.K. Hasan, M.A. Khan, G.F. Issa, A. Atta, A.S. Akram, M. Hassan, Smart waste management and classification system for smart cities using deep learning, presented at the 2022 International Conference on Business Analytics for Technology and Security (ICBATS), IEEE, 2022, pp. 1–7.

[5] S.M. Cheema, A. Hannan, I.M. Pires, Smart waste management and classification systems using cutting edge approach, Sustainability 14 (16) (2022). Available from https://doi.org/10.3390/su141610226.

[6] X. Chen, Machine learning approach for a circular economy with waste recycling in smart cities, Energy Reports 8 (2022) 3127–3140. Available from https://doi.org/10.1016/j.egyr.2022.01.193.

[7] R. Ali, N.N.L. Nik Ibrahim, A. Karim, H. Lam, N.S. Sani, Utilization of process network synthesis and machine learning as decision-making tools for municipal solid waste management, International Journal of Environmental Science and Technology vol. 19, (2021). Available from https://doi.org/10.1007/s13762-021-03250-0.

[8] B. Carrera, J.B. Mata, V.L. Piñol, K. Kim, Environmental sustainability: a machine learning approach for cost analysis in plastic recycling classification, Resources, Conservation and Recycling 197 (2023) 107095.

[9] K. Lin, Y. Zhao, L. Wang, W. Shi, F. Cui, T. Zhou, MSWNet: a visual deep machine learning method adopting transfer learning based upon ResNet 50 for municipal solid waste sorting, Frontiers of Environmental Science and Engineering 17 (no. 6) (2023) 77. Available from https://doi.org/10.1007/s11783-023-1677-1.

[10] S. Ascher, I. Watson, S. You, Machine learning methods for modelling the gasification and pyrolysis of biomass and waste, Renewable and Sustainable Energy Reviews 155 (C, 2022) (2024). Available from https://ideas.repec.org//a/eee/rensus/v155y2022ics1364032121011680.html.

[11] L. Zhao, Y. Pan, S. Wang, Z. Abbas, M. Islam, S. Yin, Skip-YOLO: domestic garbage detection using deep learning method in complex multi-scenes, International Journal of Computational Intelligence Systems 16 (2023) 139. Available from https://doi.org/10.1007/s44196-023-00314-6.

[12] Y. Zhao, H. Li, Understanding municipal solid waste production and diversion factors utilizing deep-learning methods, Utilities Policy 83 (2023) 101612. Available from https://doi.org/10.1016/j.jup.2023.101612.

[13] I. Yogeashwaran, Domestic waste segregation using deep neural networks, Masters, Dublin, National College of Ireland, 2024. Available from https://norma.ncirl.ie/6347/.

[14] P.R. Kshirsagar, et al., Artificial intelligence-based robotic technique for reusable waste materials, Computational Intelligence and Neuroscience 2022 (2022) 1–9. Available from https://doi.org/10.1155/2022/2073482.

[15] B. Erdebilli, B. Devrim-İçtenbaş, Ensemble voting regression based on machine learning for predicting medical waste: a case from Turkey, Mathematics 10 (14) (2022) 2466. Available from https://doi.org/10.3390/math10142466.

[16] M. Kannangara, R. Dua, L. Ahmadi, F. Bensebaa, Modeling and prediction of regional municipal solid waste generation and diversion in Canada using machine learning approaches, Waste Management 74 (2018) 3–15. Available from https://doi.org/10.1016/j.wasman.2017.11.057.

[17] T. Khoa, et al., Waste management system using IoT-based machine learning in university, Wireless Communications and Mobile Computing 2020 (2020) 1–13. Available from https://doi.org/10.1155/2020/6138637.

[18] H. Abdu, M.H. Mohd Noor, A survey on waste detection and classification using deep learning, IEEE Access 10 (2022) 128151–128165. Available from https://doi.org/10.1109/ACCESS.2022.3226682.

[19] P. Nowakowski, T. Pamuła, Application of deep learning object classifier to improve e-waste collection planning, Waste Management 109 (2020) 1–9. Available from https://doi.org/10.1016/j.wasman.2020.04.041.

[20] A. Namoun, B.R. Hussein, A. Tufail, A. Alrehaili, T.A. Syed, O. BenRhouma, An ensemble learning based classification approach for the prediction of household solid waste generation, Sensors 22 (9) (2022) 3506. Available from https://doi.org/10.3390/s22093506.

[21] L.R. Kambam, R. Aarthi, Classification of plastic bottles based on visual and physical features for waste management, in: 2019 IEEE International Conference on Electrical, Computer and Communication Technologies (ICECCT), Coimbatore, India: IEEE, Feb. 2019, pp. 1–6. Available from https://doi.org/10.1109/ICECCT.2019.8869191.

[22] M.D.W. Rahman, R. Islam, A. Hasan, N.I. Bithi, M.D.M. Hasan, M.M. Rahman, Intelligent waste management system using deep learning with IoT, Journal of King Saud University 34 (5) (2022) 2072–2087. Available from https://doi.org/10.1016/j.jksuci.2020.08.016.

[23] G.-W. Cha, H.J. Moon, Y.-C. Kim, A hybrid machine-learning model for predicting the waste generation rate of building demolition projects, Journal of Cleaner Production 375 (2022) 134096. Available from https://doi.org/10.1016/j.jclepro.2022.134096.

[24] M. Cubillos, Multi-site household waste generation forecasting using a deep learning approach, Waste Management 115 (2020) 8–14. Available from https://doi.org/10.1016/j.wasman.2020.06.046.

[25] Z. Kang, J. Yang, G. Li, Z. Zhang, An automatic garbage classification system based on deep learning, IEEE Access 8 (2020) 140019–140029. Available from https://doi.org/10.1109/ACCESS.2020.3010496.

[26] D. Rutqvist, D. Kleyko, F. Blomstedt, An automated machine learning approach for smart waste management systems, IEEE Transactions on Industrial Informatics 16 (1) (2020) 384–392. Available from https://doi.org/10.1109/TII.2019.2915572.

[27] S. Kumar, D. Yadav, H. Gupta, O.P. Verma, I.A. Ansari, C.W. Ahn, A novel YOLOv3 algorithm-based deep learning approach for waste segregation: towards smart waste management, Electronics 10 (1) (2020) 14. Available from https://doi.org/10.3390/electronics10010014.

[28] C. Sirawattananon, N. Muangnak, W. Pukdee, Designing of IoT-based smart waste sorting system with image-based deep learning applications, in: 2021 18th International Conference on Electrical Engineering/Electronics, Computer, Telecommunications and Information Technology (ECTI-CON), Chiang Mai, Thailand: IEEE, May 2021, pp. 383–387. Available from https://doi.org/10.1109/ECTI-CON51831.2021.9454826.

[29] D.O. Gómez, S.C. Agudelo, S.I. Cadavid, M. Toro, J.C. Ramirez, A pipeline for solid domestic waste classification using computer vision. OSFPREPRINTS (2021). http://dx.doi.org/10.31219/osf.io/rvzyc.

[30] M.M. Hossen, et al., GCDN-Net: garbage classifier deep neural network for recyclable urban waste management, Waste Management 174 (2024) 439–450.

[31] S. Jin, Z. Yang, G. Królczykg, X. Liu, P. Gardoni, Z. Li, Garbage detection and classification using a new deep learning-based machine vision system as a tool for sustainable waste recycling, Waste Management 162 (2023) 123–130.

[32] M. Zeng, X. Lu, W. Xu, T. Zhou, Y. Liu, PublicGarbageNet: a deep learning framework for public garbage classification, presented at the 2020 39th Chinese Control Conference (CCC), IEEE, 2020, pp. 7200–7205.

[33] D. Ziouzios, N. Baras, V. Balafas, M. Dasygenis, A. Stimoniaris, Intelligent and real-time detection and classification algorithm for recycled materials using convolutional neural networks, Recycling 7 (1) (2022). Available from https://doi.org/10.3390/recycling7010009.

[34] T.H. Witkowski, World War II poster campaigns–preaching frugality to American consumers, Journal of Advertising 32 (1) (2003) 69–82. Available from https://doi.org/10.1080/00913367.2003.10639053.

[35] L. Davidson Cummings, Voluntary strategies in the environmental movement: recycling as cooptation, Journal of Voluntary Action Research 6 (3–4) (1977) 153–160. Available from https://doi.org/10.1177/089976407700600307.

[36] C.R. Álvarez-Chávez, S. Edwards, R. Moure-Eraso, K. Geiser, Sustainability of bio-based plastics: general comparative analysis and recommendations for improvement, Journal of Cleaner Production 23 (1) (2012) 47–56. Available from https://doi.org/10.1016/j.jclepro.2011.10.003.

[37] C.I. Idumah, I.C. Nwuzor, Novel trends in plastic waste management, SN Applied Sciences 1 (11) (2019) 1402. Available from https://doi.org/10.1007/s42452-019-1468-2.

[38] M.A.M. Bhuiyan, R.K. Sahi, M.R. Islam, S. Mahmud, Machine learning techniques applied to predict tropospheric ozone in a semi-arid climate region, Mathematics 9 (22) (2021) 2901. Available from https://doi.org/10.3390/math9222901.

[39] A.Y. Bayahya, W. Alhalabi, S.H. Alamri, Older adults get lost in virtual reality: visuospatial disorder detection in dementia using a voting approach based on machine learning algorithms, Mathematics 10 (12) (2022) 1953. Available from https://doi.org/10.3390/math10121953.

[40] W. Liang, S. Luo, G. Zhao, H. Wu, Predicting hard rock pillar stability using GBDT, XGBoost, and LightGBM algorithms, Mathematics 8 (5) (2020) 765. Available from https://doi.org/10.3390/math8050765.

[41] U. Gopalan, D. Rajalakshmi, Y. Teekaraman, R. Kuppusamy, A. Radhakrishnan, A novel strategy for waste prediction using machine learning algorithm with IoT based intelligent waste management system, Wireless Communications and Mobile Computing 2022 (2022) 1–15. Available from https://doi.org/10.1155/2022/2063372.

[42] S.R.J. Ramson, D.J. Moni, S. Vishnu, T. Anagnostopoulos, A.A. Kirubaraj, X. Fan, An IoT-based bin level monitoring system for solid waste management, Journal of Material Cycles and Waste Management 23 (2) (2021) 516–525. Available from https://doi.org/10.1007/s10163-020-01137-9.

[43] M. Cerchecci, F. Luti, A. Mecocci, S. Parrino, G. Peruzzi, A. Pozzebon, A low power IoT sensor node architecture for waste management within smart cities context, Sensors 18 (4) (2018) 1282. Available from https://doi.org/10.3390/s18041282.

[44] W. Xia, Y. Jiang, X. Chen, R. Zhao, Application of machine learning algorithms in municipal solid waste management: a mini review, Waste Management and Research 40 (2021). Available from https://doi.org/10.1177/0734242X211033716, p. 0734242X2110337.

[45] M. Erkınay Özdemir, Z. Ali, B. Subeshan, E. Asmatulu, Applying machine learning approach in recycling, Journal of Material Cycles and Waste Management 23 (2021). Available from https://doi.org/10.1007/s10163-021-01182-y.

[46] Y. Chu, C. Huang, X. Xie, B. Tan, S. Kamal, X. Xiong, Multilayer hybrid deep-learning method for waste classification and recycling, Computational Intelligence and Neuroscience 2018 (2018).

[47] C. Wang, J. Qin, C. Qu, X. Ran, C. Liu, B. Chen, A smart municipal waste management system based on deep-learning and Internet of Things, Waste Management 135 (2021) 20–29. Available from https://doi.org/10.1016/j.wasman.2021.08.028.

[48] L.A. Akanbi, A.O. Oyedele, L.O. Oyedele, R.O. Salami, Deep learning model for demolition waste prediction in a circular economy, Journal of Cleaner Production 274 (2020) 122843.

[49] M. Abdallah, M.Abu Talib, S. Feroz, Q. Nasir, H. Abdalla, B. Mahfood, Artificial intelligence applications in solid waste management: a systematic research review, Waste Management 109 (2020) 231–246. Available from https://doi.org/10.1016/j.wasman.2020.04.057.

[50] J. Arbeláez-Estrada, et al., A systematic literature review of waste identification in automatic separation systems, Recycling 8 (2023) 86. Available from https://doi.org/10.3390/recycling8060086.

[51] J. Li, et al., Automatic detection and classification system of domestic waste via multimodel cascaded convolutional neural network, IEEE Transactions on Industrial Informatics 18 (1) (2022) 163–173. Available from https://doi.org/10.1109/TII.2021.3085669.

[52] S. Zhang, Y. Chen, Z. Yang, H. Gong, Computer vision based two-stage waste recognition-retrieval algorithm for waste classification, Resources, Conservation and Recycling 169 (2021) 105543. Available from https://doi.org/10.1016/j.resconrec.2021.105543.

[53] S. Majchrowska, et al., Deep learning-based waste detection in natural and urban environments, Waste Management 138 (2022) 274–284. Available from https://doi.org/10.1016/j.wasman.2021.12.001.
[54] W.-L. Mao, W.-C. Chen, H.I.K. Fathurrahman, Y.-H. Lin, Deep learning networks for real-time regional domestic waste detection, Journal of Cleaner Production 344 (2022) 131096. Available from https://doi.org/10.1016/j.jclepro.2022.131096.
[55] J.K. Solano Meza, D. Orjuela Yepes, J. Rodrigo-Ilarri, E. Cassiraga, Predictive analysis of urban waste generation for the city of Bogotá, Colombia, through the implementation of decision trees-based machine learning, support vector machines and artificial neural networks, Heliyon 5 (11) (2019) e02810. Available from https://doi.org/10.1016/j.heliyon.2019.e02810.
[56] M.A. Mohammed, et al., Automated waste-sorting and recycling classification using artificial neural network and features fusion: a digital-enabled circular economy vision for smart cities, Multimedia Tools and Applications 82 (25) (2023) 39617–39632. Available from https://doi.org/10.1007/s11042-021-11537-0.
[57] H.-N. Guo, Y.-J. Tian, J. Zhang, H. Liu, Application of machine learning methods for the prediction of organic solid waste treatment and recycling processes: a review, Bioresource Technology 319 (2020) 124114. Available from https://doi.org/10.1016/j.biortech.2020.124114.
[58] M. Hossen, et al., A reliable and robust deep learning model for effective recyclable waste classification, IEEE Access (2024) 1. Available from http://doi.org/10.1109/ACCESS.2024.3354774.
[59] K. Lin, et al., Toward smarter management and recovery of municipal solid waste: a critical review on deep learning approaches, Journal of Cleaner Production 346 (2022) 130943. Available from https://doi.org/10.1016/j.jclepro.2022.130943.
[60] I. Ihsanullah, G. Alam, A. Jamal, F. Shaik, Recent advances in applications of artificial intelligence in solid waste management: a review, Chemosphere 309 (2022) 136631. Available from https://doi.org/10.1016/j.chemosphere.2022.136631.
[61] A.K. Abdella Ahmed, A.M. Ibraheem, M.K. Abd-Ellah, Forecasting of municipal solid waste multi-classification by using time-series deep learning depending on the living standard, Results in Engineering 16 (2022) 100655. Available from https://doi.org/10.1016/j.rineng.2022.100655.
[62] W. Lu, J. Lou, C. Webster, F. Xue, Z. Bao, B. Chi, Estimating construction waste generation in the Greater Bay Area, China using machine learning, Waste Management 134 (2021) 78–88. Available from https://doi.org/10.1016/j.wasman.2021.08.012.
[63] M.S. Jassim, G. Coskuner, M. Zontul, Comparative performance analysis of support vector regression and artificial neural network for prediction of municipal solid waste generation, Waste Management and Research 40 (2) (2022) 195–204. Available from https://doi.org/10.1177/0734242X211008526.
[64] R. Khan, et al., Machine learning and IoT-based waste management model, Computational Intelligence and Neuroscience 2021 (2021) 5942574. Available from https://doi.org/10.1155/2021/5942574.
[65] S. Dubey, P. Singh, P. Yadav, K.K. Singh, Household waste management system using IoT and machine learning, Procedia Computer Science 167 (2020) 1950–1959. Available from https://doi.org/10.1016/j.procs.2020.03.222.
[66] H.H. Chin, P.S. Varbanov, F. You, F. Sher, J.J. Klemeš, Plastic circular economy framework using hybrid machine learning and pinch analysis, Resources, Conservation and Recycling 184 (2022) 106387.
[67] A.J. Morrissey, J. Browne, Waste management models and their application to sustainable waste management, Waste Management 24 (3) (2004) 297–308. Available from https://doi.org/10.1016/j.wasman.2003.09.005.
[68] R.M. Deus, R.A.G. Battistelle, G.H.R. Silva, Scenario evaluation for the management of household solid waste in small Brazilian municipalities, Clean Technologies and Environmental Policy 19 (1) (2017) 205–214. Available from https://doi.org/10.1007/s10098-016-1205-0.
[69] A. Gulghane, R.L. Sharma, P. Borkar, Quantification analysis and prediction model for residential building construction waste using machine learning technique, Asian Journal of Civil Engineering 24 (6) (2023) 1459–1473. Available from https://doi.org/10.1007/s42107-023-00580-x.
[70] J. Huang, D.D. Koroteev, Artificial intelligence for planning of energy and waste management, Sustainable Energy Technologies and Assessments 47 (2021) 101426. Available from https://doi.org/10.1016/j.seta.2021.101426.
[71] Y. Chen, et al., Classification and recycling of recyclable garbage based on deep learning, Journal of Cleaner Production 414 (2023) 137558.

[72] Y. Jiang, et al., Prediction for odor gas generation from domestic waste based on machine learning, Waste Management 156 (2023) 264–271. Available from https://doi.org/10.1016/j.wasman.2022.12.006.

[73] X.C. Nguyen, et al., Development of machine learning - based models to forecast solid waste generation in residential areas: a case study from Vietnam, Resources, Conservation and Recycling 167 (2021) 105381. Available from https://doi.org/10.1016/j.resconrec.2020.105381.

[74] Q. Zhang, et al., Recyclable waste image recognition based on deep learning, Resources, Conservation and Recycling 171 (2021) 105636. Available from https://doi.org/10.1016/j.resconrec.2021.105636.

CHAPTER

19

Future aspects of machine learning/automation for the waste management

Nandini Gaikwad[1], *and Girish Sambhaji Gund*[2]

[1]Department of Information Technology, Mahatma Phule Arts, Science and Commerce College, Panvel, Maharashtra, India [2]Department of Physics, Mahatma Phule Arts, Science and Commerce College, Panvel, Maharashtra, India

Introduction

As the world deals with the escalating challenges of waste management, there is a growing recognition that technology, particularly machine learning (ML), can play a pivotal role in revolutionizing this critical sector. Integrating ML into waste management practices holds the promise of helping to create sustainable solutions that enhance efficiency, reduce environmental impact, and contribute to creating a circular economy. In this era of rapid technological advancement, the future applications of ML in waste management are poised to transform traditional methods of handling waste into smart, data-driven systems [1,2].

ML, a subset of artificial intelligence (AI), empowers systems to learn from data, identify patterns, and make informed decisions without explicit programming. In the context of waste management, this technology can be harnessed across various stages of the waste lifecycle, from generation and collection to processing and disposal. Integrating ML algorithms into waste management practices can bring several key advantages [1,2].

One primary area where ML can significantly impact waste collection routes is in optimizing them. Intelligent algorithms can analyze historical data, such as patterns of waste generation, population density, and traffic conditions, to dynamically plan and optimize collection routes. This reduces fuel consumption and emissions and enhances the overall efficiency of waste collection services [3].

Furthermore, ML can revolutionize waste sorting and recycling processes. Automated sorting systems equipped with ML algorithms can efficiently identify and separate different types of materials, improving recycling rates and reducing contamination. Additionally, these systems can adapt to changing waste compositions, ensuring flexibility in handling diverse and evolving waste streams [4].

Predictive analytics powered by ML can also assist in anticipating waste generation trends, enabling municipalities and waste management organizations to allocate resources more effectively. ML models can forecast future waste generation by analyzing various

Harnessing Automation and Machine Learning for Resource Recovery and Value Creation
DOI: https://doi.org/10.1016/B978-0-443-27374-2.00019-4

factors such as demographic changes, consumption patterns, and economic indicators, allowing for proactive and sustainable waste management strategies.

The integration of Internet of Things (IoT) devices and sensors, coupled with ML, offers real-time monitoring and management capabilities. Smart waste bins equipped with sensors can signal when they are nearing capacity, optimizing collection schedules and reducing unnecessary pickups. This enhances operational efficiency and minimizes the environmental impact associated with overcollection [5,6].

Aspects of machine learning

ML is an application of AI where a computer/machine learns from past experiences (input data) and makes future predictions [6–8]. The performance of such a system should be at least human level. There are three main elements to every ML algorithm, and they include:

Representation

This component defines the model's appearance and how knowledge is represented. Examples of representations include decision trees, sets of rules, instances, graphical models, neural networks, support vector machines, model ensembles, and others [6].

Evaluation

This component defines how good models are differentiated and how programs are evaluated. The evaluation compares different models and selects the best one for a given task. Common evaluation methods include cross-validation, holdout validation, and bootstrapping.

Optimization

This component defines what the process for finding good models and how programs are generated. Optimization is used to search through the space of possible models and find the one that best fits the data. Common optimization methods include gradient descent, stochastic gradient descent, and genetic algorithms [7].

Types of machine learning

The various types of ML based on the process are summarized in Fig. 19.1 and each type is explained in detail below:

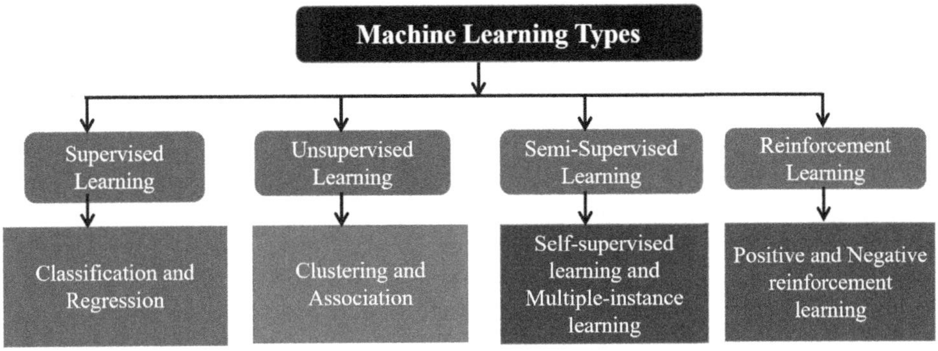

FIGURE 19.1 The flow chart of types of machine learning.

Supervised learning

Supervised learning uses labeled datasets to train algorithms for accurate classification or prediction of results. During training, the machine uses labeled data to predict future output. Labeled data helps set strategic paths for machines as they map inputs to outputs. Additionally, analysts use test datasets to check the accuracy of the analysis after continuous training. Various industries, such as healthcare, finance, and marketing, extensively use supervised learning [9].

Types of supervised learning

Supervised ML can be classified into two types of problems: classification and regression [10]. Classification algorithms are utilized when the output variable is binary or categorical, making them ideal for identifying whether emails are spam or legitimate. On the other hand, regression algorithms are utilized to make predictions like weather and market conditions for problems that involve a linear relationship between the input and output variables [11].

Applications of supervised learning

- Include medical diagnosis
- Fraud detection
- Spam detection
- Speech recognition.

Unsupervised learning

Unsupervised learning is one of the four main types of ML techniques. It utilizes unlabeled and unclassified datasets to make predictions without human intervention. This method is useful for categorizing or grouping unsorted data based on similarities and differences and discovering hidden patterns and trends in the input data. Unlike supervised learning, unsupervised learning algorithms learn from the data without any fixed output variable, making them suitable for complex tasks (no date). They identify hidden patterns or groupings in the data, which makes them ideal for clustering, anomaly detection, and exploratory data analysis.

Types of unsupervised learning

Unsupervised learning can be classified into two main techniques: clustering and association. Cluster analysis is an unsupervised learning technique that involves grouping data points based on their similarities, differences, or features without prior knowledge or labels. Businesses commonly use this technique in applications such as retail marketing, email marketing, and streaming services to identify similarities and patterns (no date; [12]). On the other hand, association is an unsupervised learning technique where machines discover interesting relations and connections among variables in large input datasets. Businesses commonly use associations for applications such as web usage mining, plagiarism checking, developing recommendations, and continuous production.

Applications of unsupervised learning

- Anomaly detection to identify fraud transactions
- Forecasting and predictive modeling
- Identifying and targeting market segments
- Recommendation systems to suggest products or services.

Semisupervised learning

Semisupervised learning is a highly efficient and cost-effective ML technique that combines labeled and unlabeled data during training. It allows machines to learn from all available data by utilizing both supervised and unsupervised learning advantages [11,12]. This method first uses unsupervised learning algorithms to group similar data. It labels previously unlabeled data afterwards, making the dataset more robust for training. Additionally, semisupervised learning enables machines to improve their accuracy and performance.

Types of semisupervised learning

Semisupervised ML includes self-supervised learning and multiple-instance learning. The self-supervised learning technique frames the problem as a supervised learning task to generate labeled data from unlabeled data. This approach is particularly useful when obtaining labeled data is relatively inexpensive. Additionally, it has shown impressive results in various applications. Notably, Google and Facebook utilize self-supervised techniques in computer vision and natural language processing [13]. On the other hand, multiple-instance learning provides weakly supervised learning where bags containing training instances are labeled instead of individual instances. This approach allows leveraging weakly labeled data often present in business problems due to the high cost of labeling. Furthermore, it is useful in scenarios where only partial information is available and has been applied in various fields, including medical diagnosis, image classification, and natural language processing ([12–14]).

Applications of semisupervised learning

Medical image analysis
Speech recognition, and natural language processing
Text classification and sentiment analysis
Fraud detection in finance.

Reinforcement learning

Reinforcement learning is an ML technique in which an agent learns to take optimal actions through environmental feedback. Unlike other types of ML, this technique does not rely on labeled data. Instead, it utilizes a trial-and-error approach with a feedback-based process that allows the agent to learn from its experiences [15]. One of the main advantages of reinforcement learning is its ability to learn from experience and improve performance over time.

Types of reinforcement learning

Reinforcement learning includes positive reinforcement learning and negative reinforcement learning. Positive reinforcement learning is a type of learning where the agent is rewarded for taking actions that lead to positive outcomes. The reward can be a numerical value, such as a score or a probability, or a symbolic value, such as a label or a tag. Negative reinforcement learning, on the other hand, is a type of learning where the agent is punished for taking actions that lead to negative outcomes. Here, the agent learns to avoid actions that result in punishment and instead takes actions that lead to positive outcomes [15].

Applications of reinforcement learning

- Building intelligent robotics
- Developing personalized treatment plans for patients
- Autonomous vehicles
- Game playing.

Types and automation in waste management

Types of waste management

Waste management encompasses various strategies and techniques for handling, disposing and minimizing waste [12]. The common processes and types of waste management methods are illustrated in Fig. 19.2 and mentioned below:

1. Landfill disposal
2. Incineration
3. Recycling

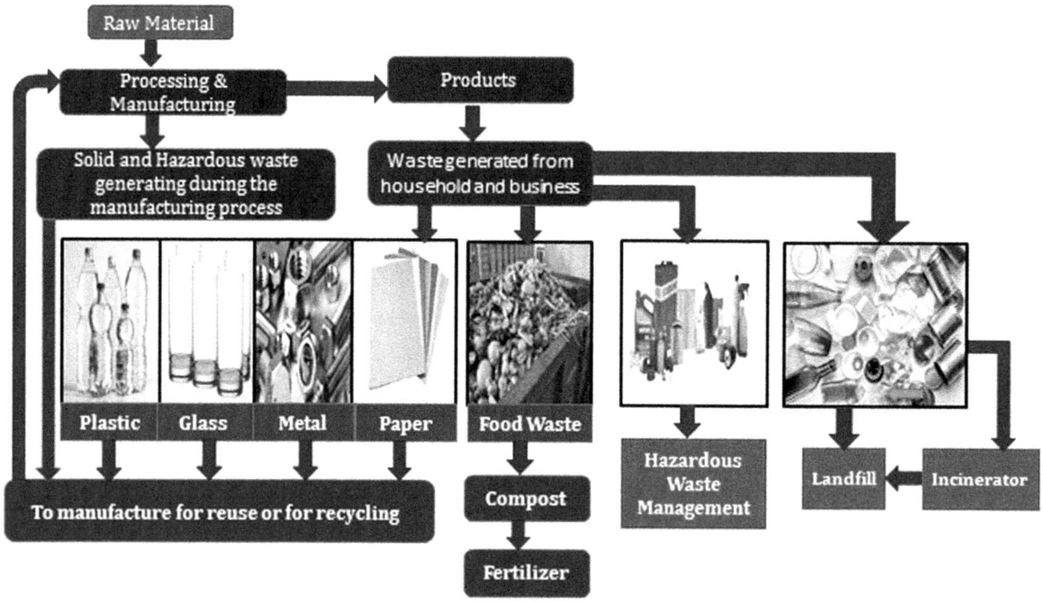

FIGURE 19.2 The schematic illustration of types of waste management.

4. Composting
5. Biological Treatment
6. Waste-to-Energy (WtE)
7. Source Reduction and Waste Minimization
8. Electronic Waste (E-waste) recycling
9. Construction and Demolition (C&D) waste management.

Automation categories in waste management

Automation is revolutionizing waste management practices, introducing efficiency, accuracy and sustainability ([12–14,16]). Here's an overview of how automation transforms various aspects of waste management.

Smart waste collection

Smart waste collection introduces innovative methods for optimizing waste management processes. Here's an in-depth look at smart waste collection [13]:

Sensor-based monitoring

- **Smart bins:** Equipped with sensors to detect fill levels, these bins provide real-time data on waste accumulation.
- **Types of sensors:** Ultrasonic, infrared, or weight-based sensors are commonly used to measure fill levels.

- **Benefits of sensor-based systems:**
- **Optimized collection routes:** Real-time data allows for efficient route planning, reducing unnecessary trips and fuel consumption.
- **Cost savings:** Optimized routes and schedules minimize operational costs for waste collection services.
- **Reduced overflow and littering:** Timely collection based on fill levels prevents overflow, minimizing littering and environmental pollution.
- **Resource efficiency:** Avoids unnecessary collection trips to partially filled bins, optimizing resource utilization.

Internet of Things integration

- IoT integration refers to the process of seamlessly connecting IoT devices and systems to work harmoniously, sharing data and insights. The goal is to achieve interoperability and enhance these interconnected devices' overall functionality and efficiency. Fig. 19.3 indicates the IoT based Radio-frequency identification (RFID) reader which can help to detect the overflow of the waste in dustbin and update it to the garbage truck [9].
- **Internet of Things:** Smart waste collection incorporates IoT devices to enable communication between bins and collection systems. These devices are outfitted with sensors that monitor the bins' fill level. The data collected is sent to central systems or directly to collection vehicles, allowing for the optimization of collection routes and schedules. This approach enhances waste collection efficiency while reducing operational costs.

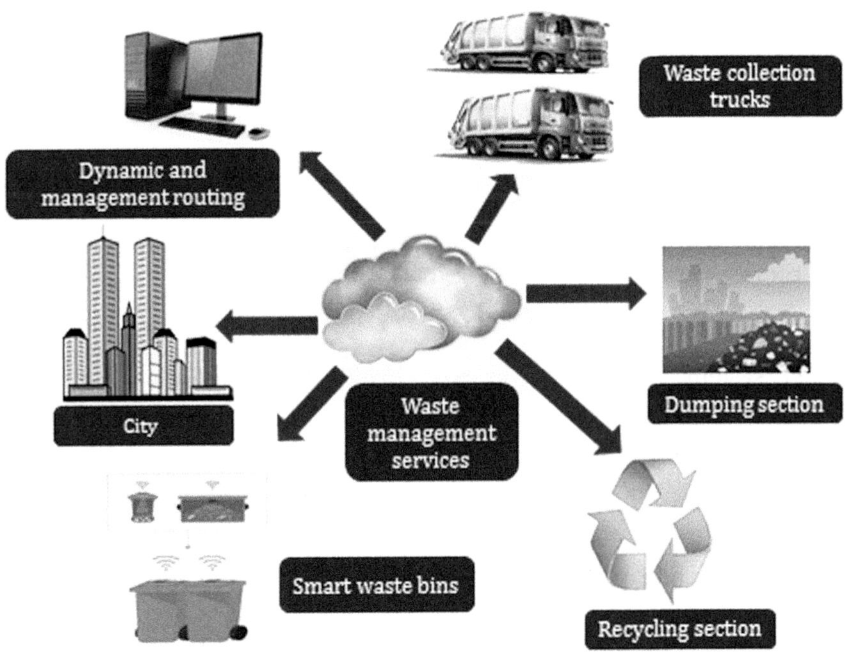

FIGURE 19.3 IoT-based RFID reader which can help to detect the overflow of the waste in dustbin and update it to the garbage truck.

- **Data transmission:** Smart bins utilize wireless connectivity to transmit fill-level data to centralized systems or collection vehicles, aiding in efficient waste management by preventing bin overflow and minimizing environmental pollution risks. These smart bins are integrated into the IoT, enabling real-time communication among bins, waste management teams, and central control systems. The data collected from these bins, including fill levels, waste types, and locations, can be analyzed to optimize waste collection routes and schedules.

An IoT-based system has been proposed to address the waste management challenges in smart cities. This system enhances waste collection efficiency and includes features for fire detection in waste materials and predicting future waste generation. The IoT devices manage and monitor the smart bins, which are wirelessly linked to a central hub, transmitting data about their fill levels and locations.

Route optimization

- **AI and data analytics:** Algorithms process sensor data to determine the most efficient collection routes.
- **Dynamic scheduling:** Routes are adjusted in real time based on current fill levels, traffic conditions, or other relevant factors.

Remote monitoring and alerts

- **Alert systems:** Notifications or alerts are generated for collection crews when bins reach predefined fill levels or require attention.
- **Remote monitoring:** Allows for remote tracking and management of waste collection operations.

Automated sorting and processing

- **Robotic sorting:** AI-powered robots and machines in recycling facilities sort and separate different types of materials with precision, enhancing recycling efficiency.
- **AI-assisted sorting:** ML algorithms identify and categorize recyclables accurately, reducing contamination and improving material recovery rates.

Predictive maintenance

- Automated systems monitor equipment health using sensors and data analytics, predicting maintenance needs to prevent breakdowns and optimize machinery performance in recycling plants.

Data analytics and decision support

- Automation collects and analyzes vast amounts of data on waste composition, generation patterns, and recycling processes, providing valuable insights for informed decision-making[23].
- AI-driven decision support systems aid in optimizing waste management strategies, such as material sourcing, recycling methods, and resource allocation.

Robotic waste collection and processing

- Autonomous robots are being developed for tasks like litter picking, sorting, and processing in waste facilities, reducing the need for manual labor and improving safety [12].

Blockchain for waste tracking

- Automation through blockchain technology facilitates transparent and traceable waste tracking, enabling the monitoring of waste movement and ensuring proper disposal or recycling.

Integration with Internet of Things and artificial intelligence

- Integration of IoT devices and AI enhances waste management systems, allowing real-time monitoring, data analysis, and decision-making for more efficient operations.

Benefits of automation in waste management

Automation brings several benefits to waste management practices, revolutionizing traditional methods. Here are the key advantages:

Enhanced efficiency

- **Streamlined processes** - Automation reduces manual intervention, speeding up waste collection, sorting, and processing [6].
- **Optimized resource allocation:** Automated systems efficiently allocate resources, such as collection routes and personnel, minimizing wastage.

Improved accuracy and precision

- **Reduced contamination -** Automated sorting systems use AI and sensors to accurately separate materials, minimizing contamination in recyclables.
- **Enhanced sorting quality -** Machines ensure consistent and precise sorting, improving the quality of recycled materials [4].

Cost savings

- **Operational efficiency -** Automated waste collection and processing lower operational costs by optimizing routes, reducing labor, and minimizing fuel consumption.
- **Long-term savings -** Despite initial setup costs, automation often leads to long-term cost savings through increased efficiency and reduced errors.

Environmental benefits

- **Resource conservation -** Effective sorting and recycling through automation conserve natural resources by reducing the need for virgin materials [15].
- **Reduced landfilling -** Automation helps divert more waste from landfills by promoting recycling and resource recovery.

Data-driven decision making

- **Insights and analytics** - Automation generates and analyzes vast amounts of data, providing valuable insights for optimizing waste management strategies [16].
- **Predictive maintenance** - Predictive analytics aid in maintaining equipment, preventing breakdowns, and minimizing downtime.

Safety and occupational health

- **Reduced human risk** - Automation in waste handling reduces risks associated with manual labor, enhancing worker safety and minimizing workplace accidents [11].

Innovation and technological advancements

- **Continuous improvement:** Advancements in technology continue to improve automation systems, enhancing their capabilities and adaptability [15].
- **Encourages innovation:** The drive towards automation fosters innovation in waste management practices, encouraging the development of sustainable solutions.

Scalability and adaptability

- **Scalable systems:** Automated systems can be designed to handle various waste volumes and adapt to different waste compositions.
- **Flexibility:** Systems can be adjusted and upgraded to accommodate changing waste management needs over time.

Challenges and opportunities in waste management

Current challenges in machine learning

The current challenges faced in the field of ML are crucial to address for improving the performance and reliability of ML models ([11,13–16]; no date; [12]). Here are some key points:

Data quality: High-quality data is essential for effective ML. Poor data quality, such as missing values, noise, and outliers, can adversely affect model accuracy and generalization.

High dimensionality: As data grows, the dimensionality increases. Handling high-dimensional data requires efficient feature selection, dimensionality reduction techniques, and scalable algorithms.

Model scalability: ML models should scale well with large datasets. Scalability challenges arise when dealing with massive amounts of data, especially in distributed computing environments.

Streaming data: Real-time data streams pose challenges due to their dynamic nature. ML models must adapt to changing data distributions and make real-time.

Model interpretability: Black-box ML models lack transparency, making it difficult to understand their decision-making process. Interpretable models are crucial for establishing trust and accountability.

Adaptability: ML models should adapt to concept drifts, changes in data distribution, and evolving user preferences. Continuous learning and adaptation are crucial.

Usability: ML tools and libraries should be user-friendly, accessible, and well-documented. Researchers and practitioners need efficient tools to build, evaluate, and deploy models.

Error rates and testing: Rigorous testing and evaluation are crucial. Researchers must consider error rates, statistical significance, and robustness to ensure reliable results.

Ethical and legal challenges: ML models can perpetuate biases, raise privacy concerns, and violate data protection laws. Addressing these challenges is essential.

Current challenges in automation

In the 21st century, the rapid growth of technology has left no option other than technology to accomplish tasks where human input can be minimized. This is how automation came into the scenario [16]. It includes enterprise applications such as IT automation, network automation, industrial automation, and automation of integration between systems and consumer applications.This execution and planning automation strategy has flawlessly covered all these spectra of the IT sector. It not only minimizes human input but also maximizes productivity and accuracy.

Opportunities in machine learning and automation

Waste management faces several pressing challenges globally, impacting environmental sustainability, public health, and resource conservation:

Rapid increase in waste generation: As populations grow and economies develop, there is a simultaneous rise in waste generation. This wave often outpaces the capacity of existing waste management infrastructure.

Inadequate waste collection and disposal infrastructure: Many regions lack efficient waste collection systems, resulting in improper disposal methods like open dumping or uncontrolled landfilling, which can pollute soil, air, and water sources.

Limited recycling and resource recovery: Despite efforts to promote recycling, challenges persist in efficient material separation and recovery. A considerable portion of recyclable materials ends up in landfills due to contamination or lack of proper sorting facilities.

Hazardous waste management: The management of hazardous waste, including electronic waste (e-waste), medical waste, and industrial chemicals, presents unique challenges due to their potential environmental and health hazards if not handled properly.

Waste composition complexity: Diverse waste streams with varying compositions and contaminants complicate recycling efforts. It requires advanced sorting technologies and processes to effectively separate materials for recycling.

Illegal dumping and waste trafficking: Illegal dumping of waste, including hazardous materials, presents environmental risks and challenges law enforcement efforts. Additionally, waste trafficking across regions or countries presents regulatory and control challenges.

Lack of public awareness and engagement: Inadequate public awareness and participation in waste reduction, recycling, and proper waste disposal practices hinder efforts to create a more sustainable waste management ecosystem.

Limited innovation and technology adoption: The adoption of innovative technologies, such as ML, IoT, and advanced sorting techniques, faces barriers due to costs, infrastructure limitations, and a lack of technical expertise in some regions.

Opportunities in waste management

Waste sorting automation: ML algorithms can be employed to automate waste materials sorting at recycling facilities. Computer vision systems can identify and categorize different types of materials, improving sorting accuracy and efficiency.

Predictive analytics for collection routes: ML algorithms can analyze historical data, weather conditions, and other relevant factors to predict the optimal collection routes for waste management vehicles. This can lead to fuel, time savings, and reduced environmental impact.

Dynamic pricing models for waste disposal: ML can be used to develop dynamic pricing models for waste disposal services. These models can consider factors such as demand, type of waste, and environmental impact, leading to more efficient pricing strategies that encourage recycling and waste reduction.

Smart bin technology: ML-enabled sensors in smart bins can monitor waste levels, identify the types of waste being disposed of, and send real-time data to waste management systems. This data can be used to optimize collection schedules and resource allocation.

v. Illegal dumping detection: ML algorithms can analyze satellite imagery and sensor data to detect illegal dumping activities. This can help authorities take timely actions to prevent environmental hazards and ensure proper waste disposal practices.

Energy recovery optimization: ML can optimize the process of energy recovery from waste by predicting the calorific value of waste materials. This information can be used to improve the efficiency of waste-to-energy plants and enhance energy generation from waste.

Customer behavior analysis: Understanding consumer behavior through ML can help design effective waste reduction campaigns. By analyzing data on consumer habits and preferences, waste management authorities can tailor educational programs and incentives to promote recycling and reduce waste generation.

Circular economy integration: ML can facilitate the transition to a circular economy by optimizing material reuse and recycling. Predictive analytics can help identify opportunities for recycling specific materials and minimize the need for new resource extraction.

Robotics in waste management: Integrating ML with robotics can enhance the capabilities of waste management robots. These robots can be deployed to collect, sort, and process waste in various environments.

Blockchain for transparent waste tracking: Blockchain technology, along with ML, can create a transparent and traceable system for tracking waste from its source to its final disposal or recycling. This can help enforce regulations, ensure accountability, and promote sustainable practices.

Future outlook and perspectives

ML for resource recovery and value creation in waste management presents exciting prospects for the future [8]:

Enhanced sorting and recycling

- **Advanced sorting technologies:** ML-driven sensors and robotics can improve sorting accuracy, enabling better separation of recycling materials, thus reducing contamination.
- **Optimized material recovery:** Algorithms can identify optimal methods to recover valuable resources from complex waste streams, promoting resource efficiency.

Predictive maintenance and efficiency

- **Smart maintenance systems:** ML can predict machinery breakdowns in recycling plants, minimizing downtime and optimizing operations.
- **Energy and resource optimization:** ML algorithms can optimize energy usage and resource allocation within waste management facilities for cost-effectiveness and sustainability.

Real-time decision-making

- **Data-driven decision support:** AI systems can analyze real-time data from waste collection and processing, aiding in quick decision-making for process optimization or resource allocation.
- **Dynamic routing optimization:** ML algorithms can dynamically adjust waste collection routes based on current traffic, weather, or real-time demand, reducing fuel consumption and time.

Waste-to-value innovations

- **Circular economy integration:** Automation and ML can facilitate the transition towards a circular economy by identifying opportunities for waste-to-value processes.
- **Innovative product design:** AI can assist in designing products for better recyclability, reducing waste generation at the source.

Environmental impact mitigation

- **Emissions reduction:** Automation can help monitor emissions in waste processing facilities, ensuring compliance and reducing environmental impact.
- **Landfill optimization:** ML can aid in optimizing landfill operations, minimizing environmental risks and maximizing space utilization.

Community engagement and education

- **Behavioral analytics:** Using ML to analyze community behavior patterns, enabling targeted educational campaigns for waste reduction and proper disposal practices.
- **Consumer feedback loop:** AI-driven systems can gather feedback from consumers on recyclability and sustainability, informing product design and waste management strategies.

Ethical considerations and transparency

- **Fairness and bias mitigation:** Ensuring that AI systems in waste management are fair and unbiased, addressing ethical concerns in decision-making processes.
- **Transparency and accountability:** Automation can improve transparency in waste management processes, enhancing accountability and trust among stakeholders.

The future of ML and automation in waste management lies in their potential to revolutionize the entire lifecycle, from waste collection to resource recovery, fostering sustainable practices, and creating value from what was once considered waste. However, it will require collaboration between technology developers, waste management industries, policymakers, and communities to fully realize this potential.

References

[1] P. Flach, Machine Learning: The Art and Science of Algorithms that Make Sense of Data, (2012).
[2] D.J. Patil, H. Mason, (2015). Data-Driven: Creating a Data Culture.
[3] B. Fang, J. Yu, Z. Chen, A.I. Osman, M. Farghali, I. Ihara, E.H. Hamza, D.W. Rooney, P.-S. Yap, Artificial intelligence for waste management in smart cities: a review, Environmental Chemistry Letters 21 (2023). Available from https://doi.org/10.1007/s10311-023-01604-3.
[4] M.E. Ozdemir, Z. Ali, B. Subeshan, E. Asmatulu, Applying machine learning approach in recycling, Journal of Material Cycles and Waste Management 23 (3) (2021) 855–871. Available from https://doi.org/10.1007/s10163-021-01182-y.
[5] J.Y. Khan, M.R. Yuce, Internet of Things (IoT): Systems and Applications, (2019).
[6] J. Brownlee, Basic Concepts in Machine Learning - Machine Learning Mastery, (2020).
[7] T. Ali, M. Irfan, A.S. Alwadie, A. Glowacz, IoT-based smart waste bin monitoring and municipal solid waste management system for smart cities, Arabian Journal for Science and Engineering 45 (2020).
[8] W. Xia, Y. Jiang, X. Chen, R. Zhao, Application of machine learning algorithms in municipal solid waste management: a mini review, Waste Management & Research 40 (2021) 609–624. Available from https://doi.org/10.1177/0734242X21103371.
[9] A Guide to 4 Important Types of Machine Learning with Use Cases by Krati Joshi, 2022. https://emeritus.org/blog/ai-and-mL-types-of-machine-learning/.
[10] M.D. Lagrega, P.L. Buckingham, J.C. Evans, Hazardous Waste Management, (2010).
[11] J. Pichtel, Waste Management Practices Municipal, Hazardous, and Industrial, (2014).
[12] S.E. Hasan, Introduction to Waste Management: A Textbook. (2022), 978–1.
[13] B. Acharya, S. Dey, M. Zidan, IoT-Based Smart Waste Management for Environmental Sustainability. (2022). Available from https://doi.org/10.1201/9781003184096.
[14] A. Sampathkumar, G.R. Kanagachidambaresan, M. Balasaraswathi, R. Maheswar, R. Rastogi, Challenges and solutions for sustainable smart city development, in: A. Sampathkumar, G.R. Kanagachidambaresan, M. Balasaraswathi, R. Maheswar, R. Rastogi (Eds.), Advances in Waste Management: Select Proceedings of Recycle, Springer, 2019.
[15] R. Kumar, U.K. Wiil, R. Kumar, U.K. Wiil (Eds.), Recent Advances in Computational Intelligence by Raman Kumar, Springer, 2019.
[16] M.N. Injadat, A. Moubayed, A.B. Nassif, A. Shami, Machine Learning Towards Intelligent Systems: Applications, Challenges, and Opportunities. 54 (2021). Available from https://doi.org/10.1007/s10462-020-09948-w.

Glossary

Accuracy A measure of the overall effectiveness of a model, indicating the ratio of correctly classified samples (both positive and negative) to the total number of samples.

Activation Functions Functions used in neural networks to introduce non-linearity, allowing the model to capture complex patterns in data

Actuators Devices that convert electrical signals into physical movement in automation systems.

Adam Optimizer An optimization algorithm used in training deep learning models, known for its efficiency in handling large datasets and its adaptability through the use of momentum and adaptive learning rates.

Additives Substances added to plastics to enhance their properties, such as flame retardants, plasticizers, and stabilizers.

AI-guided robotic sorting systems Automated systems using artificial intelligence to accurately sort and classify different types of plastic waste, improving recycling efficiency.

AI-powered collaboration platforms Platforms that use AI to facilitate cooperation and information sharing among stakeholders in the construction industry to address plastic pollution.

AlexNet A deep learning model known for its architecture with several convolutional layers, used for image classification tasks.

Artificial Intelligence (AI) A branch of computer science focused on creating systems capable of performing tasks that typically require human intelligence, such as decision-making, pattern recognition, and problem-solving.

Automated Assembly Line Machines Equipment used to automatically assemble products on a production line.

Automated Fill-Level Sensors Devices like Bintel's sensors that monitor garbage container levels, offering insights into waste management and container usage.

Automated Material Handling Systems designed to move, store, and manage materials with minimal human intervention.

Automated Quality Assurance Systems Systems that use technology to monitor and ensure the quality of products during manufacturing.

Automated Route Optimization Software that plans and adjusts garbage collection routes efficiently, minimizing manual effort and improving service coverage.

Automated Sorting AI robots that sort plastics more accurately than humans, improving recycling efficiency and reducing landfill waste.

Automation A process that converts a work procedure, machine, or system to operate with minimal human intervention, originating from the Greek word 'automaton,' referring to self-moving machines.

Batch Size The number of training examples used in one iteration of model training. A larger batch size can speed up training but requires more memory.

Best Execution Refers to the optimal performance achieved by the algorithm under the best conditions.

Bias In machine learning, bias refers to systematic errors introduced by incorrect assumptions in the learning algorithm, which can lead to models that favor certain outcomes over others.

Bioaccumulate The process by which substances, such as pollutants or toxins, accumulate in the tissues of living organisms over time, often increasing in concentration up the food chain.

Blockchain for Waste Tracking Employs blockchain technology for transparent waste tracking and ensuring proper disposal or recycling.

Blockchain Technology A decentralized digital ledger used to track and verify transactions and data, potentially applied to manage and ensure transparency in recycling processes.

Carbon Neutrality The state of balancing carbon dioxide emissions with the amount of CO_2 removed from the atmosphere to minimize net increase in CO_2 levels.

Cardboard A category of waste in the dataset representing images of cardboard items.

Centralized Sorting The process of sorting mixed waste in a central facility, where materials are separated mechanically for recycling or disposal.

Chemical Recycling Converts plastics into raw materials or chemicals for reuse, involving methods like pyrolysis.

Chemical Recycling Refibering's technology that transforms post-consumer textiles into new threads using AI and proprietary chemicals, reducing textile waste.

Circular economy An economic model that emphasizes the continuous use and recycling of materials to minimize waste and reduce reliance on new resources.

Class A category or label that an image can belong to in a classification task. For example, Glass, Paper, Cardboard, Plastic, Metal, Electronic Waste, and Medical Waste.

Classification Error Rate The complement of accuracy, used as the objective function to be minimized, calculated as 1−Accuracy1 − Accuracy1−Accuracy.

Cloud-Based Recycling Management Software that schedules and optimizes trash collection routes, manages driver assignments, tracks performance, and provides proof of service.

Clustering An unsupervised learning technique used to group similar data points into clusters based on shared characteristics or patterns.

CNN L1 Neurons The number of neurons in the first convolutional layer of the CNN.

CNN L2 Neurons The number of neurons in the second convolutional layer of the CNN.

CNN-ABC A Convolutional Neural Network optimized by the Artificial Bee Colony algorithm.

CNN-COA A Convolutional Neural Network optimized by the original version of the Chaos Optimization Algorithm.

CNN-COLSHADE A Convolutional Neural Network optimized by the COLSHADE algorithm.

CNN-GA A Convolutional Neural Network optimized by the Genetic Algorithm.

CNN-HHO A Convolutional Neural Network optimized by the Harris Hawks Optimization algorithm.

CNN-MCOA A Convolutional Neural Network optimized by a modified version of the Manta Ray Foraging Optimization algorithm.

CNN-PSO A Convolutional Neural Network optimized by the Particle Swarm Optimization algorithm.

Cohen's Kappa A metric for evaluating the agreement between observers or classifiers, accounting for chance agreement.

Composite Films Multi-layered plastic films made from different types of polymers, which are challenging to separate and recycle due to their complex composition.

Computer Vision Techniques AI methods for interpreting visual data to sort and identify plastics.

Computer Vision A field of AI that enables computers to interpret and understand visual information from the world, such as images and videos.

Computerized Systems Automation systems controlled by simple programs using graphical user interfaces to select settings and manage operations automatically.

Confusion Matrix A table used to evaluate the performance of a classification model by showing the counts of true positives, false positives, true negatives, and false negatives.

Construction and Demolition Debris Waste produced from construction, renovation, and demolition activities, including materials like wood, concrete, bricks, glass, plastics, and hazardous materials like asbestos.

Contamination Detection Identifying and removing impurities from recycled materials using AI.

Convergence The process by which an optimization algorithm approaches a solution, often visualized by plotting performance over iterations.

Conveyor Belt An early industrial automation system used to move components along an assembly line.

Crayfish Optimization Algorithm (COA) A metaheuristic optimization algorithm inspired by the behavior of crayfish in different environmental conditions, used to find optimal solutions for various problems.

Crude Oil A naturally occurring liquid fossil fuel that is refined to produce various products, including plastics.

CTR (Convolutional Trash Recognition) A model based on the ResNet18 architecture that achieved high accuracy in classifying various types of waste.

Curbside Collection A waste collection method where recyclables are picked up from homes directly, often alongside regular garbage collection.

Cyber-Physical Systems (CPS) Integrated systems involving both physical processes and computational elements, such as those used in advanced recycling technologies.

Data Analytics and Decision Support Automates data collection and analysis to aid in decision-making and optimize waste management strategies.

Data Augmentation Techniques used to artificially increase the size and diversity of training data by applying transformations such as rotations, flips, and color adjustments.

Data Cleaning The process of detecting and correcting (or removing) corrupt or inaccurate records from a dataset, ensuring data quality.

Data-driven Decision-making The use of AI to analyze large datasets and provide insights that guide decisions on reducing plastic waste and improving sustainability in construction.

Decision Trees A type of supervised learning algorithm that splits data into branches based on feature values to make predictions or decisions.

Deep Learning (DL) A subset of machine learning involving neural networks with many layers, capable of modeling complex patterns and representations.

Dense L1 Neurons The number of neurons in the first dense layer of the CNN.

Dense L2 Neurons The number of neurons in the second dense layer of the CNN.

DenseNet A type of CNN architecture that uses dense connections between layers to improve feature reuse and reduce the number of parameters.

DenseNet121 A deep learning model known for its dense connections between layers, which improves feature reuse and efficiency in image classification.

DenseNet201 A variation of DenseNet with additional layers, used for improved image classification performance compared to DenseNet121.

Density The mass per unit volume of a substance, often used to characterize materials like plastics.

Digitization The integration of digital technologies into processes and systems, enhancing efficiency and data management, including in waste management and recycling.

Discarded Plastic Plastic items that are no longer needed or useful and are thrown away, such as packaging or used electronics.

Dissolution Recycling Dissolving plastics in solvents to recover and purify polymers.

Distribution Graph A visual representation of how performance metrics are distributed across different optimizers, highlighting stability and performance variations.

Dropout Rate A regularization technique in neural networks where a fraction of neurons are randomly ignored during training to prevent overfitting.

Dropout A regularization technique used to prevent overfitting by randomly setting a fraction of input units to zero during training.

Early Stopping A regularization technique used in training neural networks to prevent overfitting by stopping the training process once the model's performance ceases to improve on the validation set for a specified number of epochs.

Ecological Services Benefits provided by ecosystems, such as clean air and water, which are affected by environmental pollution and degradation

Ecotoxicology The study of the effects of toxic substances on ecosystems and living organisms.

Eddy Current Separator A device used in magnetic sorting that repels non-magnetic, electrically conductive metals from mixed waste using rotating magnetic fields.

EfficientNet A family of CNN models designed to achieve high performance with fewer parameters by scaling depth, width, and resolution systematically.

Electronic Waste (E-Waste) Discarded electronic devices and equipment, which can contain valuable metals and hazardous elements like mercury, cadmium, and lead.

Electrostatic Separation A method that separates materials based on their electrical properties, using differences in electrostatic charge or conductivity.

Endocrine System A system of glands that release hormones into the bloodstream, which regulate various bodily functions. Disruption can lead to health issues.

Ensemble Methods Techniques that combine multiple machine learning models to improve the overall performance and accuracy of predictions.

Ensemble Voting Regression A machine learning technique that combines predictions from multiple models to improve accuracy and robustness.

Environmental Pollution The contamination of the natural environment by harmful substances or agents, often resulting from human activities.

Epoch A single pass through the entire training dataset during the training process of a neural network. Multiple epochs are typically used to improve model accuracy.

Epochs The number of complete passes through the training dataset during the learning process.

Ethical and Legal Challenges Issues related to biases, privacy, and data protection in ML models.

Evaluation Methods used to assess model performance, including cross-validation and holdout validation.

External Validation The process of evaluating a model on a dataset that was not used during training to assess its generalization capabilities to new, unseen data.

F1 Score A metric that combines precision and recall into a single value, providing a balance between the two. It is calculated as the harmonic mean of precision and recall.

False Negative An error in classification where a sample is incorrectly labeled as negative when it is actually positive.

False Positive An error in classification where a sample is incorrectly labeled as positive when it is actually negative.

Feature Engineering The process of selecting, modifying, or creating features from raw data to improve the performance of machine learning models.

Feature Map The output of a convolutional layer in a CNN, representing the spatial arrangement of features detected in the input image.

Feature Selection The process of identifying the most relevant features in a dataset to use in model training, enhancing model accuracy and reducing complexity.

Feedback and Control Quality Systems designed to monitor and adjust operations to maintain desired quality levels in automated manufacturing processes.

Ferrous Scrap Scrap derived from discarded steel, iron, and their alloys. It includes industry scrap (leftover metal from manufacturing), burr scrap (rough metal edges from production), and fuel scrap (iron powder used as a fuel source).

Fill-Level Sensors Devices that monitor waste levels in containers for optimized collection.

Film Grippers Mechanical devices used to handle and sort plastic films during recycling processes.

Fitness Function A function used to evaluate how good a solution is in terms of the optimization objective. The better the fitness, the more likely a solution is to be selected.

Fixed Automation Automation systems designed for high production volumes with dedicated equipment and fixed operations.

Flexibility The ability of a material to bend or stretch without breaking. LDPE, for example, is known for its flexibility.

Flexibility The ability of a material to bend or stretch without breaking, important for certain types of plastics like LDPE.

Fluorescent Markers Tracers added to polymers during manufacturing to enable their detection using fluorescence emissions, aiding in sorting and recycling processes.

Garbage Classification Dataset A dataset with images categorized into three high-level waste classes: organic, inorganic, and medical waste.

Garbage Detection Dataset A large dataset with images of four common waste types: plastic, paper, metal, and glass, used for object detection tasks.

Generative Adversarial Networks (GANs) A class of AI algorithms that generate new data samples by training two networks in opposition, useful for data synthesis and augmentation.

Glass A waste category in the dataset representing images of glass items.

Google Colab A cloud-based platform that provides free access to GPUs and other resources for running and training machine learning models.

GoogleNet A deep learning model characterized by its inception modules, designed to improve computational efficiency and classification performance.

Graph Theory A mathematical framework used to model and analyze relationships and networks, applied here for optimizing waste management operations.

Hazardous Waste Waste containing harmful substances that pose a risk to health or the environment, such as industrial solvents, pesticides, used oils, chemicals, batteries, and medical waste.

Heat Distortion The deformation or change in shape of a material when exposed to high temperatures.

High Dimensionality Refers to challenges in handling data with many features, requiring efficient techniques for feature selection and reduction.

Hydrometallurgical Separation A process that dissolves metals from ore or waste using aqueous or organic solutions for recovery.

Hyperparameter Tuning The process of adjusting the parameters of a machine learning model to optimize its performance on a given task.

Hyperparameter Tuning The process of adjusting the settings of a machine learning model to improve its performance. This is a complex and computationally intensive task, often addressed with metaheuristic algorithms.

Hyperparameters Parameters set before the training process begins, such as learning rate, batch size, and the number of epochs, which influence the model's learning process and performance.

IIoT (Industrial Internet of Things) The integration of internet-connected devices and sensors into industrial processes to enhance data collection and system monitoring.

Illegal Dumping and Waste Trafficking Environmental issues related to unauthorized disposal of waste and the movement of waste across regions or countries.

Image-Detecting Sorting Sensors Sensors in smart bins that use visual recognition to sort waste and detect fire hazards.

ImageNet A large-scale image dataset used for training deep learning models, particularly for transfer learning.

Imbalanced Data A situation where the classes in a dataset are not represented equally, which can lead to biased models.

InceptionV3 A convolutional neural network architecture designed for image classification, known for its inception modules that capture various feature scales.

Incinerator Ash Residue remaining after waste is burned in an incinerator, which may contain heavy metals and pollutants.

Industrial Arm Robots Robots used in manufacturing for tasks such as welding, assembling, and material handling.

Industrial Automation System Architecture Includes various machines, controllers, detectors, operator displays, and cabinets, with components such as PLCs, sensors, and actuators working together in a hierarchical structure to manage manufacturing processes.

Industrial Automation The use of technology including PLCs, AI, sensors, actuators, and machine learning to control and operate manufacturing processes with minimal human involvement.

Integrated Automation Advanced automation systems that use optimization algorithms for performing complex calculations and operations.

Intelligent waste management The application of AI sensors and data analytics to monitor, track, and optimize waste management processes, particularly for plastic waste in construction.

Internet of Things (IoT) A network of interconnected devices that can collect and exchange data, potentially used to monitor and optimize recycling systems.

Layers CNN The number of convolutional layers in the CNN architecture.

Layers Dense The number of fully connected (dense) layers in the CNN architecture.

Leachate A liquid formed when water percolates through waste material, potentially contaminating groundwater with dissolved organic and inorganic chemicals.

Learning Rate A hyperparameter that controls the size of the steps taken during optimization. It determines how quickly a model learns from the data.

Light Gradient Boosting Machine (LightGBM) A gradient boosting framework that is efficient for large datasets and performs well in terms of speed and memory usage.

Logistic Regression A statistical model used in supervised learning to predict binary outcomes (e.g., yes/no, true/false) based on one or more predictor variables.

Long Short-Term Memory (LSTM) A type of recurrent neural network (RNN) designed to model long-term dependencies in sequential data.

Long Short-Term Memory Networks (LSTMs) A type of recurrent neural network designed to model temporal sequences and dependencies, often used in combination with CNNs for tasks involving time series data.

Machine Learning (ML) A branch of artificial intelligence where algorithms learn from data to make predictions or decisions without being explicitly programmed.

Magnetic Separation A technique that uses magnetic fields to separate ferromagnetic materials from non-magnetic materials.

Manufacturing Planning and Decision-Making Systems Automated systems that assist in planning production schedules and making decisions related to manufacturing processes.

Material Recovery Facilities (MRFs) Facilities where recyclable materials are sorted and processed for reuse.

Mean Execution The average performance of the algorithm across multiple runs.

Mechanical Recycling Transforms plastics into secondary materials without changing their chemical structure.

Mechanization The use of machines to replace human muscle, impacting physical employment and displacing manual labor.

Median Execution The middle value of the performance measurements, providing a measure less affected by outliers.

Medical Waste A waste category in the dataset representing images of medical waste items.

Melting Point The temperature at which a material changes from a solid to a liquid.

Metaheuristics Algorithms Optimization algorithms inspired by natural processes or phenomena, such as genetic algorithms (GA) and particle swarm optimization (PSO), used to find near-optimal solutions for complex problems.

Metaheuristics High-level procedures designed to guide other heuristics toward better solutions by exploring and exploiting the search space effectively.

Metal A waste category in the dataset representing images of metal items.

Microplastics Tiny plastic particles resulting from the breakdown of larger plastic items due to mechanical abrasion and photo-degradation, found in marine food chains and the environment.

Mixed Collection A waste collection method where recyclables are collected together with other waste types, often requiring post-collection sorting.

MobileNet A lightweight CNN model optimized for mobile and embedded vision applications, designed to be efficient in terms of computation and memory usage.

MobileNetV3 Small A lightweight deep learning model designed for mobile and edge devices, optimizing computational efficiency and performance.

Model Evaluation The process of assessing a machine learning model's performance using metrics like accuracy, precision, recall, and F1 score.

Model Interpretability The extent to which a model's decision-making process can be understood and explained.

Model Scalability The ability of ML models to handle increasing data sizes effectively.

Modeling and Simulation AI-based tools for evaluating the environmental impact of construction projects and identifying opportunities to reduce plastic waste through various scenarios.

Monomer Recycling Breaks plastics into monomers to create new plastics with similar properties.

Multisensor-Based Separation A sorting method that uses multiple types of sensors (e.g., VIS, NIR, FIR) to collect various attributes of waste plastics, such as color and composition, for better separation.

Multi-task Learning An ML approach where multiple tasks are learned simultaneously, improving model performance through shared representations.

Municipal Solid Waste (MSW) Waste generated from residential, commercial, and institutional sources, which includes a mix of organic matter, paper, plastic, glass, and metal.

mvitv2_small A deep learning model used for image classification, known for its efficiency in handling large datasets.

Near Infrared (NIR) Sensors Sensors that detect and analyze materials based on their absorption of near-infrared light, useful for sorting different types of plastics.

Near Infrared (NIR) Technology A sorting technology that uses infrared light to identify and separate different types of plastics and other materials.

Near-Infrared (NIR) Spectroscopy Analyzes plastic materials using light to identify and classify polymers.

Neural Networks (NNs) Computational models inspired by the human brain, consisting of interconnected layers of neurons that learn to perform tasks by processing input data.

No Free Lunch (NFL) Theorem A principle stating that no single optimization algorithm performs best for every problem; the effectiveness of an algorithm depends on the specific problem.

Non-Ferrous Scrap Scrap composed of metals such as copper, aluminum, brass, and bronze, known for their malleability, light weight, and resistance to corrosion. Common sources include household items and electronic components.

Non-Renewable Resources Natural resources that cannot be replenished at the same rate they are consumed, such as fossil fuels like oil and natural gas.

Normalization The process of adjusting the pixel values of images to a standard range, typically [0, 1], by dividing by the maximum pixel value, which helps in improving the training efficiency and accuracy of the model.

Novel Disposal Techniques AI identifying effective methods for plastic waste disposal, such as discovering enzymes to break down plastics faster.

NVIDIA T4 GPU A type of graphics processing unit designed for efficient machine learning and deep learning tasks, featuring high performance and energy efficiency.

Optimization Techniques for improving model performance, such as gradient descent and genetic algorithms.

Optimization The process of finding the best solution or parameters for a given problem from a set of possible options.

Organic Recycling Breaks down biodegradable plastics into natural substances through composting or biogasification.

Overfitting A modeling error where a machine learning model performs well on training data but poorly on new, unseen data due to being too complex.

Physical and Computational Infrastructures Systems and technologies used together to improve efficiency, including those for sorting and processing recyclable materials.

Plastic Bottle Recycling Processes involving the collection, sorting, and recycling of plastic bottles using mechanical and chemical methods to create new products.

Plastic Utilization The increase in plastic production and waste over recent decades, with a focus on the environmental impact, recycling rates, and the challenges of managing plastic pollution.

Plastic Waste Waste generated from plastic products, particularly single-use plastics, which can accumulate in the environment and pose significant ecological concerns due to their slow degradation.

PLCs (Programmable Logic Controllers) Industrial digital computers used for automation of electromechanical processes, such as control of machinery on factory assembly lines.

Polyethylene (PE) A common plastic polymer used in various applications, including packaging and containers, with types like HDPE and LDPE.

Polymer A large molecule composed of repeating structural units (monomers) typically connected by covalent bonds, forming materials like plastics.

Polyolefins (POs) A group of plastics derived from olefin monomers, including polypropylene (PP) and polyethylene (PE), which are common in recycling.

Polypropylene (PP) A versatile plastic used in products such as containers and automotive parts, known for its durability and resistance to heat.

Pooling Layer A layer in a CNN that reduces the dimensionality of feature maps, helping to decrease computational complexity and reduce the risk of overfitting.

Post-Consumer Plastic Waste (PPW) Plastic waste generated by consumers after the product's initial use, such as used bottles and containers.

Post-Separation The process of sorting recyclable materials after waste collection, typically at treatment and recovery centers.

Post-Separation The sorting of recyclables after collection, typically at specialized facilities, to recover and process different materials.

Precision A metric that measures the proportion of true positive predictions among all positive predictions made by the model.

Predictive Analytics AI forecasts that adjust production levels to reduce plastic waste and improve supply chain efficiency.

Predictive Maintenance The use of machine learning models to predict when equipment will fail, allowing for timely maintenance to prevent breakdowns.

Programmable Automation Automation systems that allow for the reprogramming of operation sequences and electronic controls to accommodate different tasks or production runs.

Pyrolysis Decomposes plastics at high temperatures to produce reusable compounds.

Quasi-Reflection-Based Learning (QRL) A technique used to enhance the initialization phase of optimization algorithms by generating opposite solutions to improve exploration.

Radio-Frequency Identification (RFID) Technology Tracks waste containers and assets using radio waves for better management.

Random Forest (RF) A machine learning algorithm that constructs multiple decision trees and merges them to improve prediction accuracy and control overfitting.

Reactive Sensor Technology Technology in smart bins that sorts waste into categories like biodegradable and non-biodegradable based on sensor data.

Recall Also known as sensitivity, it measures the proportion of actual positives that were correctly identified by the model.

Recyclable Materials Materials that can be processed and used again to create new products, reducing waste and conserving resources.

Recycled plastic 3D printing The use of AI-controlled 3D printing technology to create building materials from recycled plastic, enhancing structural performance and promoting a circular economy.

Recycling Facilities Sites where materials are processed and repurposed for new products, including sorting, cleaning, and manufacturing.

Recycling Feedstock Neste Nordeste's processing of liquefied plastic waste into materials for new plastic production, substituting fossil fuels.

Recycling Metals The process of recovering and reusing metals from scrap to conserve resources and reduce environmental impact. Recycling reduces energy consumption compared to primary metal production.

Recycling of Dissolution Dissolves plastics to recover and purify polymers for reuse.

Recycling of Monomers Breaks down plastics into their basic monomers for recycling into new plastics.

Recycling Programs Initiatives designed to collect, process, and repurpose waste materials to reduce environmental impact and conserve resources.

Recycling Rate The percentage of collected materials that are successfully processed and reused in new products.

Recycling The process of converting waste materials into reusable materials to reduce the need for new raw materials and minimize environmental impact.

Recycling The process of recovering valuable materials from waste streams and reprocessing them into new products to reduce the burden on natural resources and minimize waste sent to landfills and incinerators.

RecyclingNet A deep learning model that uses convolutional neural networks to identify and sort recyclable materials.

Reduce and Reuse Strategies to minimize metal waste and extend the life of metal resources through efficient use and recycling, promoting a circular economy.

Reinforcement Learning (RL) A type of machine learning where an agent learns to make decisions by taking actions in an environment to maximize cumulative rewards.

Representation Defines the model's structure and how knowledge is represented, such as decision trees or neural networks.

ResNet A CNN architecture that uses residual connections to ease the training of deep networks and improve accuracy.

ResNet18 A deep learning model known for its residual learning framework, which helps in training deeper networks by mitigating the vanishing gradient problem.

RFID Technology Radio Frequency Identification used in smart bins to track and manage waste by reading tags that provide information about the waste type and bin status.

Robotic Sorting Systems AI-driven robots that automate the sorting of plastics for recycling.

Robotic Sorting AI-powered robots sort and process materials in recycling facilities, improving efficiency and accuracy.

Robotic Sorting The use of robots equipped with AI to automatically sort materials, potentially replacing manual sorting methods.

Robotic Waste Collection and Processing Autonomous robots perform tasks such as litter picking and sorting, reducing manual labor and enhancing safety.

ROC Curve A graphical plot that illustrates the diagnostic ability of a binary classifier system by plotting the True Positive Rate (Recall) against the False Positive Rate at various threshold settings.

Scrap Waste material composed of discarded metal pieces from objects, machines, or appliances that can be transformed into useful materials.

Semi-supervised Learning A hybrid approach that uses a small amount of labeled data combined with a large amount of unlabeled data to train models.

Sensors Devices that detect changes in the environment or system and provide feedback for automated systems to respond accordingly.

Smart Bins and Collecting Systems AI-driven bins and systems that automatically sort and handle plastic waste using sensors and machine learning, improving recycling rates and reducing contamination.

Smart Bins Waste bins equipped with sensors and machine learning algorithms to monitor fill levels and optimize waste collection processes.

Smart Waste Collection Utilizes sensors and IoT to monitor and optimize waste collection processes, improving route efficiency and reducing overflow.

Smart Waste Technologies Innovative technologies such as Pello (waste tracking), AI recycling robots, pneumatic waste pipes, solar-powered trash compactors, and e-waste kiosks that enhance waste management and recycling efficiency.

Sorting Technologies Methods used to separate and identify different types of materials, including manual and automated systems, such as air classifiers and sensor-based sorting.

Source Separation The practice of separating recyclable materials from other waste at the point of disposal, typically by the waste generator.

Source-Separated Recyclables Materials that are sorted and separated by type at the point of disposal, facilitating easier and more efficient recycling.

Specificity A metric that measures the model's ability to correctly identify negative cases, calculated as the ratio of true negatives to all actual negatives.

Spectral Analysis Examines light spectra to identify and classify plastics.

SPI Code A numerical code developed by the Society of the Plastics Industry to identify different types of plastic resins for recycling purposes.

SqueezeNet1_0 A compact deep learning model designed for efficient performance with fewer parameters and computational resources.

Standard Deviation (Std) A measure of the variation or dispersion of performance outcomes.

Stochastic Gradient Descent (SGD) An optimization algorithm used to minimize the loss function of a machine learning model by updating parameters in the direction of the gradient of the loss.

Streaming Data Data that is continuously generated and must be processed in real-time.

Supervised Learning A machine learning approach where models are trained on labeled data, with the goal of predicting outputs for new inputs.

Support Vector Regression (SVR) An ML technique that aims to find a function that approximates the relationship between variables with minimal error.

Sustainability The ability to maintain or improve environmental quality and resource use over time without depleting resources or causing harm.

Sustainable Design Predictive Analytics The use of AI to analyze and optimize building designs for reduced plastic usage and improved recyclability, contributing to more sustainable construction practices.

Synthetic Fibres Man-made fibers, such as those produced from recycled plastics, used in textiles and other applications.

Test Subset A portion of the dataset used to assess the performance of the trained model on new, unseen data.

Thermal Separation A technique involving heating metal waste to decompose or melt it, using methods such as pyrolysis, incineration, or smelting.

Transfer Learning A machine learning technique where a pre-trained model is adapted to a new, but related, task, leveraging learned features from the initial task.

Trash A waste category in the TrashNet dataset representing images of non-recyclable trash items.

TrashBox Dataset A dataset used for training and evaluating waste classification models, containing images of various types of waste.

TrashNet Dataset An external dataset used for validation, containing images of waste items categorized into classes such as glass, paper, cardboard, plastic, metal, and trash.

Triangulation Scanning A technique in optical sorting where a camera and laser are used to measure and identify the color and shape of materials.

Trommel Screen A rotating cylindrical separator used in mechanical sorting to separate materials based on size through perforated drum holes.

Trophic Levels The hierarchical levels in an ecosystem's food chain, representing different feeding positions, from primary producers to top predators.

True Negative The number of correctly identified negative samples.

True Positive The number of correctly identified positive samples.

Unsupervised Learning A machine learning approach where models learn patterns from unlabeled data, often used for clustering and association tasks.

Upcycling The process of converting waste materials or unwanted products into new materials or products of higher quality or value.

Urban Mining The extraction and recycling of valuable metals from e-waste, old buildings, and infrastructure to reduce reliance on traditional mining.

UV Sorting A method of sorting glass based on its interaction with UV light, distinguishing between different types of glass by measuring light attenuation.

Validation Subset A portion of the dataset used to evaluate the model's performance during training, helping to monitor and mitigate overfitting.

Valorization of Waste Heat Converts waste into energy or chemicals, reducing landfill use.

Variance (Var) The square of the standard deviation, indicating the degree of spread in the performance results.

Visible Light Spectroscopy (VIS) A technique that uses visible light to analyze materials, which can be used in multisensor-based sorting systems.

Visual Geometry Group with 16 layers (VGG16) A CNN model known for its simplicity and depth, used for object detection and classification tasks.

Waste Classification Data A dataset with images spanning six waste categories similar to TrashNet, but with no reported benchmark results.

Waste Classification The process of categorizing waste into different types (e.g., organic, recyclable) to manage and dispose of it properly.

Waste Heat Valorization SBS Thermal Technologies' method for converting waste heat into biofuel and clean energy, enhancing recycling and reducing energy costs.

Waste Management Regulations Laws and policies governing the collection, disposal, and recycling of waste, aiming to improve efficiency and reduce environmental impact.

Waste Management Strategies and methods for handling, disposing of, and minimizing waste. Includes landfill disposal, incineration, recycling, composting, and more.

Waste Recovery The process of retrieving and reprocessing valuable materials from waste for reuse, aiming to improve recycling rates and efficiency.

Waste Segregation The process of separating different types of waste materials to facilitate recycling and proper disposal.

Waste Stream The flow of waste materials from the point of generation through collection, sorting, and recycling or disposal.

Waste-to-Energy (WtE) A process of generating energy, usually in the form of electricity or heat, from the treatment of waste materials, reducing the amount of waste sent to landfills.

Waste-to-Energy Solutions Technologies that convert waste into energy, contributing to sustainable energy production.

Waxy Texture A smooth, somewhat greasy surface texture often associated with certain plastics like HDPE.

Worst Execution Refers to the performance of the algorithm under the least favorable conditions.

Xception A deep learning model with a depthwise separable convolution architecture, designed to improve classification performance while reducing computational complexity.

YOLOv3 A real-time object detection system that can classify and locate objects within images or video frames quickly.

YOLOv4 An object detection model that achieved notable performance in detecting waste items within images, used for garbage detection task

Zero Waste A waste management approach aimed at reducing waste by reusing, recycling, and composting materials to minimize landfill use.

Index

Note: Page numbers followed by "*f*" and "*b*" refer to figures and boxes, respectively.

P

Q

R

T

9780443273742